Asymmetric Domino Reactions

RSC Catalysis Series

Series Editor:
Professor James J Spivey, *Louisiana State University, Baton Rouge, USA*

Advisory Board:
Krijn P de Jong, *University of Utrecht, The Netherlands*, James A Dumesic, *University of Wisconsin-Madison, USA*, Chris Hardacre, *Queen's University Belfast, Northern Ireland*, Enrique Iglesia, *University of California at Berkeley, USA*, Zinfer Ismagilov, *Boreskov Institute of Catalysis, Novosibirsk, Russia*, Johannes Lercher, *TU München, Germany*, Umit Ozkan, *Ohio State University, USA*, Chunshan Song, *Penn State University, USA*

Titles in the Series:

How to obtain future titles on publication:
A standing order plan is available for this series. A standing order will bring delivery of each new volume immediately on publication.

For further information please contact:
Book Sales Department, Royal Society of Chemistry, Thomas Graham House, Science Park, Milton Road, Cambridge, CB4 0WF, UK
Telephone: +44 (0)1223 420066, Fax: +44 (0)1223 420247
Email: booksales@rsc.org
Visit our website at www.rsc.org/books

Asymmetric Domino Reactions

Hélène Pellissier
CNRS and Université Paul Cézanne Aix-Marseille III, Marseille, France
Email: h.pellissier@univ-cezanne.fr

RSCPublishing

RSC Catalysis Series No. 10

ISBN: 978-1-84973-651-0
ISSN: 1757-6725

A catalogue record for this book is available from the British Library

Published by The Royal Society of Chemistry,
Thomas Graham House, Science Park, Milton Road,
Cambridge CB4 0WF, UK

Registered Charity Number 207890

For further information see our web site at www.rsc.org

Printed in the United Kingdom by CPI Group (UK) Ltd, Croydon, CR0 4YY, UK

Preface

A domino reaction has been defined by Tietze as a reaction which involves two or more bond-forming transformations, taking place under the same reaction conditions, without adding additional reagents and catalysts, and in which the subsequent reactions are a consequence of the functionality formed by bond formation or fragmentation in the previous step.[1] It must be recognised that a relatively narrow distinction exists between domino and consecutive cascade or tandem reactions. From the point of view of an operator, the only difference between the two lies in the point along the sequence at which one or more catalysts or reagents had to be added to effect either the initiation of a sequence (that is, domino reaction) or propagation to the next step (that is, consecutive reaction). It must be noted that the descriptors domino, cascade, and tandem are often used indistinguishably from one another in the literature,[2] and a variety of opinions exist on how such reactions should be classified. According to Tietze, a domino reaction is strictly defined as a process in which two or more bond-forming transformations occur based on functionalities formed in the previous step and, moreover, no additional reagents, catalysts or additives can be added to the reaction vessel, nor can reaction conditions be changed.[1] Denmark further posited, however, that most domino reactions, as defined by Tietze, fell under the broader category of tandem processes.[3] Other tandem reactions which are not cascades involve the isolation of intermediates, a change in reaction conditions, or the addition of reagents or coupling partners. Furthermore, other authors classified domino reactions with even stricter conditions.[4,5] This book strictly follows the definition according to Tietze, and the reactions have been classified according to the mechanism of the single step(s). The quality and importance of a domino reaction can be correlated to the number of bonds generated in such a process and the increase in molecular complexity. Its goal is the resembling of nature in its highly selective sequential transformations. The domino reactions can be performed as single-, two-, and

RSC Catalysis Series No. 10
Asymmetric Domino Reactions
By Hélène Pellissier
© The Royal Society of Chemistry 2013
Published by the Royal Society of Chemistry, www.rsc.org

multicomponent transformations. Multicomponent reactions are defined as domino reactions involving at least three substrates and, consequently, constitute a subgroup of domino reactions.[5,6] It must be noted that there are also some confusing ideas among chemists about the definition of a multi-component reaction. According to Yus,[6c] this type of reaction should be clearly differentiated from other one-pot processes, such as tandem or cascade reactions, and in general from all processes that involve the reaction between two reagents to yield an intermediate which is captured by the successive addition of a new reagent (sequential component reactions).

This book is based on the definition of multicomponent reactions as convergent chemical processes which involve the well defined condensation of more than two reactants to form a product that contains significant portions of all reactants, ideally all atoms.[6b] Moreover, no additional reagents, catalysts, or additives can be added to the reaction vessel, nor can reaction conditions be changed, as in domino processes according to Tietze. The use of one-, two-, and multicomponent domino reactions in organic synthesis is increasing constantly, because they allow the synthesis of a wide range of complex molecules, including natural products and biologically active compounds, in an economically favourable way by using processes that avoid the use of costly and time-consuming protection–deprotection processes, as well as purification procedures for intermediates.[7] Indeed, decreasing the number of laboratory operations required and the quantities of chemicals and solvents used have made domino and multicomponent reactions unavoidable processes.[8] The proliferation of these reactions is evidenced by the number of recent reviews covering the literature through 1992.[1,4,6,9]

Although asymmetric synthesis is sometimes viewed as a subdiscipline of organic chemistry, actually this topical field transcends any narrow classification and pervades essentially all chemistry.[10] Indeed, the preparation of chiral compounds is an important and challenging area of contemporary synthetic organic chemistry, mainly in connection with the fact that most natural products are chiral and their physiological or pharmacological properties depend upon their recognition by chiral receptors, which will interact only with molecules of the proper absolute configuration. The use of chiral drugs in enantiopure form is now a standard requirement for virtually every new chemical entity, and the development of new synthetic methods to obtain enantiopure compounds has become a key goal for pharmaceutical companies. The growing economic importance of chiral compounds has spurred major research efforts towards the selective preparation of chiral compounds. The synthesis of optically active chiral compounds, which play an important role in medicine and materials, is one of the most fascinating aspects of modern organic synthesis. Over the last three decades an explosive growth of research in the field of asymmetric synthesis has occurred. Asymmetric synthesis constitutes one of the main strategies to gain access to enantio-enriched compounds, involving the use of either chiral auxiliaries or catalysts derived preferentially from cheap chiral pool sources. In particular, asymmetric catalysis of organic reactions to provide enantiomerically enriched products is of central importance to modern synthetic and pharmaceutical chemistry. In this context, the ability to prepare chiral compounds by joining two or more reactions in one asymmetric domino process has rapidly become one challenging goal for chemists, due to economic advantages, such as avoiding costly protecting groups and time-consuming purification procedures after each step. In particular, the combination of these fascinating one-pot reactions with asymmetric catalysis has quickly become one of the most interesting fields in organic chemistry.

The goal of this book is to provide to researchers and professionals in academic and industrial laboratories a broad overview of the most recent efforts of the chemical community in the development of novel asymmetric domino reactions published since the beginning of 2006. Indeed, the complete field of asymmetric domino reactions, including both the asymmetric domino reactions using chiral substrates and the enantioselective domino reactions using chiral catalysts, was most recently reviewed in its globality in 2006 by Tietze in a book, and by this author in two reviews.[1c,9j–k] This book is divided into three chapters dealing successively with asymmetric domino reactions based on the use of chiral substrates including chiral auxiliaries, enantioselective domino reactions based on the use of chiral metal catalysts, and enantioselective domino reactions based on the use of chiral organocatalysts. In order to facilitate presentation, each chapter is subdivided into two sections, the first of which includes one- and two-component domino reactions, while the second describes multicomponent reactions. The sections are subdivided according to the types of reaction involved in the first synthetic step(s).

Hélène Pellissier

References

1. (a) L. F. Tietze and U. Beifuss, *Angew. Chem., Int. Ed.*, 1993, **32**, 131–163; (b) L. F. Tietze, *Chem. Rev.*, 1996, **96**, 115–136; (c) L. F. Tietze, G. Brasche and K. Gericke, *Domino Reactions in Organic Synthesis*, Wiley-VCH, Weinheim, 2006.
2. K. C. Nicolaou, D. J. Edmonds and P. G. Bulger, *Angew. Chem., Int. Ed.*, 2006, **45**, 7134–7186.
3. S. E. Denmark and A. Thorarensen, *Chem. Rev.*, 1996, **96**, 137–165.
4. C. J. Chapman and C. G. Frost, *Synthesis*, 2007, 1–21.
5. D. E. Fogg and E. N. dos Santos, *Coord. Chem. Rev.*, 2004, **248**, 2365–2379.
6. (a) L. F. Tietze and A. Modi, *Med. Res. Rev.*, 2000, **20**, 304–322; (b) J. Zhu and H. Bienaymé (eds.), *Multicomponent Reactions*, Wiley-VCH, Weinheim, 2005; (c) D. J. Ramon and M. Yus, *Angew. Chem., Int. Ed.*, 2005, **44**, 1602–1634; (d) R. V. A. Orru and E. Ruijter (eds.), *Synthesis of Heterocycles via Multicomponent Reactions, Topics in Heterocyclic Chemistry*, vols I and II, Springer, Berlin, 2010.
7. (a) C. Grondal, M. Jeanty and D. Enders, *Nature Chem.*, 2010, **2**, 167–178; (b) K. C. Nicolaou and J. S. Chen, *Chem. Soc. Rev.*, 2009, **38**, 2993–3009;

(c) B. B. Touré and D. G. Hall, *Chem. Rev.*, 2009, **109**, 4439–4486; (d) M. Colombo and I. Peretto, *Drug Discov. Today*, 2008, **13**, 677–684; (e) C. Hulme and V. Gore, *Curr. Med. Chem.*, 2003, **10**, 51–80; (f) A. Padwa and S. K. Bur, *Tetrahedron*, 2007, **63**, 5341–5378.

8. Special Issue on Rapid formation of molecular complexity in organic synthesis, *Chem. Soc. Rev.*, **38**(11), 2009.

9. (a) G. H. Posner, *Chem. Rev.*, 1986, **86**, 831–844; (b) H. Waldmann, *Nachr. Chem. Tech. Lab.*, 1992, **40**, 1133–1140; (c) K. Fukumoto, *Synth. Org. Chem. Jpn.*, 1994, **52**, 2–18; (d) R. A. Bunce, *Tetrahedron*, 1995, **51**, 13103–13159; (e) P. J. Parsons, C. S. Penkett and A. J. Shell, *Chem. Rev.*, 1996, **96**, 195–206; (f) P. C. F. Balaure and P. I. A. Filip, *Rev. Roum. Chim.*, 2002, **46**(8), 809–833; (g) E. Capdevila, J. Rayo, F. Carrion, I. Jové, J. I. Borrell and J. Teixido, *Afinidad*, 2003, **506**, 317–337; (h) L. F. Tietze and N. Rackelmann, *Pure Appl. Chem.*, 2004, **76**, 1967–1983; (i) P. I. Dalko and L. Moisan, *Angew. Chem., Int. Ed.*, 2004, **43**, 5138–5175; (j) H. Pellissier, *Tetrahedron*, 2006, **62**, 2143–2173; (k) H. Pellissier, *Tetrahedron*, 2006, **62**, 1619–1665; (l) D. Enders, C. Grondal and M. R. M. Hüttl, *Angew. Chem., Int. Ed.*, 2007, **46**, 1570–1581; (m) G. Guillena, D. J. Ramon and M. Yus, *Tetrahedron: Asymmetry*, 2007, **18**, 693–700; (n) D. M. D'Souza and T. J. J. Müller, *Chem. Soc. Rev.*, 2007, **36**, 1095–1108; (o) A.-N. Alba, X. Companyo, M. Viciano and R. Rios, *Curr. Org. Chem.*, 2009, **13**, 1432–1474; (p) J. E. Biggs-Houck, A. Younai and J. T. Shaw, *Curr. Opin. Chem. Biol.*, 2010, **14**, 371–382; (q) M. Ruiz, P. Lopez-Alvarado, G. Giorgi and J. C. Menéndez, *Chem. Soc. Rev.*, 2011, **40**, 3445–3454; (r) L. Albrecht, H. Jiang and K. A. Jorgensen, *Angew. Chem., Int. Ed.*, 2011, **50**, 8492–8509; (s) H. Pellissier, *Adv. Synth. Catal.*, 2012, **354**, 237–294; (t) C. De Graaff, E. Ruijter and R. V. A. Orru, *Chem. Soc. Rev.*, 2012, **41**, 3969–4009.

10. E. N. Jacobsen, A. Pfaltz and H. Yamamoto (eds.), *Comprehensive Asymmetric Catalysis*, Springer, Berlin, 1999.

Contents

RSC Catalysis Series No. 10
Asymmetric Domino Reactions
By Hélène Pellissier
© The Royal Society of Chemistry 2013
Published by the Royal Society of Chemistry, www.rsc.org

Abbreviations

Ac	acetyl
Acac	acetylacetone
Alloc	allyloxycarbonyl
AP	activator protein
AQN	anthraquinone
Ar	aryl
BDHP	1,1'-binaphth-2,2'-diyl hydrogen phosphate
bdpp	2,4-bis(diphenylphosphino)pentane
BH	Bronsted acid
BINAP	2,2'-bis(diphenylphosphino)-1,1'-binaphthyl
BINAPO	2-diphenylphosphino-2'-diphenylphosphinyl-1,1'-binaphthalene
BINEPINE	phenylbinaphthophosphepine
BINIM	binapthyldiimine
BINOL	1,1'-bi-2-naphthol
BIPHEP	2,2'-bis(diphenylphosphino)-1,1'-biphenyl
BIPHEPHOS	6,6'-[(3,3'-Di-*tert*-butyl-5,5'-dimethoxy-1,1'-biphenyl-2,2'-diyl)bis(oxy)bis(dibenzo[d,f][1,3,2]dioxaphosphepin)
Bn	benzyl
Boc	*tert*-butoxycarbonyl
BOM	π-benzyloxymethyl
Box	bisoxazoline
BOXAX	2,2'-bis(oxazolyl)-1,1'-binaphthyl
BPTV	N-benzene-fused phthaloyl–valine
Bu	butyl
Bz	benzoyl
Cat	catechol

RSC Catalysis Series No. 10
Asymmetric Domino Reactions
By Hélène Pellissier
© The Royal Society of Chemistry 2013
Published by the Royal Society of Chemistry, www.rsc.org

Cbz	benzyloxycarbonyl
Chiraphos	2,3-bis(diphenylphosphine)butane
cod	cyclooctadiene
CSA	camphorsulfonic acid
Cy	cyclohexyl
DABCO	1,4-diazabicyclo[2.2.2]octane
dba	(*E,E*)-dibenzylideneacetone
DBU	1,8-diazabicyclo[5.4.0]undec-7-ene
DCE	dichloroethane
de	diastereomeric excess
DFT	density functional theory
DHQ	hydroquinine
DHQD	dihydroquinidine
DIFLUORPHOS	5,5′-bis(diphenylphosphino)-2,2,2′,2′-tetrafluoro-4, 4′-bi-1,3-benzodioxole
DIPEA	diisopropylethylamine
DKR	dynamic kinetic resolution
DMAD	dimethyl acetylenedicarboxylate
DME	dimethoxyethane
DMF	dimethylformamide
DMSO	dimethylsulfoxide
DOSP	*N*-*p*-dodecylbenzenesulfonylprolinate
DPEN	1,2-diphenylethylenediamine
DPPB	1,4-bis(diphenylphosphino)butane
DPPF	1,1′-bis(diphenylphosphanyl)ferrocene
dtb	ditertbutyl
dtbm	ditertbutylmethoxy
E	electrophile
ee	enantiomeric excess
ESI	electrospray ionisation
Et	ethyl
FBIP	ferrocene bis-imidazoline bis-palladacycle
Fc	ferrocenyl
FMOC	9-fluorenylmethoxycarbonyl
FOXAP	ferrocenyloxazolinylphosphine
Fu	furyl
Hex	hexyl
HFIP	hexafluoroisopropanol
HIV	human immunodeficiency virus
HMDS	hexamethyldisilazide
HMPA	hexamethylphosphoramide
HOMO	highest occupied molecular orbital
IBX	*o*-iodoxybenzoic acid
INNs	*N*-(*p*-nitrobenzenesulfonyl)iminophenyliodinane
i-Pr-DuPhos	1,2-Bis(2,5-diisopropylphospholano)benzene

Josiphos	1-[2-(diphenylphosphino)ferrocenyl]ethyl dicyclohexylphosphine
L	ligand
LA	Lewis acid
LB	Lewis base
LDA	lithium diisopropylamide
LUMO	lowest occupied molecular orbital
MAO	monoamine oxidase
MCPBA	*meta*-chloroperbenzoic acid
Me	methyl
MEDAM	bis(dimethylanisyl)methyl
Me-DuPhos	1,2-Bis(2,5-dimethylphospholano)benzene
MES	mesyl
MOM	methoxymethyl
MS	mass spectroscopy
MTBE	methyl *tert*-butyl ether
NADH	nicotinamide adenine dinucleotide
Naph	naphthyl
NBS	*N*-bromosuccinimide
Nf	nonaflate
NHC	*N*-heterocyclic carbene
NMI	N-methylimidazole
NMM	*N*-methyl morpholine
NMO	*N*-methylmorpholine-*N*-oxide
NMP	*N*-methylpyrrolidinone
NMR	nuclear magnetic resonance
Norphos	2,3-bis(diphenylphosphino)-bicyclo[2.2.1]hept-5-ene
Ns	nosyl
Nu	nucleophile
Oct	octyl
PCC	pyridinium chlorochromate
Pent	pentyl
PFBA	pentafluorobenzoic acid
Ph	phenyl
PHAL	1,4-phthalazinediyl
Phth	phthalimido
Pin	pinacolato
PINAP	4-[2-(diphenylphosphino)-1-naphthalenyl]-N-[1-phenylethyl]-1-phthalazinamine
PMB	*para*-methoxybenzyl
PMP	pentamethylpiperidine
Pr	propyl
Pro	prolinium
PTSA	*p*-toluenesulfonic acid
Py	pyridine

PYBOX	2,6-bis(2-oxazolyl)pyridine
QUINAP	1-(2-diphenylphosphino-1-naphthyl)isoquinoline
QUOX	quinoline-oxazolineSegphos: 5,5′-Bis (diphenylphosphino)-4,4′-bi-1,3-benzodioxole
rt	room temperature
Segphos	5,5′-Bis(diphenylphosphino)-4,4′-bi-1,3-benzodioxole
Solphos	7,7′-bis(diphenylphosphino)-3,3′,4,4′-tetrahydro-4,4′-dimethyl-8,8′-bis-2H-1,4-benzoxazine
SPRIX	spiro bis(isoxazoline)
SYNPHOS	6,6′-bis(diphenylphosphino)-2,2′,3,3′-tetrahydro-5,5′-bi-1,4-benzodioxin
Taniaphos	[2-diphenylphosphinoferrocenyl](*N*,*N*-dimethylamino)(2-diphenylphosphinophenyl)methane
TBA	tribromoacetic acid
TBAB	tetra-*n*-butylammonium bromide
TBAF	tetra-*n*-butylammonium fluoride
TBAI	tetra-*n*-butylammonium iodide
TBAT	tetrabutylammonium triphenyldifluorosilicate
TBCHD	2,4,4,6-tetrabromocyclohexa-2,5-dienone
TBDPS	*tert*-butyldiphenylsilyl
TBS	*tert*-butyldimethylsilyl
TC	thiophene carboxylate
TCBA	2,4,6-trichlorobenzoic acid
TCPTTL	*N*-tetrachlorophthaloyl-*tert*-leucinate
TEA	triethylamine
TEAB	triethylammonium bicarbonate
TES	triethylsilyl
Tf	trifluoromethanesulfonyl
TFA	trifluoroacetic acid
THF	tetrahydrofuran
Thio	thiophene
TIPS	triisopropylsilyl
TMEDA	tetramethylethylenediamine
TMP	2,2,6,6-tetramethylpiperidide
TMS	trimethylsilyl
Tol	tolyl
TosMIC	(*p*-toluenesulfonyl)methylisocyanide
TPAP	tetrapropylammonium peruthenate
Ts	4-toluenesulfonyl (tosyl)
C_3-TUNEPHOS	1,13-Bis(diphenylphosphino)-7,8-dihydro-6H-dibenzo[f,h][1,5]dioxonin
VAPOL	2,2′-diphenyl-[3,3′-biphenanthrene]-4,4′-diol
XANTPHOS	4,5-Bis(diphenylphosphino)-9,9-dimethylxanthene

CHAPTER 1

Asymmetric Domino Reactions Based on the Use of Chiral Substrates

1.1 Introduction

Of the methods available for preparing chiral compounds, asymmetric synthesis from chiral substrates still attracts a lot of attention.[1] Indeed, it remains the method most commonly employed in total synthesis of optically active compounds, playing an important role in medicine and materials as well as natural products, in spite of the explosive growth of organocatalysis in the last decade,[2,3] and its application in the synthesis of a number of chiral products.[4] The last six years have witnessed an explosive growth in the field of asymmetric domino reactions starting from chiral substrates and auxiliaries, with an impressive number of novel methodologies and applications in total synthesis. The goal of this chapter is to cover the recent efforts of the chemical community in the development of novel domino reactions of chiral substrates, including chiral auxiliaries and their wide applications in total synthesis, published since the beginning of 2006. This field was most recently reviewed in 2006 by Tietze in a book,[5] and by this author in a review, covering the literature until the beginning of 2006.[6] The domino reactions are catalogued on the basis of the reaction types involved in the first synthetic step(s). They have been selected according to Tietze's definition, qualifying a domino reaction as a reaction involving two or more bond-forming transformations which take place under the same reaction conditions, without adding additional reagents and catalysts, and in which the subsequent reactions are a consequence of the functionality formed by bond formation or fragmentation in the previous step.[5,7] In order to facilitate presentation, the chapter is divided into two

RSC Catalysis Series No. 10
Asymmetric Domino Reactions
By Hélène Pellissier
© The Royal Society of Chemistry 2013
Published by the Royal Society of Chemistry, www.rsc.org

principal sections, dealing successively with one- and two-component domino reactions, and multicomponent domino reactions. The first section, describing the asymmetric one- and two-component domino reactions, is subdivided into eight sections which successively deal with domino reactions with an anionic primary step, domino reactions based on cationic sequences, domino reactions initiated by a pericyclic primary step, domino reactions based on carbene sequences, palladium-catalysed domino reactions, ruthenium-catalysed domino reactions, gold-catalysed domino reactions, and finally miscellaneous domino reactions. The second section of the chapter, describing the asymmetric multicomponent reactions, defined as domino reactions involving at least three substrates and forming products that contain significant portions of all reactants, ideally all atoms,[8] is subdivided into 10 paragraphs, dealing successively with multicomponent reactions initiated by the Michael addition, multicomponent reactions based on the Hantzsch reaction, multicomponent reactions based on the Ugi reaction, multicomponent reactions based on the Strecker reaction, multicomponent reactions based on the Mannich reaction, multicomponent reactions initiated by an allylation reaction, multicomponent reactions based on the Passerini reaction, multicomponent reactions based on the Biginelli reaction, multicomponent reactions based on the Petasis reaction, and finally miscellaneous multicomponent reactions.

1.2   One and Two-Component Domino Reactions

1.2.1   Anionic Primary Step

1.2.1.1   Domino Reactions Initiated by the Michael Reaction

The nucleophilic 1,4-addition of stabilised carbon nucleophiles to electron-poor olefins, generally α,β-unsaturated carbonyl compounds, is known as the Michael addition, although it was first reported by Komnenos in 1883.[9] Michael-type reactions can be considered as one of the most powerful and reliable tools for the stereocontrolled formation of carbon–carbon and carbon–heteroatom bonds,[10] as has been demonstrated by the huge number of examples in which it has been applied as a key strategic transformation in total synthesis. As a consequence, in recent years, many different versions of this important transformation have been reported, using a wide variety of nucleophiles and conjugate acceptors.[11] In particular, the Michael–Michael domino reaction is a powerful tool in forging ring systems common to many natural products. A number of asymmetric double Michael reactions have been reported in the last few years, often applied to the total synthesis of important products. As an example, Garrido *et al.* have developed an expeditious asymmetric synthesis of a precursor of natural product (–)-pumiliotoxin C, based on an asymmetric domino aza-Michael–Michael reaction.[12] This process was initiated by a highly diastereoselective Michael addition of a chiral lithium amide to nona-2,7-diendioic diester, followed by a 6-*exo-trig* cyclisation of the thus formed enolate. As shown in Scheme 1.1, the expected chiral cyclohexane

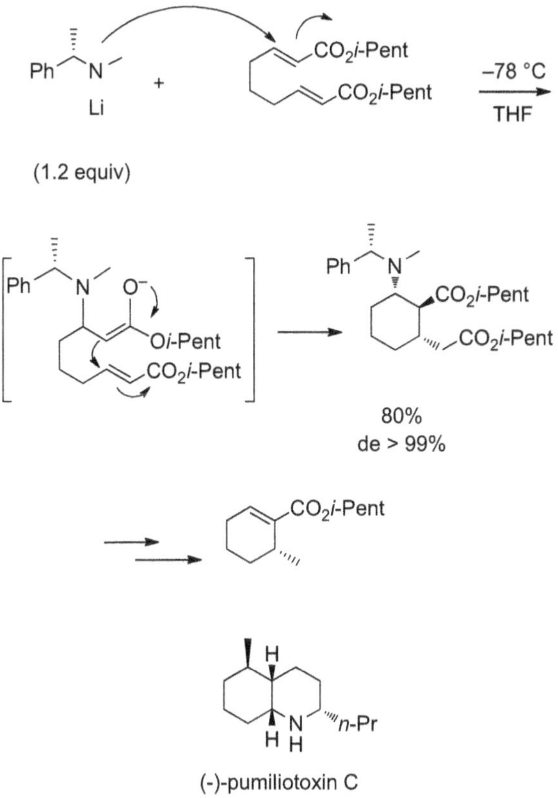

Scheme 1.1 Synthesis of precursor of (–)-pumiliotoxin C through domino aza-Michael–Michael reaction.

domino product was achieved in 80% yield as a single diastereomer which bore three contiguous stereocentres. This chiral product was subsequently converted into the required chiral precursor of (–)-pumiliotoxin C through a three-step sequence, involving successively a Cope elimination, a selective hydrolysis of the less steric demanding ester, and an efficient Barton decarboxylation.

Later, Scheerer *et al.* described the first asymmetric synthesis of the naturally occurring neoclerodane diterpene salvinorin A, which is a potent κ opioid receptor agonist, the only non-alkaloid psychoactive substance, and the most potent natural hallucinogen.[13] The key step of the synthesis was a transannular double Michael reaction cascade of chiral bisenone macrocycle promoted by tetra-*n*-butylammonium fluoride (TBAF). The corresponding tricycle was obtained in quantitative yield as a single diastereomer, as shown in Scheme 1.2. The process delivered two quaternary methyl stereocentres at C5 and C9 in a 1,3-diaxial alignment from the corresponding β, β-disubstituted enones, moieties commonly known to possess poor reactivity toward conjugate addition. Scheme 1.2 provides a rationale for the observed stereoselectivity in the domino reaction. Conformational analysis of the starting bisenone

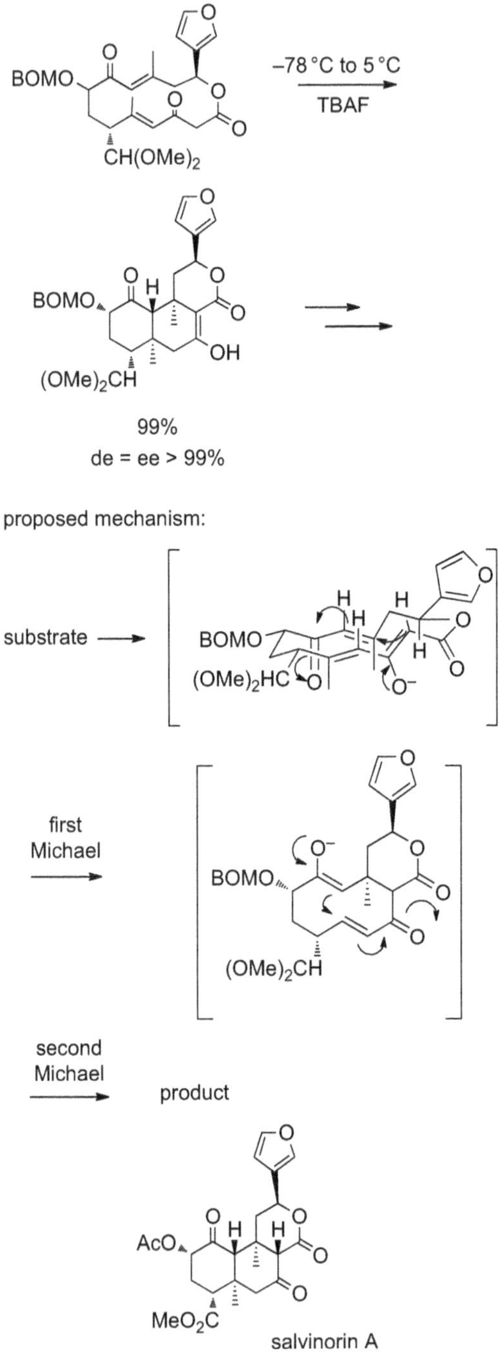
BOMO
CH(OMe)2
–78 °C to 5 °C
TBAF
(OMe)2CH
OH
99%
de = ee > 99%
proposed mechanism:
substrate
first
Michael
second
Michael
product
AcO
MeO2C
salvinorin A

Scheme 1.2 Synthesis of salvinorin A through domino Michael–Michael reaction.

macrocycle suggested that the three stereocentres, in pseudo-equatorial positions, mutually reinforced the diastereoselectivity. This analysis also suggested that enolisation favoured the Z-enolate. The final enantiopure tricycle was further converted into expected salvinorin A through seven supplementary steps in 41% overall yield.

In 2008, the first and long-awaited total synthesis of the natural cardioactive glycosylated steroid ouabain was developed by Deslongchamps *et al.* on the basis of a polyanionic cyclisation strategy, providing a tricyclic domino inter-mediate which is depicted in Scheme 1.3.[14] This intermediate arose from an asymmetric domino Michael–Michael reaction occurring between a chiral cyclohexenone and a chiral Nazarov substrate in the presence of Cs_2CO_3. This key tricyclic domino product was obtained in 85% yield and subsequently submitted to decarboxylation to give the corresponding key tricyclic product as a single diastereomer in 92% yield. Finally, this polyfunctionalised tricyclic compound was converted into the expected ouabain.

On the other hand, Kataoka *et al.* have previously investigated the reaction of a chiral 3-cinnamoyloxazolidine-2-thione with aromatic aldehydes in the presence of $BF_3 \cdot Et_2O$.[15] The reaction evolved through an asymmetric domino thia-Michael–aldol process, providing diastereoselectively the corresponding tricyclic products, which incorporated a bridgehead carbon bound to four heteroatoms. These products were produced as mixtures of two diastereomers with moderate to high diastereoselectivities of up to 90% diastereomeric excess (de), as shown in Scheme 1.4. It must be noted that this process generated four stereocentres. The chiral domino products achieved in good to high yields could be further transformed into the corresponding enantiopure propane-1,3-diols bearing three contiguous stereocentres through acid hydrolysis, allowing the chiral auxiliaries to be recovered. These structurally rare compounds were presumably formed along the reaction pathways shown in Scheme 1.4. $BF_3 \cdot Et_2O$ coordinated with the carbonyl oxygen of the N-cinnamoylthio-carbamate, allowing the enone moiety to be activated. The intramolecular Michael addition of the thione group to the enone moiety proceeded *via* intermediate **1**, and afforded the boron enolate–iminium salt **2**. An aldol reaction between the boron enolate moiety and aldehyde yielded the aldol product **3**, which intramolecularly cyclised to afford the final tricyclic products.

In 2009, Davies *et al.* demonstrated that the condensation of a chiral lithium (R)-N-(α-methylbenzyl)amide to a range of benzylidene imines provided the corresponding almost diastereo- and enantiopure domino aza-Michael–cyclisation products in good to excellent yields, as shown in Scheme 1.5.[16] This process opened a novel and powerful access to enantiopure 2-aryl-4-aminotetrahydroquinoline-3-carboxylic acid derivatives bearing three contiguous stereogenic centres with general diastereo- and enantioselectivities of >98% de and >98% enantiomeric excess (ee), respectively. In addition to being readily applicable to the preparation of libraries of chiral 4-aminotetrahydroquinolines for biological screening, this approach has been shown to be useful in the preparation of corresponding chiral diamino esters

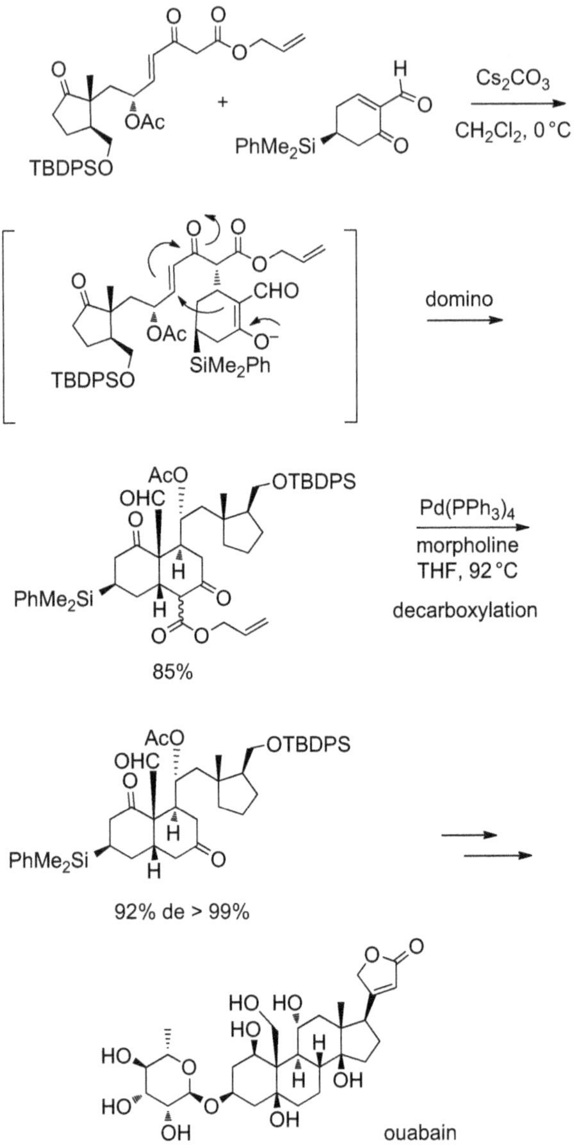

Scheme 1.3 Synthesis of ouabain through domino Michael–Michael reaction.

through simple hydrogenolysis of the domino products, which occurred without compromising the diastereo- and enantiopurity.

In 2006, a novel asymmetric domino Michael–Dieckmann cyclisation reaction was employed by Groth *et al.* as key step in a total synthesis of naturally occurring fungitoxic (–)-chokol A.[17] As shown in Scheme 1.6, the condensation of a higher-order cuprate derived from the corresponding vinyl bromide to a chiral α,β-unsaturated ester derived from (–)-phenylmenthol

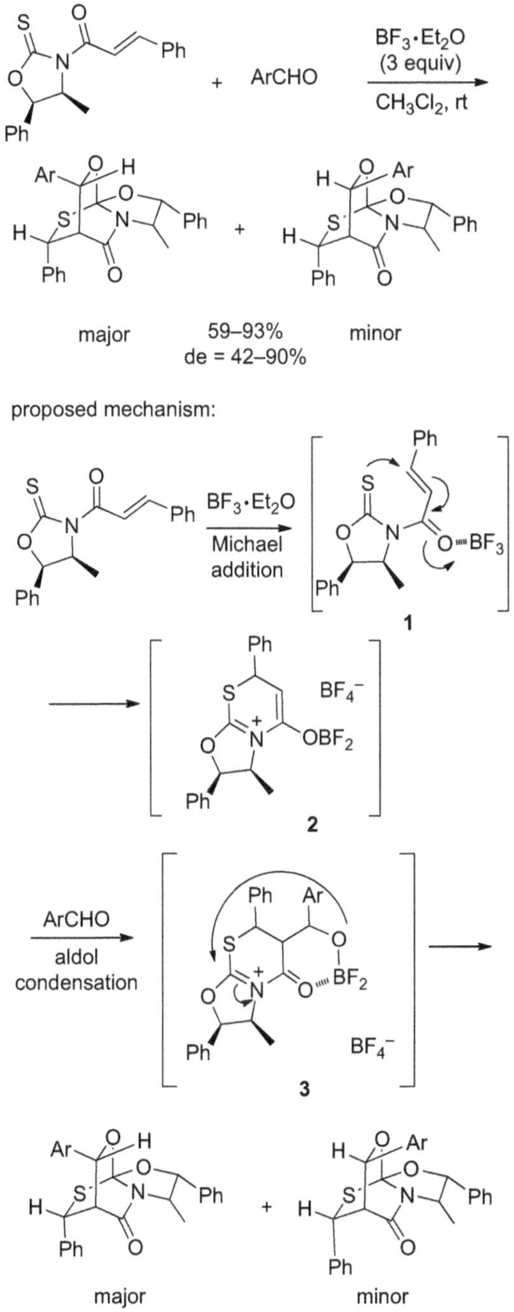

proposed mechanism:

Scheme 1.4 Domino thia-Michael–aldol reaction of chiral 3-cinnamoyloxazolidine-2-thione with aromatic aldehydes.

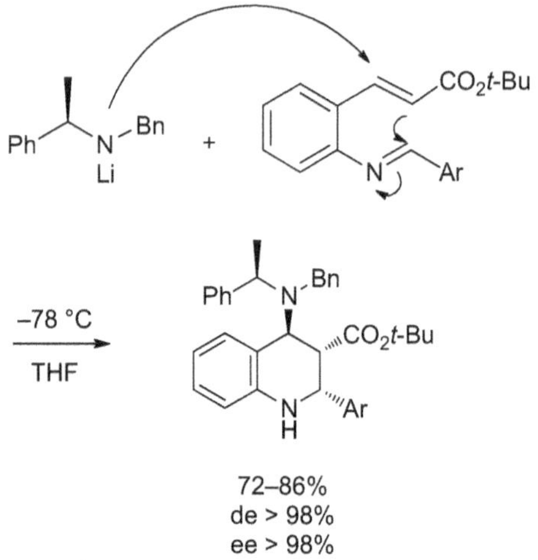

Scheme 1.5 Domino aza-Michael–cyclisation reaction of lithium (*R*)-*N*-(α-methyl-benzyl)amide with benzylidene imines.

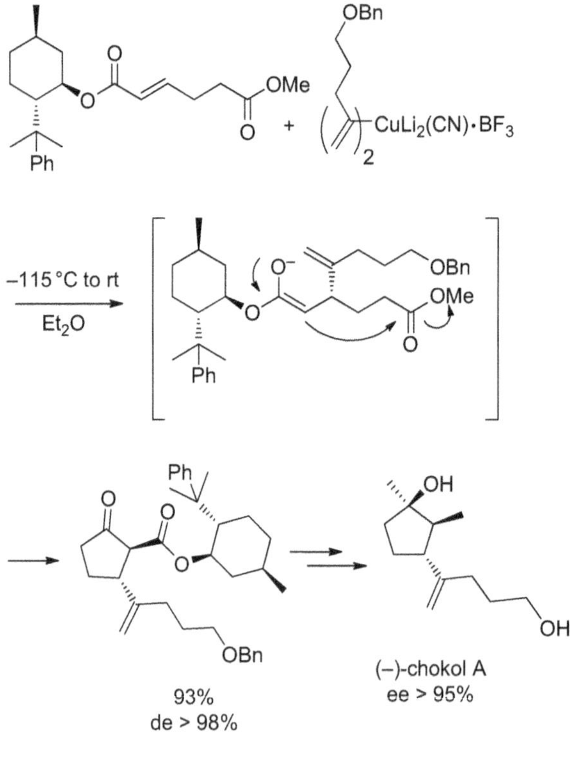

Scheme 1.6 Synthesis of (–)-chokol A through domino Michael–Dieckmann reaction.

provided, after a Michael addition followed by a Dieckmann cyclisation, the corresponding cyclic β-keto ester in 93% yield, combined with an excellent diastereoselectivity of >98% de. This chiral product was further converted into the expected (−)-chokol A through a five-step sequence, with an overall yield of 29%.

In the course of developing a novel synthesis of natural and biologically active himbacine, McCarthy *et al.* have found that a domino Michael–Dieckmann reaction of an enantioenriched α,β-unsaturated furanone (ee = 70%) with a racemic diester led to the corresponding tricyclic product, which constituted a key intermediate for the proposed synthesis (Scheme 1.7).[18] In fact, the process generated the domino product as a mixture of three

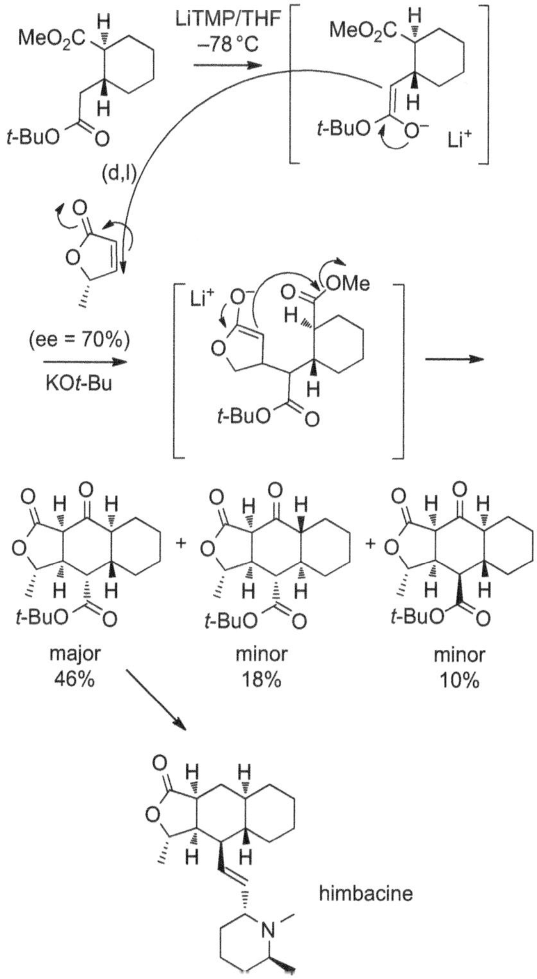

Scheme 1.7   Synthesis of a key intermediate in the synthesis of himbacine through domino Michael–Dieckmann reaction.

diastereomers, among which the major one was separated by chromatography with 46% yield. This enantiopure tricyclic product constituted a useful intermediate for a short enantioselective synthesis of himbacine and derivatives.

In 2011, another asymmetric domino Michael–Dieckmann reaction was demonstrated by Avenoza *et al.* to be an efficient entry to both enantiomers of α-(hydroxymethyl)glutamic acid.[19] As shown in Scheme 1.8, the reaction of methyl acrylate with an L-serine-derived bicyclic (3*S*,7*R*,7a*S*)-*N*,*O*-acetal provided, through a sequence of a Michael addition followed by a Dieckmann reaction promoted by the participation of the cyclic carbamate group, the corresponding domino product as a 39:61 mixture of (3*S*,6*R*,7a*S*)- and (3*S*,6*S*,7a*S*)-diastereomers. This mixture was further submitted to 6N HCl aqueous solution under reflux to achieve the (*S*)-enantiomer of α-(hydroxymethyl)glutamic acid. In the same way as that described for the synthesis of this (*S*)-enantiomer, the authors have achieved the corresponding (*R*)-enantiomer

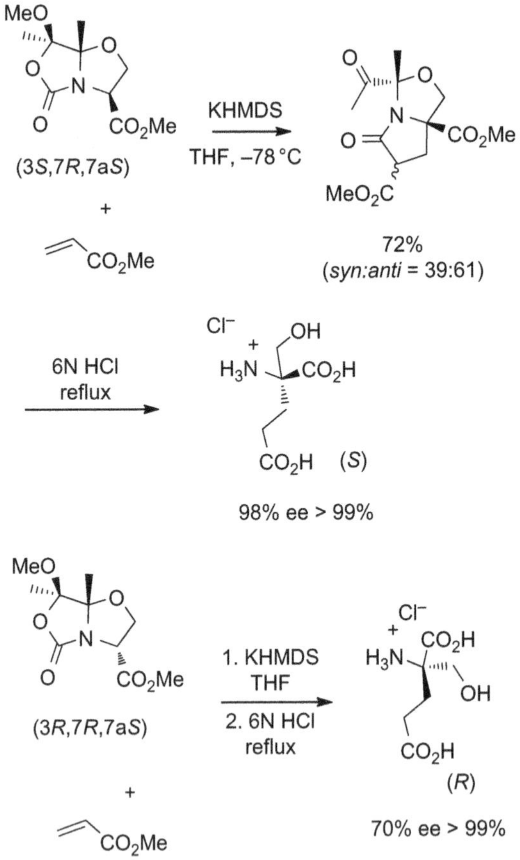

Scheme 1.8 Synthesis of both enantiomers of α-(hydroxymethyl)glutamic acid through domino Michael–Dieckmann reaction.

of α-(hydroxymethyl)glutamic acid starting from the corresponding bicyclic (3*R*,7*R*,7a*S*)-*N*,*O*-acetal derived from D-serine, as shown in Scheme 1.8.

In 2007, Nagasaka *et al.* reported the first domino Michael addition–Mannich-type reaction using a TiCl$_4$–tetra-*n*-butylammonium iodide (TBAI) system and occurring intramolecularly between α,β-unsaturated carbonyl compounds which bore an Evans oxazolidinone as a chiral auxiliary and *in situ* generated *N*-acyliminium ion intermediates.[20] The intramolecular cyclisation of the chelated iodo titanium enolate intermediates, arising from the Michael addition, afforded the corresponding chiral indolizidines bearing three stereogenic centres in moderate to good diastereoselectivities of up to 80% de, as shown in Scheme 1.9. In this study, the use of mixed solvents, such as AcOEt and CH$_2$Cl$_2$, was shown to be the most effective.

In 2011, Qin *et al.* reported a novel domino Michael–Mannich–Mannich reaction of a tryptamine derivative with a chiral amidinobenzodiene, which was generated *in situ* from the corresponding isatin-derived chloride.[21] Surprisingly, the process led, in the presence of two equivalents of AgBF$_4$, to a 3:1 diastereomeric mixture of complex products possessing a polycyclic skeleton of 2,3,4,5-diindolinohexahydropyrrole in 76% yield. The authors assumed that the formation of these products was explained through a three-step cascade reaction beginning with the attack of indole on the benzodiene, which gave the corresponding indolinium **4** (Scheme 1.10). This intermediate **4** was susceptible to addition of its imidate group to afford the corresponding iminium intermediate **5**. A final Mannich addition of the amide group in **5** to the iminium moiety provided the final products. The stereochemistry of the chiral formed products was established by X-ray crystal-structural analysis.

In another context, the Michael reaction has also been combined with intramolecular alkylation reaction. As an example, the synthesis of a pheromone was achieved by Reddy *et al.* on the basis of a domino Michael–intramolecular alkylation reaction of a chiral α,β-unsaturated ester

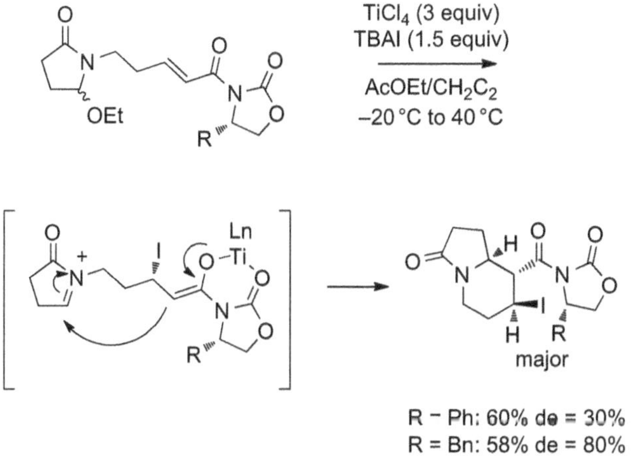

R = Ph: 60% de = 30%
R = Bn: 58% de = 80%

Scheme 1.9   Intramolecular domino Michael–Mannich-type reaction.

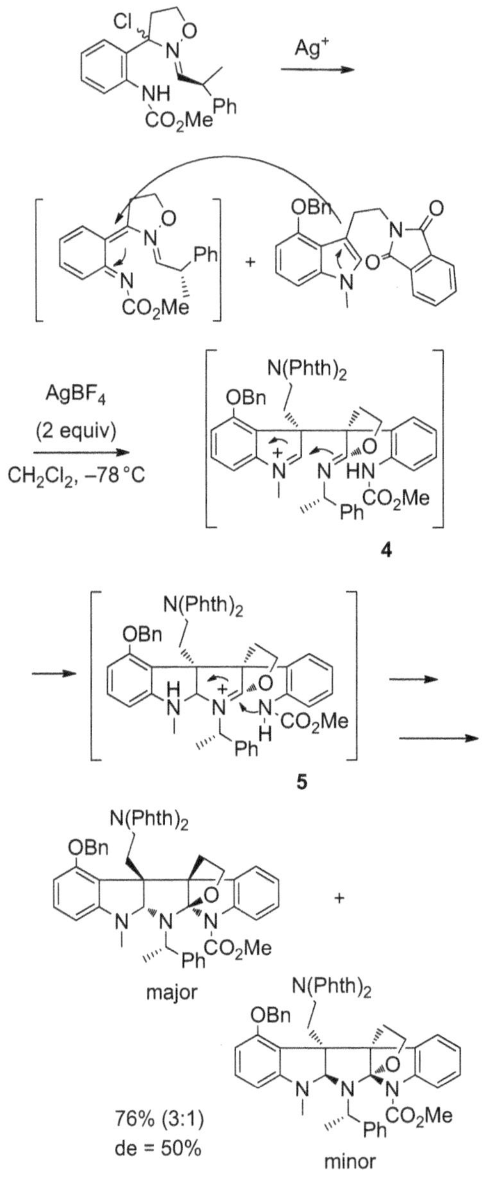

Scheme 1.10   Domino Michael–Mannich–Mannich reaction.

derived from (*R*)-pantolactone with Me$_2$CuLi.[22] Indeed, the resulting enolate arising from the Michael addition of this cuprate to the α,β-unsaturated ester subsequently cyclised to furnish the corresponding chiral cyclopentane derivative as a single diastereomer in 94% yield, as shown in Scheme 1.11. This product was further converted into the expected sex pheromone depicted in the same scheme.

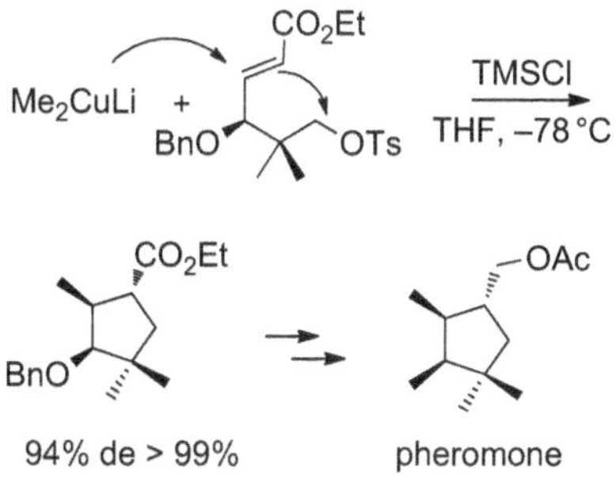

Scheme 1.11　Synthesis of a pheromone through domino Michael–intramolecular alkylation reaction.

In 2010, another type of asymmetric domino Michael–intramolecular alkylation reaction was developed by Pathak *et al.* to achieve a range of chiral six-membered heterocycles and carbocycles bearing three contiguous stereocentres.[23] In this case, the domino reaction occurred between a chiral vinyl sulfone derived from ribose and Na_2S, which provided the corresponding enantiopure cyclic sulfide in 72% yield, as shown in Scheme 1.12. This highly efficient methodology was applied to a series of other nucleophiles, such as primary amines and dialkyl malonates, leading to the corresponding piperidine derivatives and cyclohexane derivatives, respectively. Remarkably, in all cases of the substrates studied, the products were produced in good to high yields and as single diastereomers. Furthermore, the scope of the reaction was applied to malonitrile as nucleophile, yielding the corresponding enantiopure carbocycle in 68% yield, as shown in Scheme 1.12. The synthetic utility of many of these chiral products was demonstrated by converting the sulfonylated piperidine derivatives into functionalised olefinic piperidines, for example.

In 2012, Csaky *et al.* developed a nice route to enantiopure poly-functionalised tetrahydropyrans based on an asymmetric domino Michael–acetal ring-opening reaction promoted by trifluoroacetic anhydride.[24] This sequence consisted in the metal-free conjugate addition of boronic acids to chiral acetals performed in the presence of 3 equivalents of trifluoroacetic anhydride, providing the corresponding tetrahydropyrans in good to high yields and high to excellent diastereoselectivities of up to >96% de. A mechanism accounting for the diastereoselective formation of these products is depicted in Scheme 1.13. Reaction of trifluoroacetic anhydride with boronic acid could give a mono- or a diacylboronate intermediate **6**, in which the Lewis acidity of the boron atom was enhanced with respect to boronic acid. Co-ordination of this species with the γ-oxygen of acetal led to intermediate **7**. The subsequent intramolecular delivery of the R^1 group was facilitated by the lone pair on the δ-oxygen, providing intermediate **8**. Intramolecular ring-closure accounted for the formation of the final products. The stereochemistry has been found to depend on both the substrate and reagent structures. Indeed, a highly selective *trans* relative disposition between R^1 and OH groups in the final

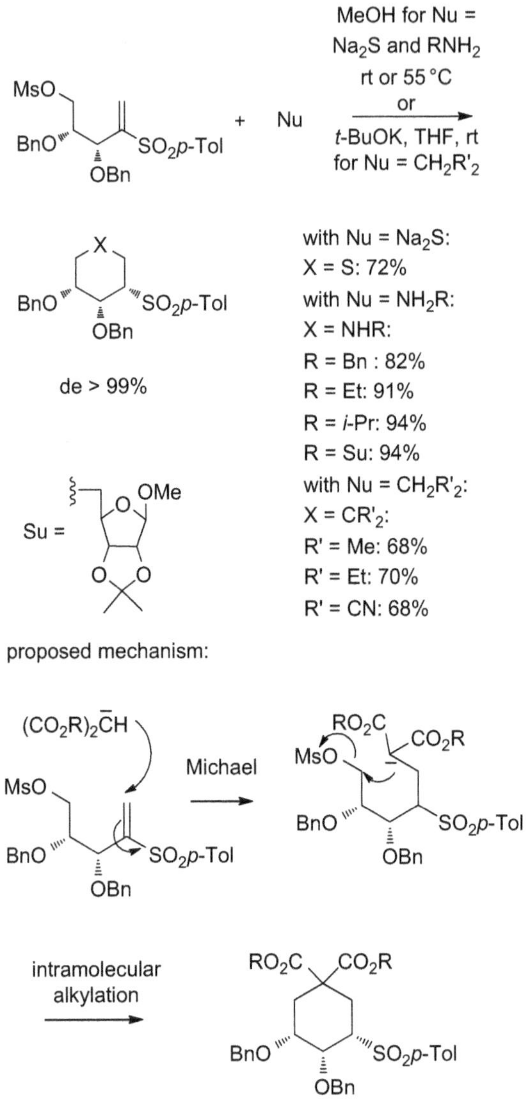

MeOH for Nu =
Na$_2$S and RNH$_2$
rt or 55 °C
or
t-BuOK, THF, rt
for Nu = CH$_2$R'$_2$

MsO

BnO''''   SO$_2$p-Tol
OBn

+   Nu

with Nu = Na$_2$S:
X = S: 72%
with Nu = NH$_2$R:
X = NHR:
R = Bn : 82%
R = Et: 91%
R = i-Pr: 94%
R = Su: 94%
with Nu = CH$_2$R'$_2$:
X = CR'$_2$:
R' = Me: 68%
R' = Et: 70%
R' = CN: 68%

X

BnO''''   '''SO$_2$p-Tol
OBn

de > 99%

Su =

OMe
O
O   O

proposed mechanism:

(CO$_2$R)$_2\overline{\text{C}}$H

MsO

BnO''''   SO$_2$p-Tol
OBn

Michael

RO$_2$C   CO$_2$R
MsO

BnO''''   SO$_2$p-Tol
OBn

intramolecular
alkylation

RO$_2$C   CO$_2$R

BnO''''   '''SO$_2$p-Tol
OBn

Scheme 1.12   Domino Michael–intramolecular alkylation reaction.

products was found, which was consistent with a substrate-controlled chelated *syn* Michael addition step, as depicted on intermediate **7**. The final cyclisation could be envisioned by an approach of the enolate to the sp^2 carbon of the electrophilic moiety in a chairlike transition state **9**. The pseudoequatorial disposition of the substituents minimised the steric interaction between R^1 and benzyl groups in the final tetrahydropyrans.

In 2008, Tang *et al.* reported a remarkable synthesis of a collection of enantiopure highly functionalised cyclohexadiene epoxide derivatives through

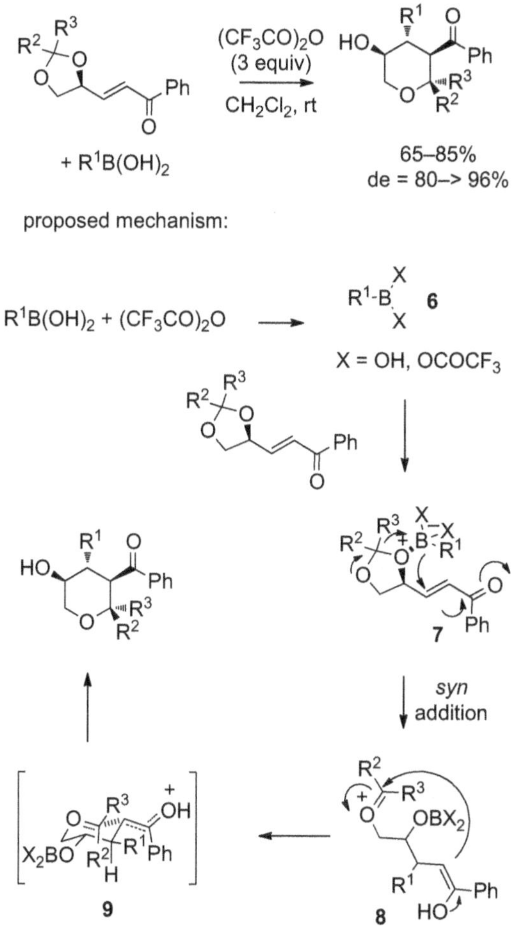

Scheme 1.13 Domino Michael–acetal ring-opening reaction.

asymmetric intramolecular as well as intermolecular domino Michael–ylide epoxidation reactions.[25] As shown in Scheme 1.14, when the two types of camphor-derived chiral allylic sulfonium salt depicted in Scheme 1.14 were treated with K_2CO_3, they afforded the corresponding cyclohexadiene epoxides as single products in good yields and generally excellent enantio-selectivities of >91% ee in both intramolecular and intermolecular reactions with α,β-unsaturated enones. In addition to being highly functionalised and enantiopure, it must be noted that these products exhibited three stereocentres.

Further studies by the same authors have revealed that the intermolecular reaction of a chiral camphor-derived sulfonium salt with α,β-unsaturated enones depicted above could be switched to give rise to enantioenriched vinylcyclopropanes.[26] Indeed, when the reaction was performed in the presence of a strong base, such as *t*-BuOK in tetrahydrofuran (THF), it selectively

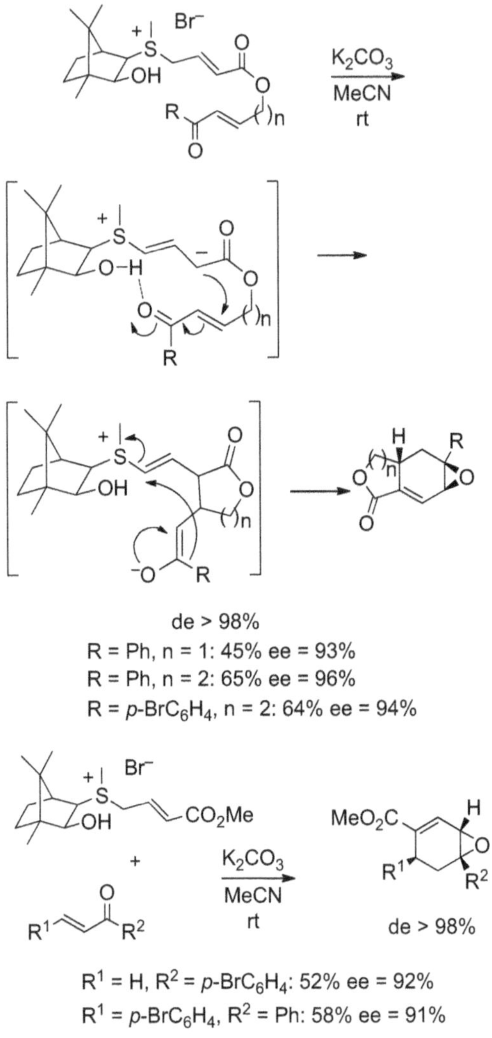

de > 98%
R = Ph, n = 1: 45% ee = 93%
R = Ph, n = 2: 65% ee = 96%
R = p-BrC₆H₄, n = 2: 64% ee = 94%

R¹ = H, R² = p-BrC₆H₄: 52% ee = 92%
R¹ = p-BrC₆H₄, R² = Ph: 58% ee = 91%

Scheme 1.14 Intra- and intermolecular domino Michael–ylide epoxidation reactions.

afforded the corresponding vinylcyclopropanes instead of the cyclohexadiene epoxides obtained by using a weak base such as K_2CO_3 in MeCN. As shown in Scheme 1.15, a range of trisubstituted chiral vinylcyclopropanes were achieved in good yields and excellent enantioselectivities of up to >99% ee in almost all cases of the substrates studied. Moreover, the diastereoselectivity of the process was found to be in a range of >92 to >98% de for all the substrates studied. A possible mechanism to account for the results observed with the asymmetric tunable ylide-initiated reaction is depicted in Scheme 1.15. The sulfur ylide has two major contributing resonances, **10** and **10′**, and the latter underwent Michael addition with unsaturated kenone to form intermediate **11**. In the

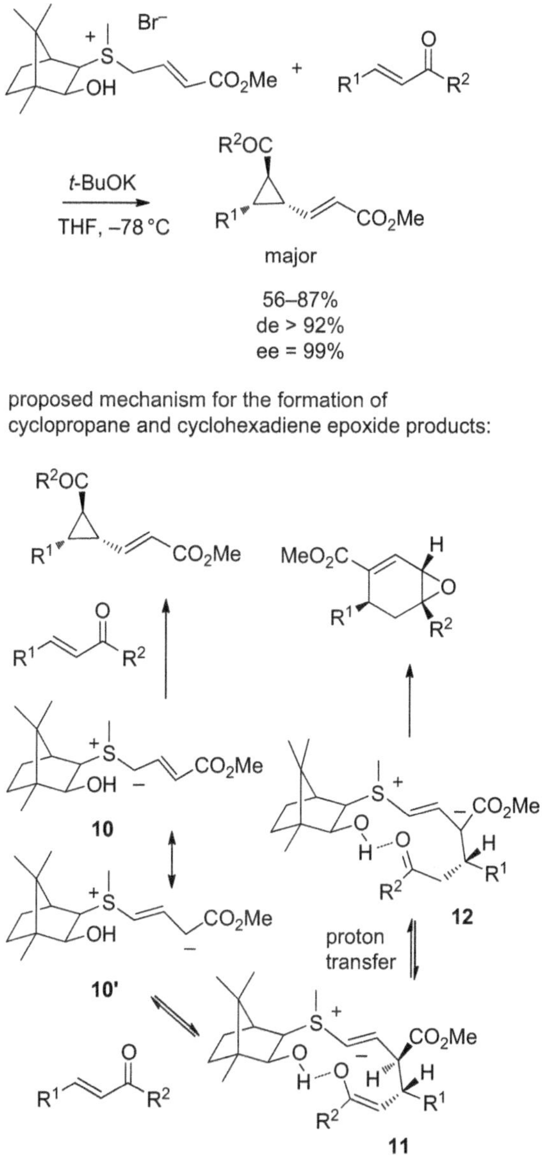

proposed mechanism for the formation of
cyclopropane and cyclohexadiene epoxide products:

Scheme 1.15 Domino Michael–ylide cyclopropanation reaction.

presence of a weak base such as Cs_2CO_3, the base could not remove the proton
smoothly on the α-position of the ester group in intermediate **11** and,
consequently, the proton transferred to enolate to form **12**, allowing the
intramolecular ylide epoxidation to afford a cyclohexadiene epoxide as a major
product. In the presence of a strong base and under low temperature condition,
intermediate **11** disfavoured a proton-transfer process to form ketone **12** and,

as a consequence, a normal ylide cyclopropanation between ylide **10** and unsaturated ketone was preferred to afford a cyclohexadiene epoxide as a major product.

In 2009, the same authors developed an asymmetric domino Michael–ylide olefination reaction of crotonate-derived chiral phosphonium salts with α,β-unsaturated ketones in the presence of a base such as Cs_2CO_3.[27] This domino process provided the corresponding enantiomerically enriched functionalised cyclohexa-1,3-diene derivatives in moderate to good yields and enantioselectivities (25–90% ee) when using (R)-2-MeO-MeOBIPHEP-derived phosphonium salt as chiral auxiliary. As depicted in Scheme 1.16, the chiral phosphonium salt generated the corresponding chiral phosphorus ylide **13**, which underwent Michael addition to the α,β-unsaturated enone to give an enolate **14**. This enolate generated a novel phosphorus ylide **15** which underwent intramolecular Wittig olefination to give the final cyclohexadiene.

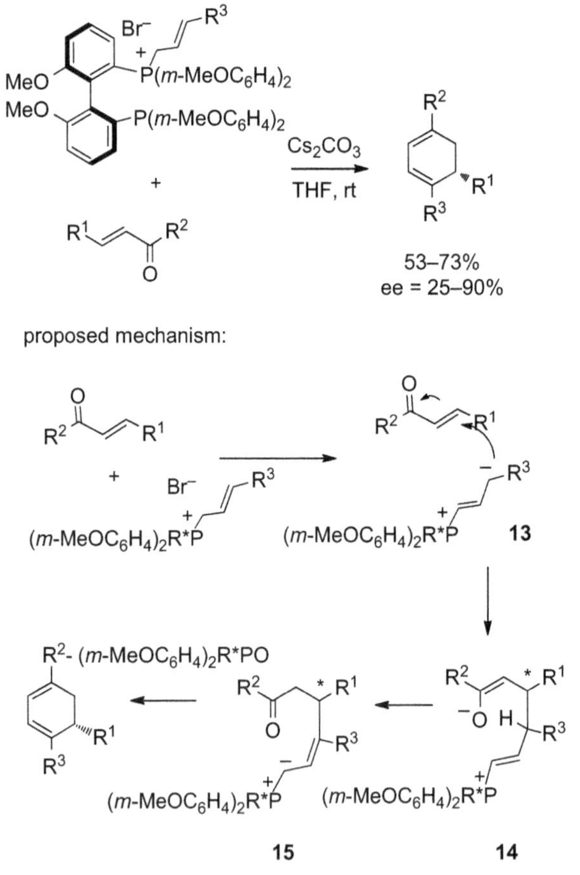

Scheme 1.16   Domino Michael–Wittig reaction.

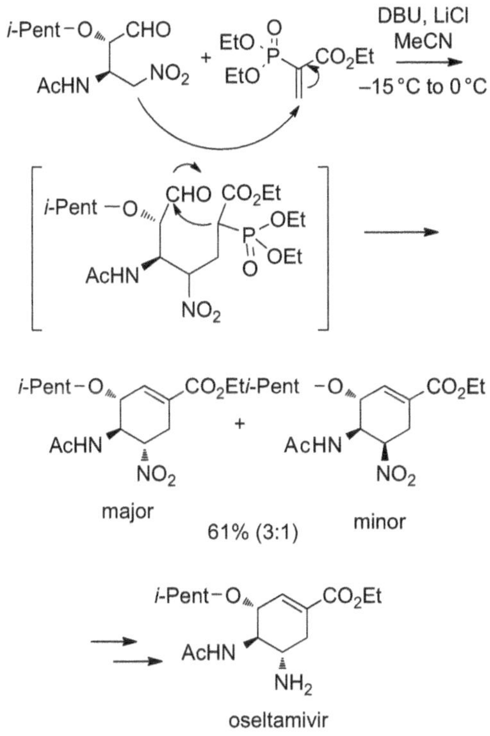

Scheme 1.17 Synthesis of oseltamivir through domino Michael–Horner–Wadsworth–Emmons reaction.

In another context, Lu *et al.* have described a practical and azide-free synthetic approach to the antiviral agent oseltamivir (Tamiflu) from diethyl D-tartrate.[28] The key step of this synthesis consisted in a domino aza-Michael–Horner–Wadsworth–Emmons reaction of a chiral nitroalkane, which allowed the cyclohexene ring of the antiviral product to be achieved. The corresponding domino product was obtained in 61% yield as a 3:1 mixture of diastereomers, as shown in Scheme 1.17. This diastereomeric mixture was further converted into the expected oseltamivir in three supplementary steps.

In the same area, another asymmetric domino Michael–Horner–Wadsworth–Emmons reaction was employed as key step in the synthesis of dipeptidyl peptidase IV-selective inhibitor ABT-341, which was reported by Hayashi *et al.* in 2011.[29] In this case, the chiral starting material of the domino reaction was another chiral nitroalkane, which is depicted in Scheme 1.18. The process evolved through the formation of a Michael product which was subsequently submitted to Horner–Wadsworth–Emmons reaction, affording the corresponding chiral *cis*-cyclohexene as a single diastereomer in 87% yield. This product could be further converted into ABT-341 in four steps.

In another context, an asymmetric domino aza-Michael–imine formation reaction was reported by Taylor *et al.*, allowing the synthesis of the alkaloid

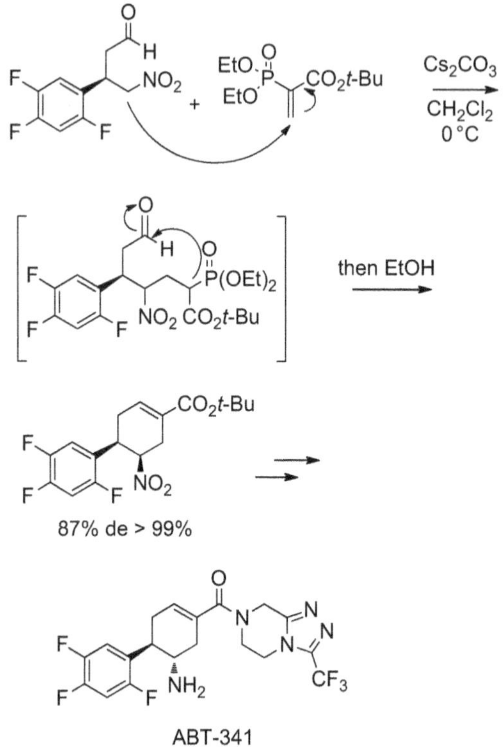

Scheme 1.18   Synthesis of ABT-341 through domino Michael–Horner–Wadsworth–Emmons reaction.

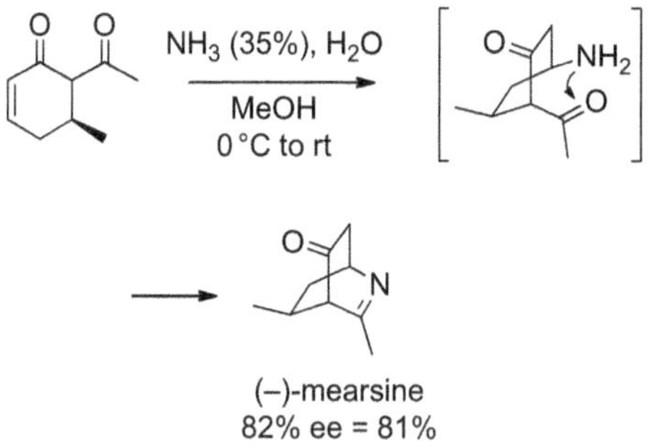

Scheme 1.19   Synthesis of (–)-mearsine through domino aza-Michael–imine formation reaction.

(–)-mearsine to be achieved.[30] As shown in Scheme 1.19, treatment of a chiral diketone with 35% aqueous ammonia in methanol initiated the domino reaction with the Michael addition of ammonia to the enone, providing an amine intermediate which further underwent intramolecular imine formation reaction to give the expected (–)-mearsine in 82% yield combined with enantioselectivity of 81% ee.

1.2.1.2 Domino Reactions Initiated by Other Anionic Reaction

Domino reactions involving the formation of anionic intermediates in their first step constitute the largest family of domino processes. In reactions of this type, the primary step is the formation of an anion or a nucleophile, as for the domino reactions initiated by a Michael addition discussed in Section 1.2.1.1. This family of domino reactions has been used extensively in total synthesis. A number of anionic reactions other than the Michael addition have been incorporated as the first steps in asymmetric domino reactions in the past few years. As an example, Fustero *et al.* have developed a new domino reaction consisting in the addition of fluorinated nucleophiles to (*R*)-*N*-(*tert*-butanesulfinyl)imines, followed by an intramolecular aza-Michael addition.[31] As shown in Scheme 1.20, this asymmetric process has been applied to the synthesis of various chiral fluorinated 1,3-disubstituted isoindolines in moderate to high yields and excellent diastereoselectivities of up to 98% de. The removal of the *tert*-butanesulfinyl group was achieved by treatment of these isoindolines with 4 M HCl in dioxane, yielding the corresponding free amines after basification with 2.2 M NaOH. In order to make the domino methodology synthetically more useful, its scope was expanded to partially fluorinated nucleophiles, such as the nucleophilic fluoroalkylating agent

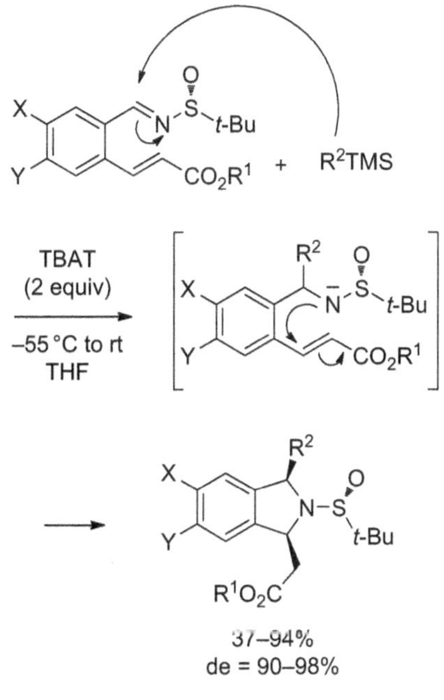

Scheme 1.20 Domino nucleophilic addition–aza-Michael reaction.

PhSO$_2$CF$_2$H. In this case, the reaction was performed at $-78\,°$C in THF in the presence of 1.1 equivalents of lithium hexamethyldisilazide (LiHMDS) as a base and provided the corresponding domino product in 63% yield as a single diastereomer.

On the other hand, Ma *et al.* have reported a novel domino cross-Rauhut–Currier–acetalisation reaction, occurring between 2-bromo-cyclohexenal and (–)-menthyl ester, which provided in the presence of 1,8-diazabicyclo[5.4.0]undec-7-ene (DBU) the corresponding functionalised spiro-3,4-dihydropyran in 74% yield, albeit combined with a low diastereo-selectivity of 11% de.[32] In another context, Brimble *et al.* have developed a flexible asymmetric synthesis of the tetracyclic core of berkelic acid, an extre-mophile natural product with selective activity against ovarian cancer, which was based on a domino Horner–Wadsworth–Emmons–oxa-Michael reaction of a α-ketophosphonate with a chiral lactone.[33] The domino process provided the corresponding isochroman as an approximately 1 : 1 mixture of *cis* : *trans* chiral diastereomers in 80% yield. This mixture constituted the key inter-mediate in the synthesis of the tetracyclic benzannulated spiroketal core of berkelic acid. In the same area, Kawasaki *et al.* have employed an asymmetric domino Horner–Wadsworth–Emmons–isomerisation–Claisen rearrangement reaction as a key step in the synthesis of biologically active marine indole alkaloids, such as (–)-flustramines A, B, (–)-flustramides A, B, (–)-fructigenine A, and (–)-5-*N*-acetylardeemin.[34] As shown in Scheme 1.21, the domino process occurred between diethyl cyanomethylphosphonate and a diastereoisomer mixture of substituted *N*-acetylindolin-3-ones in the presence of *t*-BuOK to provide the corresponding Horner–Wadsworth–Emmons products, which subsequently isomerised into intermediates **16**. These intermediates sub-sequently underwent a Claisen rearrangement to afford the final almost enantiopure oxindoles. These key products were further converted into natural pyrrolo[2,3-*b*]indole alkaloids, including (–)-flustramines A and B, (–)-flustramides A and B, (–)-fructigenine A, and (–)-5-*N*-acetylardeemin.

In another context, a highly efficient synthesis of enantiopure highly func-tionalised piperidines from chiral sulfinyl imines has been achieved by Ghorai *et al.* on the basis of an asymmetric domino imino-aldol–aza-Michael reaction with α-arylmethylidene-β-keto esters.[35] As shown in Scheme 1.22, the reaction of chiral 2-aryl-*N*-sulfinylaldimines with (*E*)-ethyl 2-benzylidene-3-oxobutanoate led to the corresponding 2,6-disubstituted piperidines as single diastereomers in good yields and remarkable enantioselectivities of $>98\%$ ee.

In 2008, enantioselective total syntheses of biologically active lyconadin alkaloids A and B were achieved by Smith *et al.* on the basis of an asymmetric domino intramolecular aldol–Michael reaction of a chiral diketo aldehyde, allowing the formation of an enantiopure tricyclic key product bearing three stereogenic centres in 84% yield.[36] This key product could be further converted into (+)-lyconadin A and (–)-lyconadin B. In 2009, Tietze *et al.* reported an efficient formal total synthesis of the erythrina alkaloid (+)-erysotramidine by using a domino process consisting of an amidation of a chiral ketoester with a primary amine to provide the corresponding amide **17**. This amide reacted with

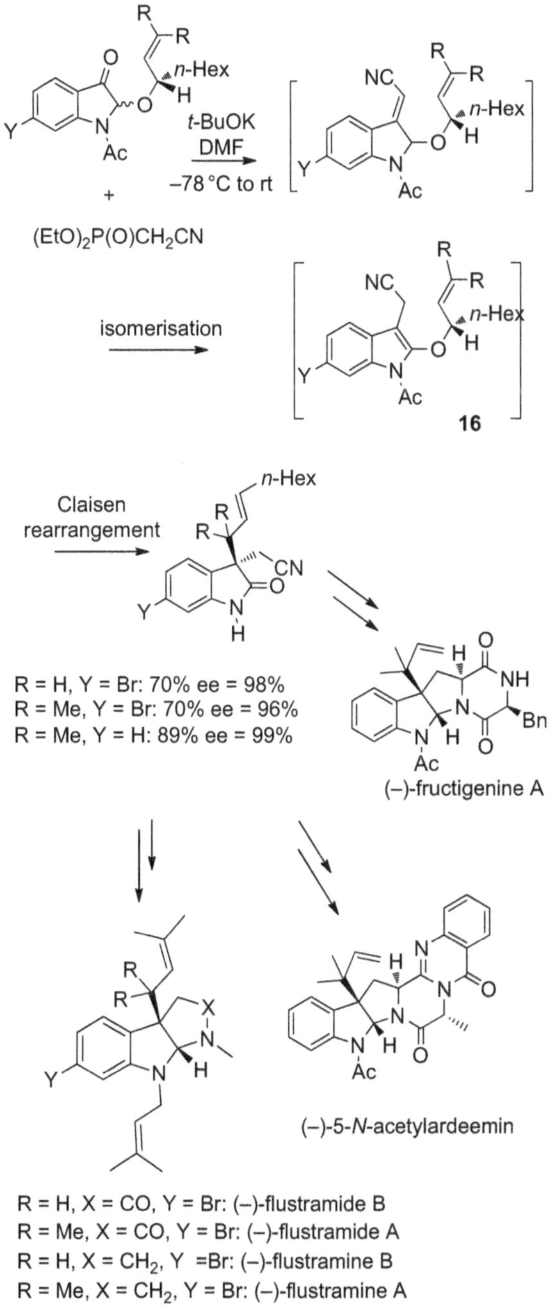
t-BuOK
DMF
−78 °C to rt
$(EtO)_2P(O)CH_2CN$
isomerisation
16
Claisen
rearrangement
R = H, Y = Br: 70% ee = 98%
R = Me, Y = Br: 70% ee = 96%
R = Me, Y = H: 89% ee = 99%
(−)-fructigenine A
(−)-5-N-acetylardeemin
R = H, X = CO, Y = Br: (−)-flustramide B
R = Me, X = CO, Y = Br: (−)-flustramide A
R = H, X = CH_2, Y =Br: (−)-flustramine B
R = Me, X = CH_2, Y = Br: (−)-flustramine A

Scheme 1.21 Synthesis of alkaloids through domino Horner–Wadsworth–Emmons–isomerisation–Claisen rearrangement reaction.

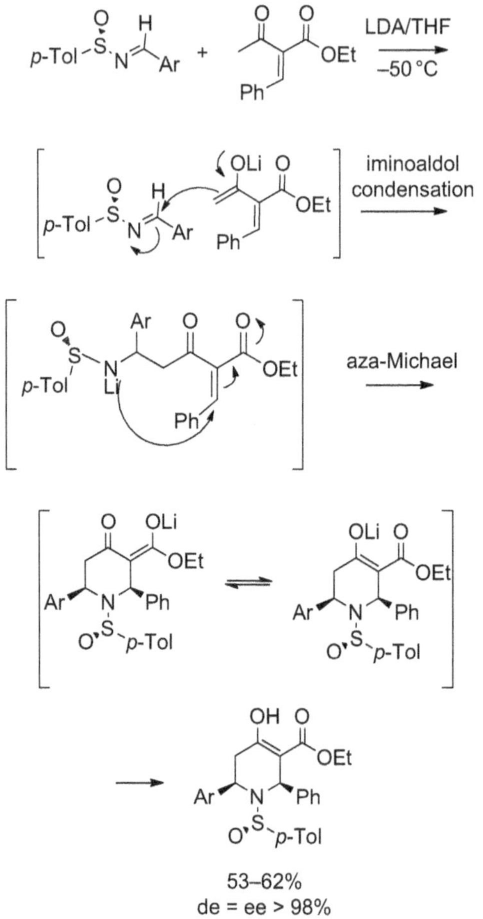

Scheme 1.22 Domino imino-aldol–aza-Michael reaction.

the carbonyl moiety through intramolecular cyclisation to give an iminium ion **18** which constituted the domino amidation-cyclisation product.[37] Under treatment with trifluoromethanesulfonyl hydroxide (TfOH), the iminium ion **18** underwent an intramolecular electrophilic aromatic substitution to give the corresponding spirocyclic skeleton of (+)-erysotramidine as a 4:1 mixture of diastereomers, as shown in Scheme 1.23.

On the other hand, Romo *et al.* have reported a concise total synthesis of the biologically active natural product (–)-belactosin C and derivatives based on a diastereoselective domino Mukaiyama aldol–lactonisation reaction of chiral dipeptide glyoxamide or a novel tartrate-derived chiral glyoxylate with chiral silyl ketene acetals.[38] These processes allowed the construction of the pharmacophoric β-lactone moiety of these proteasome inhibitors to be achieved in a single step, albeit in zero to moderate diastereoselectivities. On the other hand, Hurd *et al.* have developed a practical two-step asymmetric synthesis of the

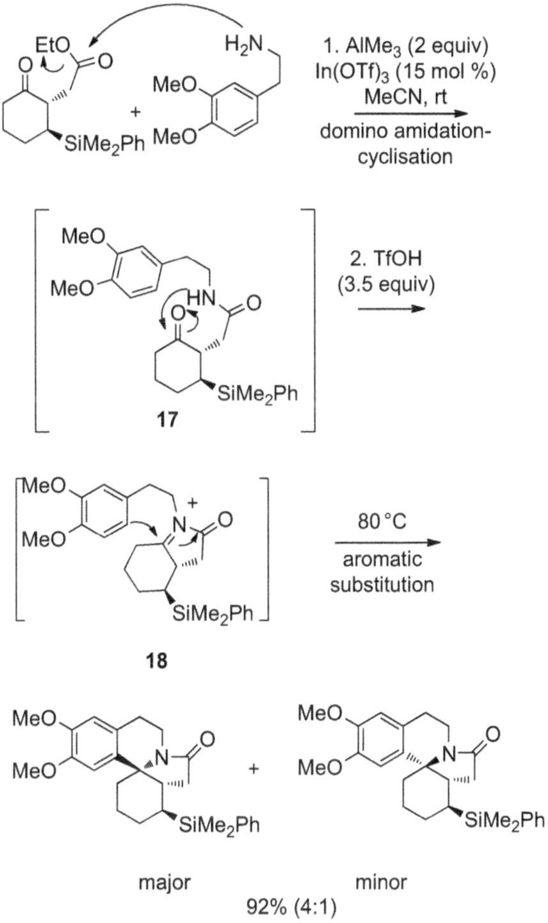

Scheme 1.23 Domino amidation–cyclisation reaction followed by intramolecular electrophilic aromatic substitution.

(−)-enantiomer of PNU-286607, which is a promising member of a novel class of antibacterial agents.[39] The key step of the synthesis was a domino Knoevenagel–[1,5]-hydrogen shift–cyclisation reaction, occurring between a chiral aldehyde and barbituric acid when heated at 117 °C in *n*-butanol. As shown in Scheme 1.24, the process began with the Knoevenagel condensation between the two substrates, producing alkylidene intermediate **19**. This intermediate then underwent a [1,5]-hydrogen shift to give zwitterion **20**, which finally cyclised to afford, in 74% yield, the enantiopure (−)-PNU-286607 as a single stereoisomer.

In 2011, Metz *et al.* described the first total synthesis of the macrodiolide antibiotic pamamycin-649B by using as the key step an asymmetric domino elimination–alkoxide-directed 1,6-addition of the ethyllithium reaction of a chiral sulfone induced by BF₃·Et₂O.[40] In spite of a moderate yield of 36%, the

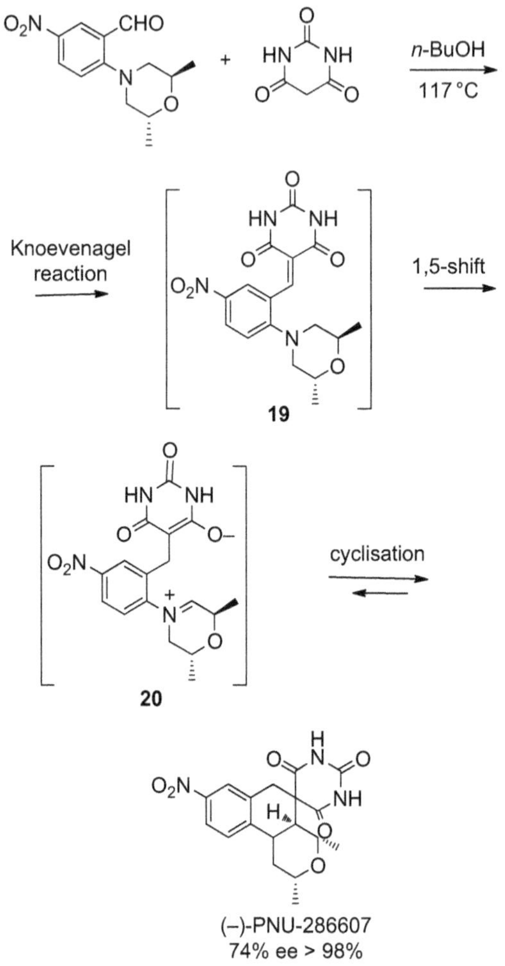

Scheme 1.24   Synthesis of (–)-PNU-286607 through domino Knoevenagel–[1,5]-
              hydrogen shift–cyclisation reaction.

corresponding enantiopure bicyclic sulfone exhibiting four stereogenic centres
was produced as a single stereoisomer, as shown in Scheme 1.25.

   Another important natural and biologically active product, (+)-chinensiolide
B, was recently synthesised on the basis of an asymmetric domino
allylboration–lactonisation reaction starting from a chiral aldehyde derived
from (*R*)-carvone, as depicted in Scheme 1.26.[41] When this aldehyde was
treated by a *Z/E* mixture of allylboronate in the presence of BF$_3$·Et$_2$O, it
provided the corresponding *trans*-γ-lactone product in remarkable yield of 87%
and diastereoselectivity of >90% de. This chiral product was further converted
into (+)-chinensiolide B in nine supplementary steps.

   In 2008, Fustero *et al.* developed a one-pot synthesis of optically pure
fluorinated indolines on the basis of a novel asymmetric domino reaction

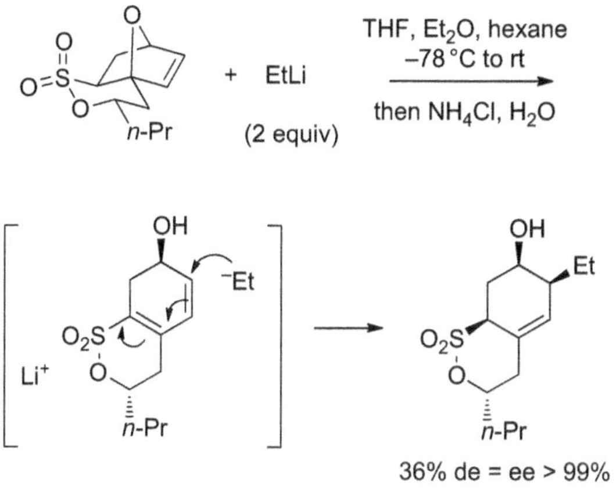

Scheme 1.25 Domino elimination–1,6-addition reaction.

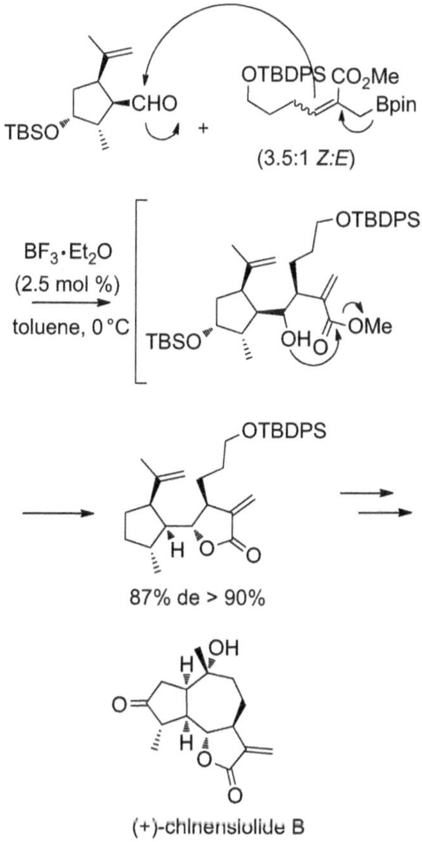

Scheme 1.26 Synthesis of (+)-chinensiolide B through domino allylboration–lactonisation reaction.

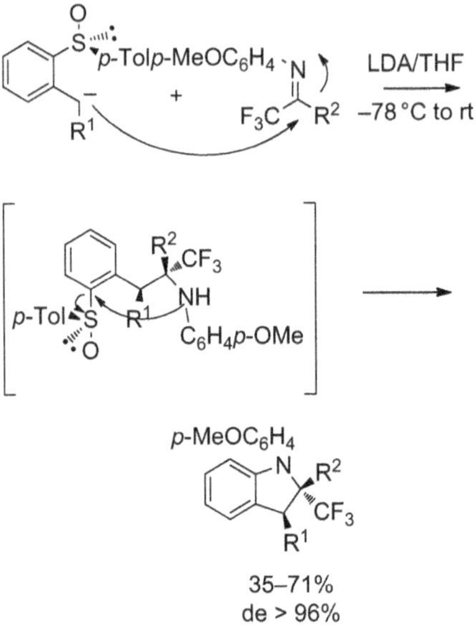

Scheme 1.27 Domino nucleophilic addition–nucleophilic aromatic substitution reaction.

including a nucleophilic addition of the carbanion of chiral 2-*p*-tolylsulfinyl alkylbenzenes, generated in the presence of lithium diisopropylamide (LDA) as a base, to fluorinated aldimines and ketimines. This was followed by an intramolecular nucleophilic aromatic substitution of the *p*-tolylsulfinyl group by the amine anion, providing the corresponding fluorinated indolines containing one or two stereogenic centres in good yields and diastereoselectivity of >96% de in all cases of the substrates studied (Scheme 1.27).[42]

In the same year, Garrido *et al.* reported a novel asymmetric domino process, involving a diastereoselective Ireland–Claisen rearrangement followed by a Michael addition, allowing the first practical and efficient one-pot route to optically active γ-substituted δ-amino acids to be achieved.[43] As shown in Scheme 1.28, the reaction of Baylis–Hillman adducts with a chiral lithium amide afforded the corresponding chiral γ-substituted δ-amino acids in moderate to good yields, combined with both high diastereo- and enantio-selectivities of up to >95% de and ee, respectively. It is reasonable to think that deprotonation of the Baylis–Hillman adduct was favoured over initial Michael addition to a trisubstituted double bond. Consequently, a rearrangement, probably through transition state **21**, gave rise to transition state **22**, which further led to intermediate **23**. Then, intermediate **23** underwent Michael addition with the chiral lithium amide to afford the final product. This methodology was applied to the total synthesis of the non-peptidic neurokinin NK1 receptor antagonists (+)-L-733,060, and (+)-CP-99,994.[44] Moreover, these authors have shown that, when benzaldehyde was used as the starting material in the Baylis–Hillman adduct (R^1 = Ph-CH=CH, R^2 = OMe), the

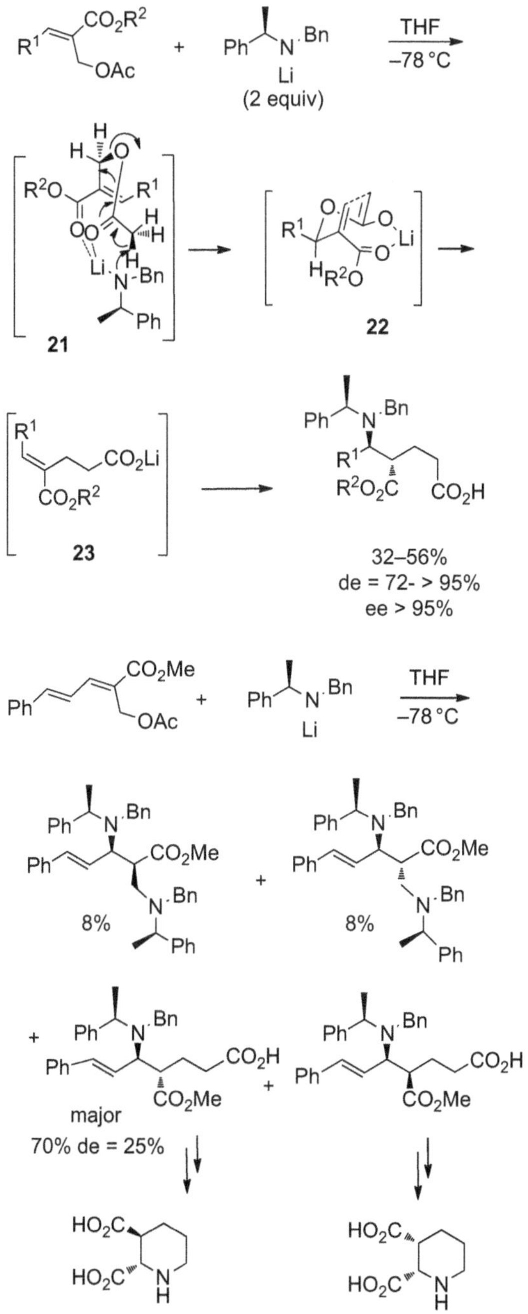

Scheme 1.28 Domino Ireland–Claisen rearrangement–Michael reactions.

domino nucleophilic addition–nucleophilic aromatic substitution reaction of the corresponding domino substrate was preceded by an allylic acetate rearrangement to give the corresponding chiral δ-amino acids as a 63 : 37 diastereoisomeric mixture, along with diastereoisomeric byproducts arising from the double addition reaction (Scheme 1.28). These products were further converted into biologically active chiral *cis*- and *trans*-piperidine dicarboxylic acids.[45]

In addition, a novel total synthesis of (–)-oseltamivir was reported by Fukuyama *et al.*, in which the building of the cyclohexene skeleton was achieved through an asymmetric domino reaction, evolving successively through ethanolysis of a chiral *N*-Boc (*tert*-butoxycarbonyl) lactam, dehydrobromination, and aziridine formation *via* an S_N2 reaction to provide the corresponding enantiopure highly functionalised cyclohexene in 87% yield, as shown in Scheme 1.29.[46] This key product was further converted into (–)-oseltamivir in four steps.

A S_N2 reaction was also employed by Gharpure *et al.* as first step in a diastereoselective domino reaction allowing the construction of 2,3,3,6-tetrasubstituted tetrahydropyrans to be achieved.[47] As shown in Scheme 1.30, the domino S_N2–Michael reaction of active methylene compounds with vinylogous carbonates performed in the presence of cesium carbonate as a base led to the

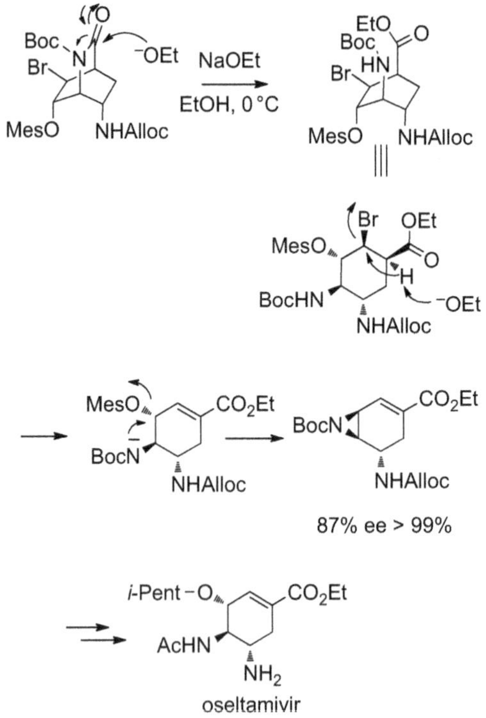

87% ee > 99%

oseltamivir

Scheme 1.29 Synthesis of oseltamivir through domino ethanolysis–dehydrobromation–aziridination reaction.

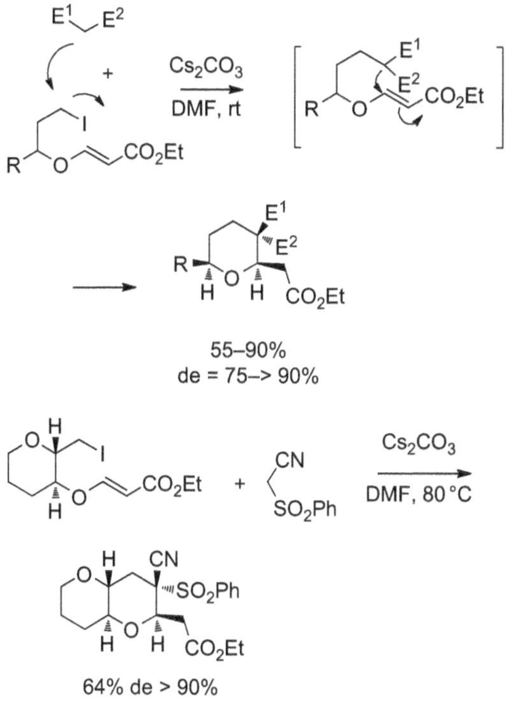

55–90%
de = 75–> 90%

64% de > 90%

Scheme 1.30 Domino S_N2–Michael reactions.

corresponding substituted tetrahydropyran derivatives in good to high yields and relative diastereoselectivities of up to >90% de. Symmetrical active methylene compounds, such as dimethyl malonate or malononitrile, as well as unsymmetrical active methylene compounds, such as sulfone nitrile, sulfone ester or ethyl cyanoacetate, gave good results. Interestingly, better diastereo-selectivities were obtained when using unsymmetrical nucleophiles than symmetrical nucleophiles. The scope of this methodology was extended to the synthesis of fused bicyclic tetrahydropyrans starting from an iodide derived from tri-*O*-acetyl-(D)-glucal, which was reacted with a sulfone nitrile to give the corresponding bis-tetrahydropyran in 64% yield and excellent relative diastereo-selectivity of >90% de, as shown in Scheme 1.30. It must be noted that this type of product is applicable in polycyclic ether fragments of ladder toxins.

In another area, Wolfe *et al.* have described asymmetric domino Wittig rearrangement–aldol reactions, occurring between chiral *O*-benzyl- and *O*-allylglycolate esters of chiral *trans*-2-phenylcyclohexanol and aldehydes.[48] The reactions proceeded through *Z*-boron ester enolates, and afforded the corresponding *syn*-α-alkyl-α,β-dihydroxy esters in good yields, enantio-selectivities of up to 95% ee, and >95:5 *syn* : *anti* diastereoisomeric ratio, after cleavage of the 2-phenylcyclohexanol chiral auxiliary achieved by treatment with LiAlH₄, as shown in Scheme 1.31. The scope of this methodology was

domino Wittig rearrangement-aldol reaction:

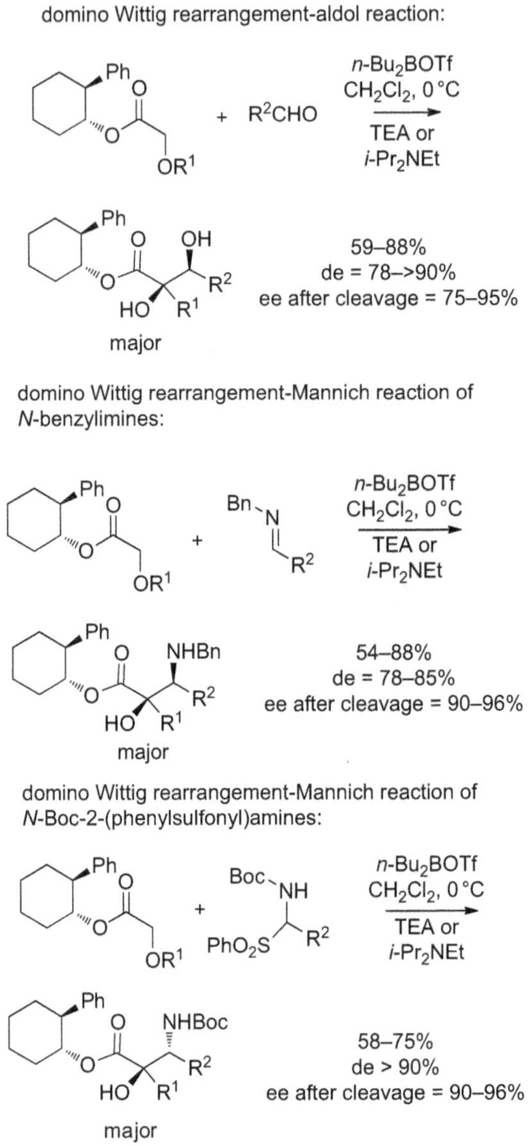

domino Wittig rearrangement-Mannich reaction of
N-benzylimines:

domino Wittig rearrangement-Mannich reaction of
N-Boc-2-(phenylsulfonyl)amines:

Scheme 1.31 Domino Wittig rearrangement–aldol reaction and domino Wittig rearrangement–Mannich reactions.

extended by the same authors to asymmetric domino Wittig rearrangement–Mannich reactions by using imines instead of aldehydes.[49] As shown in Scheme 1.31, the reaction of the same chiral auxiliaries with a range of imines provided the corresponding α-alkyl-α-hydroxy-β-amino esters in good yields. It must be noted that the reaction of *N*-benzylimines derived from aromatic aldehydes led stereoselectively to the corresponding *syn*-domino products, while the reaction of *N*-benzylimines derived from unbranched aliphatic aldehydes failed to undergo the Mannich reaction. Moreover, the reaction of *N*-Boc imines or enamides derived from aliphatic aldehydes also failed to undergo the domino process. On the other hand, the authors have demonstrated that *N*-Boc-2-(phenylsulfonyl)amines stereoselectively reacted with *O*-benzyl- and *O*-allylglycolate esters of *trans*-2-phenylcyclohexanol to give the corresponding *anti*-α-alkyl-α-hydroxy-β-amino esters in good yields and diastereoselectivity of >90% de in all cases of the substrates studied. After cleavage of the chiral auxiliaries by treatment with LiAlH$_4$, the corresponding *anti*-aminoalcohols were achieved with enantioselectivities ranging from 90 to 96% ee, as shown in Scheme 1.31.

Finally, Compain *et al.* have developed the synthesis of spirocyclopropyl γ-lactams on the basis of a diastereoselective domino intramolecular azetidine ring-opening–closing reaction.[50] The key step of this process was an intra-molecular SN$_2$-type ring-opening of trimethylsilyl (TMS)OTf-activated azetidine rings by silyl ketene acetals generated by treatment with TMSOTf and triethylamine. As shown in Scheme 1.32, the reaction of azetidine in the presence of TMSOTf and triethylamine led to the corresponding functionalised 5-azaspiro[2.4]heptane derivative in good yield and relative diastereoselectivity of >98% de. The key step of the process was an SN$_2$-type ring-opening of the TMSOTf-activated azetidine ring by the silyl ketene acetal generated by treatment with TMSOTf and triethylamine (TEA). The amino ester **24** formed finally underwent an intramolecular cyclisation to afford the corresponding five-membered lactam through the reaction of the amine function with the ester

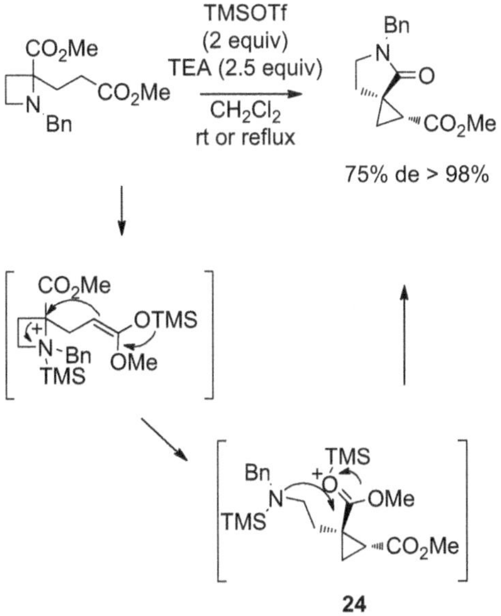

Scheme 1.32 Domino intramolecular azetidine ring-opening–closing reaction.

group in the γ position, as depicted in Scheme 1.32. Remarkably, in this process TMSOTf played a triple role by generating the reactive nucleophilic intermediate (silyl ketene acetal), by activating the azetidine for the nucleophilic ring-opening, and by activating the carbonyl group of the tertiary ester group for the final amide bond formation. The scope of this methodology was found, however, to be limited because substrates such as corresponding *tert*-butyl ester or corresponding lactam, *N*-Tos or *N*-Boc related azetidines, and azetidines bearing substituents in the α or β position did not provide the expected corresponding spirocyclopropyl γ-lactams.

1.2.2   Cationic Sequences

Cationic-mediated reactions constitute one of the oldest known subsets of domino reactions. In these processes, a carbocation is formed, either formally or in reality. This carbocation can be formed by elimination, or by addition of a positive particle such as a proton. The carbocation then reacts with a nucleophile to form a new carbocation that undergoes one or more comparable further transformations in a cationic–cationic process, finally being trapped by a nucleophile or stabilised by elimination of a proton. As an example, a diastereoselective cationic domino cyclisation was employed by Blaauw *et al.* as the key step in a total synthesis of the natural product (–)-dysibetaine PP.[51] As shown in Scheme 1.33, treatment of various enantiopure dipeptides by a catalytic amount of 4-toluenesulfonyl (Ts)OH underwent the formation of the corresponding bicyclic *N,N*-acetals through cascade cationic cyclisations in good yields and diastereoselectivities of up to 88% de in favour of the *trans*-isomer. This methodology was applied to a chiral dipeptide derived from L-allysine to give the corresponding amine, which was further converted into (–)-dysibetaine PP.

   In 2008, a $BF_3 \cdot Et_2O$-mediated asymmetric domino cationic epoxide-opening–cyclisation reaction was developed by Wiener *et al.* in the course of synthesising natural and biologically active tetracyclic schweinfurthins.[52] As shown in Scheme 1.34, the cascade sequence was initiated by a $BF_3 \cdot Et_2O$-promoted opening of a chiral epoxide, and terminated by a cationic cyclisation of one of the phenolic oxygens to give the corresponding chiral tricyclic product in 52% yield, along with another tricyclic product in 30% yield, both exhibiting the hexahydroxanthene core of schweinfurthins. The usefulness of this methodology was demonstrated in a total synthesis of schweinfurthin G starting from the major tricyclic product.

   In the same area, the first enantioselective total synthesis of a potential anti-influenza A virus agent (+)-stachyflin was achieved by Katoh *et al.* on the basis of an asymmetric domino epoxide-opening–rearrangement–cyclisation reaction, which stereoselectively formed the requisite pentacyclic ring system.[53] As shown in Scheme 1.35, treatment of a tetracyclic chiral epoxide by $BF_3 \cdot Et_2O$ led to the corresponding domino product as a mixture of two C3 epimers, which could be readily separated by chromatography. The desired epimer (C3 β-OH) was isolated in 9% yield along with the undesired epimer

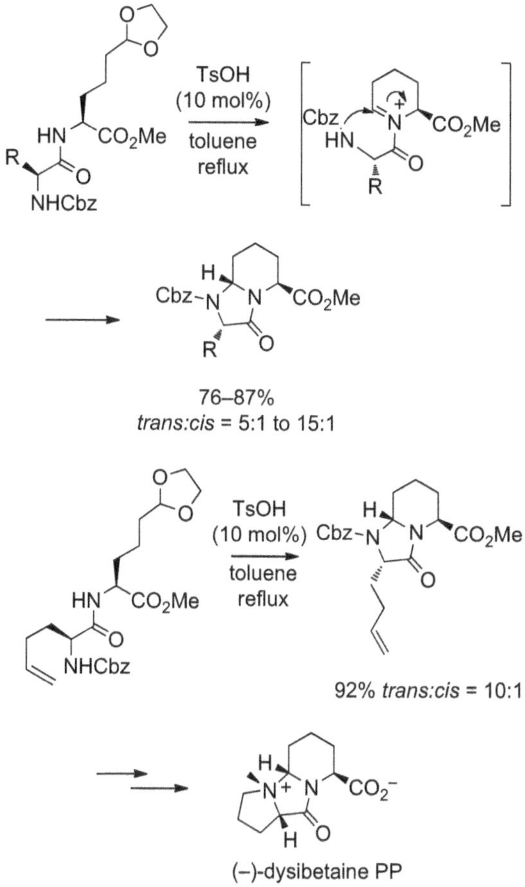

Scheme 1.33 Synthesis of (–)-dysibetaine PP through domino cyclisation reactions.

(C3 α-OH), which was isolated in 66% yield. The major epimer (C3 α-OH) could be converted in 96% yield into the desired β-OH epimer by successive oxidation with Dess–Martin periodinane and reduction with LiAlH(O*t*-Bu)$_3$. This domino sequence occurred in a stepwise manner through the carbocation intermediates **25**, **26** and **27** (Scheme 1.35). Indeed, the first coordination–activation between the Lewis acid and the epoxide moiety of the epoxide led to an epoxide ring-opening and the formation of intermediate **25**, which further produced intermediate **26** through the migration of the C5 methyl group to the C4 carbocation centre. Intermediate **26** underwent a 1,2-hydride shift from the C10 position to the C5 carbocation centre on the α-face of the molecule to provide intermediate **27**, wherein the C10 carbocation centre was trapped by the inner phenolic hydroxyl group to deliver, after elimination of the Lewis acid, the final desired cyclised product. The separated chiral minor product was transformed into (+)-stachyflin through two supplementary steps.

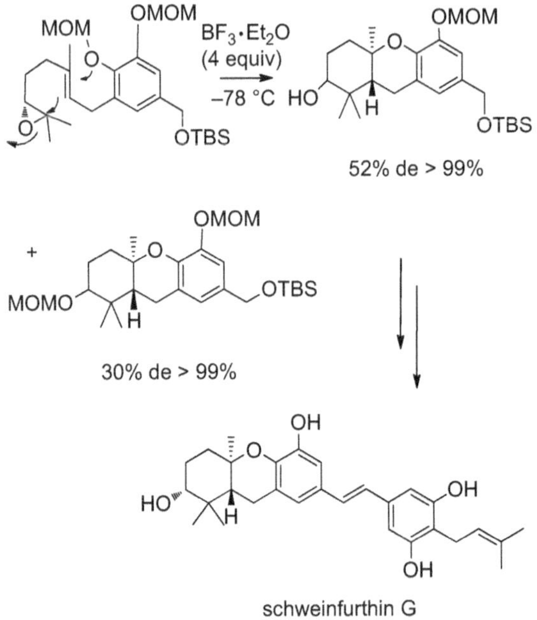

Scheme 1.34 Synthesis of schweinfurthin G through domino epoxide-opening–
cyclisation reaction.

In 2011, a domino Prins–Friedel–Crafts reaction of a chiral enal derived
from L-serine was successfully employed by Aubé *et al.* to achieve a key chiral
intermediate in a formal synthesis of the biologically active alkaloid
(–)-haouamine A.[54] This domino product was produced in 65% yield as a
1.5 : 1 mixture of two C2 α/β OH epimers. This mixture was further oxidised
into the corresponding ketone, which was subsequently submitted to
hydrogenolysis to provide the expected key product, as shown in Scheme 1.36.
Consequently, the required indeno-tetrahydropyridine was constructed in 13
steps from L-serine.

Finally, Canesi *et al.* have reported a diastereoselective oxidative domino
Prins–pinacol process mediated by a hypervalent iodine reagent, such as
PhI(OAc)$_2$, performed in a mixture of hexafluoroisopropanol (HFIP) and
dichloromethane as the solvent.[55] As shown in Scheme 1.37, the reaction
involved various phenols bearing a terminal alkene as an internal nucleophile to
trigger the oxidative Prins process, followed by a semipinacol-type rearrangement
to produce the main corresponding polyfunctionalised spiro[4.5]decanyl
compounds in moderate to good yields and relative diastereoselectivities of
>90% de. It was assumed that during the umpolung activation, which was
mediated by a single-electron transfer, the phenoxonium ion **28** generated was
trapped *via* an oxidative Prins process by the double bond, possibly through a
cyclic chairlike transition state, as depicted in Scheme 1.37. Then, a stereocontrolled ring contraction occurred with retention of the configuration of the

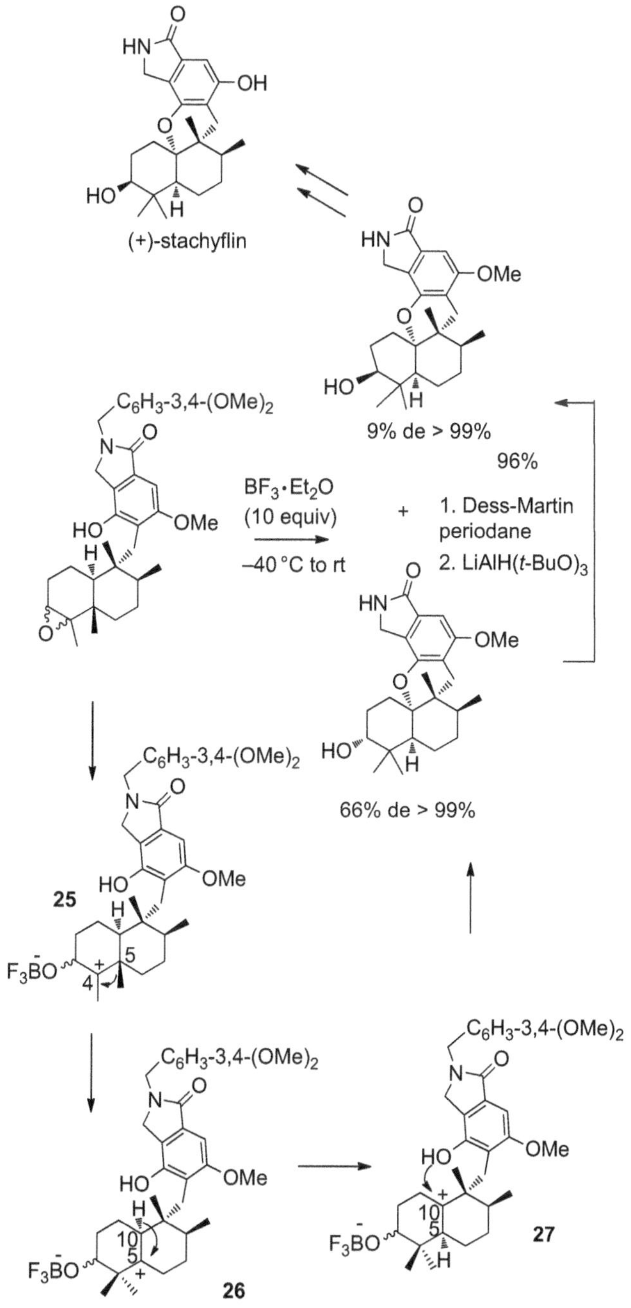

Scheme 1.35 Synthesis of (+)-stachyflin through domino epoxide-opening–rearrangement–cyclisation reaction.

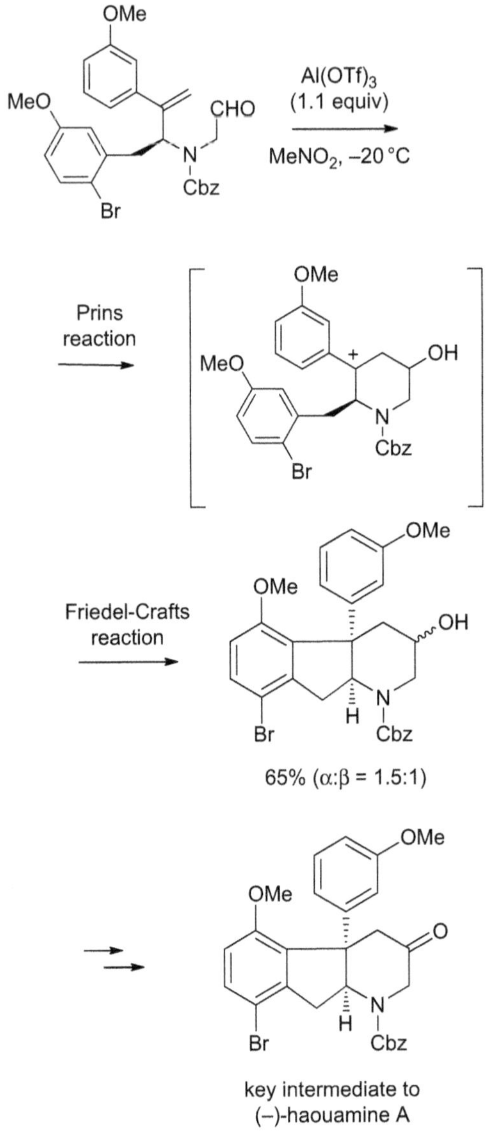

Scheme 1.36   Synthesis of a key intermediate in the synthesis of (–)-haouamine A through domino Prins–Friedel–Crafts reaction.

emerging quaternary centre. Moreover, a single diastereomer of a tricyclic product bearing contiguous tertiary and quaternary centres was obtained in 42% overall yield starting from the corresponding *trans*-cycloether. The domino product was isolated in 70% yield as a mixture of anomers, which was subsequently oxidised by treatment with Dess–Martin periodinate into the corresponding keto-aldehyde in 60% yield as a single diastereomer. As a first

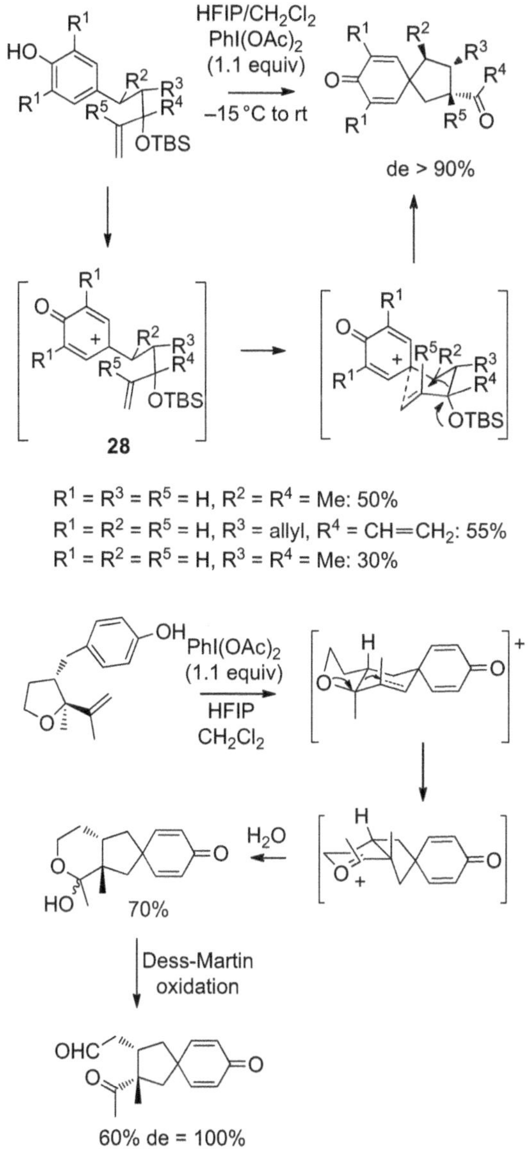

Scheme 1.37 Oxidative domino Prins–Pinacol reactions.

application of this novel diastereoselective domino process, a formal synthesis of the important antibiotic agent (–)-platensimycin was achieved.

1.2.3 Domino Reactions Initiated by a Pericyclic Primary Step

Pericyclic reactions, such as the Diels-Alder, ene, Claisen, Cope, or electrocyclic reactions, are by themselves extremely useful transformations. By combining two or more pericyclic reactions, however, the effect can be multiplied. There have been considerable advances in the use of pericyclic processes to initiate both inter- and intramolecular sequences.[56] In particular, asymmetric domino sequences involving cycloaddition reactions are highly effective processes for the rapid elaboration of complex polycyclic systems, because each cyclo-addition event generates a new ring and two new covalent bonds. Most of the asymmetric pericyclic sequences include a Diels–Alder reaction, generally in the first step. As an example, Boger *et al.* have developed an asymmetric domino reaction of a chiral 1,3,4-oxadiazole bearing a tethered trisubstituted olefin with an electron-donating substituent to activate dienophile for participation in an inverse electron-demand intramolecular Diels–Alder reaction with the electron-deficient 1,3,4-oxadiazole as the first step of the domino process.[57] The stereochemistry of this [4+2] cycloaddition was controlled by the chiral substituent on the tether linking the dienophile, providing a diazo compound, which subsequently reacted through 1,3-dipolar cycloaddition to give the corresponding cascade cycloadduct possessing six stereocentres, as shown in Scheme 1.38. This complex chiral product was further applied as a key inter-mediate in the total synthesis of vindorosine, vindoline, and key vinblastine analogues.

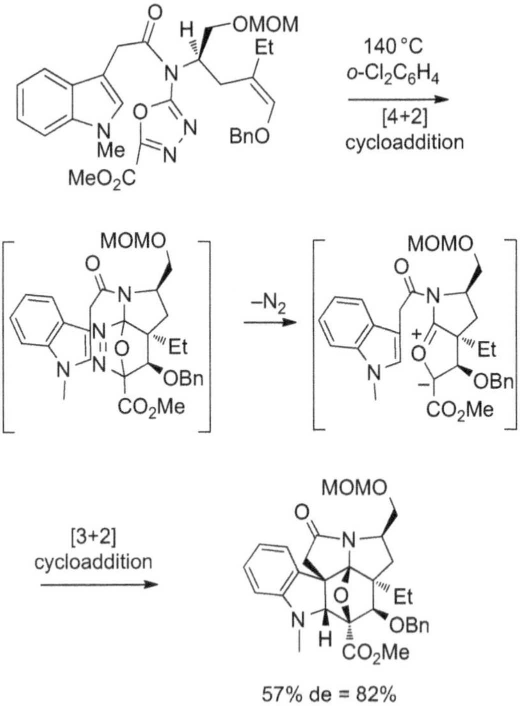

Scheme 1.38   Domino intramolecular Diels–Alder cycloaddition–1,3-dipolar cyclo-addition reaction.

On the other hand, Danheiser *et al.* have reported the synthesis of polycyclic benzofused nitrogen heterocycles *via* a strategy which involved a domino ynamide benzannulation reaction followed in a second step by a ring-closing metathesis.[58] The domino benzannulation step was based on the reaction of cyclobutenones with ynamides. This cascade process proceeded *via* a sequence of four successive pericyclic reactions, such as a 4 electron electrocyclic cleavage, a [2 + 2] cycloaddition, a 4 electron electrocyclic cleavage, and a 6 electron electrocyclic closure, as depicted in Scheme 1.39. This sequence furnished multiply substituted aniline derivatives. The authors have applied this powerful methodology to a chiral ynamide, which afforded, by reaction with 3-methoxymethylcyclobutenone, the corresponding enantiopure aniline in 88–94% yield. This product was subsequently submitted to ring-closing metathesis to give a key benzazocine core of the anticancer agent (+)-FR900482.

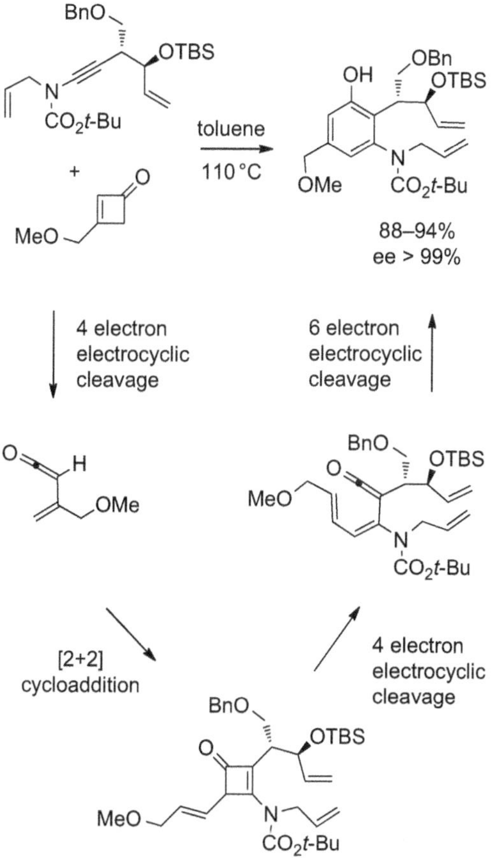

Scheme 1.39   Domino intramolecular ynamide benzannulation reaction.

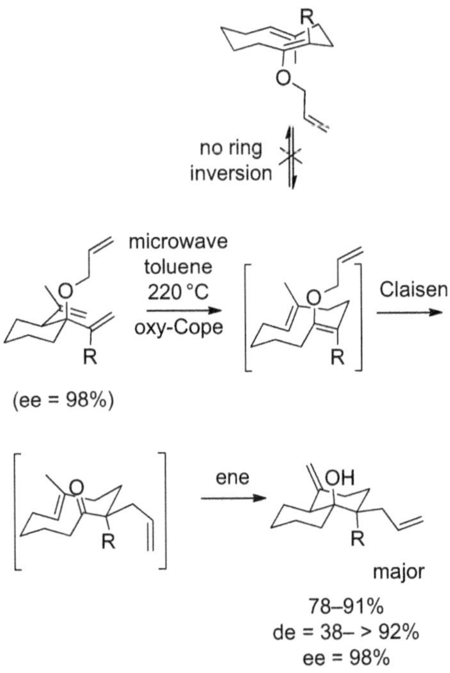

Scheme 1.40   Domino oxy-Cope–Claisen–ene reaction.

In 2007, Barriault *et al.* investigated the asymmetric domino oxy-Cope–Claisen–ene reaction of chiral allyl 1,2-divinylcyclohexanol ethers, providing through microwave irradiation and heating the corresponding domino decalin cycloadducts in high yields and moderate to high diastereoselectivities of up to 92% de, combined with excellent enantioselectivities of >98% ee, as shown in Scheme 1.40.[59] The observed conservation of enantiomeric excess was taken by the authors as evidence that the ring inversion of the intermediary enol ether did not occur before the Claisen reaction.

In 2009, Porco *et al.* accomplished an enantioselective synthesis of (+)-chamaecypanone, which is an anticancer agent, on the basis of an asymmetric domino retro-Diels–Alder cycloaddition–Diels–Alder cycloaddition reaction.[60] This key step of the synthesis involved a Diels–Alder cycloaddition between a diarylcyclopentadienone, which was generated *in situ*, and a chiral *ortho*-quinol derived from a retro-Diels–Alder reaction of its dimeric form. Later, these authors improved and extended the scope of this type of domino reaction by using microwave irradiation.[61] In this work, the authors have accomplished a parallel screen of cycloaddition partners for *ortho*-quinols using a plate-based microwave system. A range of alkene partners including dienes and dienophiles were demonstrated to react with a chiral *o*-quinol dimer, providing the corresponding domino products as single diastereomers in high to excellent yields in all cases of the substrates studied, as shown in Scheme 1.41. Indeed, complete diastereoselectivity was observed when using a

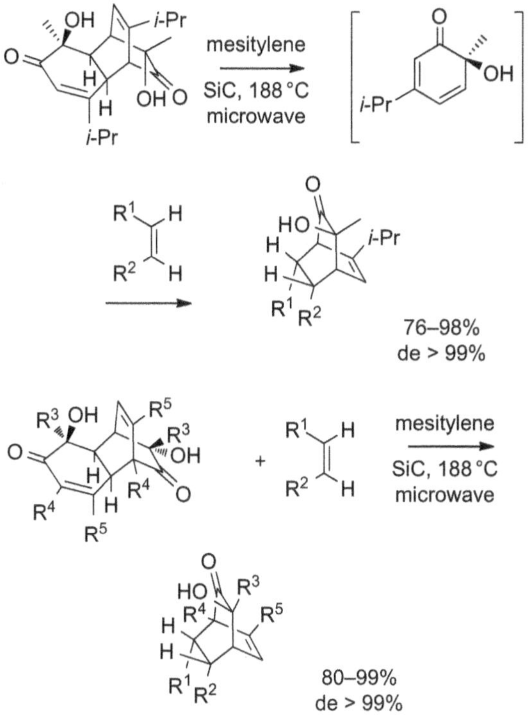

Scheme 1.41 Domino retro-Diels–Alder cycloaddition–Diels–Alder cycloaddition reactions.

range of alkene partners, which could be normal-demand dienophiles, such as methylvinylketone, indene and 4-methoxystyrene, as well as inverse-demand dienophiles, such as dihydrofuran and vinylene carbonate. In addition, the authors have extended the scope of this methodology to other *ortho*-quinol precursors, which led, under the same reaction conditions, to the corresponding cycloadducts as single diastereomers in good to excellent yields, as shown in Scheme 1.41. Among a number of chiral products generated using this powerful process, a bicyclo[2.2.2]octenone was found to be a novel inhibitor of activator protein-1 (AP1), an oncogenic transcription factor.

Another asymmetric domino sequence, including a Diels–Alder cyclo-addition followed by sulfoxide elimination, was used by Urbano *et al.* as the key step in a total synthesis of rubiginones A_2 and C_2, and their 11-methoxy regioisomers, which are angucyclinone-type natural products.[62] The domino reaction occurs between an enantiopure vinyl cyclohexene and a racemic methoxy-substituted sulfinylnaphthoquinone, such as 5-methoxy-2-(*p*-tolylsulfinyl)-1,4-naphthoquinone, providing, by heating in dichloromethane, the corresponding enantiopure tetracyclic quinine as a sole regioisomer and pure diastereomer in 52% yield, as shown in Scheme 1.42. This tetracyclic product resulted from a regioselective Diels–Alder cycloaddition, followed by

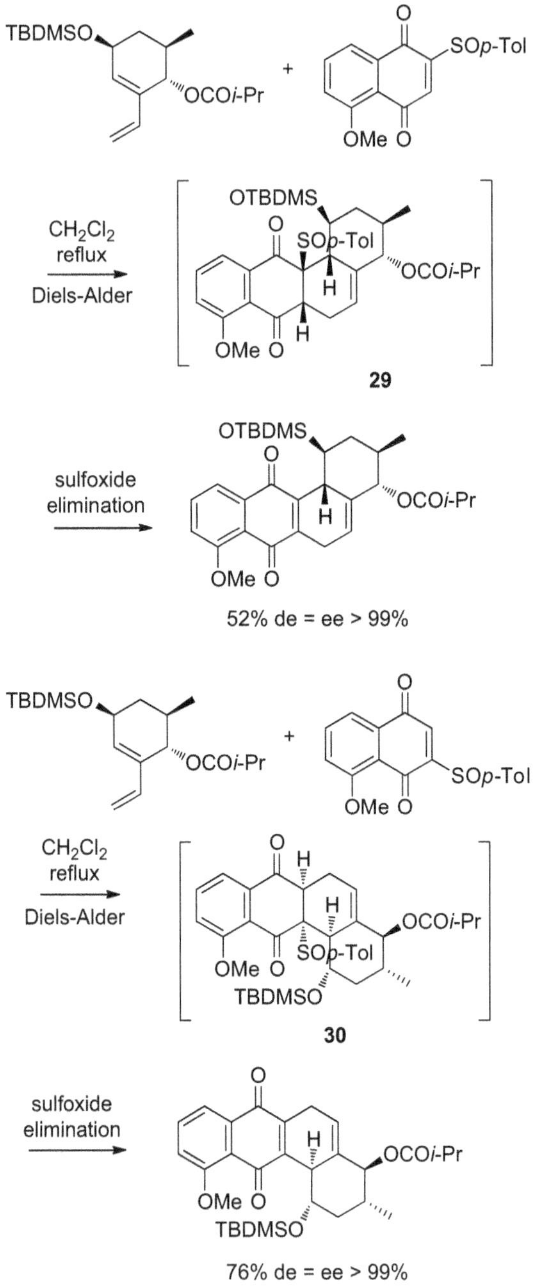

Scheme 1.42 Domino Diels–Alder cycloaddition–sulfoxide elimination reactions.

the spontaneous elimination of *p*-tolylsulfenic acid, which regenerated the quinonic double bond. It was further converted into C4 oxygenated angucyclinones rubiginones C_2 and A_2. In addition, the same enantiopure vinyl cyclohexene reacted with racemic 5-methoxy-3-(*p*-tolylsulfinyl)-1,4-naphthoquinone to give, under similar conditions, the corresponding tetracyclic quinine in 76% yield. This domino product also resulted from the spontaneous elimination of the sulfoxide in the initially formed Diels–Alder cycloadduct **30** (Scheme 1.42), which was formed in a completely regio- and diastereoselective way. The regiochemistry of the initial cycloadduct **30**, resulting from 5-methoxy-3-(*p*-tolylsulfinyl)-1,4-naphthoquinone bearing the sulfoxide at C3, must be the opposite to that of cycloadduct **29**, arising from reaction of 5-methoxy-2-(*p*-tolylsulfinyl)-1,4-naphthoquinone with the sulfoxide at C2. The tetracyclic product derived from intermediate **30** was further converted into regioisomeric angucyclinones.

1.2.4 Carbene Sequences

Cascade reactions initiated from carbene intermediates have been a productive area of discovery during the past 15 years. Carbenes and carbenoids can react with a series of functional groups, and a more reactive intermediate (ylide) is frequently formed that can undergo further reactions, allowing the synthesis of many complex molecules to be achieved.[63] Of these processes, the catalytic domino carbonyl ylide formation–1,3-dipolar cycloaddition offers an elegant route to highly substituted oxygen-containing heterocycles.[63b,64] This powerful methodology has been extensively advanced by the Padwa group, in particular.[63c,65] In 2006, a diastereoselective version of this type of domino reaction was used by Hashimoto *et al.* as the key step in the first total syntheses of squalene synthase inhibitors zaragozic acids A and C.[66] As shown in Scheme 1.43, the rhodium-catalysed domino carbonyl ylide formation–1,3-dipolar cycloaddition reaction of a chiral α-diazo ester carbonyl ylide precursor derived from di-*tert*-butyl D-tartrate with 3-butyn-2-one afforded the corresponding cycloadduct as a single diastereomer in 72% yield. This chiral domino product was further converted into zaragozic acids A and C. In addition, an intramolecular version of a comparable strategy was applied by the same authors to synthesise biologically active natural products (–)-polygalolides A and B.[67]

Later, Schaus *et al.* developed a nice intermolecular domino carbonyl ylide formation–1,3-dipolar cycloaddition reaction of a chiral dihydropyrimidone with a maleimide which afforded, in the presence of rhodium(II) acetate, the corresponding cycloadduct in 90–93% yield as a single stereoisomer, as shown in Scheme 1.44.[68] The process evolved through the *in situ* formation of dipole **31** by treatment of substrate with rhodium(II) acetate dimer in refluxed benzene. In the presence of maleimine as a dipolarophile, the dipole intermediate **31** underwent 1,3-dipolar cycloaddition to generate the final domino product (Scheme 1.44).

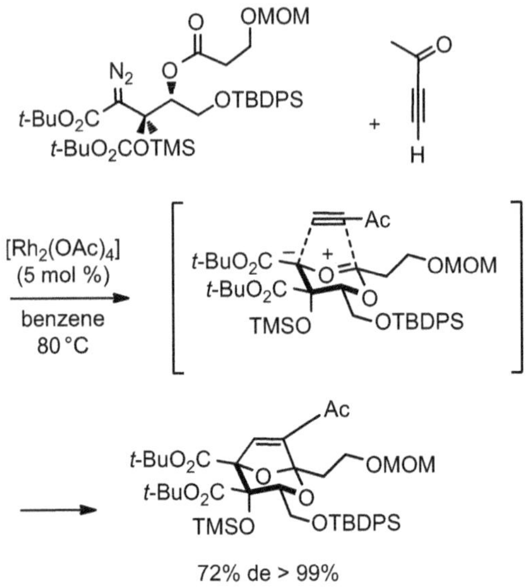

Scheme 1.43 Rh-catalysed domino carbonyl ylide formation–1,3-dipolar cyclo-addition reaction of a di-*tert*-butyl D-tartrate-derived diazo ester with 3-butyn-2-one.

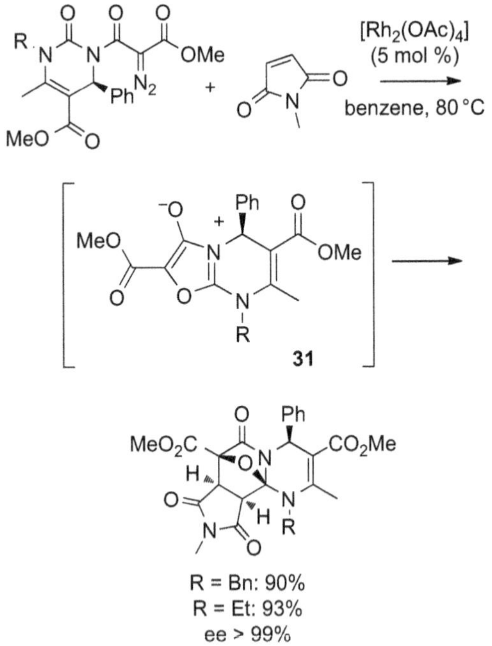

R = Bn: 90%
R = Et: 93%
ee > 99%

Scheme 1.44 Rh-catalysed domino carbonyl ylide formation–1,3-dipolar cycloaddition reaction of a chiral dihydropyrimidone diazo ester with a maleimide.

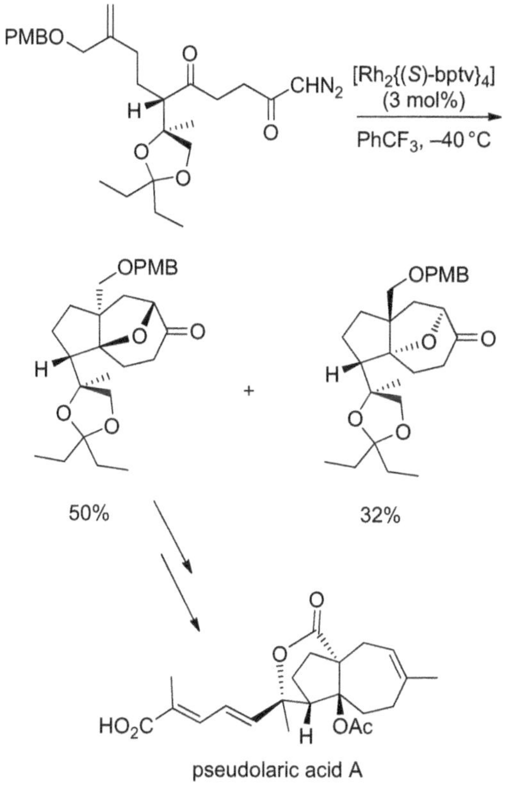

Scheme 1.45 Synthesis of pseudolaric acid A through Rh-catalysed domino carbonyl ylide formation–1,3-dipolar cycloaddition reaction.

In the same area, Chiu *et al.* have reported total syntheses of natural and biologically active pseudolaric acid A,[69] and (–)-indicol, on the basis of rhodium-catalysed intramolecular domino carbonyl ylide formation–1,3-dipolar cycloaddition reactions of chiral α-diazoketones derived from commercially available glycidol derivatives, which allowed the core bicyclo[5.4.0]undecane skeleton to be assembled.[70] As shown in Scheme 1.45, treatment of a chiral α-diazoketone with a catalytic amount of chiral catalyst [Rh₂{(S)-bptv}₄] in benzotrifluoride at –40 °C afforded the corresponding domino product in 82% yield as a 1.6 : 1 mixture of two diastereoisomers. The major diastereomer was separated and then converted into pseudolaric acid A through 10 steps. It must be noted that the use of achiral dirhodium catalysts, such as rhodium(II) acetate, favoured the formation of the undesired diastereomer, since in this case the ratio of diastereomers was 1 : 3, combined with a 61% yield.

In 2009, Saba *et al.* developed concise syntheses of novel chiral tricyclic alkaloids, starting from a readily available (S)-1,2,3,4-tetrahydroisoquinoline carboxylic acid methyl ester as chiral auxiliary, which underwent a domino

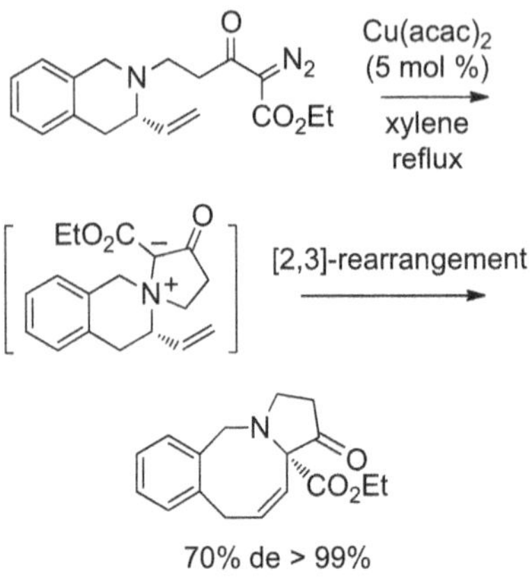

Scheme 1.46   Cu-catalysed domino ammonium ylide formation–[2,3]-rearrangement
               reaction.

ammonium ylide formation–[2,3]-rearrangement reaction, affording the corresponding key pyrrolo benzoazacyclononenone, as shown in Scheme 1.46.[71] The
process, mediated by copper, first provided a Cu(acac)$_2$-generated metallo
carbenoid species, which cyclised to the corresponding spiricyclic ammonium
ylide intermediate. This ylide then underwent a [2,3]-rearrangement to afford
the final chiral tricyclic product as a single stereoisomer in 70% yield. The
domino process resulted in a three-carbon expansion of the starting amine ring
moiety. Finally, Li *et al.* have reported an expedient synthesis of Δ^4-oxocene
cores of (+)-laurencin, and (+)-prelaureatin on the basis of diastereoselective
domino ring cyclisation–fragmentation–expansion reaction of a chiral
tetrahydrofuran with ethyl diazoacetate.[72] This cascade reaction mediated by
SnCl$_2$ involved an intramolecular oxo-carbenoid insertion and a β-silyl fragmentation sequence, which provided the corresponding enantiopure eight-
membered cyclic ether in low yields (20–25%).

1.2.5   Palladium-Catalysed Domino Reactions

Over the past 40 years, a large number of novel transition metal-catalysed
reactions have been discovered, and activity in this area has remained high.
Among the increasing number of asymmetric domino processes initiated by a
transition metal-catalysed reaction, the asymmetric palladium-catalysed
domino transformations have seen an astounding development over the past
few years. Indeed, given that palladium has the advantage of being compatible
with many functional groups, it constitutes an ideal catalyst for domino
reactions.[73] Although palladium-catalysed domino processes have only recently
been extensively reported in the literature,[74] the concept of sequential
palladium-mediated transformations was actually pioneered some time before
the word "domino" was coined. The discovery of the ability of transition
metals to interact with organic moieties, to connect inter- or intramolecularly

alkenes, alkynes, carbon monoxide, *etc.* in cascading processes, is certainly a breakthrough in organometallic synthesis. It must be remembered that the astonishing simplicity of achieving many complex polycyclisations is sometimes directly proportional to the labour required for the synthesis of the cyclisation precursor. The possible modes by which an organopalladium complex can be engaged in consecutive bond formations, or the manner in which two sequential palladium-catalysed processes can be coupled using a single catalytic system, is only limited by the chemist's imagination. The Heck reaction is an important way to couple aryl and vinyl systems in the presence of palladium,[75] and has been recently exploited as the key step of many total syntheses of natural products.[76] Furthermore, it forms the keystone of many domino reactions.[77] As a recent example, Tietze *et al.* have reported the synthesis of new chiral photochromic switches based on helical alkenes through asymmetric domino processes consisting either of a carbopalladation of an alkyne and a Heck reaction, or a combination of a carbopalladation and a Stille reaction.[78] As shown in Scheme 1.47, the palladium-catalysed domino carbopalladation–Heck reaction of chiral alkynes provided the corresponding tetrasubstituted helical alkenes as single diastereomers in high yields. While enantiopure *syn*-diastereomers provided good results, the reaction of *anti*-diastereomers did not achieve the corresponding domino helical alkenes. Moreover, these authors have developed a domino palladium-catalysed carbopalladation–Stille reaction of enantiopure organotin compounds, which afforded the corresponding chiral tetrasubstituted helical alkenes in good yields, as shown in Scheme 1.47.

In 2009, Roy *et al.* reported a convenient regioselective synthesis of enantiopure 3-*C*-linked mannopyranosyl coumarins on the basis of a domino Heck–lactonisation reaction.[79] As shown in Scheme 1.48, the reaction of a chiral β-*C*-mannopyranosyl acrylate with 2-iodophenol led to the corresponding chiral 3-*C*-mannopyranosyl coumarin derivative in 86% yield through a domino Heck–lactonisation process, while the corresponding chiral α-*C*-mannopyranosyl acrylate provided a mixture of the corresponding domino product in 77% yield, along with the corresponding intermediary Heck product as minor product (see Scheme 1.48). This intermediary Heck product could be further converted, by treatment with NaOMe in methanol, into the corresponding domino product in 83% yield. An analogous chiral α-*C*-mannopyranosyl acrylate depicted in was also submitted to the same conditions, providing a 1 : 2 mixture of the corresponding Heck product and the corresponding chiral domino 3-*C*-mannopyranosyl coumarin in 71% yield (Scheme 1.48). This intermediary Heck product was also further transformed by treatment with tetra-*n*-butylammonium bromide (TBAB) and NaHCO₃ in dimethylformamide (DMF) into the corresponding domino product in 38% yield.

Another asymmetric domino reaction based on the Heck reaction was reported by Pfeffer *et al.*, in 2012.[80] In this process, chiral 1,3-disubstituted tetrahydroisoquinolines were achieved in good yields from the domino Heck reaction of an enantiopure phenylalanine derivative with α,β-unsaturated

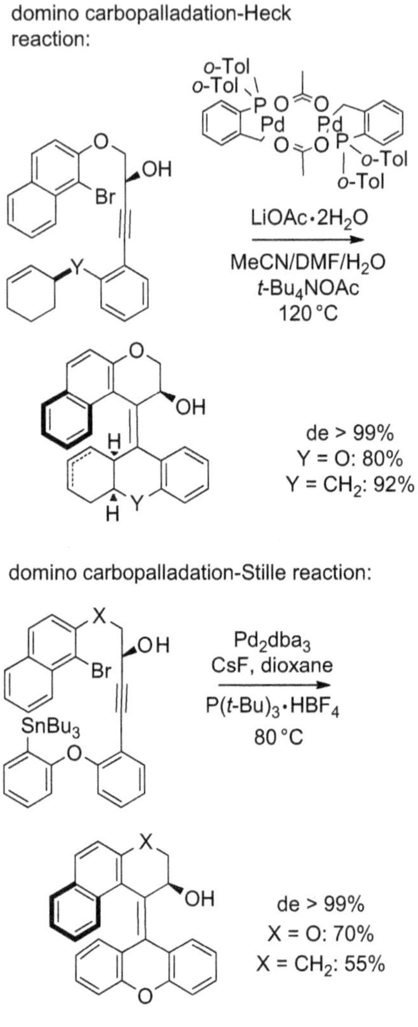

domino carbopalladation-Heck
reaction:

de > 99%
Y = O: 80%
Y = CH$_2$: 92%

domino carbopalladation-Stille reaction:

de > 99%
X = O: 70%
X = CH$_2$: 55%

Scheme 1.47   Domino   carbopalladation–Heck   reaction   and   domino   carbopalladation–Stille reaction.

carbonyl compounds, followed by an intramolecular aza-Michael addition.
Moderate to high diastereoselectivities (85–92% de; Scheme 1.49) were
obtained for the corresponding domino products, while the extension of this
methodology to indole substrates provided the corresponding C1-substituted
tetrahydro-β-carbolines in moderate diastereoselectivities (≤60% de).

In 2006, a novel palladium-catalysed oxy-carbopalladation process was
developed by Gouverneur *et al.*, allowing the orchestrated union of hydroxy
ynones with ethyl acrylate.[81] With enantiopure β-hydroxy ynones, the domino
Wacker–Heck reaction provided the corresponding dihydropyranones with
excellent enantioselectivities of 96% ee with no detectable racemisation, as

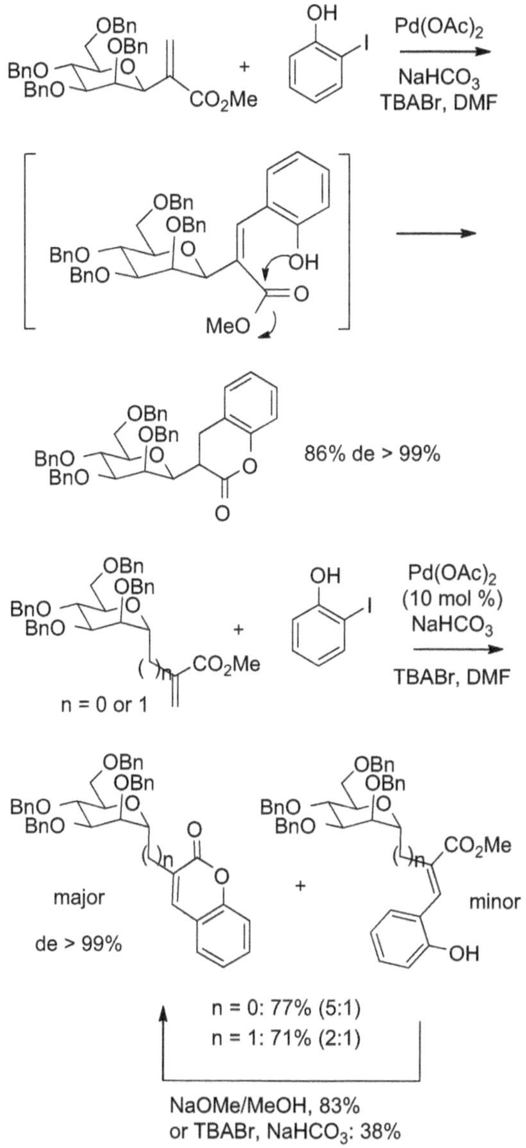

Scheme 1.48 Domino Heck–lactonisation reactions.

shown in Scheme 1.50. The importance of this efficient methodology was highlighted by the fact that these compounds were unknown. A plausible mechanism for this domino process is depicted in Scheme 1.50. It involved the initial coordination of the metal to the triple bond, resulting in intramolecular nucleophilic attack by the proximal hydroxy group (oxypalladation). This led to the formation of a σ-alkenyl palladium intermediate, which could not

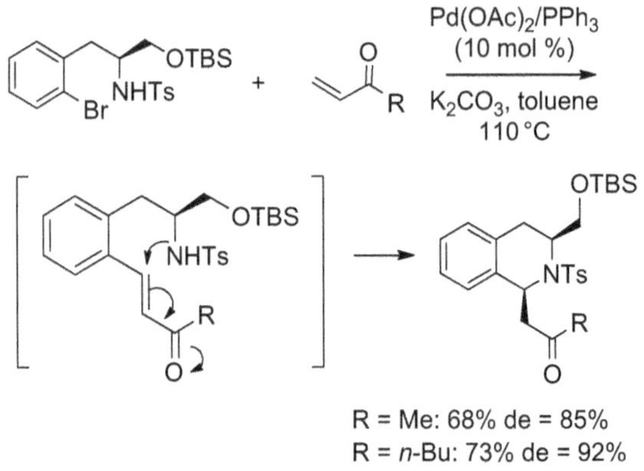

R = Me: 68% de = 85%
R = *n*-Bu: 73% de = 92%

Scheme 1.49   Domino Heck–aza-Michael reaction.

undergo β-hydride elimination. This intermediate further reacted in a subsequent carbopalladation process with ethyl acrylate present in large excess. After β-hydride elimination, the trisubstituted dihydropyranone was formed with concomitant formation of a palladium hydride species which could, upon reductive elimination, allow Pd(0) to re-enter the catalytic cycle after oxidation by molecular oxygen.

The first total synthesis of natural biologically active (–)-panacene was accomplished by Snieckus *et al.* on the basis of a key asymmetric domino alkoxycarbonylation–lactonisation reaction of a chiral phenol with retention of enantioselectivity, as shown in Scheme 1.51.[82] The tricyclic domino product formed was further converted into (–)-panacene in seven steps.

In 2011, Ohno *et al.* described the enantioselective total syntheses of (+)-lysergic acid, (+)-lysergol, and (+)-isolysergol through palladium-catalysed domino cyclisation of chiral allenes bearing amino and bromoindolyl groups.[83] This key domino reaction enabled the direct construction of the C/D ring system of the ergot alkaloid skeleton, as well as the creation of the C5 stereogenic centre, with transfer of the allenic axial chirality to the central chirality. As shown in Scheme 1.52, the domino process of a chiral allenic amide (88% de) provided the corresponding expected tetracyclic domino product in 76% yield and diastereoselectivity of 84% de. This chiral key product was further converted into the alkaloids (+)-lysergic acid, (+)-lysergol, and (+)-isolysergol. In order to explain these results, the authors have proposed the mechanism depicted in Scheme 1.52. The process began with the amino-palladation of indolylpalladium bromide **32**, formed by oxidative addition of the substrate to Pd(0), which proceeded through conformation **33** to produce the alkenylpalladium(II) intermediate **34** stereoselectively. This was followed by reductive elimination, leading to the final product as the major isomer.

In 2010, an expeditious route to 1α,25-dihydroxyvitamin D$_3$ was reported by Mourino *et al.* through an aqueous domino palladium-catalysed cyclisation–Suzuki coupling reaction.[84] As depicted in Scheme 1.53, the process

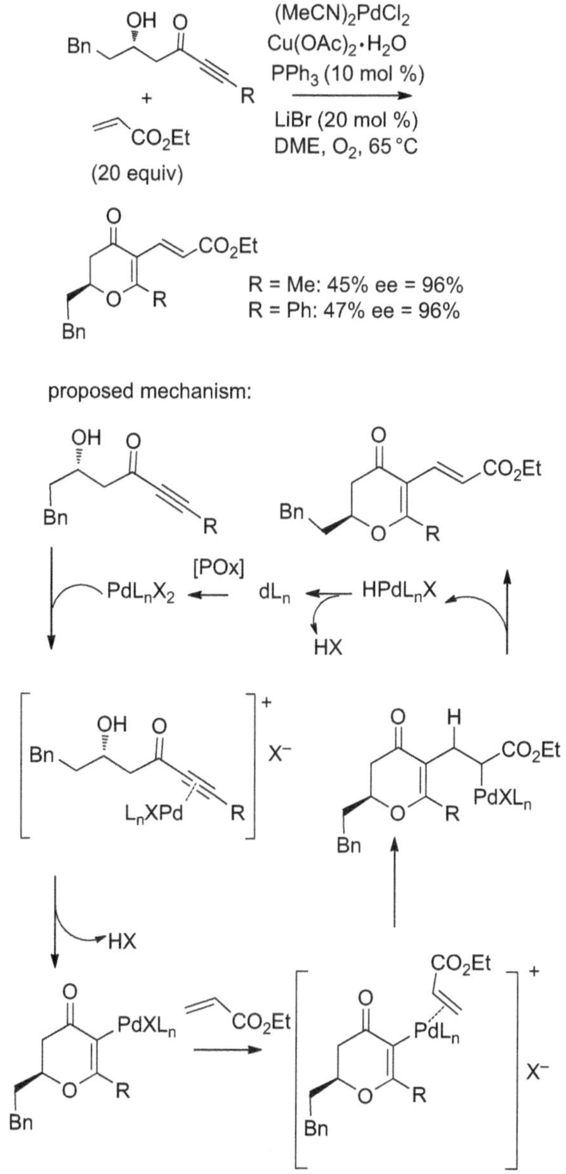

Scheme 1.50 Domino Wacker–Heck reaction.

consisted in a highly stereoselective intramolecular cyclisation of a chiral enol triflate, allowing ring A of $1\alpha,25$-dihydroxyvitamin D_3 to be built, followed by a Suzuki–Miyaura coupling of the resulting palladium intermediate with a chiral alkenyl boronic ester. The corresponding domino product was further submitted to deprotection of the silyl ether to give the expected

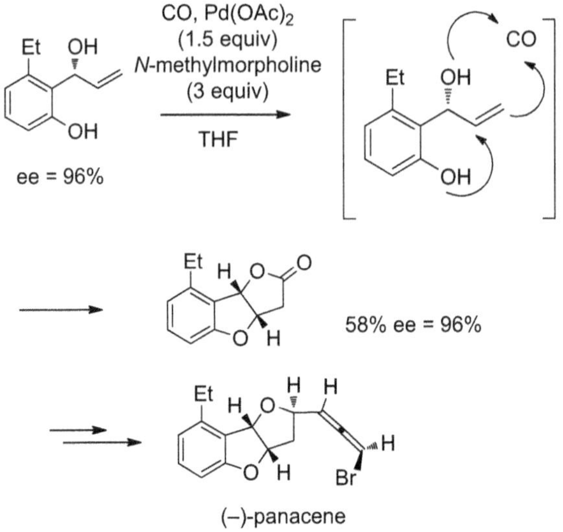

Scheme 1.51 Synthesis of (–)-panacene through domino Wacker–Heck reaction.

α,25-dihydroxyvitamin D$_3$ in 81% yield (for the two steps, domino and desilylation reactions).

Finally, Menche *et al.* have developed a novel diastereoselective palladium-catalysed domino reaction allowing a concise synthesis of tetrahydropyrans to be achieved.[85] The process involved, as the first step of the sequence, an oxa-Michael addition of a chiral homoallylic alcohol to an α-substituted nitroalkene, which provided the corresponding intermediate enolate **35**. In the presence of a catalytic amount of [{Pd(allyl)Cl}$_2$] combined with a base such as LiHMDS, this enolate produced a π-allyl complex **36**, which was subsequently trapped in an intramolecular fashion through an allylic substitution reaction to afford, finally, the corresponding polysubstituted tetrahydropyran as a mixture of two diastereomers. The authors carried out the domino process with several chiral homoallylic alcohols differentiated by the nature of the carbonate function borne at position C5. Methyl and *tert*-butyl carbonates proved to be the best leaving groups of those evaluated, providing low to good yields of up to 72% combined with moderate to good relative diastereoselectivities of up to 82% de (Scheme 1.54). In spite of these moderate diastereoselectivities, this domino oxa-Michael–Tsuji–Trost reaction demonstrates a conceptually novel methodology for the synthesis of functionalised tetrahydropyrans.

1.2.6   Ruthenium-Catalysed Domino Reactions

Olefin metathesis represents one of the most powerful and attractive tools in polymer science and organic synthesis for the formation of carbon–carbon double bonds.[86] Moreover, the combination of two or more metathesis reactions has increased its efficiency immensely; in particular, the domino

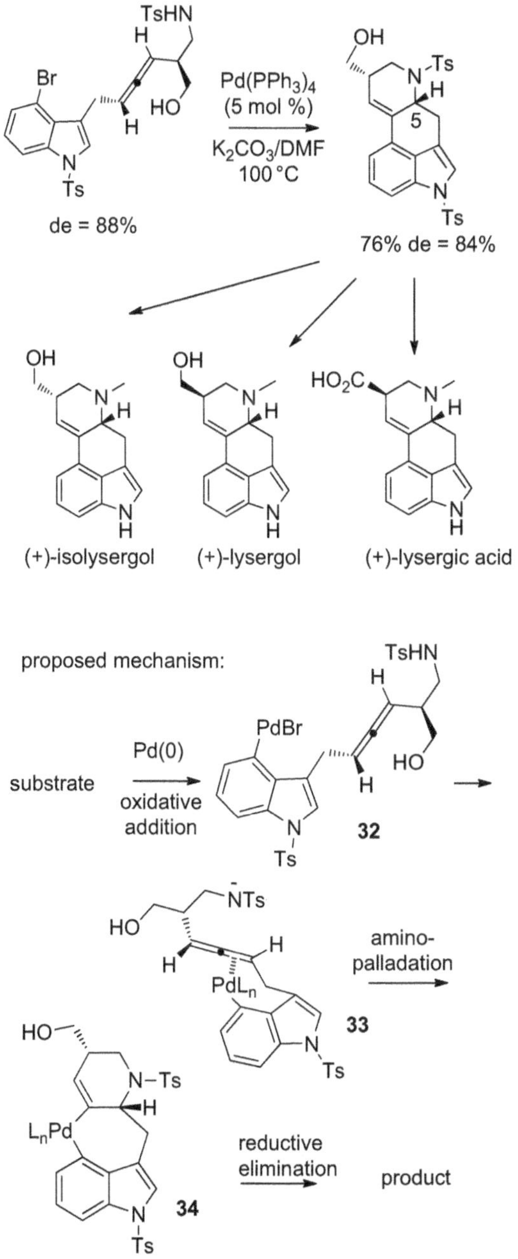

Scheme 1.52 Syntheses of (+)-lysergic acid, (+)-lysergol, and (+)-isolysergol through domino cyclisation reaction.

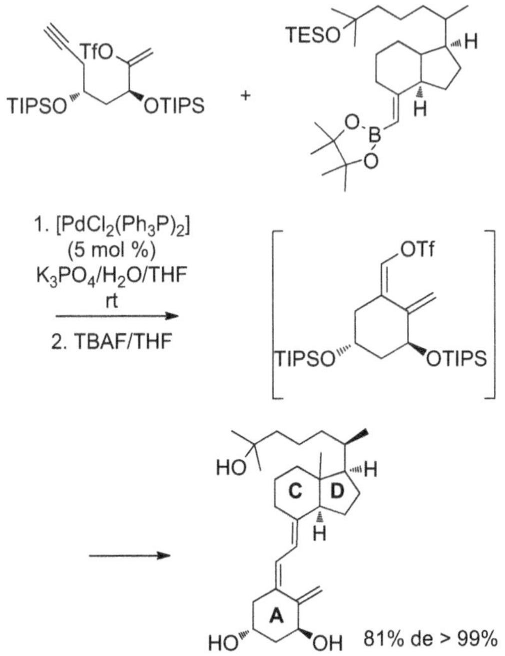

Scheme 1.53 Synthesis of 1α,25-dihydroxyvitamin D$_3$ through domino cyclisation–Suzuki coupling reaction.

ring-opening–ring-closing metathesis has been widely used. Indeed, the large number of asymmetric domino reactions include a ruthenium-catalysed olefin ring-closing metathesis, ring-opening metathesis, enyne metathesis, or cross-metathesis, which can be associated in domino processes with various other reactions. In 2007, Fustero *et al.* reported the first example of domino olefin cross-metathesis–intramolecular aza-Michael reaction.[87] The scope of this novel methodology was extended to an asymmetric variant by using chiral Cbz-protected secondary amines as starting materials. As shown in Scheme 1.55, the reaction of chiral amines with methyl vinyl ketone in the presence of a Hoveyda–Grubbs second-generation catalyst led to the corresponding chiral protected 2,5-substituted pyrrolidines in good to high yields (76–98%) combined with moderate diastereoselectivities (50–72% de). These authors have also demonstrated the utility of a closely related methodology by achieving the total synthesis of the piperidine natural product (–)-pinidinol.[88] In this case, the nitrogen nucleophiles were chiral *N*-sulfinyl secondary amines which provided, through domino olefin cross-metathesis–intramolecular aza-Michael reaction, the corresponding 2-substituted pyrrolidines and piperidines in low to high yields (10–92%) and moderate to good diastereoselectivities (40–84% de). Another diastereoselective domino olefin cross-metathesis–aza-Michael reaction was more recently reported by Cho *et al.*, and applied to the

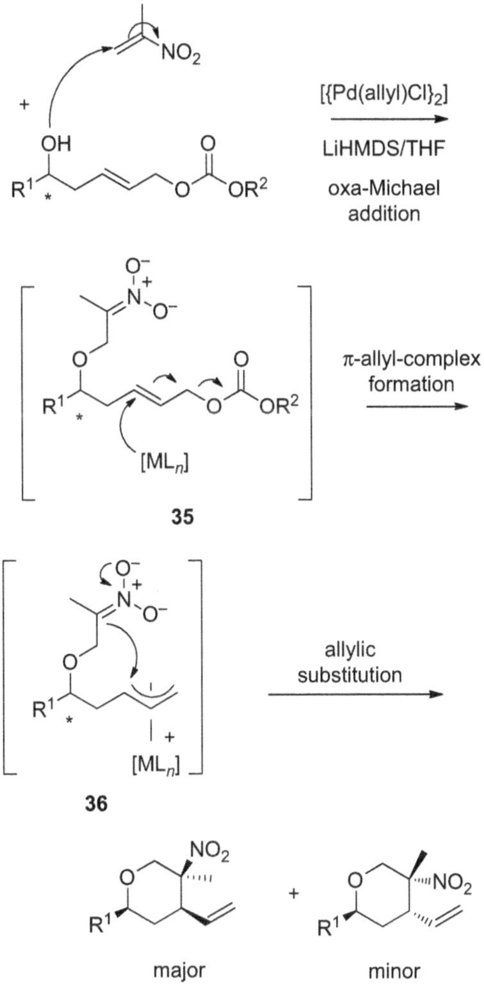

R^1 = Ph, R^2 = Me: 17% de = 80%
R^1 = Ph, R = *t*-Bu: 62% de = 80%
R^1 = *p*-Tol, R^2 = *t*-Bu: 51% de = 78%
R^1 = *p*-MeOC$_6$H$_4$, R^2 = *t*-Bu: 57% de = 78%
R^1 = *o*-MeOC$_6$H$_4$, R^2 = *t*-Bu: 59% de = 72%
R^1 = Cy, R^2 = *t*-Bu: 72% de = 66%
R^1 = CH$_2$Bn, R^2 = *t*-Bu: 48% de = 75%
R^1 = 2-Naph, R^2 = *t*-Bu: 67% de = 82%

Scheme 1.54 Domino oxa-Michael–Tsuji–Trost reaction.

asymmetric formal synthesis of pyrrolopiperazinone natural products.[89] In this case, the domino reaction occurred between acrolein and chiral tertiary allyl amides derived from the condensation of 4,5-dibromo-1*H*-pyrrole-2-carboxylic acid to (*R*)-*N*-allyl-1-arylethylamine. It furnished the corresponding

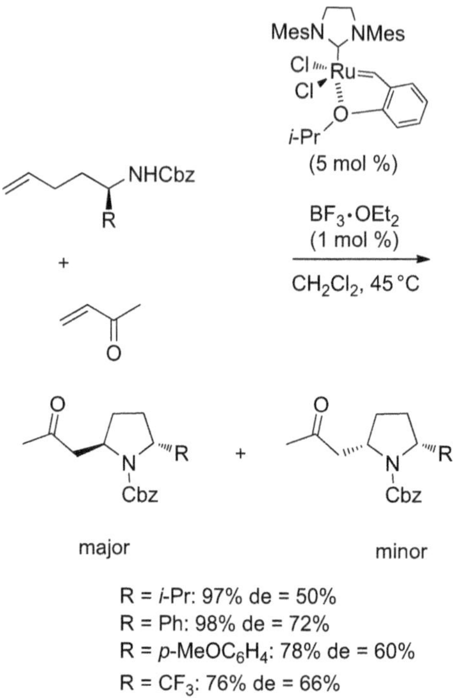

Scheme 1.55   Domino olefin cross-metathesis–intramolecular aza-Michael reaction.

2-((*S*)-6,7-dibromo-1-oxo-2-((*R*)-1-arylethyl)-1,2,3,4-tetrahydropyrrolo[1,2-*a*]pyrazin-4-yl)acetaldehydes as mixtures of two diastereomers with low to moderate diastereoselectivities ranging from 14 to 54% de.

On the other hand, Fuwa *et al.* have developed highly diastereoselective synthesis of chiral 2,6-*cis*-disubstituted tetrahydropyrans *via* asymmetric domino olefin cross-metathesis–intramolecular oxa-Michael reaction of chiral 6-hydroxy alkenes with α,β-unsaturated aliphatic and aromatic ketones.[90] As shown in Scheme 1.56, a range of substituted tetrahydropyrans could be achieved on the basis of this powerful methodology using a Hoveyda–Grubbs second-generation catalyst. These chiral products were produced in good to excellent yields (77–97%) combined with diastereoselectivity of >90% de in all cases of the substrates studied. A closely related methodology was later employed by Hong *et al.* to achieve an efficient formal synthesis of biologically active SCH 351448.[91]

Asymmetric domino enyne ring-closing metathesis–cross metathesis processes have been developed by several groups. As an example, Martin *et al.* have reported an enantioselective synthesis of the natural product (+)-8-*epi*-xanthatin on the basis of the enyne metathesis–cross metathesis reaction of a chiral enyne with methyl vinyl ketone in the presence of a phosphine-free ruthenium catalyst, which provided the corresponding enantiopure seven-membered bicyclic (+)-8-*epi*-xanthatin in 83% yield

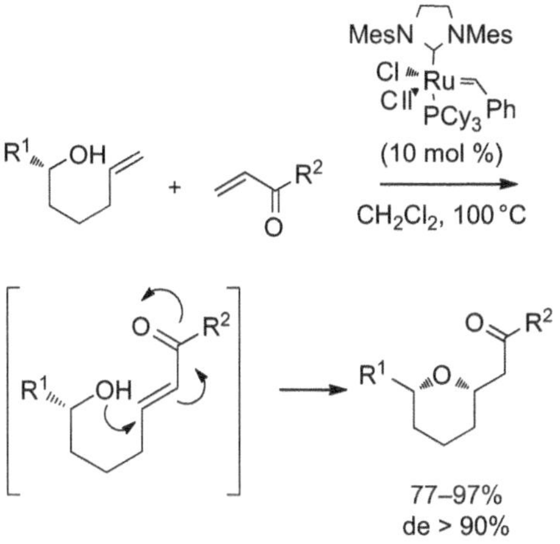

Scheme 1.56 Domino olefin cross-metathesis–intramolecular oxa-Michael reaction.

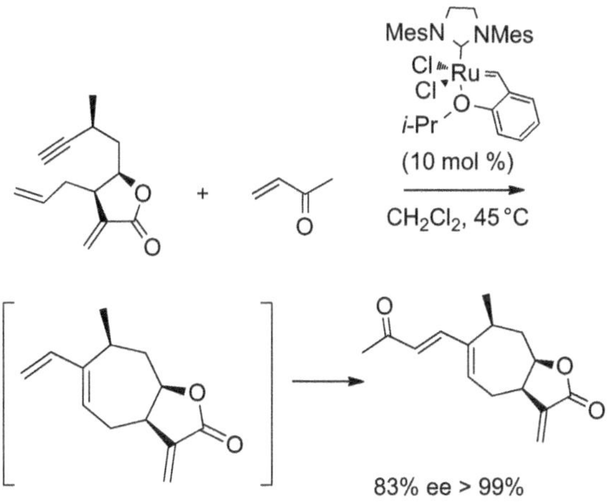

Scheme 1.57 Domino enyne metathesis–cross metathesis reaction.

(Scheme 1.57).[92] In the same area, Lee *et al.* have developed a total synthesis of another naturally occurring and biologically active product, (+)-panepophen-anthrin, by using an asymmetric domino enyne ring-closing metathesis–cross metathesis reaction, which occurs between a chiral linear enyne and 2-buten-3-ol.[93] The process was catalysed by a Hoveyda – Grubbs second-generation catalyst and yielded the corresponding key chiral cyclohexene in 51% yield combined with moderate diastereoselectivity of 34% de.

Another type of asymmetric ruthenium-catalysed domino reaction, including domino ring-opening–ring-closing metathesis, has been reported by Phillips *et al.* in application to the syntheses of several natural tetramic acid-containing macrolactams, such as aburatubolactam A,[94] and cylindramide A.[95] In the

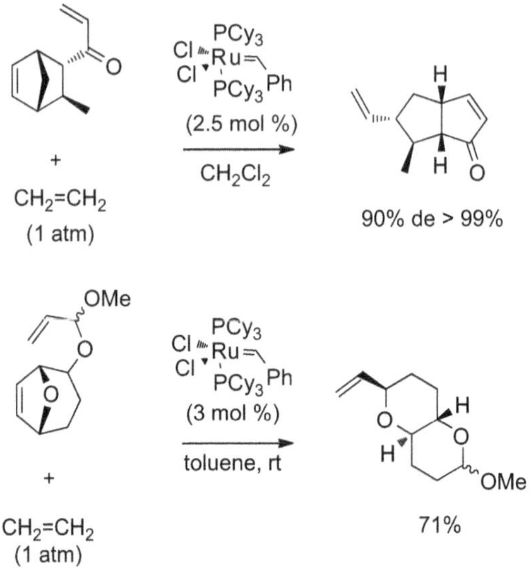

Scheme 1.58 Domino ring-opening–ring-closing metathesis reactions.

case of the synthesis of aburatubolactam A, the process involved the domino ring-opening–ring-closing metathesis of a chiral functionalised bicyclo[2.2.1]heptene, performed in the presence of ethylene upon catalysis by a first-generation Grubbs' catalyst. The reaction afforded the corresponding bicyclo[3.3.0]octene in 90% yield as a single stereoisomer, as shown in Scheme 1.58. Moreover, the same authors have reported the total synthesis of the biologically active natural product norhalichondrin B, which also involved a domino ring-opening–ring-closing metathesis of a chiral highly functionalised ether under very similar conditions, providing the corresponding chiral key pyranopyran in 71% yield, as shown in Scheme 1.58.[96]

A novel asymmetric approach to a densely functionalised lactarane skeleton in enantiomerically pure form, involving a domino ring-opening–ring closing metathesis of a chiral norbornene derivative, was reported by Ghosh *et al.*, in 2008.[97] As shown in Scheme 1.59, the reaction of a chiral auxiliary with ethylene in the presence of a first-generation Grubbs' catalyst afforded the corresponding enantiopure hydroazulene derivative in 65% yield as a single product.

Another chiral norbornene derivative was used by the same authors to produce the corresponding enantiopure tricycle through an asymmetric domino ring-opening–ring-closing metathesis reaction under comparable reaction conditions.[98] As shown in Scheme 1.60, the corresponding enantiopure domino product was nicely produced in 94% yield as a single stereoisomer. This product constituted the CDE core of nortriterpenoid schintrilactones A and B, which are naturally occurring anti-human immunodeficiency virus (HIV) agents.

In 2011, an intramolecular domino metathesis reaction of a chiral dienyne was used by Metz *et al.* as the key step in a total synthesis of diterpenes, such as

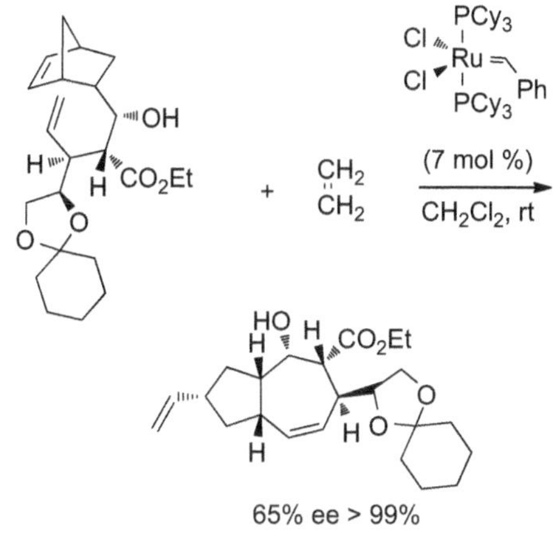

65% ee > 99%

Scheme 1.59   Domino ring-opening–ring-closing metathesis reaction.

94% ee > 99%

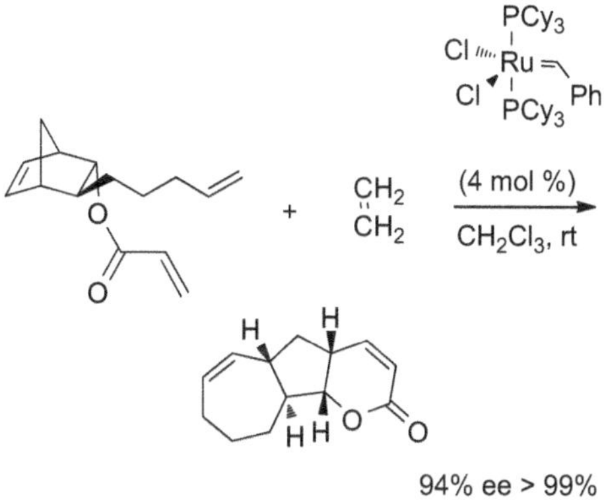

Scheme 1.60   Domino ring-opening–ring-closing metathesis reaction.

kempene-2, kempene-1, and 3-*epi*-kempene-1.[99] The process, catalysed by a Hoveyda–Grubbs second-generation catalyst, provided the corresponding enantiopure tetracyclic product in 92% yield, as shown in Scheme 1.61. On the other hand, Trost *et al.* have reported a ruthenium-catalysed domino alkyne-enone coupling–Michael reaction.[100] Indeed, the reaction of chiral propargylic alcohols with β,γ-enones in the presence of [CpRu(MeCN)$_3$][PF$_6$] provided the corresponding 2,6-*cis*-dihydropyrans as major diastereomers in moderate to good yields (31–80%). The *cis : trans* ratios ranged from 5:1 to 8:1. The process was demonstrated to be compatible with a variety of alcohol protecting groups. One of the products formed was used as a key chiral intermediate in the syntheses of ring A and B subunits of the bryostatins.

With the aim of synthesising the chiral benzo[*d*]xanthene tetracyclic core of anti-influenza active sesquiterpene natural products, Cramer *et al.* have

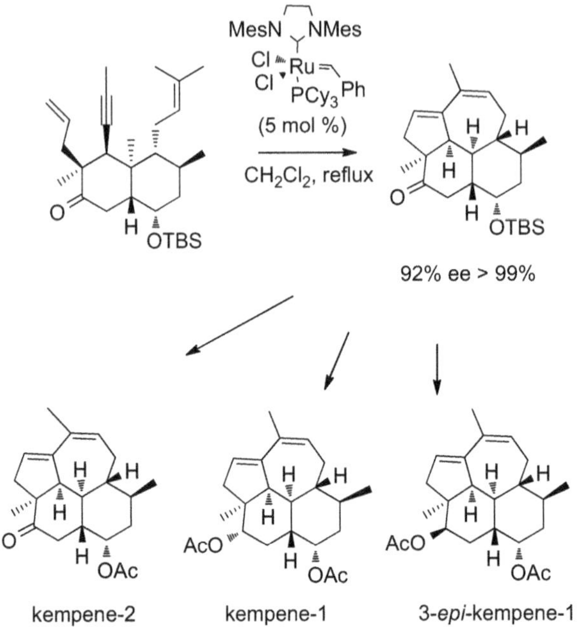

Scheme 1.61 Synthesis of kempene-2, kempene-1, and 3-*epi*-kempene-1 through domino enyne metathesis reaction.

investigated asymmetric ruthenium(III)-catalysed domino cyclisation reaction of a chiral phenol.[101] As shown in Scheme 1.62, treatment of this functionalised phenol with cationic ruthenium(III) complexes promoted its cascade cyclisation to form the corresponding tetracycle as a 93:7 mixture of *trans*- and *cis*-diastereomers in 74% yield. The authors have shown, however, that it was possible to favour the diastereoselective formation of the *cis*- required product through treatment of the substrate with triflic acid in nitromethane, albeit in low yield (20%). The authors have not proposed a mechanism for the formation of the domino product; nevertheless, a plausible one is depicted in Scheme 1.62. First, ruthenium precursor reacted with AgOTf to give a cationic ruthenium species. This cationic Ru(III) species had some TfO⁻, ligand, and/or solvent molecules. The C=C double bond of cyclohexene coordinated to this cationic ruthenium to give a Ru–olefin complex, which allowed reaction with phenolic oxygen in a nucleophilic fashion intramolecularly to give a cyclised intermediate. At this stage, the counter anion held the phenolic proton, and the resulting TfOH reacted with the carbon–ruthenium σ-bond by cyclisation to the second C=C double bond to give, after protonolysis, the final product.

1.2.7   Gold-Catalysed Domino Reactions

The use of gold catalysts for organic transformations has become increasingly common over the past decade, leading to the development of a number of

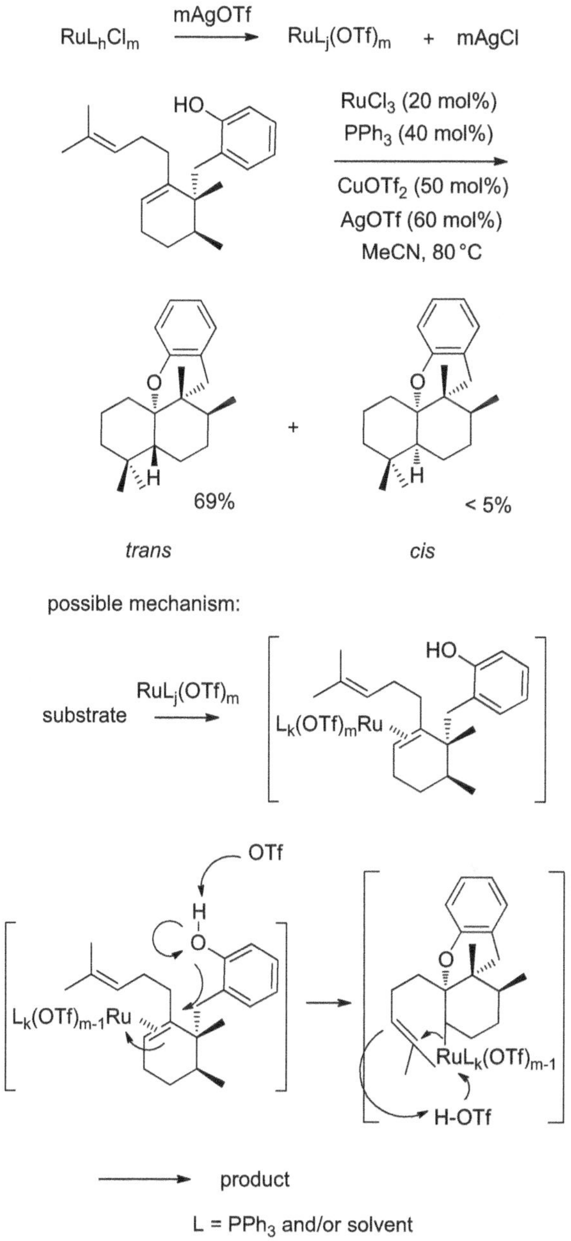

Scheme 1.62 Domino cyclisation reaction.

useful carbon–carbon and carbon–heteroatom bond-forming processes, allowing the synthesis of a wide variety of carbo- and heterocycles presenting a high degree of structural complexity.[102] In particular, a number of asymmetric gold-catalysed domino reactions have been successfully developed in recent

years. As an example, Chen *et al.* have reported a novel gold-catalysed
diastereoselective domino allylation–enyne cycloisomerisation reaction
occurring between allylic acetates and nonterminal propargylic alcohols.[103]
Remarkable levels of relative diastereoselectivities of >94% de in combination
with good yields were achieved in all cases of the substrates studied in the
formation of a range of densely functionalised oxygen heterocycles, as shown in
Scheme 1.63. The reaction was catalysed by 5 mol% of Ph$_3$PAuNTf$_2$ and was
found to follow another route when employing terminal propargylic alcohols
because, in the case of these substrates, corresponding tetrahydrofurans were
produced as single diastereomers in high yields, as shown in Scheme 1.63. In
order to explain the two different routes to these two types of product, the
authors have proposed the mechanism depicted in Scheme 1.63. Enyne ether
intermediates **37**, generated *in situ* from the intermolecular allylation, underwent
subsequent cycloisomerisation to yield a series of oxygen heterocycles. In the
second step, trapping of the gold-activated triple bond by the alkene group was
favoured to proceed through transition state **38**, in which the allylic R^1 group
pointed away from the alkyne bond to avoid the steric hindrance between R^1
and gold complexes. The stereochemistry of bicyclo[4.1.0]heptylidene gold(I)
carbene **40** (route I) and bicyclo[3.1.0]hexylidene gold(I) carbene **41** (route II)
was then determined by the favoured transition state **38**. The final six-membered
product was obtained from intermediate **40** by β-hydrogen elimination.
Trapping intermediate **41** derived from transition state **39**, with terminal
propargylic alcohol, afforded the five-membered product.

In 2011, Chan *et al.* demonstrated gold(I)-catalysed domino 1,3-migration–
[2 + 2] cycloaddition reaction of 1,7-enyne benzoates to be a regioselective and
stereoconvergent strategy for the construction of highly functionalised chiral
azabicyclo[4.2.0]oct-5-enes.[104] Indeed, when chiral 1,7-enyne benzoates were
treated with a catalytic amount of an Au(I) complex at 80 °C, they led to the
corresponding enantiopure domino products as single diastereomers, irre-
spective of whether the reaction started from a single or a diastereomeric
mixture of the substrate. As shown in Scheme 1.64, the process was shown
to tolerate a diverse set of 1,7-enyne substrates and furnished stereochemically
a range of enantiopure cyclobutane-fused pyridines bearing up to four
stereogenic centres, which constituted key starting materials for the synthesis of
natural and biologically interesting products. A tentative mechanism for this
novel domino process was proposed by the authors, in which activation of the
alkyne moiety of the 1,7-enyne substrate by the gold(I) catalyst resulted in *syn*
1,3-migration of the carboxylic ester group and formation of allenene **42**. To
avoid unfavourable steric interactions between the gold complex and the
substituents on the allene moiety, the catalyst selectively coordinated to
the alkene bond of the newly formed allenene **42** to give intermediate **43**. This
intermediate then underwent a [2 + 2] cycloaddition to give the piperidine
intermediate **44**. A subsequent nucleophilic addition of the Au–C(sp^3) bond to
the carbonyl carbon centre of the benzoyl cationic moiety generated from this
initial intramolecular cyclisation step then delivered the final product, as
depicted in Scheme 1.64.

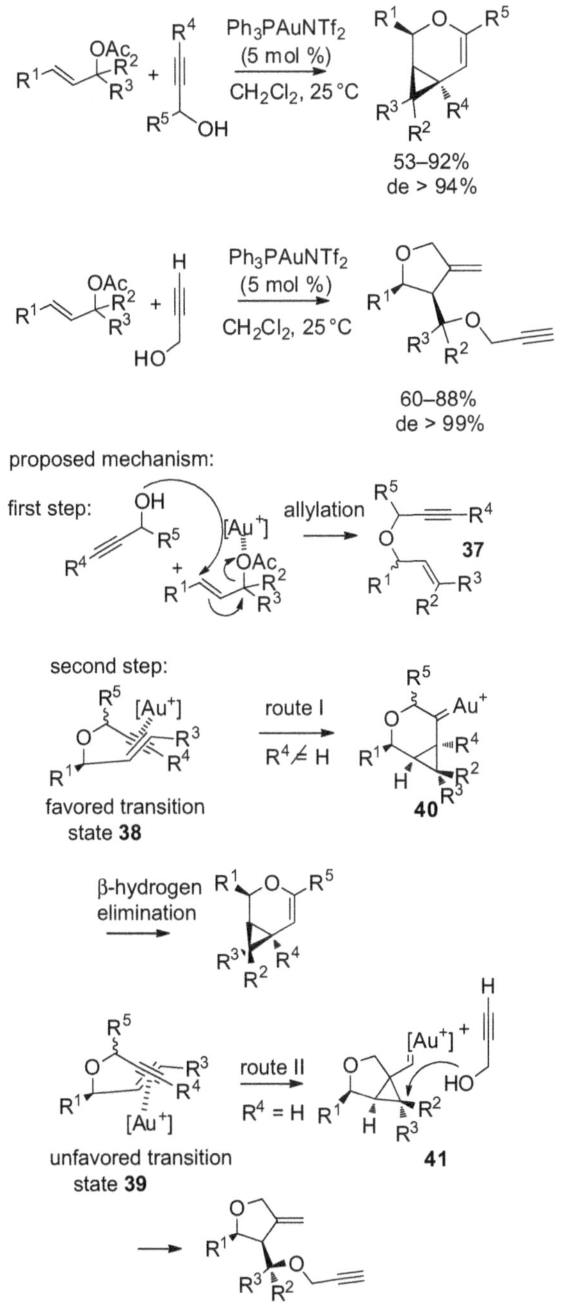

Ph3PAuNTf2
(5 mol %)
CH2Cl2, 25 °C
53–92%
de > 94%
60–88%
de > 99%
proposed mechanism:
first step:
allylation
37
second step:
route I
favored transition state 38
40
β-hydrogen elimination
route II
unfavored transition state 39
41

Scheme 1.63 Domino allylation–enyne cycloisomerisation reactions.

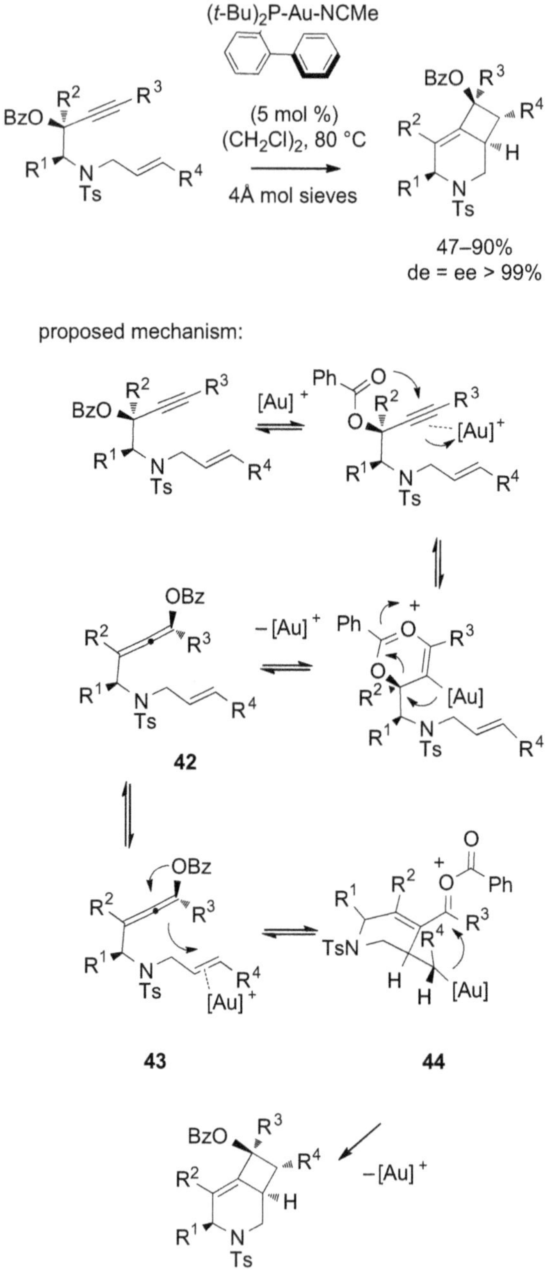

Scheme 1.64   Domino 1,3-migration–[2 + 2] cycloaddition reaction.

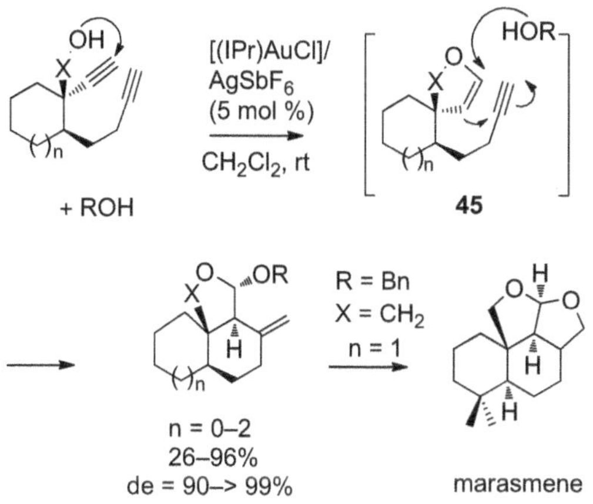

Scheme 1.65 Synthesis of marasmene through domino cyclisation reaction.

In addition, Li *et al.* have developed total syntheses of drimane-type sesquiterpenoids on the basis of a gold-catalysed domino reaction of chiral 1,7-diynes, involving both internal and external nucleophiles.[105] This domino process allowed the formation of three bonds, two rings, and two stereogenic centres. The reaction began with a 5-*endo*-dig addition of oxygen to an alkyne to give a polarised olefin functionality **45**, which functioned as the nucleophile in the following 6-*exo*-dig cyclisation (Scheme 1.65). The reaction was terminated by an external nucleophile as an alcohol, affording the corresponding tricyclic domino product. Indeed, when 1,7-diynes were treated by 5 mol% of the catalyst system [(IPr)AuCl]/AgSbF₆ in the presence of an alcohol as external nucleophile, they led to the corresponding tricyclic domino products in low to excellent yields (26–96%) and relative diastereoselectivities of >90% de in all cases of the substrates studied. The utility of this methodology was applied to the synthesis of the natural C15 oxygenated drimane-type sesquiterpenoid marasmene.

1.2.8   Miscellaneous Domino Reactions

In addition, miscellaneous asymmetric domino reactions have been reported in recent years, a number of which were catalysed by metals other than gold, ruthenium and palladium. Among them are several asymmetric Michael-terminated processes involving chiral auxiliaries. As an example, Hong *et al.* have recently combined an oxidation to an oxa-Michael addition in an asymmetric domino sequence, which constituted the key step of a formal synthesis of leucascandrolide A.[106] Indeed, the domino oxidation–oxa-Michael reaction of a chiral allylic alcohol (Scheme 1.66) led, in the presence of MnO₂, to the corresponding 2,6-*cis*-tetrahydropyran aldehyde in 86% yield and relative diastereoselectivity of >90% de. This product constituted a potent key precursor of leucascandrolide A. In addition, these authors have extended the

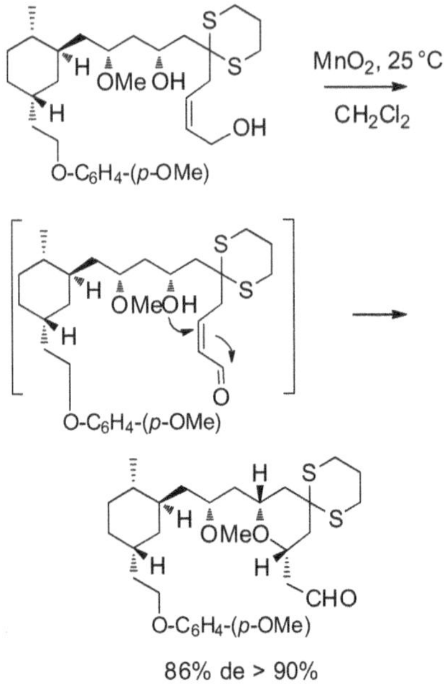

Scheme 1.66 Mn-catalysed domino oxidation–oxa-Michael reaction.

scope of this methodology to another chiral allylic alcohol which afforded, through domino oxidation–oxa-Michael reaction, the corresponding chiral 2,6-*cis*-tetrahydropyran aldehyde in 90% yield and >90% de. This product was used as the key intermediate in a formal synthesis of SCH 351448, which is a potent agent against hypercholesterolaemia.

Efficient novel syntheses of the most biologically active members of the cedranoic family, α-cedrene, α-pipitzol, and *sec*-cedrenol, have been achieved by Pettus *et al.*, beginning with the oxidative dearomatization of curcuphenol, which was followed by an intramolecular [5 + 2] cycloaddition of the phen-oxonium intermediate across the tethered olefin.[107] In this asymmetric domino reaction, the benzylic stereocentre effectively guided the formation of the first two stereocentres during the [5 + 2] cycloaddition. The sequence then terminated with the selective incorporation of acetic acid to generate a third stereocentre, setting it apart from other previous cationic [5 + 2] cycloaddition reactions. As shown in Scheme 1.67, the scope of this domino process could be extended to several chiral substrates, providing the corresponding domino products as single diastereomers in moderate to good yields, and opening the route to the enantiopure sesquiterpenes.

In 2012, Bull *et al.* reported a novel method of preparing enantiopure hydroxyl-γ-butyrolactones containing up to four stereocentres in high yields and moderate to excellent diastereoselectivities.[108] This process, performed in the presence of 10 mol% of osmium tetroxide (OsO_4) and *N*-methyl-morpholine-*N*-oxide (NMO), consisted in the diastereoselective domino

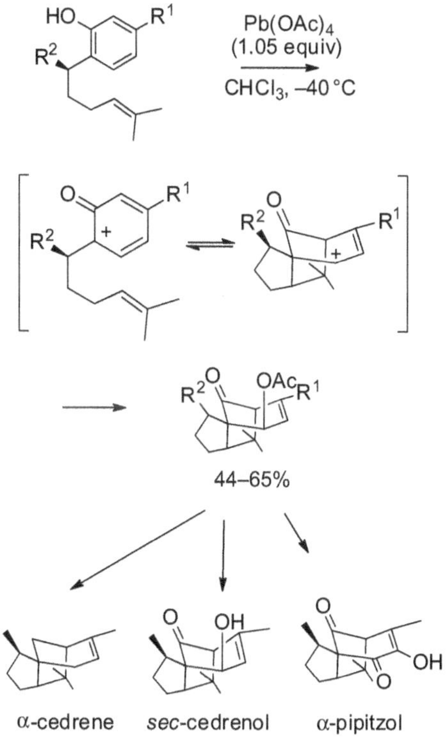

Scheme 1.67 Syntheses of α-cedrene, *sec*-cedrenol, and α-pipitzol through Pb-catalysed domino oxidation–oxa-Michael reaction.

dihydroxylation–lactonisation reaction of a range of chiral β-alkenyl-β-hydroxy-*N*-acyloxazolidin-2-ones, which provided the corresponding enantiopure hydroxyl-γ-butyrolactones in high yields and diastereoselectivities of up to >96% de, as shown in Scheme 1.68. It was found that 1-substituted, 1,1-disubstituted, (*E*)-1,2-disubstituted, (*Z*)-1,2-disubstituted, and 1,1,2-tri-substituted alkenes underwent dihydroxylation with *anti*-diastereoselectivity to their β-hydroxyl groups, whereas a 1,2,2-trisubstituted alkene led to the corresponding *syn*-diastereomer. The synthetic utility of this novel methodology has been demonstrated with a short synthesis of 2-deoxy-D-ribonolactone.

On the other hand, an asymmetric domino epoxidation–cyclisation reaction was used by Chandrasekhar *et al.* as the key step of a highly stereoselective approach to the total synthesis of the biologically active alkaloid (–)-codo-nopsinine, in 2006.[109] In this process, a chiral allylic alcohol, bearing a Boc-protected amino group in the α-position of the hydroxy group, was submitted to epoxidation by treatment with *meta*-chloroperbenzoic acid (MCPBA) to furnish directly the corresponding pyrrolidine diol in 89% yield and dia-stereoselectivity of 80% de in favour of the *anti*-diol, as shown in Scheme 1.69. The first epoxide formed was subsequently submitted to a regioselective opening with the internal nitrogen nucleophile in an *endo* fashion facilitated by

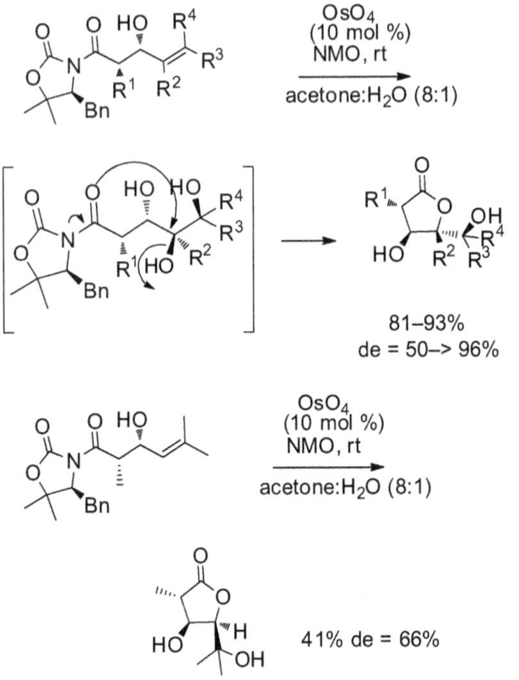

Scheme 1.68 Os-catalysed domino dihydroxylation–lactonisation reactions.

the 4-methoxy phenyl group, which allowed facile benzylic cleavage to produce the final diol. The major *anti*-diol was subsequently converted into (−)-codonopsinine through reduction of the Boc group into a methyl group in 83% yield.

An efficient synthesis of enantiopure lactols was described by Kita *et al.* on the basis of a domino reaction of chiral *cis*-2,3-epoxy-1-alcohols with a hypervalent iodine(III) reagent, such as PhI(OCOCF$_3$)$_2$, in the presence of H$_2$O.[110] In this process, the nucleophilic addition of water causing the oxirane ring-opening first occurred at the C3 position (Scheme 1.70). In the reaction, PhI(OCOCF$_3$)$_2$ reacted with the hydroxy function, which accelerated the reaction rate and the nucleophilic addition at the C3 position. Cleavage at the C1–C2 bond by forming a five-membered transition state then gave the hydroxy keto aldehyde, in which automatic lactol formation occurred to produce the final lactol. As shown in Scheme 1.70, several lactols were achieved in good to high yields and complete diastereoselectivity, and were subsequently converted into enantiopure lactones by treatment with Jones reagent. The utility of this efficient novel asymmetric domino epoxide-opening–oxidative cleavage–cyclisation reaction was applied to the total synthesis of (+)-tanikolide, an antifungal natural marine product.

Moreover, PhI(OAc)$_2$ was also employed by Aseniyadis *et al.* to mediate another asymmetric domino reaction, beginning with the oxidation of a chiral

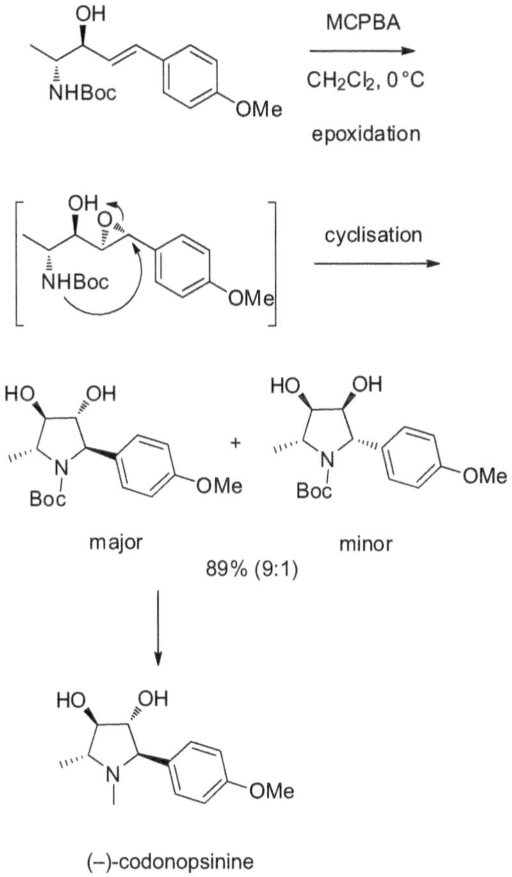

Scheme 1.69 Synthesis of (−)-codonopsinine through domino epoxidation–cyclisation reaction.

1,2-diol (Scheme 1.71) into the corresponding dialdehyde **46**, which underwent an intramolecular hetero-Diels–Alder cycloaddition to give the corresponding cyclic ene-acetal in 72% yield as a single stereoisomer, as shown in Scheme 1.71.[111] This chiral product was applied as key intermediate in an enantioselective total synthesis of a natural Ca^{++} antagonist, 1-*epi*-pathy-lactone A. As an extension of this methodology, these authors have demonstrated that using Pb(OAc)$_4$ as the oxidant instead of PhI(OAc)$_2$ under microwave irradiation and heating meant that the intermediate Diels–Alder cycloadduct **47** further evolved through ring expansion during the domino process, providing the iodonium intermediate **48**. This was subsequently submitted to an S$_N$2 opening with the reagent's acetate resulting in a ring system interchange to lead to the final product.[112] As shown in Scheme 1.71, a range of chiral ring-expanded domino products could be prepared in high yields by employing this methodology.

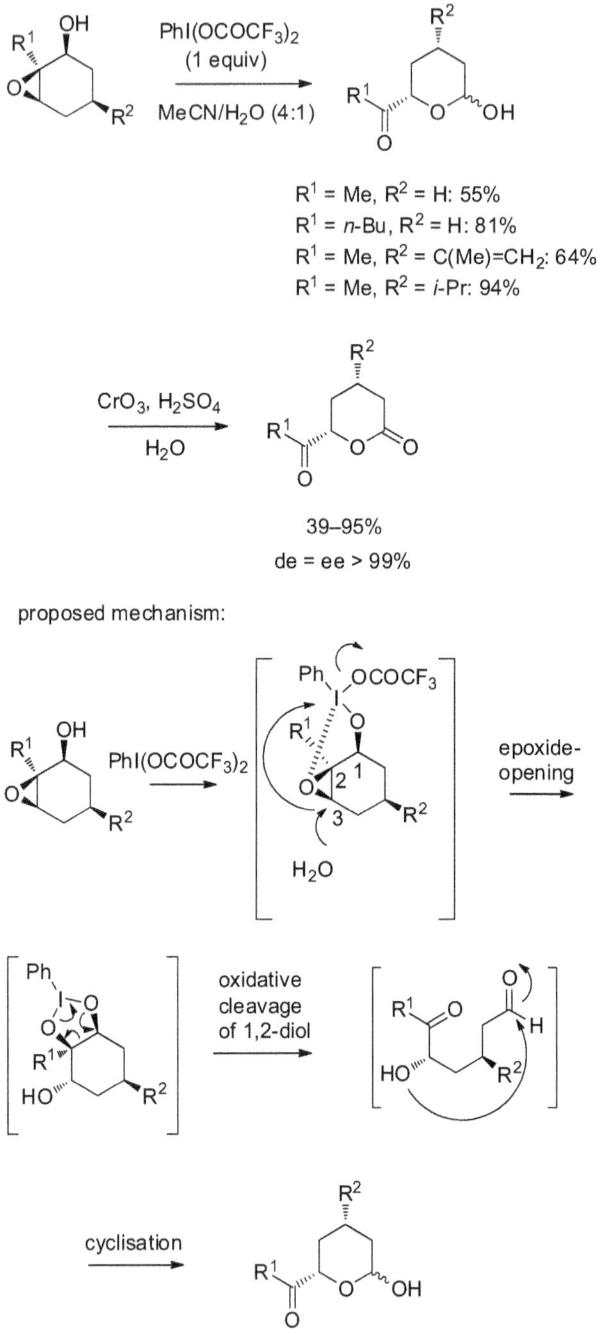
PhI(OCOCF3)2
(1 equiv)
MeCN/H2O (4:1)
R1 = Me, R2 = H: 55%
R1 = n-Bu, R2 = H: 81%
R1 = Me, R2 = C(Me)=CH2: 64%
R1 = Me, R2 = i-Pr: 94%
CrO3, H2SO4
H2O
39–95%
de = ee > 99%
proposed mechanism:
PhI(OCOCF3)2
H2O
epoxide-opening
oxidative cleavage of 1,2-diol
cyclisation

Scheme 1.70 Domino epoxide-opening–oxidative cleavage–cyclisation reaction.

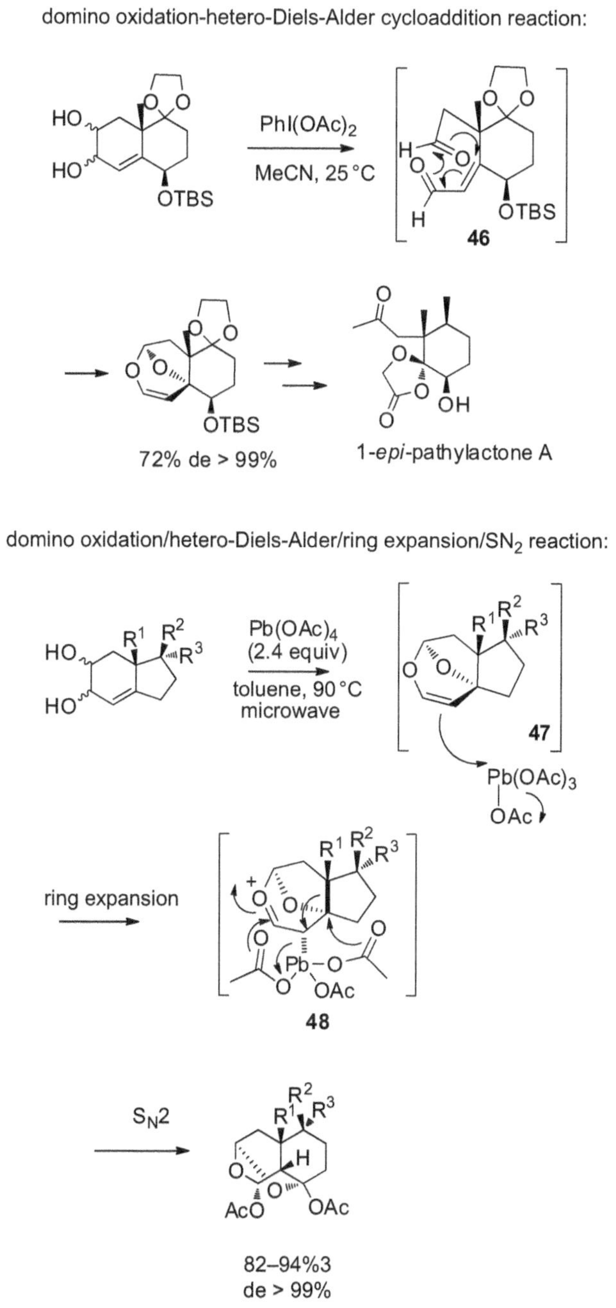

domino oxidation-hetero-Diels-Alder cycloaddition reaction:

72% de > 99%   1-*epi*-pathylactone A

domino oxidation/hetero-Diels-Alder/ring expansion/SN₂ reaction:

82–94%3
de > 99%

Scheme 1.71 Synthesis of 1-*epi*-pathylactone A through domino oxidation–intramolecular hetero-Diels–Alder cycloaddition reaction, and domino oxidation–intramolecular hetero-Diels–Alder cycloaddition–ring expansion–S$_N$2 reaction.

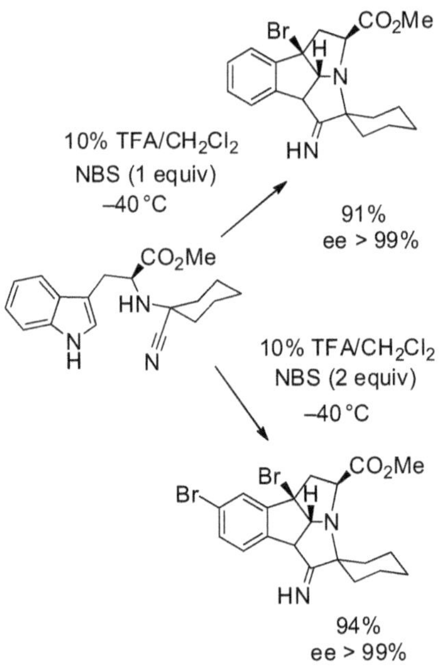

Scheme 1.72   Domino (di)bromination–cyclisation reactions.

In 2007, Herranz *et al.* described a novel stereoselective domino electrophile bromation–cyclisations reaction of tryptophan-derived α-amino nitriles.[113] As shown in Scheme 1.72, treatment of an enantiopure tryptophan-derived α-amino nitrile by one equivalent of *N*-bromosuccinimide (NBS) in 10% aqueous trifluoroacetic acid (TFA) in dichloromethane at –40 °C led to the corresponding chiral monobrominated product in 91% yield, whereas the use of two equivalents of NBS provided the corresponding enantiopure dibrominated derivative in 94% yield. Attempts to produce the corresponding iodocyclisation of this α-amino nitrile were unsuccessful, while the corresponding chlorocyclisation led to a 4:1 mixture of diastereoisomeric mono-chlorides. Moreover, these authors have shown that this α-amino nitrile could also be allylated at the same position as that for the monobromination through treatment with prenyl bromide and $Mg(NO_3)_2 \cdot 6H_2O$ in AcOH/AcONa buffer with complete stereoselectivity in 45% yield. These novel domino processes allowed an easy entry to chiral indole alkaloid analogues to be achieved.

In another context, Hsung *et al.* have reported a remarkable asymmetric quadruple domino process, consisting in a 1,3-H–1,7-H shift–6π electron pericyclic ring-closure–[4 + 2] cycloaddition reaction, allowing a rapid assembly of chiral complex tricycles from simple chiral allenamides to be achieved.[114] As shown in Scheme 1.73, heating enantiopure α-substituted allenamides bearing an *E*-alkene moiety provided the corresponding chiral amide-substituted 1,3,5-hexatrienes through 1,3-H–1,7-H shift, which underwent a 6π-electron ring-closure to give the corresponding cyclohexadienes as single stereoisomers.

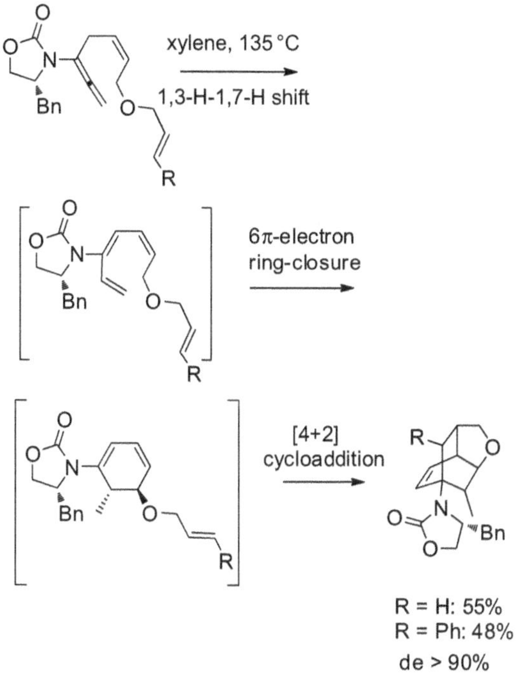

Scheme 1.73 Domino 1,3-H–1,7-H shift–6π electron pericyclic ring-closure–[4 + 2] cycloaddition reaction.

Furthermore, the authors have shown that this sequence could be followed by a [4 + 2] cycloaddition, affording the corresponding tricyclic products as almost single stereoisomers (de > 90%) in good yields (48–55%).

In 2009, Nicolaou *et al.* developed a stereoselective aluminium-catalysed domino epoxide-opening–Diels–Alder cycloaddition reaction, which constituted the key step in a total synthesis of naturally occurring and biologically active hirsutellone B.[115] Indeed, the treatment of an enantiopure epoxide derived from (R)-(+)-citronellal by Et₂AlCl led to the corresponding tricyclic domino product in 50% yield as a single diastereomer, as shown in Scheme 1.74. The cascade of reactions depicted in Scheme 1.74 could explain the formation of this key enantiopure product, which was subsequently converted into hirsutellone B in 17 steps.

In 2010, Ghorai *et al.* described a remarkable synthesis of enantiopure functionalised γ-lactams on the basis of a highly stereoselective copper-catalysed domino aziridine-opening–cyclisation reaction.[116] As shown in Scheme 1.75, monosubstituted aziridines reacted, in the presence of Cu(OTf)₂ as catalyst and NaH as base, with active methylene carbon nucleophiles, such as diethyl malonate or ethyl acetoacetate, to afford the corresponding enantio- and diastereopure γ-lactams in good to excellent yields of up to >99%. These products arose from the Cu-catalysed SN2-type ring-opening of aziridines with

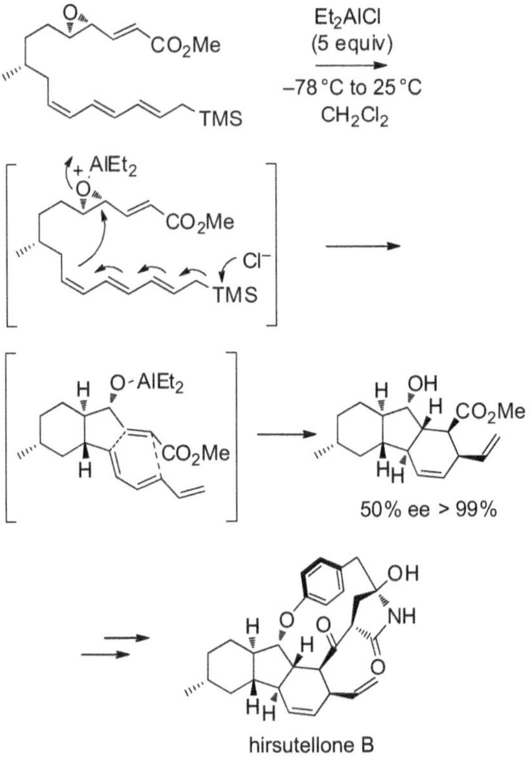

Scheme 1.74   Synthesis of hirsutellone B through Al-catalysed domino epoxide-opening–Diels–Alder cycloaddition reaction.

enolates followed by intramolecular cyclisation. The scope of this methodology could be extended to *trans-N*-tosyl-2,3-disubstituted aziridines, which led to the corresponding enantiopure trisubstituted γ-lactams as single diastereomers in comparable high yields (Scheme 1.75).

On the other hand, White *et al.* have reported total syntheses of natural solandelactones A, B, E and F by exploiting an asymmetric titanium-catalysed domino Petasis–Claisen lactonisation reaction.[117] As shown in Scheme 1.76, Petasis methylenation of an enantiopure cyclic carbonate in tandem with a Claisen rearrangement generated the corresponding chiral octenalactone portion of solandelactones as a single stereoisomer in 60–65% yield. This chiral product was further converted into a mixture of expected solandelactones A, B, E and F.

In 2010, Helmchen *et al.* reported a novel route to highly enantioenriched complex heterocycles on the basis of asymmetric domino platinum-catalysed enyne isomerisation–Diels–Alder reaction of enantiopure 1,6-enynes.[118] As shown in Scheme 1.77, heating a mixture of these 1,6-enynes and various dienophiles in the presence of $PtCl_2$ in toluene led to the corresponding domino bi- or tricyclic cycloadducts in moderate to good yields as single diastereomers in all cases of the substrates studied.

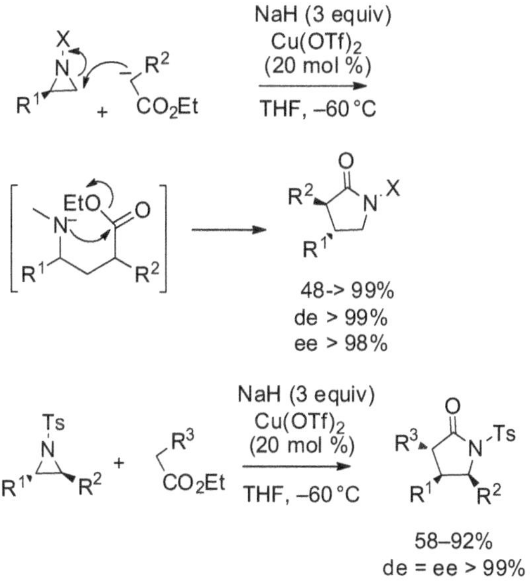

Scheme 1.75 Cu-catalysed domino aziridine-opening–cyclisation reaction.

A novel route towards chiral aminocyclopentanols was reported by Grée et al., in 2011.[119] It was based on an iron-catalysed domino isomerisation–Mannich reaction of chiral N-tert-butanesulfinamide as chiral auxiliary. As shown in Scheme 1.78, this process was performed using Fe(CO)$_5$ as the catalyst with full stereocontrol, because the (S)-configured N-tert-butanesulfinamide yielded the corresponding (R)-configured cyclopentanone as a single diastereomer in 85% yield, and the (R)-configured N-tert-butanesulfinamide led to the corresponding (S)-configured cyclopentanone in 84% yield, as shown in Scheme 1.78. These products were further converted into biologically interesting chiral aminocyclopentitols, such as mannostatin A analogues.

Nickel catalysts have also been employed to promote asymmetric domino reactions.[120] As an example, Grée et al. have developed a stereoselective synthesis of functionalised 1,3-diols on the basis of the domino isomerisation–aldolisation reaction of a 4:1 diastereomeric mixture of an allylic alcohol bearing a chiral imidazolidine group with benzaldehyde.[121] The process was catalysed by NiHCl(dppe) associated with MgBr$_2$ and led to the corresponding bicyclic aldol product as a mixture of three diastereomers in 70% yield and moderate diastereoselectivity of ≤40% de, as shown in Scheme 1.79. It began with the isomerisation of the starting allylic alcohol into its corresponding Z-enol **49** which subsequently reacted with benzaldehyde through aldol condensation to afford the final product.

In 2010, Chmielewski et al. reported a copper-catalysed domino Kinugasa cycloaddition–rearrangement reaction of chiral acetylenes derived from D-glyceraldehyde and propargyl aldehyde with nitrones, affording the

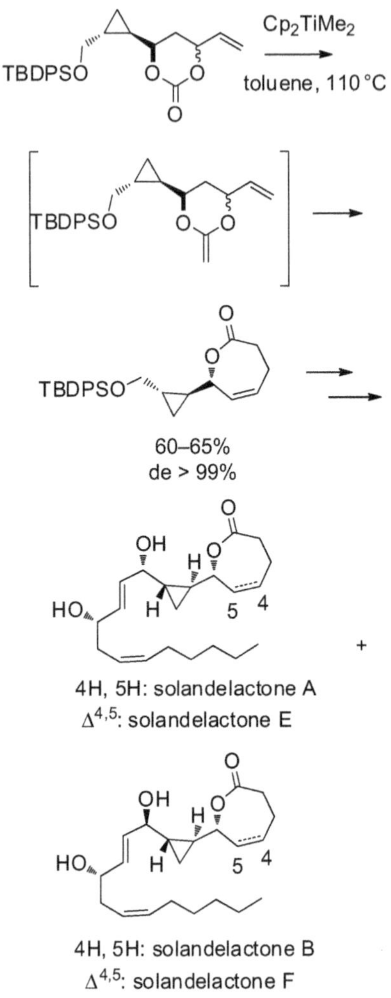

Scheme 1.76 Synthesis of solandelactones A, B, E, and F through Ti-catalysed domino Petasis–Claisen lactonisation reaction.

corresponding chiral *cis*-β-lactams in moderate to good yields and *cis : trans* ratios of up to >90% de.[122] One of the best results is depicted in Scheme 1.80. The authors assumed that the process evolved *via* formation of a rigid dinuclear copper(I) complex in which each copper ion was coordinated to one of both oxygen atoms in the acetylene molecule and to both triple bonds. They have proposed the catalytic cycle depicted in Scheme 1.80, which began with the coordination of the alkyne to Cu(I) species **50** to form π-complex **51** in which the copper ion was also linked to other nucleophilic centres (oxygen atom/atoms, phenyl ring). The deprotonation of **51** led to the dinuclear copper(I) complex **52** that, in the presence of nitrone, underwent stepwise or concerted

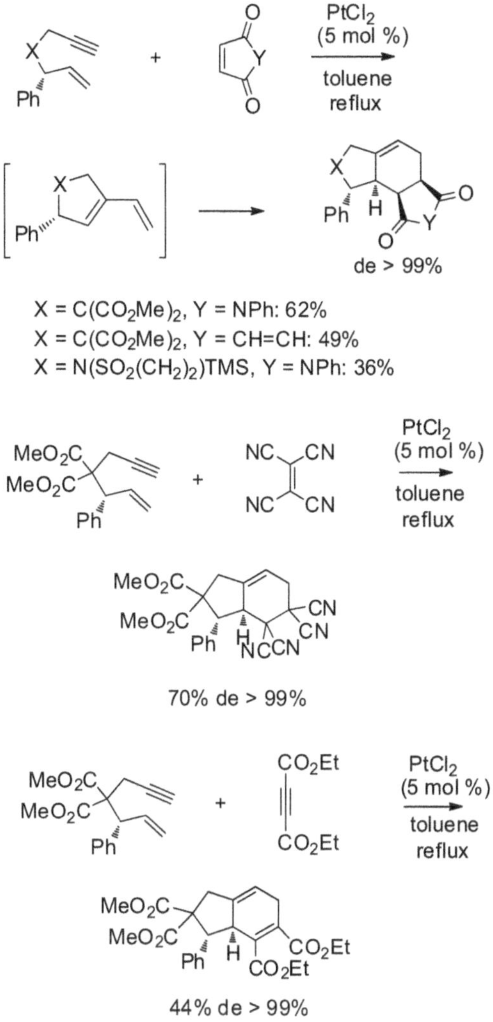
PtCl2
(5 mol %)
toluene
reflux
de > 99%
X = C(CO2Me)2, Y = NPh: 62%
X = C(CO2Me)2, Y = CH=CH: 49%
X = N(SO2(CH2)2)TMS, Y = NPh: 36%
70% de > 99%
44% de > 99%

Scheme 1.77 Pt-catalysed domino enyne isomerisation–Diels–Alder reactions.

cycloaddition leading to **55**. A stepwise process, which should be more plausible in cases of metal-catalysed reaction, proceeded through intermediates **53** and **54**. Then, six-membered copper metallocycle **54** contracted to five-membered isoxazoline **55** and rearranged to enolate **56**. Finally, protonation of **56** gave the final 2-azetidinone and released the copper catalyst. The domino process has allowed a range of chiral carbapenams to be achieved in acceptable yields even in the presence of a catalytic amount of copper salt.

Finally, a gold(I)-catalysed domino cyclisation approach to tetracyclic indolines was reported by Wang *et al.*[123] The scope of this methodology was applied to an enantioenriched alkynylindole bearing a secondary propargyl

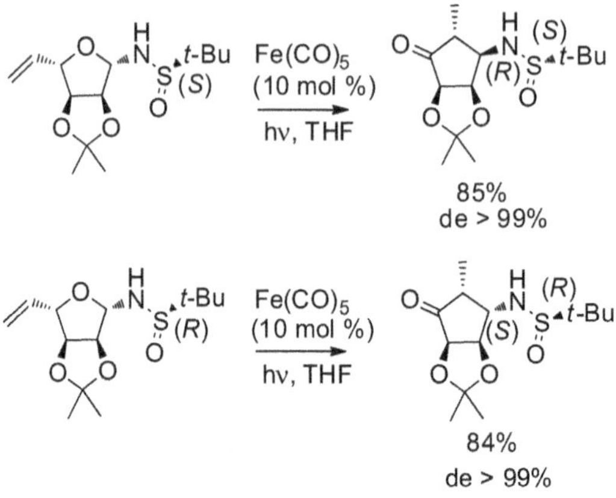

Scheme 1.78 Fe-catalysed domino isomerisation–Mannich reaction.

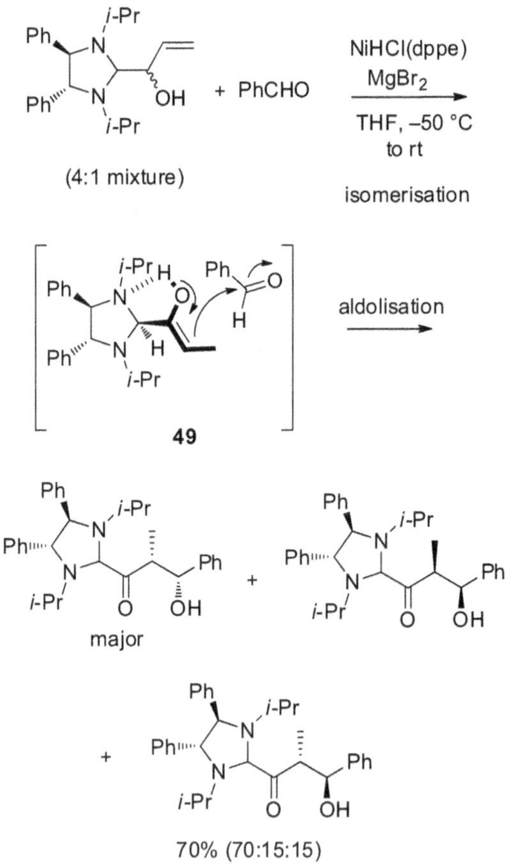

Scheme 1.79 Ni-catalysed domino isomerisation–aldolisation reaction.

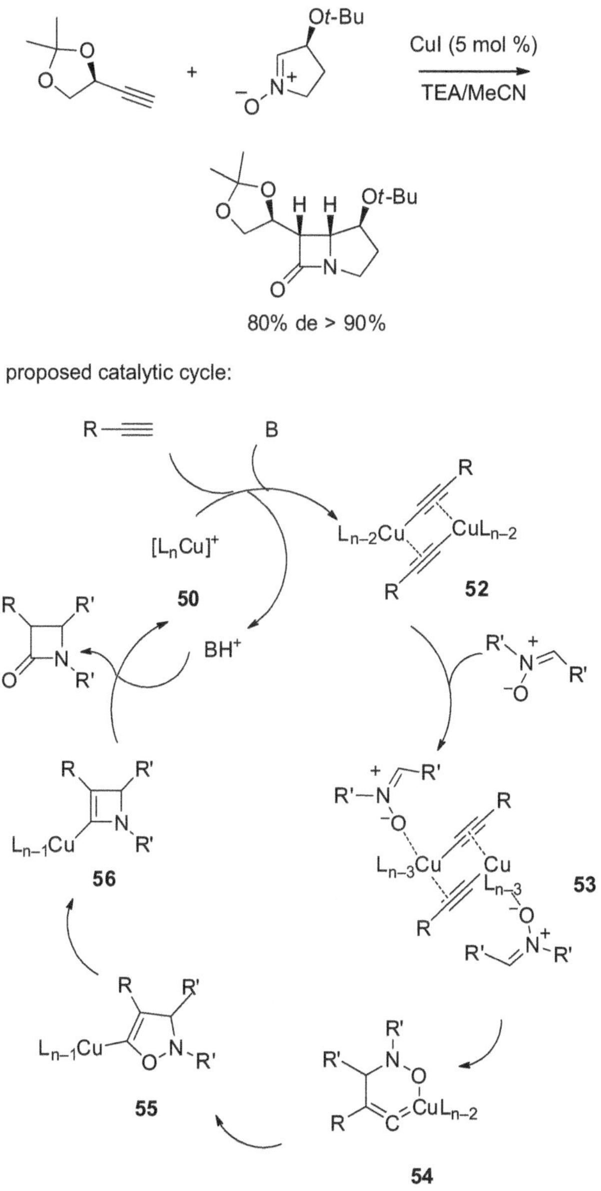

80% de > 90%

proposed catalytic cycle:

Scheme 1.80 Cu-catalysed domino Kinugasa cycloaddition–rearrangement reaction.

alcohol, which provided the corresponding tetracyclic indoline with quantitative retention of chirality (81% ee) at the secondary propargyl alcohol, which should be useful for the synthesis of complex indoline alkaloids. As shown in Scheme 1.81, the first cyclisation catalysed by Ph₃AuSbF₆ led to the formation of intermediate **57** containing a secondary alcohol, which served as

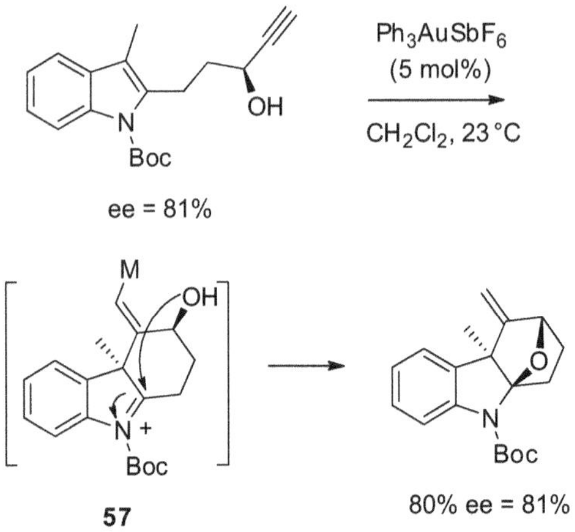

Scheme 1.81   Au-catalysed cyclisation reaction.

an internal nucleophile in the second cyclisation step to afford the final tetra-cyclic indoline.

1.3   Multicomponent Reactions

One of the challenges in organic synthesis is to implement various reaction strategies in a multicomponent reaction, which is a domino reaction involving at least three substrates,[5,7,8,124] to achieve multi-bond formation in a single reaction vessel, forming a new product which contains portions of all the components. Multicomponent reactions convert more than two educts directly into their product by one-pot reactions. The starting materials for this kind of chemical transformation are rich in functional groups. Typically, multi-component reactions lead to very complex products by reacting structurally simple starting materials. The multicomponent reaction proceeds according to the domino principle, because subsequent transformations are a consequence of the functionalities produced in the previous transformation. These reactions are highly flexible, (chemo)-selective, convergent and atom efficient processes of high exploratory power. Inspired by the mode of action of nature, they have brought the concept of ideal synthesis closer to reality.[125] Indeed, this strategy is atom economical and avoids the necessity of protecting groups and isolation of intermediates. Its goal is the resemblance of nature in its highly selective sequential transformations. Even though the history of multicomponent reactions dates back to the second half of the 19th century, with the reactions of Strecker, Hantzsch and Biginelli, it was only in the last decades with the work of Ugi that the concept of the multicomponent reaction has emerged as a powerful tool in synthetic chemistry.[126]

1.3.1 Multicomponent Reactions Initiated by the Michael Reaction

Several asymmetric multicomponent reactions employing chiral auxiliaries and initiated by a Michael addition have been recently reported, as described below. For example, Davies *et al.* reported, in 2007, asymmetric three-component couplings of α,β-unsaturated esters and alkylidene malonates initiated by chiral lithium amide conjugate addition, which proceeded with high levels of 2,3-*anti*-diastereoselectivity of >90% de.[127] As shown in Scheme 1.82, the domino reaction began with the Michael addition of lithium (S)-N-benzyl-N-α-methylbenzylamide to α,β-unsaturated esters, providing the corresponding enolates which further reacted with alkylidene malonates according to a second Michael addition, furnishing the final β-amino esters in moderate to good yields and *cis*-diastereoselectivities. Investigating the scope of the reaction has shown that higher yields and diastereoselectivities were observed with β-aryl substituents in both ester and malonate components, although the reaction successfully tolerated both β-alkyl and β-alkenyl functionality. The utility of this process was demonstrated by the conversion of the chiral formed products into polysubstituted piperidinones through hydrogenolysis.

In the same area, a novel diastereoselective aza-allyl conjugate addition–Michael addition–ring closure reaction sequence was developed by Andrews *et al.* in the same year.[128] This three-component process consisted in the reaction of the sodium anion of α-(S)-N-(α-methylbenzyl)allylamine employed as chiral auxiliary with two equivalents of *tert*-butyl cinnamate, resulting in a remarkable domino double Michael addition–ring closure reaction, affording the corresponding chiral aminocyclohexane containing six new vicinal stereogenic centres with an excellent level of stereocontrol of >90%

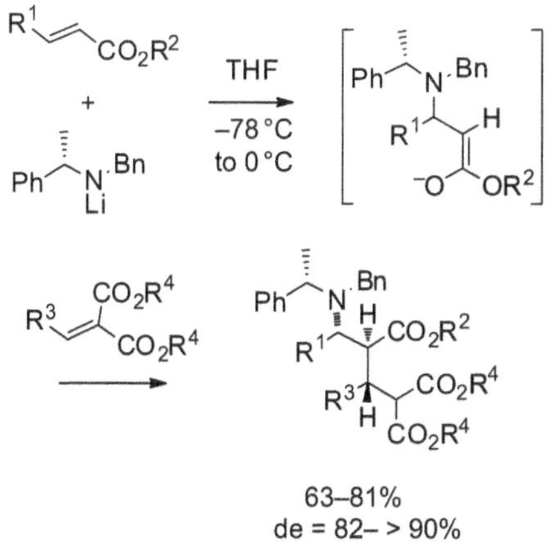

63–81%
de = 82– > 90%

Scheme 1.82 Three-component domino double Michael reaction of α,β-unsaturated esters, alkylidene malonates, and lithium (S)-N-benzyl-N-α-methyl-benzylamide.

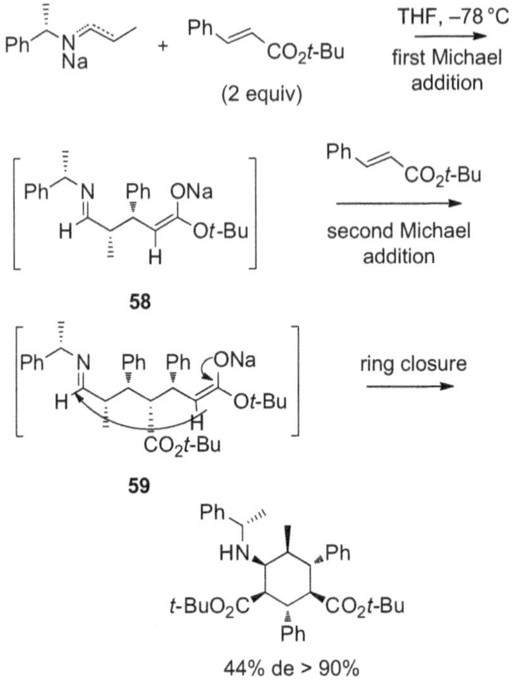

THF, −78 °C

first Michael
addition

(2 equiv)

Ph CO_2t-Bu

second Michael
addition

58

ring closure

59

44% de > 90%

Scheme 1.83 Three-component domino double Michael–ring closure reaction of sodium anion of (*S*)-*N*-(α-methylbenzyl)allylamine and two equivalents of *tert*-butyl cinnamate.

de and 44% yield. The authors have proposed the mechanism depicted in Scheme 1.83 in which the sequence began with the first Michael addition of (*S*)-*N*-(α-methylbenzyl)allylamine to one equivalent of *tert*-butyl cinnamate, providing the corresponding imine (*Z*)-enolate **58**. This enolate subsequently added to another equivalent of *tert*-butyl cinnamate through a second Michael addition to give a novel imine (*Z*)-enolate **59**, which finally underwent an intramolecular 6-*exo*-trig cyclisation of the enolate fragment onto its imino functionality, resulting in the formation of the final cyclohexane derivative.

In 2008, Rodriguez *et al.* reported a diastereoselective domino Michael–aldol–retro-Dieckmann reaction of β-ketoesters with α,β-unsaturated aldehydes in methanol as solvent and third component, allowing functionalised seven-membered rings to be achieved under mild reaction conditions.[129] Thus, a range of substituted cycloheptanols was regio- and diastereoselectively synthesised through the reaction of β-substituted acroleins with 2-oxo-cyclopentanecarboxylate methyl esters in the presence of DBU as a base in methanol at room temperature. These products, containing up to five stereogenic centres, were produced in moderate to high yields and relatively high diastereoselectivities of >92% de, as shown in Scheme 1.84. Furthermore, by starting from enantiopure 5-methyl-2-oxo-cyclopentanecarboxylate methyl

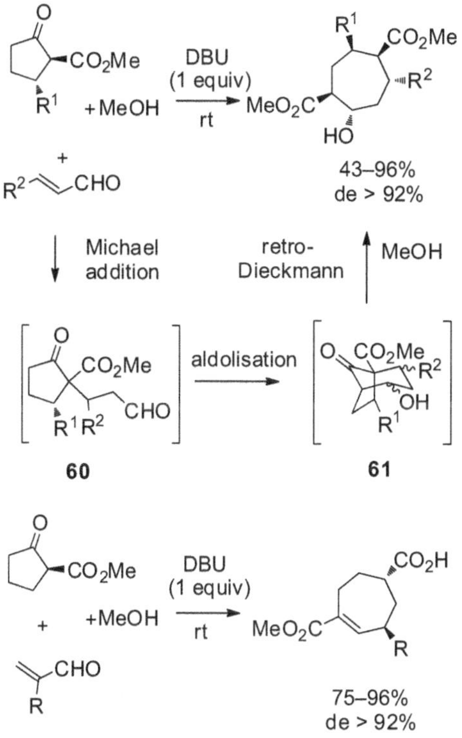

Scheme 1.84 Three-component domino Michael–aldol–retro-Dieckmann reaction of 2-oxo-cyclopentanecarboxylate methyl esters, α,β-unsaturateds aldehydes, and methanol.

ester ($R^1 = R^2 = Me$) derived from (+)-pulegone and crotonaldehyde in methanol, the reaction led to the corresponding chiral cycloheptanol as a single stereoisomer in 43% yield (Scheme 1.84). Investigating the scope of the reaction, the authors have found that the use of of α-substituted acroleins provided, by reaction with Dieckmann ester and methanol, the corresponding dehydrated cycloheptenes in good to high yields and complete relative diastereo-selectivity, as shown in Scheme 1.84. Remarkably, the crude did not require purification because a simple acidic work up allowed pure products to be achieved. As depicted in Scheme 1.84, the sequence began with the Michael addition of the β-ketoesters to the α,β-unsaturated aldehydes, providing intermediate aldehydes **60**, which underwent an intramolecular aldol reaction to give the corresponding 8-oxo-bicyclo[3.2.1]octanes **61**. Then, a retro-Dieckmann-type fragmentation of these bicyclic compounds led to seven-membered products or after a supplementary dehydration to the corresponding alkenes. The further extension of the scope of the process to related five-membered β-ketosulfones allowed the corresponding cycloheptanols to be afforded in good to high yields albeit moderate relative diastereoselectivities (34–66% de).

Later, a one-pot synthesis of chiral substituted piperidines was developed by Shi *et al.* on the basis of a three-component reaction occurring between a nitroalkene, a chiral aminoester, and an enone.[130] As depicted in Scheme 1.85, the process began with the aza-Michael addition of the chiral amine to the nitroalkene to give nitroalkane **62**, which underwent a second aza-Michael addition to the enone, providing novel nitroalkane **63**. This nitroalkane then cyclised according to a Henry-aldol reaction to give the corresponding substituted piperidine. The reaction provided the corresponding enantiopure piperidines in good yields and moderate diastereoselectivities (50–72% de) with only C4 isomers observed, as shown in Scheme 1.85.

On the other hand, Node *et al.* have investigated asymmetric domino three-component reactions initiated by the aza-Michael addition of a chiral amine,

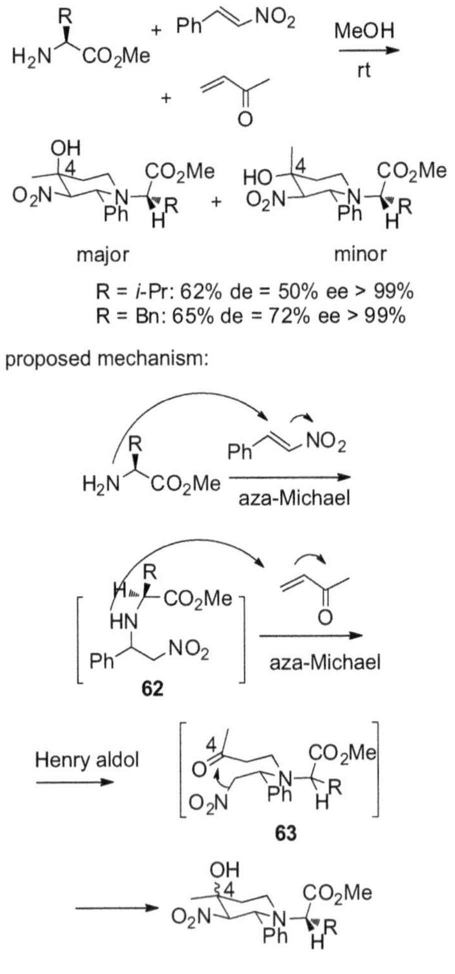

Scheme 1.85   Three-component domino double aza-Michael–Henry-aldol reaction of a chiral amino ester, a nitroalkene, and an enone.

such as *N*-benzyl-2(*R*)-methoxy-(+)-10-bornylamine, to α,β-unsaturated esters.[131] For example, the asymmetric domino Michael–aldol reaction of this chiral amine with an α,β-unsaturated ester and an aldehyde occurred in the presence of a base, such as *n*-butyl (Bu)Li, to give the corresponding domino products in moderate to good yields and diastereoselectivities of up to 80% de. Another asymmetric domino three-component reaction initiated by the aza-Michael addition of the same chiral amine to α,β-unsaturated esters was developed by these authors. In that case, the three components were the chiral amine, an aldehyde, and a diester, such as di-*tert*-butyl 2,6-octadien-1,8-dioate, which reacted through a domino double-Michael–aldol reaction, providing the corresponding highly functionalised domino products bearing five contiguous chiral centres. The process was performed in the presence of a base, such as *n*-BuLi, and provided moderate to good yields (40–67%) and diastereoselectivities of up to 50% de. In spite of these moderate stereoselectivities, this process constituted the first method of amination that could build up to five contiguous chiral centres in a one-pot reaction. The best results of both these domino processes are collected in Scheme 1.86.

In 2007, Kamimura *et al.* showed that optically active *N*-sulfinimines underwent stereoselective domino aza-Michael–nucleophilic addition reaction triggered by magnesium thiolate to give the corresponding α-phenylthio-methyl-β-(*N*-sulfinylamino) esters in good to high diastereoselectivities (62–94% de).[132] As shown in Scheme 1.87, the major *syn*-products were obtained in good to quantitative yields. These products could be further converted into useful corresponding aza-Baylis–Hilman adducts.

In the same area, Lhommet *et al.* have reported the asymmetric three-component condensation of various dicarbonyl compounds, acrolein, and (*S*)-2-phenylglycinol, providing the corresponding chiral 6-carbonyl-3-phenyl-2,3,8,8a-tetrahydro-7*H*-[1,3]oxazolo[3,2-*a*]pyridines in low to moderate yields (19–63%) and zero to good diastereoselectivities of up to 80% de.[133] This domino reaction involved the Michael addition of dicarbonyl compounds to acrolein, followed by condensation of the chiral amine (*S*)-2-phenylglycinol with the resulting tricarbonyl compounds to provide the final chiral bicyclic functionalised tetrahydropyridines. It must be noted that this simple procedure allowed the concomitant formation of four bonds and two stereogenic centres. Although only limited success was achieved in terms of reactivity and selectivity, various β-ketoesters and β-diketones were tolerated in the process. Furthermore, the utility of this domino process was illustrated by its application in a total synthesis of quinolizidine alkaloid (−)-lupinine. Other asymmetric multicomponent domino reactions initiated by a Michael addition have been developed, such as domino Michael–Mannich reactions of dialkylzincs, cyclic enones and chiral *N*-sulfinimines, which were reported by Yus *et al.*, in 2008.[134] In this work, a double induction by using a chiral copper catalyst in addition to the chiral *N*-sulfinimine auxiliary was necessary to obtain the corresponding chiral β-aminocyclohexanones as single stereoisomers. Indeed, the use of both a chiral catalyst and a chiral auxiliary as a chiral *N*-sulfinimine allowed the enantiopure domino Michael–Mannich products to be achieved in

domino Michael-aldol reaction:

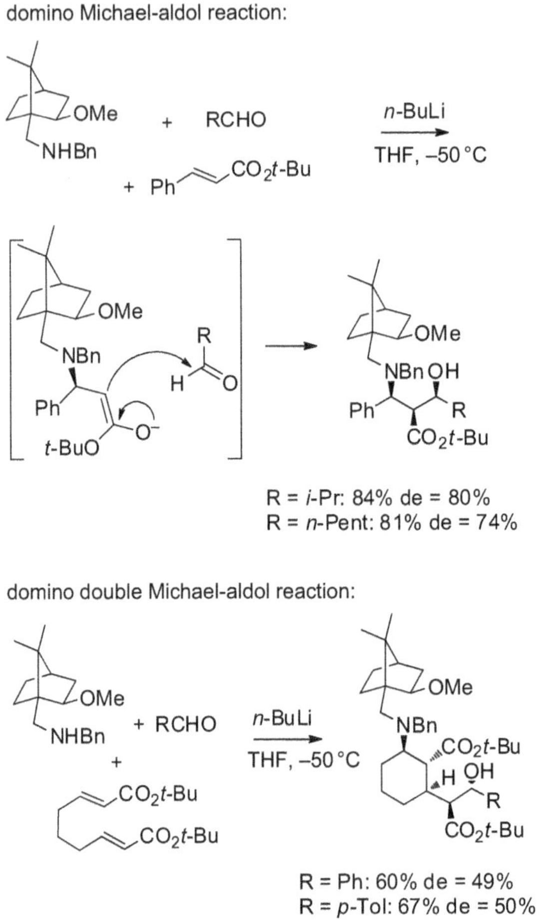

R = *i*-Pr: 84% de = 80%
R = *n*-Pent: 81% de = 74%

domino double Michael-aldol reaction:

R = Ph: 60% de = 49%
R = *p*-Tol: 67% de = 50%

Scheme 1.86   Three-component domino aza-Michael–aldol reaction, and three-component domino double Michael–aldol reaction of *N*-benzyl-2(*R*)-methoxy-(+)-10-bornylamine.

good to high yields, as shown in Scheme 1.88. This process could be applied to a wide range of substrates and was especially efficient when enolisable *t*-BuS imines were used.

In 2010, Wang *et al.* reported the first example of dynamic kinetic aza-Michael addition, allowing the convenient synthesis of an important class of chiral heterocyclic compounds which are imidazolidin-4-ones.[135] Indeed, it is known that these products exhibit a range of biological activities and constitute key chiral building blocks for the total synthesis of medically important compounds.[136] The pyridyl moiety of the chiral auxiliary amino amide activated the Michael acceptor under acidic conditions, while acting as a remote directing group to position the enone and the secondary amine in a favourable orientation for the aza-Michael addition, as depicted in

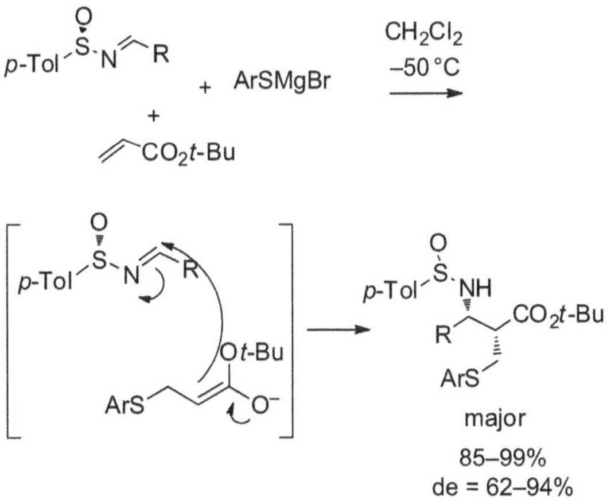

Scheme 1.87 Three-component domino aza-Michael–nucleophilic addition reaction of chiral *N*-sulfinimines, *t*-butyl acrylate, and magnesium thiolate.

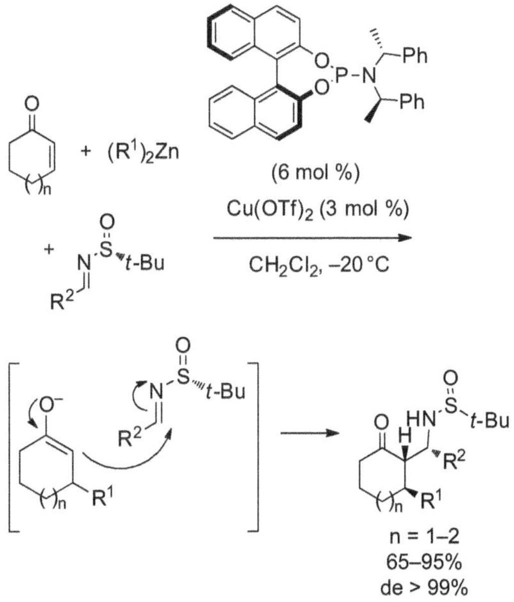

Scheme 1.88 Three-component domino Michael–Mannich reaction of chiral *N*-sulfinimines, cyclic enones, and dialkylzincs.

Scheme 1.89. The importance of the nitrogen position in the protonated pyridyl directing group was illustrated by failure of the Michael addition reaction with regioisomeric pyridyl derivatives of the chiral auxiliary amino amide. Both *trans*- and *cis*-intermediates provided the corresponding *cis*-products as the reaction outcome of the final Michael addition reaction. The authors assumed

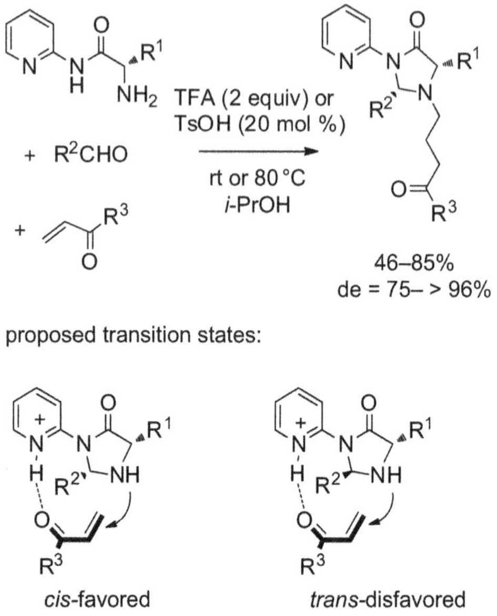

Scheme 1.89   Remote group-aza-Michael–based three-component reaction of chiral amino amides, aldehydes, and enones.

the occurrence of a dynamic kinetic pathway in which the reaction of the *cis*-intermediate was faster owing to reduced steric hindrance. While electron-poor aromatic aldehydes afforded the Michael products in moderate to good yields (64–90%) and good to excellent diastereoselectivities (84 to >96% de) in the presence of trifluoroacetic acid, electron-rich aromatic aldehydes required the use of a stronger Bronsted acid, such as *p*-toluenesulfonic acid (TsOH), to provide acceptable results (63–70% yield, 72–86% de). Furthermore, electron-rich aliphatic aldehydes required elevated temperatures (80 °C) to achieve better yields (46–53%) and diastereoselectivities (92–94% de) in the aza-Michael addition-based three-component process.

1.3.2   Multicomponent Reactions Based on the Hantzsch Reaction

A venerable and old multicomponent reaction is the so-called Hantzsch reaction, which was first reported in 1882,[137] which allows the synthesis of 1,4-dihydropyridines through the reaction of enamines, aldehydes, and 1,3-dicarbonyl compounds. Dihydropyridines are well known for exhibiting activity against calcium channels, multidrug inflammatory targets, in addition to their usefulness as tools for reducing imines to amines. The harsh reaction conditions usually applied to carry out the Hantzsch reaction have significantly decelerated the development of asymmetric versions of this reaction. In comparison with the other asymmetric multicomponent reactions, the asymmetric Hantzsch process has probably known the greatest progress in the last six years. In 2006, Dondoni *et al.* reported the synthesis of chiral 1,4-dihydropyridines through the three-component Hantzsch reaction between chiral aldehydes bearing a *N*-Boc benzyl glycinate group, β-keto esters, and enamino esters, which led to the corresponding chiral 1,4-dihydropyridines.[138] The scope of this methodology was later extended by the same authors to the synthesis of a variety of enantiopure *C*-glycosylmethyl pyridylalanines, starting from the corresponding chiral aldehydes.[139] A group of eight novel *C*-glycosyl-methyl pyridine amino acids were achieved in 55–68% yields with total preservation of the stereocentre integrity and functional group protection. Furthermore, the same authors developed, in 2009, the first organocatalysed three-component Hantzsch reaction between a chiral *C*-glycosyl aldehyde, a β-diketone, and an enamine, to provide the corresponding enantiopure substituted 1,4-dihydropyridine *C*-glycoconjugate in 50% yield and excellent diastereoselectivity of >98% de, as shown in Scheme 1.90.[140] This method, based on a double asymmetric induction, and arising from the use of a chiral auxiliary associated with a chiral organocatalyst, allowed the synthesis of biologically relevant *C*-nucleosides to be achieved; these were not accessible through uncatalysed procedures.

1.3.3 Multicomponent Reactions Based on the Ugi Reaction

The modern concept of multicomponent reaction is intimately related to the reactions developed with isocyanide reagents.[141] The Ugi four-component reaction is the reaction of a carbonyl compound (usually an aldehyde), an amine, an isocyanide, and a carboxylic acid (or an alcohol) to yield α-amino acid derivatives. Its general mechanism involves *in situ* formation of an imine from the aldehyde or ketone and the primary amine, followed by α-addition of the isocyanide component to this imine and carboxylic acid and subsequent rearrangement to furnish diversely substituted α-amino acid derivatives. This reaction, first described in 1959, has been more widely studied and used than any other multicomponent reaction.[126a] In their early work, Ugi *et al.* determined that the use of a chiral acid or isonitrile in the reaction did not provide any degree of stereoselectivity.[142] In contrast, chiral ferrocenylamine inputs resulted in the synthesis of non-racemic amino acid derivatives with low to modest levels of diastereoselectivity.[143] Later, Kunz *et al.* developed more versatile chiral auxiliaries for the Ugi reaction using carbohydrate derivatives.[144] High enantioselectivities (>90% ee) of (*R*)-amino acids were obtained in reactions employing a galactosylamine derivative.[144a] A drawback of this asymmetric Ugi reaction was that high levels of stereoselectivity were only observed for reactions using *tert*-butyl isonitrile. The asymmetric synthesis of (*S*)-amino acids *via* the Ugi reaction was achieved using an arabinosylamine derivative, but the stereoselectivity was not as high as that observed in the synthesis of the corresponding (*R*)-enantiomer.[144b] A single variant of the chemistry developed by Kunz has been reported by Goebel *et al.*[145] In 1991, these authors showed that 2,3,4,6-tetra-*O*-alkyl-β-D-glucopyranosylamines

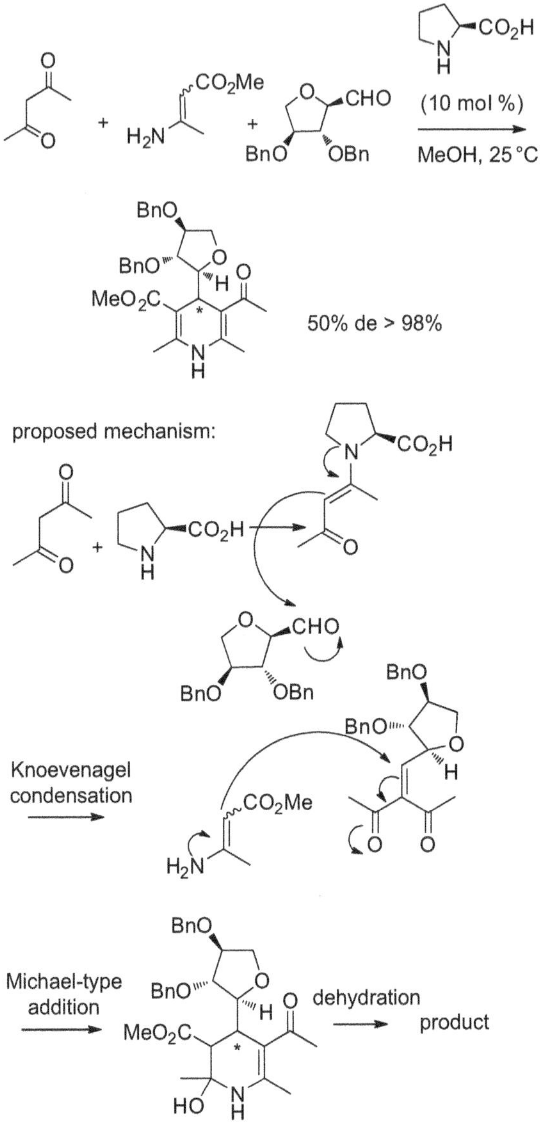

Scheme 1.90 Synthesis of 1,4-dihydropyridine *C*-glycoconjugate through three-component Hantzsch reaction.

used as chiral amine components were as favourable as the *O*-acyl-aldopyranosylamides used by Kunz *et al.*[144a] On the other hand, the Ugi three-component is a variant of the general reaction in which either two of the usual groups are included in the same reagents, or the condensation of the carbonyl compound with the amine takes place before the addition of the isocyanide and acid derivatives. Even though the imine condensation could be performed in a one-pot process and its isolation was not necessary, it must be included in this category. A number of diastereoselective versions of both three-component and four-component Ugi reactions have been successfully developed. In 2006, Dyker *et al.* reported an interesting and novel route to highly functionalised chiral dihydroisoquinolines and isoindoles based on a two-step sequence including an asymmetric four-component Ugi reaction using an amino acid, such as L-valine, as chiral auxiliary.[146] Indeed, the Ugi reaction between L-valine, *t*-butyl isocyanide, methanol, and benzaldehydes as the source of the alkyne moiety proceeded somewhat sluggishly at room temperature within 6 to 9 days, building up the corresponding highly functionalised amines, finally with satisfactory yields (68–71%) and diastereoselectivities ≤82% de, as shown in Scheme 1.91. These chiral products were further submitted to a gold-catalysed

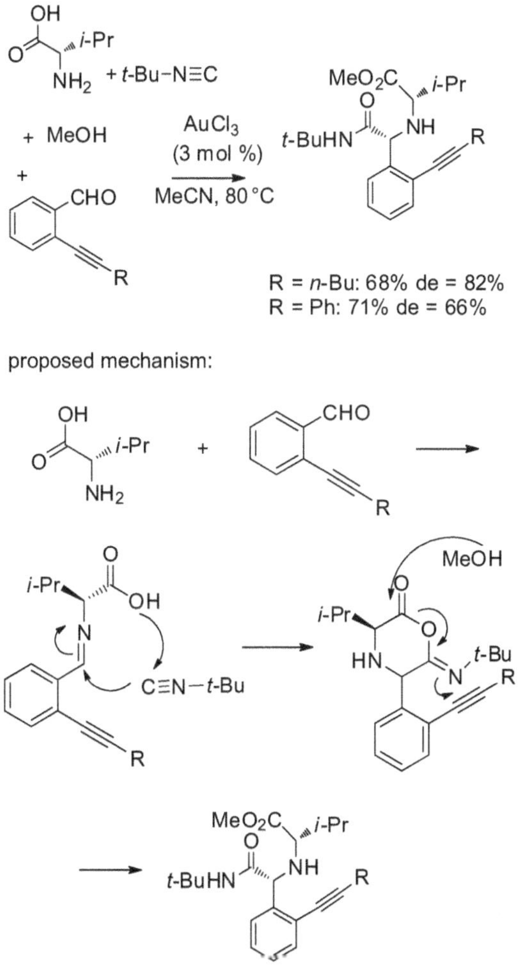

Scheme 1.91 Synthesis of highly functionalised amines through four-component Ugi reaction.

intramolecular hydroamination to provide a series of important chiral heterocycles, such as isoindoles and dihydroisoquinolines.

In the same year, Guanti *et al.* developed the synthesis of complex chiral fused polycyclic scaffolds containing up to eight stereogenic centres on the basis of a sequence of reactions beginning with a four-component asymmetric Ugi reaction of optically pure (−)-*N*-allyl-3-amino-7-oxa-[2.2.1]-bicyclohept-5-ene-2-carboxylic acid.[147] As shown in Scheme 1.92, the reaction of this chiral substrate with different combinations of aldehydes and isocyanides in methanol at room temperature for two to four days resulted in the formation of the corresponding Ugi products in moderate to good yields (46–71%) and as single diastereomers. These products were subsequently submitted to a ring-opening–ring-closing metathesis to give a range of enantiopure highly func-tionalised tricyclic products bearing five stereocentres. In addition, the scope of the Ugi methodology was extended to another optically pure bicyclic amino acid bearing a propargyl group instead of an allylic group as in previous chiral auxiliary. In spite of the fact that a propargyl group is less bulky than an allylic group, the corresponding Ugi products resulting from the reaction of the chiral propargylic amino acid with various aldehydes and isocyanides were un-expectedly achieved as diastereomeric mixtures. In one case of a product derived from *tert*-butyl isocyanide and *p*-chlorobenzaldehyde, however, the stereoselection was satisfactory (80% de) and the major diastereomer was separated by chromatography. This enantiopure product was then submitted to

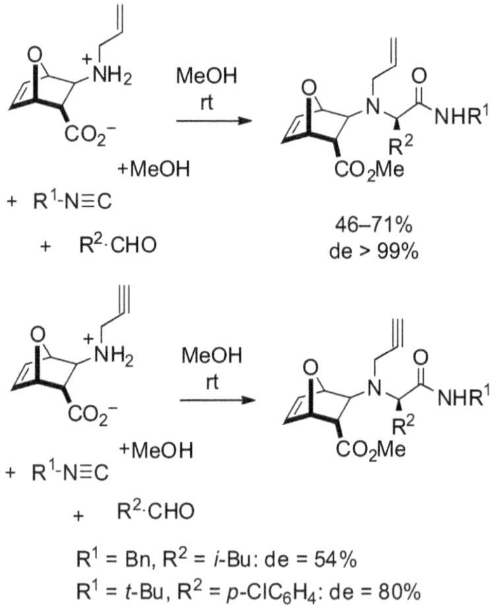

Scheme 1.92 Four-component Ugi reaction of (−)-*N*-allyl- and *N*-propargyl-3-amino-[2.2.1]-bicyclohept-5-ene-2-carboxylic acids with isocyanides, aldehydes, and methanol.

a sequence of ring-opening–ring-closing metathesis and Diels–Alder cyclo-addition to afford complex highly functionalised tri- and tetracyclic enantiopure products bearing up to eight stereocentres. More recently, the same authors combined this four-component Ugi reaction with a subsequent palladium-catalysed ring-opening to transform oxabicycloheptene-based β-amino acids into two families of regioisomeric polyfunctionalised cyclohexenols.[148] The whole process was found to be completely stereoselective, and enantiomerically pure products were obtained in high overall yields. Interestingly, this diversity-oriented approach has been recently exploited to discover novel inhibitors of protein–protein interactions involved in apoptotic processes.[149]

Water as a solvent is not only inexpensive and environmentally begnin, but may allow good reactivity. In this context, Fülöp *et al.* have developed a modified three-component Ugi reaction of various β-amino acids with isocyanides and aldehydes in water to construct β-lactam libraries.[150] In addition to requiring shorter reaction times, these experimental procedures presented the advantage that the final products insoluble in water could be isolated by simple filtration. As shown in Scheme 1.93, this benefit could be exploited when less water-soluble entities, such as norbornene- and norbornane-based β-amino acids, were employed as substrates to give the corresponding Ugi products in good yields and moderate to complete relative diastereoselectivity. The mechanism of the reaction is depicted in Scheme 1.93. In the first step, the β-amino acid was reacted with the appropriate aldehyde, resulting in the formation of a protonated Schiff base, followed by addition of the isocyanide to afford the final β-lactam *via* intramolecular cyclisation and rearrangement. Furthermore, other cyclic β-amino acids, such as the *cis*-aminocyclopentane carboxylic acid and *cis*-aminocyclohexane carboxylic acid depicted in Scheme 1.93, provided under the same conditions the corre-sponding Ugi products in moderate to good yields and complete relative diastereoselectivity when cyclohexylisocyanide was employed. The authors have studied the same reactions performed in methanol instead of water, and found that the yields were slightly better in water than in methanol, while the diastereoselectivities were comparable.

In 2007, Nenajdenko *et al.* reported the first example of a diastereoselective thio-Ugi reaction using (*S*)-α-methylbenzylamine as chiral auxiliary, affording thioamides which constitute amide bond surrogates in a number of biologically active peptides and incorporated into various natural molecules.[151] As shown in Scheme 1.94, (*S*)-α-methylbenzylamine reacted with thiobenzoic acid, iso-butyraldehyde, and an isocyanide to give the corresponding thioamide as a 2:1 mixture of two diastereomers from which the major (*R*)-product was separated by chromatography in 35% yield. Unexpectedly, the authors observed inversion of the stereoselectivity of the reaction upon replacement of thiobenzoic acid by benzoic acid, which furnished the corresponding (*S*)-configured Ugi product as the major diastereomer in 80% yield and 55% de. In order to explain the preferential formation of the (*R*)-diastereomer in the thio-Ugi reaction with (*S*)-α-methylbenzylamine, and that of the (*S*)-diastereomer in the Ugi reaction with the same substrate, the authors have proposed that the

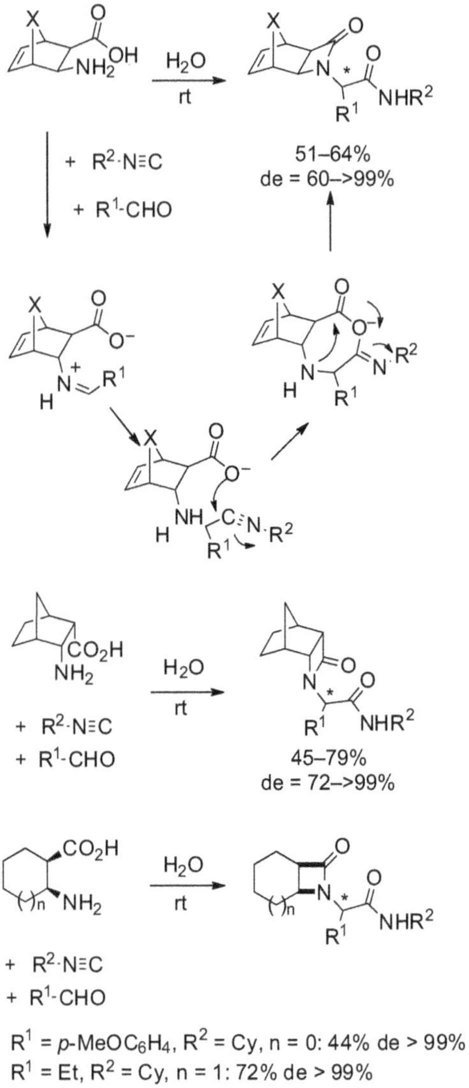

Scheme 1.93 Three-component Ugi reactions of chiral β-amino acids, isocyanides, and aldehydes in water.

thio-Ugi and Ugi reactions followed different mechanisms, A and B respectively, which are depicted in Scheme 1.94. Carboxylic acids formed salts **64** with the *in situ* generated imine (Mechanism B) without further nucleophilic addition to the iminium centre. Thereafter, an isocyanide attack occurred from the less sterically hindered side, and the corresponding Ugi product having the *S* configuration was formed as major diastereomer. In the case of the thio-Ugi reaction (Mechanism A), the much more nucleophilic thiocarboxylate attacked imine **64** from the less sterically hindered side, and the intermediate (*R*)-**65** was

thio-Ugi reaction:

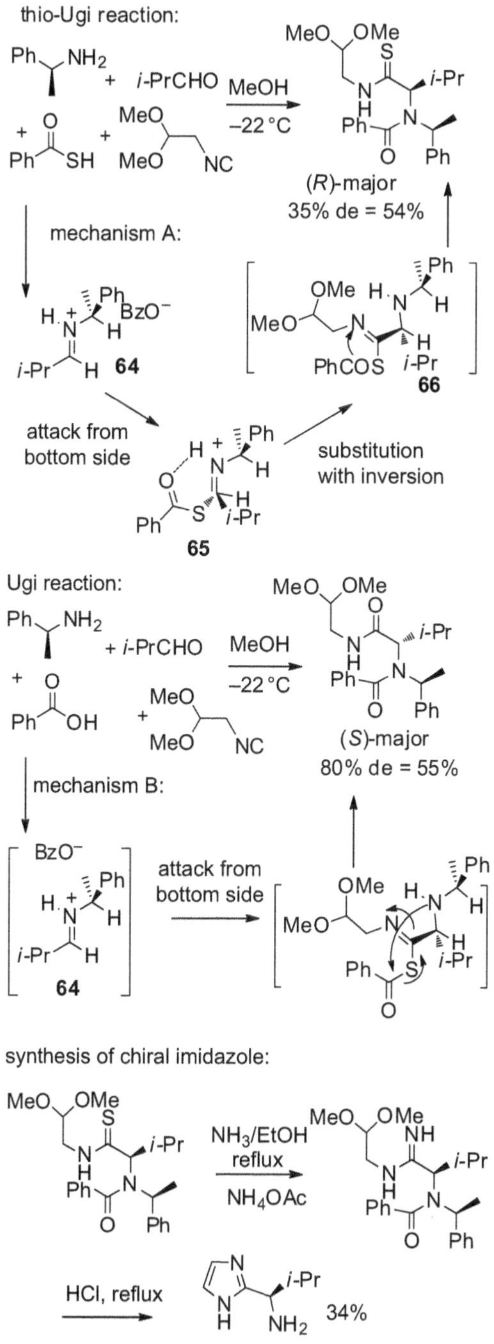

mechanism A:

Ugi reaction:

mechanism B:

synthesis of chiral imidazole:

Scheme 1.94 Four-component (thio)-Ugi reaction of (*S*)-α-methylbenzylamine, isobutyraldehyde, isocyanide, and (thio)benzoic acid.

formed. Furthermore, an S_N2-type nucleophilic substitution with inversion of the configuration of the formed stereocentre resulted in intermediate **66** as the (*R*)-isomer. Therefore, the (*R*)-configured thioamide was formed as the major diastereomer. Moreover, these chiral thioamides were treated with ammonia to afford the corresponding substituted amidines, which could be cyclised to imidazole derivatives in aqueous HCl (Scheme 1.94). Among these chiral imidazoles, one constituted a key synthon in the synthesis of SB203386, which is a naturally occurring orally bioactive HIV-1 protease inhibitor.

In the same year, the first diastereoselective three-component Ugi reaction performed without chiral amines was reported by Nenajdenko *et al.*[152] Indeed, the use of chiral 2-(2-formyl-1-*H*-pyrrol-1-yl)acetic acids derived from natural L-amino acids as chiral auxiliaries in the Ugi reaction with *tert*-butylisocyanide and primary amines provided the corresponding chiral pyrroloketopiperazines in good yields and moderate diastereoselectivities of up to 60% de, as shown in Scheme 1.95. The low diastereoselectivity observed could be explained by the almost planar structure of the heterocyclic fragment in the target molecules. In order to improve the diastereoselectivity, the authors have carried out the reaction in the presence of chiral amines, but this did not improve the diastereoselectivity of the process significantly owing to the structural peculiarities of pyrroloketopiperazines.

Only modest diastereoselectivities of $\leq36\%$ de were achieved by Guanti *et al.* for a three-component Ugi reaction occurring between a chiral pyrroline as preformed imine, various isocyanides and carboxylic acids.[153] On the other hand, the corresponding enantiomerically pure *N*-acyl-2,5-disubstituted pyrrolidines were produced in moderate to high yields (47–85%). These products were employed as key intermediates in the synthesis of two series of enantiopure bicyclic derivatives, namely hexahydro pyrrolo-oxazocinediones and -diazepinediones. A slightly more consistent diastereoselective three-component Ugi reaction of carboxylic acids, isocyanides and a D-*lyxo*-derived pyrroline as the chiral imine component was reported by Overkleeft *et al.*[154] During this sequential one-pot procedure, this imine component was generated

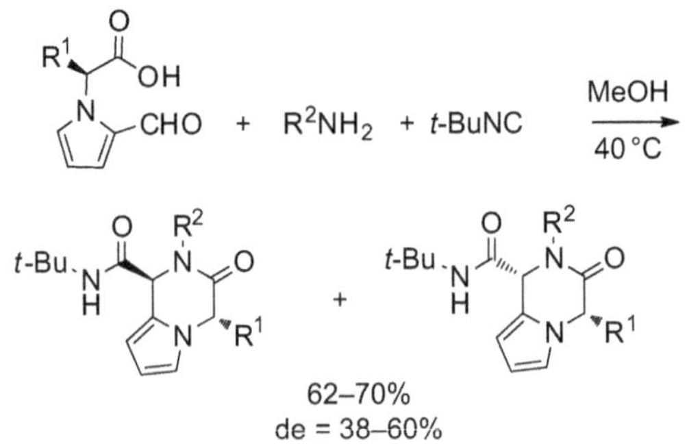

Scheme 1.95   Three-component Ugi reaction of chiral 2-(2-formyl-1-*H*-pyrrol-1-yl)acetic acids, *tert*-butylisocyanide, and primary amines.

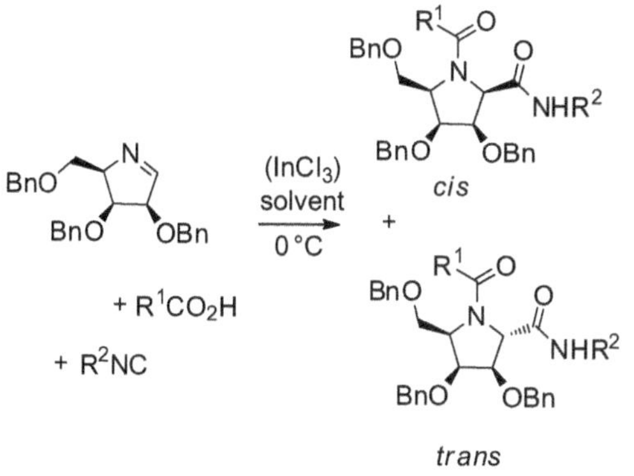

in MeCN, with $InCl_3$ (5 equiv):
R^1 = allyl, R^2 = $(Me)_2C\text{-}CH_2\text{-}t\text{-}Bu$:
63% *trans:cis* = 84:16
R^1 = $CH_2NH(Fmoc)$, R^2 = $(Me)_2C\text{-}CH_2\text{-}t\text{-}Bu$:
20% *trans:cis* > 90:10
R^1 = H, R^2 = $(Me)_2C\text{-}CH_2\text{-}t\text{-}Bu$:
68% *trans:cis* = 71:29
R^1 = H, R^2 = Cy:
56% *trans:cis* = 86:14
R^1 = $p\text{-}NO_2C_6H_4$, R^2 = $(Me)_2C\text{-}CH_2\text{-}t\text{-}Bu$:
38% *trans:cis* = 18:82

in MeOH, without $InCl_3$:
R^1 = allyl, R^2 = $(Me)_2C\text{-}CH_2\text{-}t\text{-}Bu$:
72% *cis:trans* > 90:10

Scheme 1.96 Three-component Ugi reaction of chiral D-*lyxo*-derived pyrroline, isocyanides, and carboxylic acids.

in situ through a Staudinger reduction of an L-ribose-derived 4-azido aldehyde with trimethylphosphine, followed by an intramolecular aza-Wittig reaction with the aldehyde. The subsequent three-component Ugi reaction was directed towards the formation of 2,3-*trans*-configured pyrrolidines through the addition of three equivalents of a Lewis acid such as $InCl_3$, performing the reaction at 0 °C and working in dilute solution (Scheme 1.96). On the other hand, the formation of the corresponding 2,3-*cis*-configured pyrrolidines was promoted in apolar solvents, such as MeCN, as well as by omitting the Lewis acid. The control of the diastereoselectivity of the process was found to be highly dependent on the nature of the chiral imine, the carboxylic acid, and the isocyanide involved, providing the corresponding carbohydrate-derived pyrrolidines in poor to reasonable yields (20–72%) and diastereoselectivities of 0 to >80% de for the *trans*-adduct.

7-Aza-norbornene *trans*-dicarboxylic acids were used as chiral auxiliaries by Basso *et al.* in a diastereoselective intramolecular four-component Ugi reaction,

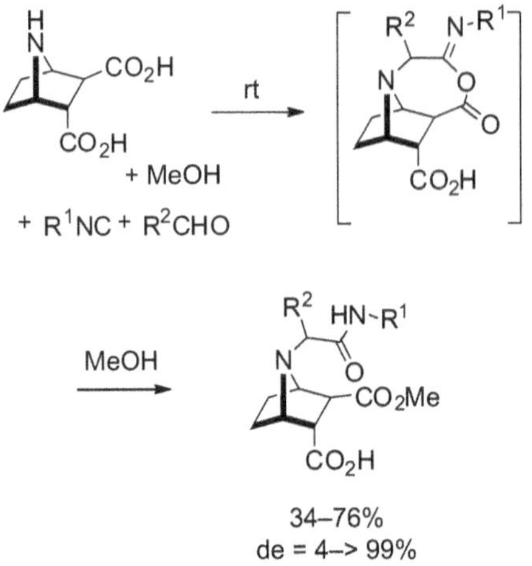

Scheme 1.97   Four-component Ugi reaction of chiral 7-aza-norbornene *trans*-dicarboxylic acids, isocyanides, aldehydes, and methanol.

in 2009.[155] The resulting unnatural amino acid derivatives constitute valuable building blocks in drug discovery, because they can be applied for the synthesis of various peptidomimetics.[156] Interestingly, the alkylated bridgehead nitrogen atom showed abnormally high barriers for nitrogen inversion,[157] due to the so-called bicyclic effect.[158] As expected, the seven-membered ring intermediate (Scheme 1.97) in this Ugi process was attacked by methanol to afford the corresponding azabicycloheptanes in moderate to good yields (34–76%) and low to excellent diastereoselectivities (4 to >99% de), as shown in Scheme 1.97. The *endo*-configured carboxylic acid group was unable to participate in the Ugi reaction and provided an interesting handle for post-condensation transformations. Various isocyanides were tolerated, showing no particular difference in reactivity, while the influence of the nature of the aldehydes was found to be more important. Acetone was demonstrated to be incompatible with the process. As there is a lack of methods for the introduction of complex alkyl substituents onto bridgehead nitrogen atoms, this methodology is very useful for the preparation of polyfunctionalised azabicyclic peptidomimetics.

In 2010, Orru *et al.* developed a highly stereoselective synthesis of chiral substituted prolyl peptides by employing as the key step an asymmetric three-component Ugi reaction.[159] As shown in Scheme 1.98, the three-component Ugi reaction of two types of chiral cyclic imine provided under mild conditions the corresponding almost enantiopure prolyl peptides in high yields and high to complete diastereoselectivities. The starting chiral 3,4-*cis*-substituted 1-pyrrolines arose from the biocatalytic desymmetrisation of the corresponding 3,4-*cis*-substituted *meso*-1-pyrrolines with engineered monoamine oxidase-N (MAO-N) from *Aspergillus niger*. The utility of this powerful three-component Ugi reaction was demonstrated by its application in a rapid asymmetric synthesis of alkaloid-like polycyclic compounds through a tandem MAO-N

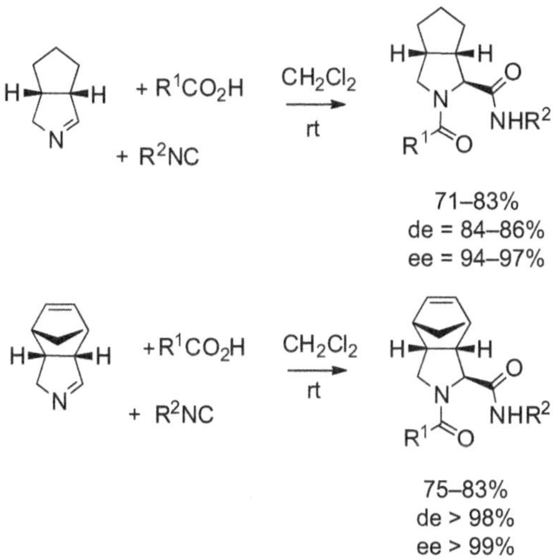

71–83%
de = 84–86%
ee = 94–97%

75–83%
de > 98%
ee > 99%

Scheme 1.98 Synthesis of substituted prolyl peptides through three-component Ugi reactions.

oxidation–three-component Ugi reaction–Pictet–Spengler-type cyclisation sequence.[160] The process occurred between a biocatalytically desymmetrised pyrroline, α-ketocarboxylic acids, and homoveratryl isocyanide or 2-(indol-3-yl)ethyl isocyanide, providing the corresponding Ugi products in good yields (75–83%) and both remarkable diastereo- and enantioselectivities of >98% de and >99% ee, respectively. These products were subsequently submitted to a Pictet–Spengler-type cyclisation to give polycyclic 2,5-diketopiperazines in generally good yields (40–92%) and poor to excellent diastereoselectivities (14 to >96% de). It must be noted that these products are structurally similar to naturally occurring alkaloids displaying a broad range of biological activities.[161] Moreover the biocatalytic desymmetrisation–three-component Ugi reaction sequence of the same bicyclic pyrroline was also investigated by Turner *et al.*[162] The scope of the Ugi reaction was explored with several carboxylic acids and isocyanides, affording the corresponding prolyl peptides in high diastereoselectivities (84–86% de) and excellent enantioselectivities (94–97% ee). It was demonstrated that the stereochemical outcome of the process, providing stereoselectively the 2,3-*trans* isomers in all cases of substrates studied, was solely determined by the configuration of the chiral imine.

In addition, these authors have employed a related sequence constituted by a biocatalytic desymmetrisation followed by a three-component Ugi reaction as the key step in a highly efficient synthesis of telaprevir, which is a potent peptidic hepatitis C virus NS3 protease inhibitor.[163] As shown in Scheme 1.99, the reaction occurred between a readily available enantiopure carboxylic acid, an isocyanide, and a chiral cyclic imine previously desymmetrised by treatment with MAO-N, to afford the corresponding key intermediate in 76% yield. The latter compound was further converted in two steps into the expected drug

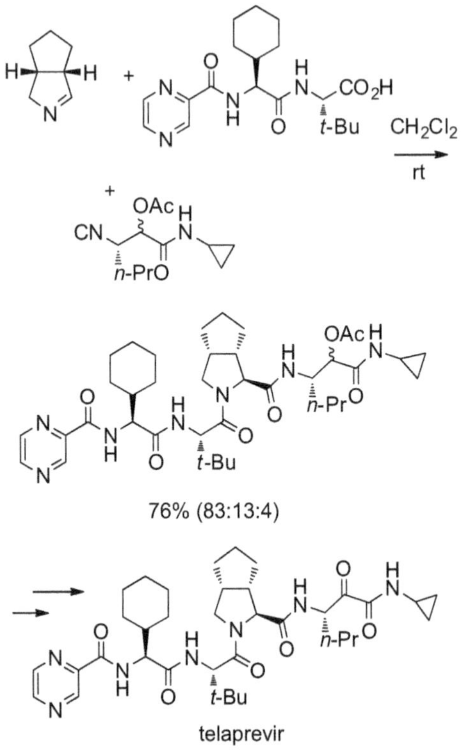

Scheme 1.99   Synthesis of telaprevir through three-component Ugi reaction.

candidate telaprevir which constituted the major isomer of an 83 : 13 : 4 mixture of diastereomers in 80% overall yield. This novel synthetic route to telaprevir represented impressive improvement over previous strategies, because minimisation of the use of protective groups as well as a greatly shortened procedure was achieved.

On the other hand, Dömling *et al.* have reported a four-component version of the thiazole-Ugi reaction, employing, for the first time in this reaction, (*R*)-4-methoxyphenylethylamine as the chiral auxiliary to react with acetaldehyde, thioacetic acid, and 2,4-dimethoxybenzylamine. The reaction led, however, to a 1 : 1 mixture of two diastereomers in 65% yield, which could be further separated through chromatography. After cleavage of the chiral auxiliary, the corresponding (*R*)-amide could be converted into naturally occurring (−)-bacillamide C.[164] In addition, Carreira *et al.* have recently reported the total synthesis of the core fragment of natural and biologically active massadine in which the key step was an asymmetric four-component Ugi reaction using a chiral norbornenone as the chiral auxiliary.[165] As shown in Scheme 1.100, the reaction of this chiral norbornene with 2-nitrophenylisonitrile, trifluoroacetic acid, and 2,4-dimethoxybenzylamine led to the corresponding Ugi-product as a single stereoisomer in 79% yield. This key

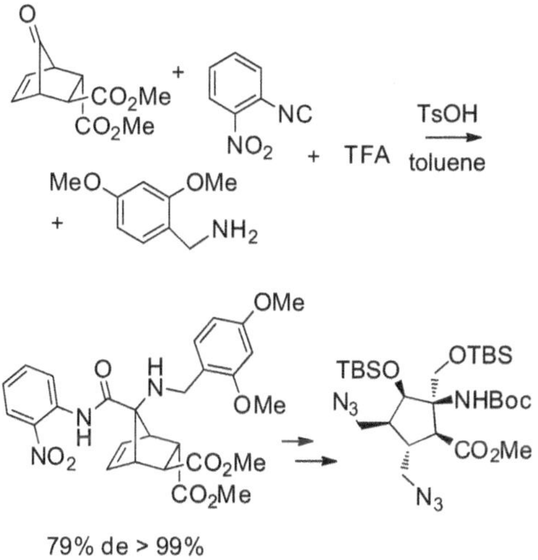

79% de > 99%

Scheme 1.100 Synthesis of core fragment of massidine through four-component Ugi reaction.

product was further converted into the chiral core fragment of massadine in 14 steps.

1.3.4 Multicomponent Reactions Based on the Strecker Reaction

The asymmetric Strecker reaction, discovered in 1850,[166] is one of the most widely employed methods for synthesizing α-aminonitriles that can be further hydrolysed to give α-amino acids.[167] The Strecker reaction is a three-component coupling among carbonyl derivatives, amines, and a cyanide source, such as hydrogen cyanide, to give the corresponding α-aminonitriles. The mechanism of the Strecker reaction involves the initial formation of an imine by condensation of the carbonyl and amine components, after which addition of the cyanide component to the imine intermediate follows. While many enantioselective catalytic protocols employing either metal-based catalysts or organocatalysts have been shown to afford enantiopure α-aryl-aminonitriles, they often use expensive chiral catalysts or require multistep syntheses of chiral ligands. Moreover, many of these catalytic methodologies require the use of trimethylsilyl cyanide as a nucleophile, which is expensive in addition to being a toxic reagent. Finally, the substrate specificity profile of a number of catalytic processes is often limited. Consequently, an alternative approach to asymmetric catalysis is to use enantiopure amines as chiral auxiliaries to perform diastereoselective Strecker reactions. As a recent example, James *et al.* have developed the synthesis of enantiopure α-aryl glycines through asymmetric three-component Strecker reaction of various aryl aldehydes, sodium cyanide in solution, and (*S*)-1-(4-methoxyphenyl)ethylamine as the chiral auxiliary.[168] As shown in Scheme 1.101, the process afforded the

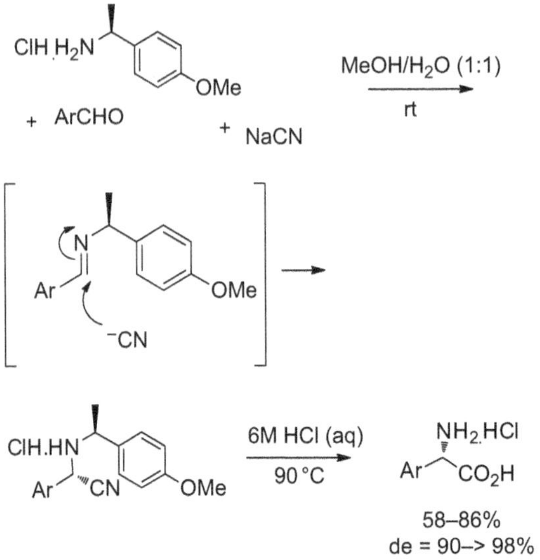

Scheme 1.101   Synthesis of α-arylglycines through three-component Strecker reaction.

corresponding crystalline α-aminonitriles in good yields and as single stereoisomers after fractional recrystallisation of the crude from Et_2O in almost all cases of substrates studied. A further heating of these products in 6 M aqueous HCl at reflux resulted in the cleavage of their chiral auxiliary fragments and hydrolysis of their nitrile groups to afford the corresponding enantiopure (*S*)-α-arylglycines.

1.3.5   Multicomponent Reactions Based on the Mannich Reaction

The classic direct Mannich reaction, discovered in 1912,[169] is an aminoalkylation of carbonylic compounds involving ammonia (or a primary or secondary amine derivative), a non-enolizable aldehyde (usually formaldehyde) or a ketone, and an enolizable carbonyl compound, leading to β-aminocarbonyl derivatives.[170] In addition to enantioselective catalytic Mannich reactions,[171] all of the possibilities of using chiral starting materials for this asymmetric multicomponent reaction have been reported. In their aim of discovering novel glycopeptide-based drugs for the treatment of bacterial and viral infections, cancer and inflammatory processes,[172] Dondoni *et al.* have employed a range of chiral *C*-glycosyl aldehydes as chiral auxiliaries in the Mannich reaction with *p*-methoxybenzylamine, and commercially available ketene silyl acetal 1-methoxy-2-methyl-1-trimethylsilyloxypropene.[138] This process was promoted by $InCl_3$, and provided the corresponding chiral *C*-glycosyl α,α-dimethyl

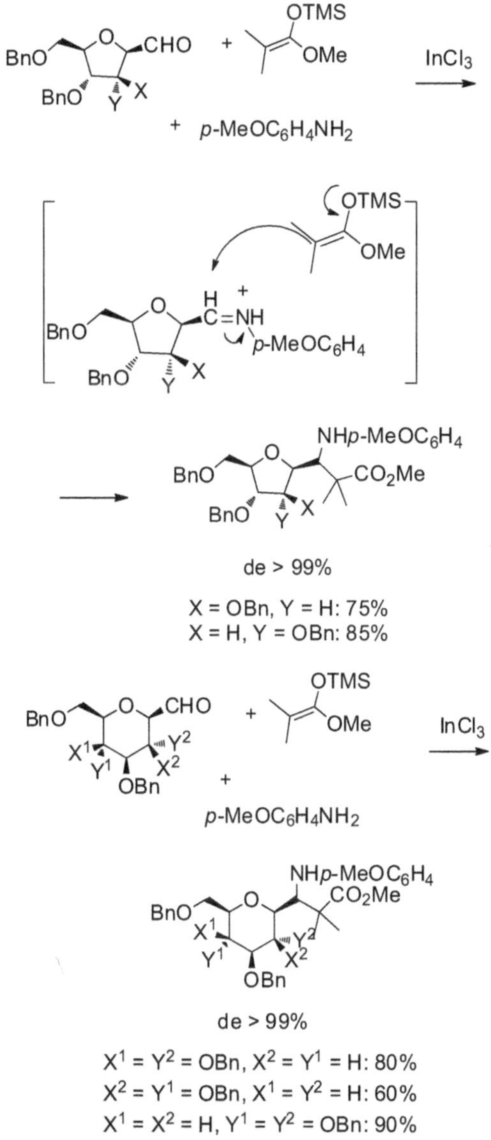

Scheme 1.102 Synthesis of *C*-glycosyl α,α-dimethyl β-amino esters through three-component Mannich reactions.

β-amino esters in high yields and complete diastereoselectivity in all cases of substrates studied, as shown in Scheme 1.102.

In 2007, a solvent-free diastereoselective Mannich-type reaction was described by Petrini *et al.*[173] This process occurred among β-naphthol, amines, and chiral functionalised aldehydes to afford the corresponding chiral aminonaphthols in moderate yields (24–62%) and diastereoselectivities

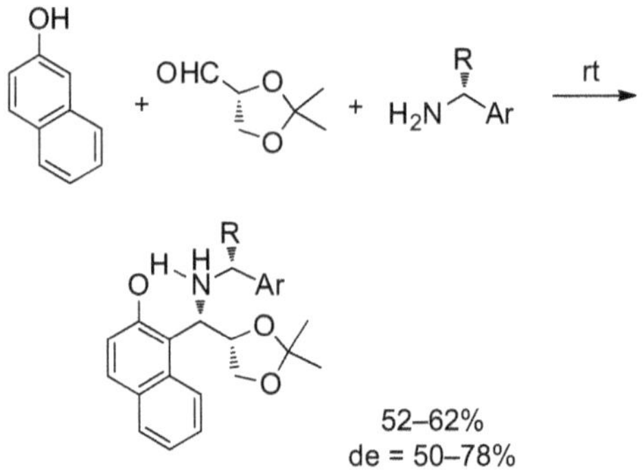

52–62%
de = 50–78%

Scheme 1.103   Synthesis of aminonaphthols through Mannich-type reaction.

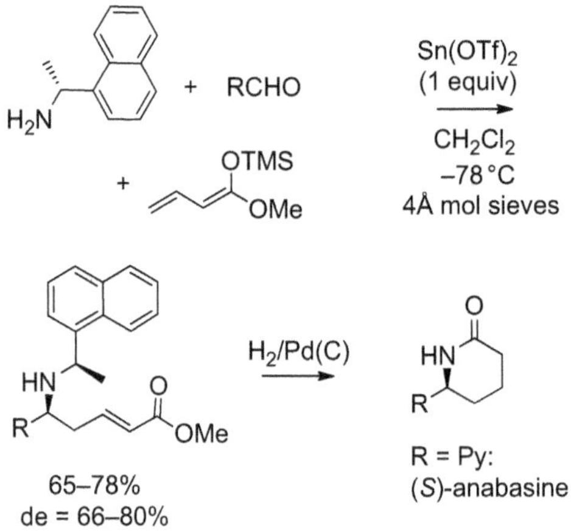

65–78%
de = 66–80%

R = Py:
(S)-anabasine

Scheme 1.104   Synthesis of (S)-anabasine through three-component vinylogous Mannich reaction.

(24–78% de). As shown in Scheme 1.103, the best diastereoselectivities were reached when using a chiral amine in addition to the chiral aldehyde. A subsequent deprotection of the formed products by acid hydrolysis (aqueous HCl/THF) allowed the corresponding enantiopure aminodiols to be achieved in high yields.

In 2011, a novel stereoselective approach to chiral 6-alkylated piperidinones and 2-piperidines was reported by Yang *et al.* on the basis of a three-component vinylogous Mannich reaction involving a chiral amine as chiral auxiliary.[174] As shown in Scheme 1.104, the reaction of (R)-(+)-1-naphthalen-1-yl-ethylamine with various aldehydes and a silylketene acetal, such as (1-methoxy-buta-1,3-dienyloxy)-trimethylsilane, provided, in the presence of 1 equivalent of Sn(OTf)$_2$, the corresponding chiral amines in good yields and moderate to good diastereoselectivities ranging from 66 to 80% de.

These products could be readily converted through hydrogenation into the corresponding cyclised 6-alkylated piperidinones, including the biologically active and natural alkaloid (*S*)-anabasine.

1.3.6 Multicomponent Reactions Initiated by an Allylation Reaction

Stereoselective allylations of carbonyl compounds, such as aldehydes and ketones, are useful but challenging reactions in organic chemistry. The resulting chiral secondary and tertiary homoallylic alcohols or ethers are valuable building blocks in the synthesis of a number of biologically active natural compounds. Although researchers have developed several methods for the stereoselective allylation of aldehydes, allylation of ketones still poses a severe problem. In this context, Tietze *et al.* have developed a highly diastereoselective domino three-component allylation reaction of ketones and allyltrimethyl silane, using the trimethylsilyl ether of norpseudoephedrine as the chiral auxiliary, which provided the corresponding ethers in good yields and diastereoselectivities of up to 96% de, as shown in Scheme 1.105.[175] The reaction was performed in the presence of a catalytic amount of trifluoromethanesulfonic acid and led to the corresponding tertiary ethers. In order to determine the origin of the stereoselectivity of the reaction, these authors have undertaken density functional theory (DFT) calculations.[176] Computational investigations on the allylation of butanone have suggested an S_N1-type mechanism *via* the attack by allyltrimethylsilane of an intermediately formed oxocarbenium ion

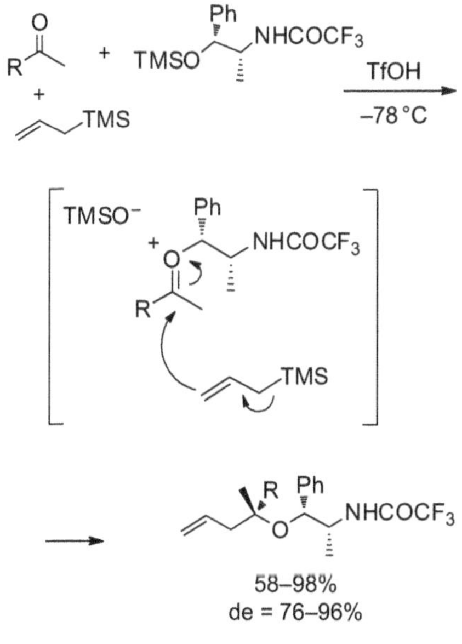

58–98%
de = 76–96%

Scheme 1.105 Three-component allylation of ketones.

from butanone and silyl ether. The identification of preferred transition states led to a straightforward rationalisation of the observed selectivity. A screening process based on B3LYP/AM1 energies allowed narrowing down of the number of potentially relevant transition states. The predicted selectivities were in good agreement with experimentally determined ones.

The procedure could also be used for the allylation of aliphatic aldehydes with diastereoselectivities of up to >98% de, as shown in Scheme 1.106. It must be noted that ketones led to the corresponding 4,1'-*syn* products, while

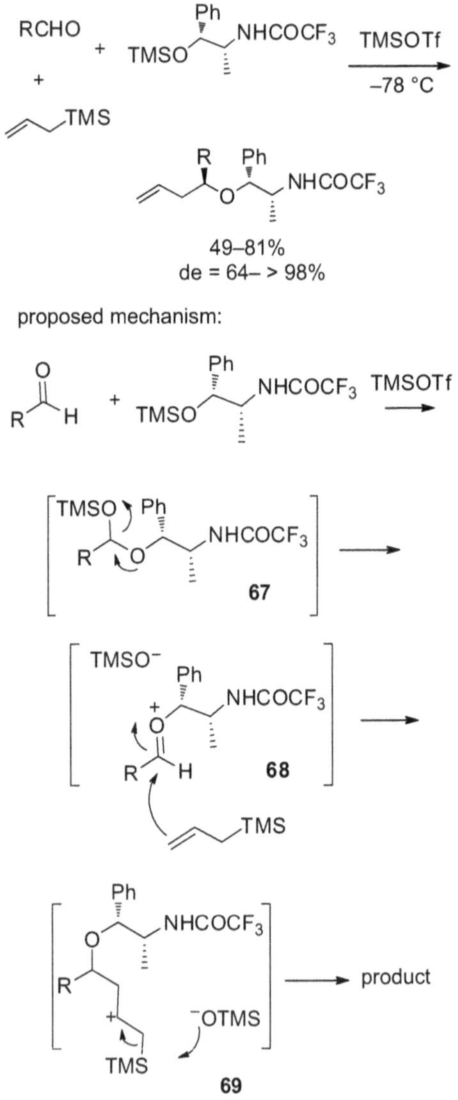

Scheme 1.106   Three-component allylation of aldehydes.

aldehydes yielded the corresponding ethers bearing two stereogenic centres with *anti*-diastereoselectivity of >98% de. The homoallylic ethers formed in the domino multicomponent processes can be used in further transformations. Indeed, the auxiliary can serve as a protecting group or can be cleaved reductively to give the corresponding homoallylic alcohols. The reaction was catalysed by TMSOTf which was sufficient to initiate the reaction, whereas reactions of ketones required the use of TfOH as catalyst. The authors have proposed that the domino reaction evolved *via* the formation of a mixed acetal species **67**, which was transformed into an oxocarbenium ion **68** bearing the organic part of the silyl ether residue at its oxygen atom (Scheme 1.106). Intermediate **68** was then intercepted by the weak nucleophile trimethylallyl silane to give the carbenium ion **69**, which in the following step was attacked by the trimethylsilyloxide anion to yield the final homoallylic product.

The utility of these methodologies was illustrated by their application as key steps of total syntheses of several natural products, such as (−)-hydroxy-myoporone, 5,6-dihydrocineromycin B, polyoxygenated cembrene,[175a] and (*R*)-(+)-orizaterpenyl benzoate.[177] As shown in Scheme 1.107, the key step of the synthesis of (*R*)-(+)-orizaterpenyl benzoate consisted in an asymmetric three-component allylation reaction of 6-methyl-2-heptanone, employing a chiral trimethylsilyl ether which was easier to prepare than the corresponding norpseudoephedrine derivative. The corresponding tertiary homoallylic ether was obtained in 85% yield as a 90 : 10 mixture of two diastereomers. The major isomer was further converted into (*R*)-(+)-orizaterpenyl benzoate.

In another context, Marko *et al.* have reported the first total synthesis of the natural product jerangolid D by using an asymmetric three-component Sakurai reaction.[178] This domino process occurred between allyl trimethyl silane, a chiral aldehyde, and a chiral ether in the presence of a catalytic amount of

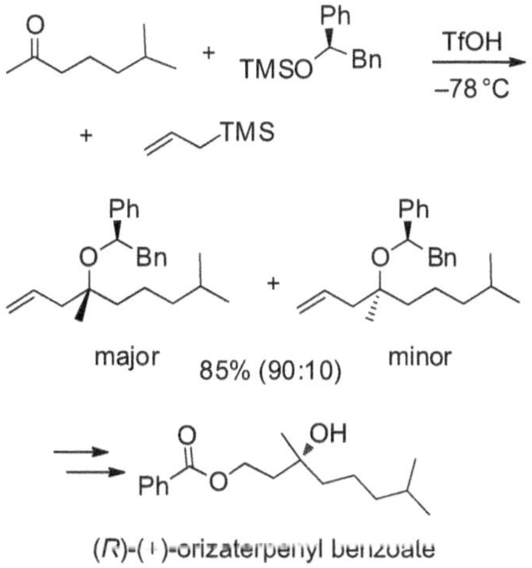

Scheme 1.107   Synthesis of (*R*)-(+)-orizaterpenyl benzoate through three-component allylation reaction.

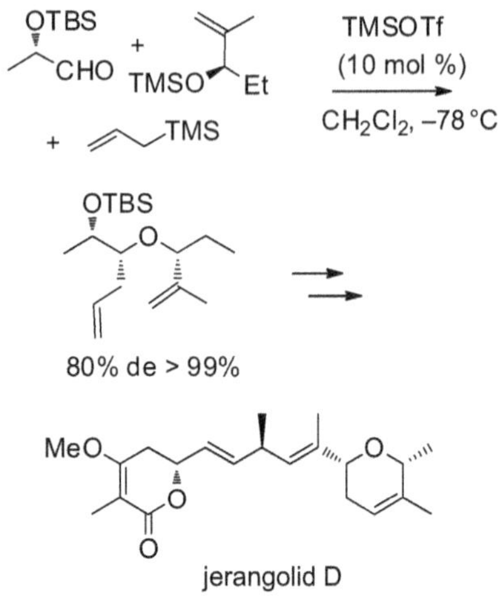

80% de > 99%

jerangolid D

Scheme 1.108   Synthesis of jerangolid D through three-component Sakurai reaction.

TMSOTf, which afforded the corresponding *syn–syn* ether as a single stereoisomer in 80% yield, as shown in Scheme 1.108. This product was further converted into jerangolid D.

1.3.7   Multicomponent Reactions Based on the Passerini Reaction

The Passerini three-component reaction, discovered in 1921, involves the condensation of carbonyl compounds, carboxylic acids, and isocyanides to afford the corresponding α-acyloxy carboxamides.[179] Among several advantages are the mildness of the reaction conditions, the broad scope, and the high variability of the inputs.[180] In spite of a number of investigations of the mechanism of the Passerini reaction, the exact intermediates involved in this reaction are still discussed.[141,181] Several chiral auxiliaries have been applied to control the stereochemical outcome of the diastereoselective Passerini reaction. This reaction was performed with the chiral version of each of the components before 2006.[182] It must be noted that, in the last six years, only a few novel examples of diastereoselective Passerini processes have been described. Among them, an interesting Passerini reaction, using (*p*-toluenesulfonyl)methyl-isocyanide (TosMIC) for the first time in such a reaction as the isonitrile component, was reported by Krishna *et al.*, in 2006.[183] As shown in Scheme 1.109, the process employed carbohydrate-derived aldehydes as the chiral auxiliaries to react with TosMIC and carboxylic acids, providing the corresponding products as mandelamides in moderate to good yields and *anti*-diastereoselectivities (30–90%). The highest diastereoselectivities of up to 90% de were reached by employing Garner's aldehyde as the chiral auxiliary. The use of a chiral carboxylic acid in this reaction did not allow the stereo-selectivity of this reaction to be enhanced.

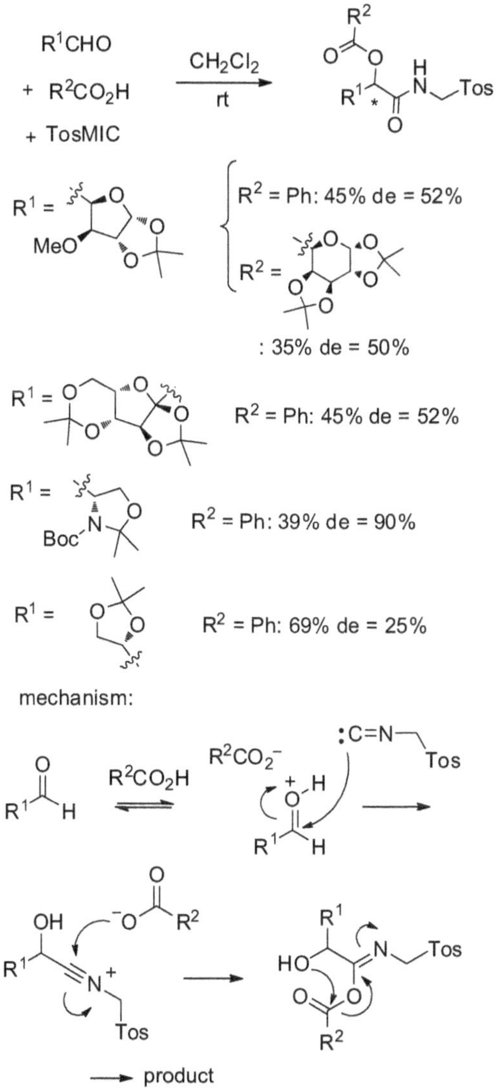

Scheme 1.109 Three-component Passerini reaction of chiral carbohydrate-derived aldehydes, carboxylic acids, and TosMIC.

Later, the same authors investigated the use of chiral 2,3-epoxy aldehydes as chiral auxiliaries in the Passerini reaction with TosMIC and carboxylic acids as the two other components.[184] As shown in Scheme 1.110, the corresponding densely substituted products were achieved in moderate to good yields and lower diastereoselectivities (40–86% de) than those obtained by using carbohydrate-derived aldehydes. The best results were obtained through double asymmetric induction using a chiral epoxide in combination with chiral

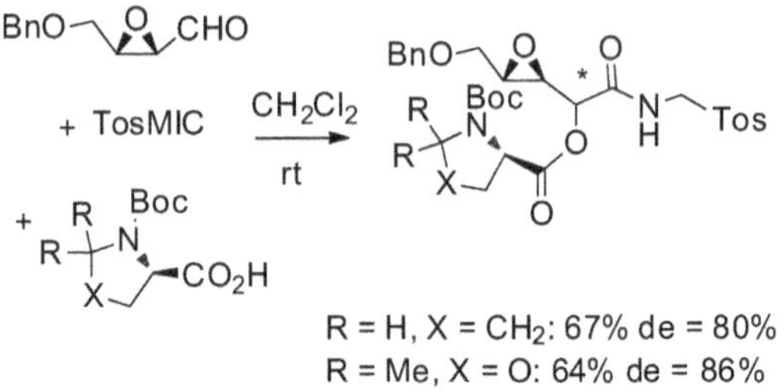

Scheme 1.110   Three-component Passerini reaction of chiral 2,3-epoxy aldehyde, chiral carboxylic acids, and TosMIC.

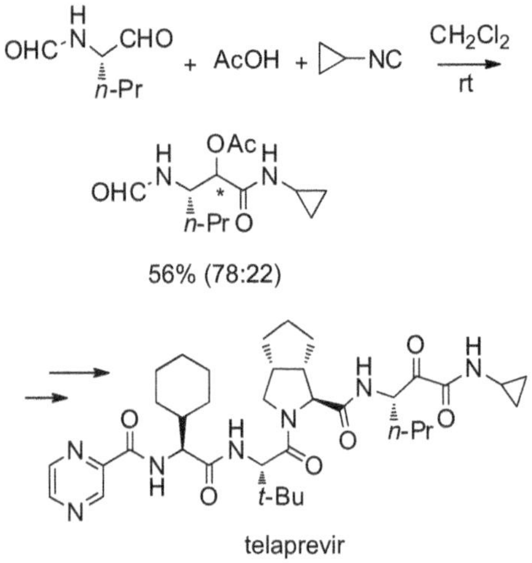

Scheme 1.111   Synthesis of telaprevir through three-component Passerini reaction.

carboxylic acids. Interestingly, the use of *trans*-epoxy aldehydes resulted in the formation of *syn*-adducts, whereas *anti*-isomers were obtained by employing *cis*-epoxy aldehydes.

Finally, a diastereoselective three-component Passerini reaction was used in a total synthesis of the potent peptidic hepatitis C virus NS3 protease inhibitor telaprevir, in addition to the diastereoselective Ugi reaction depicted in Scheme 1.111.[163] This Passerini reaction occurred among a chiral aldehyde, acetic acid, and cyclopropyl isocyanide to provide the corresponding Passerini adduct in 56% yield as a 78:22 mixture of diastereomers, as shown in Scheme 1.111.

1.3.8   Multicomponent Reactions Based on the Biginelli Reaction

The Biginelli reaction, discovered in 1893, is a three-component reaction allowing the synthesis of 3,4-dihydropyrimidin-2-(1*H*)-ones or -thiones by reacting urea or thiourea, a 1,3-dicarbonyl derivative, and an aldehyde.[185] The heterocyclic pyrimidinone products are known to exhibit a wide range of important pharmacological properties and make up a large family of medically relevant compounds. Consequently, the asymmetric version of the Biginelli reaction is of significant contemporary interest. The accepted mechanism for the Biginelli multicomponent reaction involves intermediate imine formation between the aldehyde and urea components, followed by a Mannich-type reaction with the enol derivative of the 1,3-dicarbonyl derivative.[186] The production of enantiopure dihydropyrimidines has mostly been based on chemical resolution and diastereoselective Biginelli reactions of chiral auxiliaries.[187] As a recent and nice example, Yadav *et al.* have developed a novel efficient version of diastereoselective Biginelli reaction which involved a mercaptoacetylating active methylene building block, unprotected aldoses as biorenewable aldehyde components, and ureas or thioureas.[188] As shown in Scheme 1.112, the process allowed a range of thio-sugar-annulated multi-functionalised dihydropyridine scaffolds of pharmaceutical potential, arising from the use of D-xylose and D-glucose as aldoses. This reaction was performed without solvent under microwave irradiation at 90 °C in the presence of Montmorillonite K-10 nanoclay and provided these important chiral products in high general yields combined with high diastereoselectivities of up to 96% de.

1.3.9 Multicomponent Reactions Based on the Petasis Reaction

The Petasis multicomponent reaction was discovered in 1993,[189] and involves the condensation of amines, carbonyl derivatives, and aryl- or vinylboronic acids for the synthesis of amine derivatives,[190] such as α-amino acids, when using glyoxylic acid as the aldehyde component.[191] Advantages such as mild reaction conditions and high accessibility of the reagents have contributed to render the Petasis reaction a powerful synthetic tool in the last decade. Generally, secondary amines give better yields than primary amines, and vinylboronic acids are more reactive than arylboronic acids. In 2006, Hutton *et al.* investigated the effect of chirality of the amine and organoboron species on the stereochemical outcome of the Petasis reaction.[192] The best results were achieved by using chiral secondary amines, such as (*S*)-α-methylbenzyl amine, which was reacted with a chiral boronate and glyoxylic acid monohydrate to afford the corresponding Petasis product in moderate yields (50–60%) and excellent diastereoselectivity of >90% de, as shown in Scheme 1.113. Since both enantiomers of *anti*-boronate produced the same diastereomer of product, it can be concluded that the stereochemical outcome of this reaction was dominated by the stereochemical configuration of the amine.

In the same year, Schreiber *et al.* reported a remarkable combination of an ʟ-phenylalanine derivative with a boronate and an enantiopure lactol as the aldehyde component.[193] The corresponding Petasis product was achieved as almost enantiopure stereoisomer in 85% yield, as shown in Scheme 1.114. Opposite enantiopreference was observed by using either a (*S*)-lactol or its

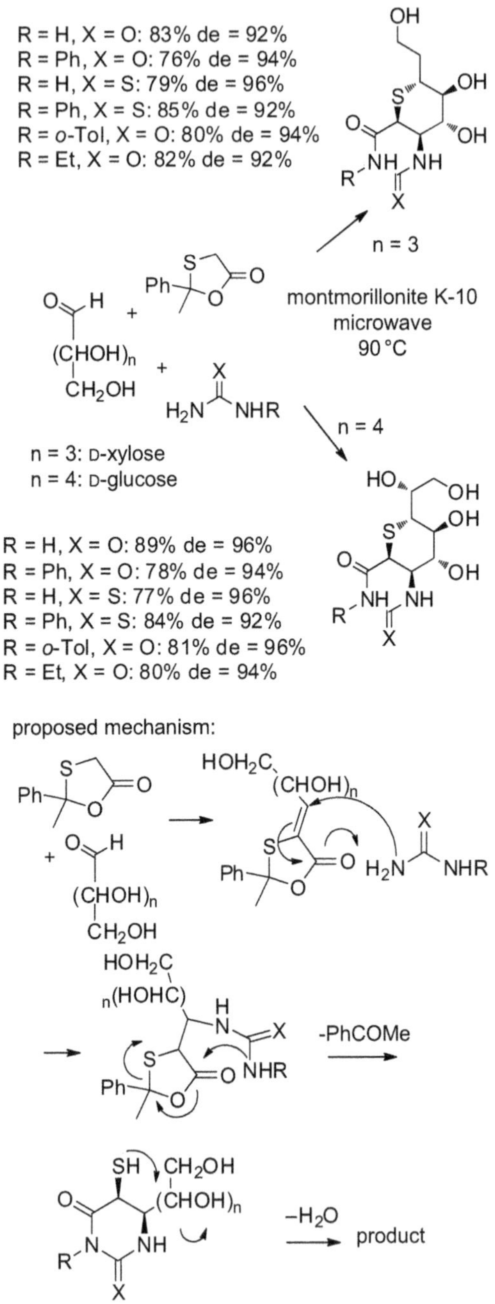

R = H, X = O: 83% de = 92%
R = Ph, X = O: 76% de = 94%
R = H, X = S: 79% de = 96%
R = Ph, X = S: 85% de = 92%
R = o-Tol, X = O: 80% de = 94%
R = Et, X = O: 82% de = 92%

n = 3

montmorillonite K-10
microwave
90 °C

n = 3: D-xylose
n = 4: D-glucose

n = 4

R = H, X = O: 89% de = 96%
R = Ph, X = O: 78% de = 94%
R = H, X = S: 77% de = 96%
R = Ph, X = S: 84% de = 92%
R = o-Tol, X = O: 81% de = 96%
R = Et, X = O: 80% de = 94%

proposed mechanism:

-PhCOMe

$-H_2O$

product

Scheme 1.112 Three-component Biginelli reaction of unprotected chiral aldoses, (thio)ureas, and mercaptoacetylating active methylene building block.

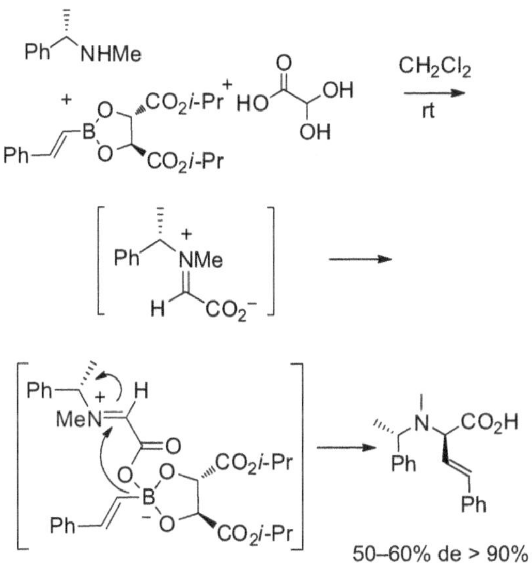

Scheme 1.113 Petasis reaction of (S)-α-methylbenzyl amine, chiral boronate, and glyoxylic acid monohydrate.

(R)-enantiomer, while an L-phenylalanine derivative was used in both cases. In order to account for this result, a secondary hydroxyl group adjacent to the intermediate imine was assumed to direct the stereochemical outcome of the process. Therefore, the directing effect of the stereocentre of the L-phenylalanine derivative was inferior to that of the aldehyde component in this case. The scope of this powerful methodology could be extended to other amino acid derivatives, which were reacted under the same reaction conditions with boronate and (S)-lactol, providing the corresponding Petasis products as almost single stereoisomers in high yields, as shown in Scheme 1.114.

Later, the scope of the boronate component was further broadened by Kukhar *et al.* in the Petasis reaction of glyoxylic acid monohydrate, various boronic acids, and chiral aminophosphonic esters instead of amino esters.[194] Electron-rich boronic acid components were successful in this Petasis process, particularly in combination with aminophosphonate, as depicted in Scheme 1.115, because the corresponding Petasis products were obtained in good to excellent yields (69–95%) and high diastereoselectivities of up to >90% de, as shown in Scheme 1.115. On the other hand, the use of a primary amine-containing phosphonic acid resulted in both lower yields (43–76%) and lower diastereoselectivities (≤80% de) for the corresponding Petasis products, as summarised in Scheme 1.115.

The asymmetric Petasis reaction has been applied by Ritthiwigrom *et al.* as the key step in total syntheses of natural products, such as (–)-uniflorine A,[195] and related casuarine, australine and 3-*epi*-australine, which are biologically active pyrrolizidine alkaloids.[196] All these syntheses employed as key intermediate a chiral tetraol which arose from the asymmetric Petasis reaction of

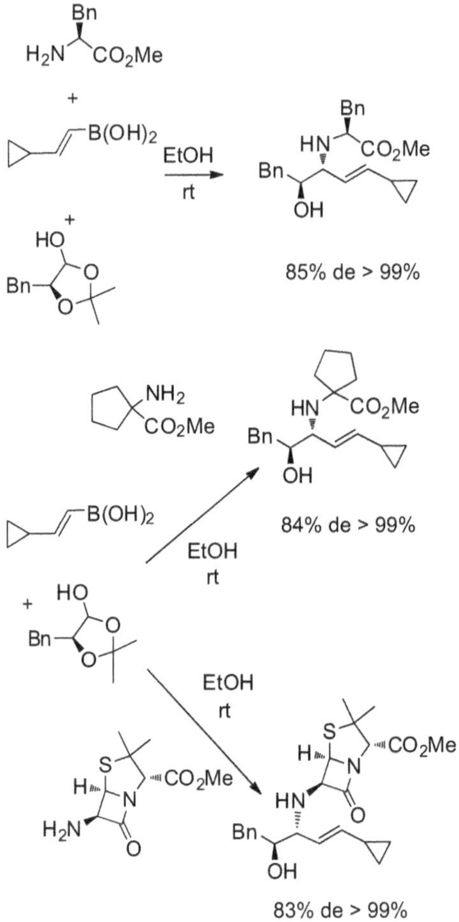

Scheme 1.114   Petasis reactions of chiral lactol, chiral amino esters, and boronate.

L-xylose with allylamine and (*E*)-styrene boronic acid. As shown in Scheme 1.116, the Petasis key product **70** was achieved in 92% yield as a single stereoisomer. This enantiopure tetraol was further converted into corresponding chiral 2-substituted-2,5-dihydropyrrole, which was subsequently transformed into the desired pyrrolizidine alkaloids uniflorine A, casuarine, australine, and 3-*epi*-australine.

In addition, Wissinger *et al.* have developed the synthesis of various other heterocyclic chiral frameworks on the basis of a diastereoselective Petasis reaction of chiral β-amino amides with boronic acids and hydroxyaldehydes.[197] This key step furnished the corresponding Petasis products in high yields (83–96%), but no precision for the diastereoselectivity of the process was given, because this crude product was directly transformed to provide a range of chiral bi- and tetraheterocyclic products. A Petasis-type aminocyclisation

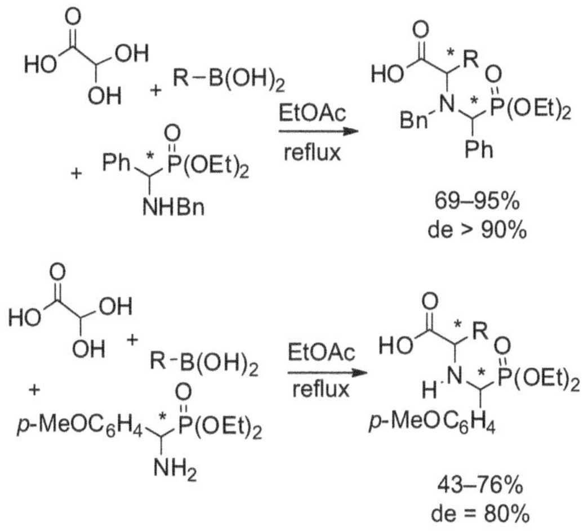

Scheme 1.115 Petasis reactions of chiral aminophosphonic esters, glyoxylic acid monohydrate, and boronic acids.

reaction can be applied to generate biologically interesting iminocyclitols, as reported by Wong *et al.*, in 2009.[198] The oxocarbenium ion-like transition state of carbohydrate-processing enzymes is mimicked by iminocyclitols, which are metabolically inert carbohydrates.[199] The various biological properties of iminocyclitols as anticancer, antiviral and antidiabetic agents render the synthesis of these key products through one-step Petasis-type reaction highly desirable.[200] An unexpected diastereoselectivity of up to >98% de was achieved by Wong *et al.* in their elegant Petasis-type reaction of a chiral sugar-derived tartaric acid with styrylboronic acid and ammonia as the amine donor, which provided the corresponding almost enantiopure pyrrolidine in 70% yield, as shown in Scheme 1.117. Iminocyclitol and analogues could be further produced through ozonolysis of these domino products in acidic solution and subsequent reduction with sodium borohydride. Furthermore, the scope of this methodology could be extended to the synthesis of chiral six-membered iminocyclitols by using commercially available 1,2- or 2,3-*O*-isopropylidene-protected D-glucose, D-mannose, D-galactose, and D-allose as starting materials. The intermediate corresponding six-membered Petasis-type products of these sugar derivatives were achieved in comparable almost complete diastereoselectivity, however in slightly lower yields, ranging from 55 to 65%.

1.3.10 Miscellaneous Multicomponent Reactions

1.3.10.1 Metal-catalysed Multicomponent Reactions

In 2006, Somfai *et al.* reported a rhodium(II)-catalysed diastereoselective three-component 1,3-dipolar cycloaddition of *in situ* generated carbonyl ylides with chiral imines as chiral auxiliaries to construct chiral highly functionalised

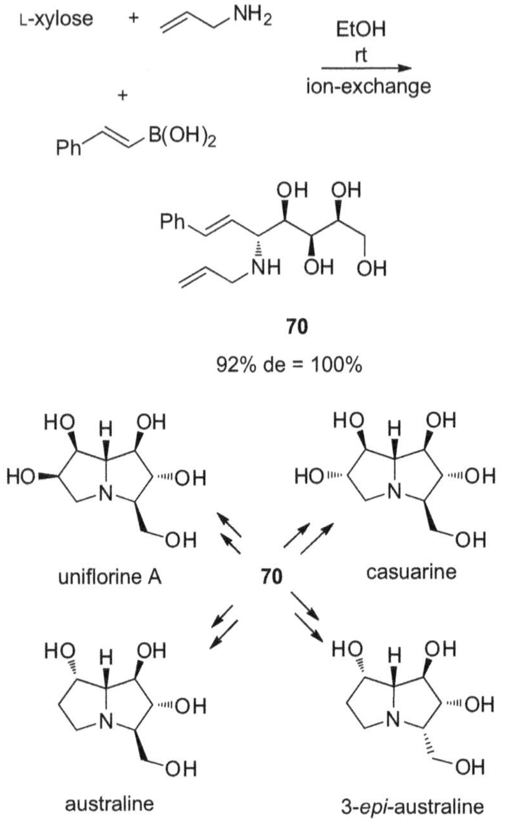

Scheme 1.116 Synthesis of uniflorine A, casuarine, australine, and 3-*epi*-australine through Petasis reaction of L-xylose, allylamine, and (*E*)-styrene boronic acid.

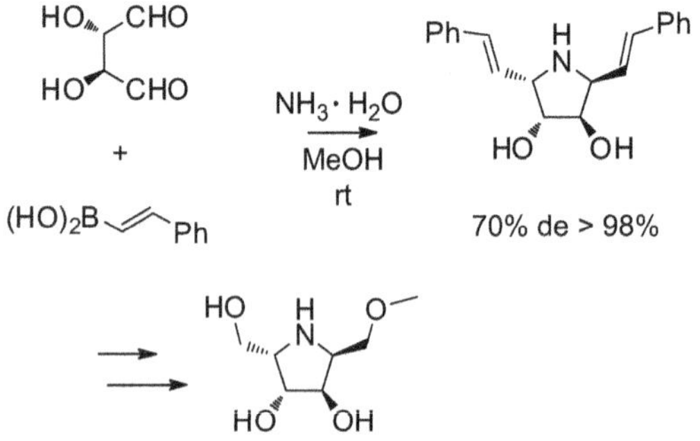

Scheme 1.117 Petasis-type reaction of chiral tartaric acid, styrylboronic acid, and ammonia.

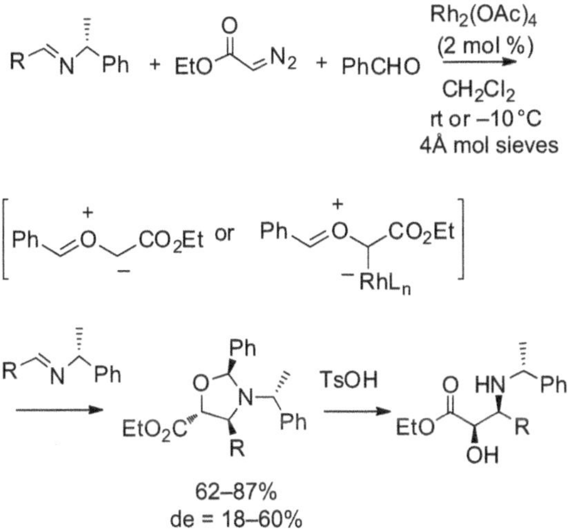

Scheme 1.118 Rh-catalysed three-component carbonyl ylide-formation–1,3-dipolar cycloaddition reaction.

oxazolines.[201] When chiral α-methylbenzylimines were reacted with benzaldehyde and ethyl diazoacetate in the presence of a catalytic amount of Rh$_2$(OAc)$_4$, they provided the corresponding oxazolines in good yields and moderate to good diastereoselectivities of up to 60% de, as shown in Scheme 1.118. These products could readily be hydrolysed by treatment with TsOH to afford *syn*-α-hydroxy-β-amino esters, which constitute interesting building blocks in the synthesis of natural products and biologically important compounds.[202] Furthermore, these motifs can be applied to the synthesis of various chiral ligands and auxiliaries.[203] This methodology was also employed as the key step in a short asymmetric synthesis of paclitaxel side-chain.

In 2007, Hu *et al.* described an efficient trapping of oxonium ylides, generated *in situ* from phenyldiazoacetates and alcohols, by imines through a highly diastereoselective three-component domino reaction.[204] As shown in Scheme 1.119, the reaction of chiral N-(*tert*-butylsulfinyl)imines with methyl phenyldiazoacetate and a range of alcohols provided the corresponding almost enantiopure β-amino-α-hydroxyesters in moderate to good yields combined with exceptionally high diastereoselectivity of >96% de in all cases of substrates studied.

Enantioselective syntheses of 2-substituted pyrrolidines from allylamines through rhodium-catalysed domino hydroformylation–reductive amination reaction were reported by Helmchen *et al.*, in 2009.[205] It must be noted that the outcome of the hydroformylation reaction was found to be controlled by the nature of the substituent at nitrogen, and not by the substituent at carbon. In the case of N-alkylallylamines, an *in situ* reduction to the corresponding almost enantiopure pyrrolidines occurred (first reaction of Scheme 1.120), while with N-sulfonyl- and N-acyl derivatives, the corresponding hemiaminals were

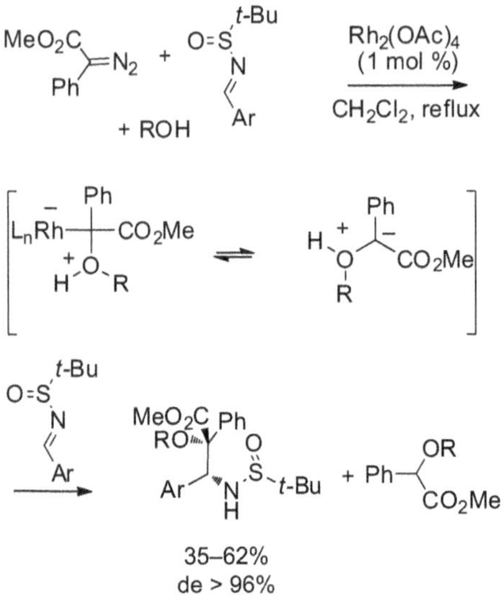

Scheme 1.119   Rh-catalysed three-component domino oxonium ylide formation–nucleophilic addition reaction.

generated (second reaction of Scheme 1.120) and, with unprotected primary amines, the corresponding cyclic imines were formed (third reaction of Scheme 1.120). In all these cases of substrates studied, the yields were moderate to good and the enantioselectivities generally excellent, up to 98% ee, as shown in Scheme 1.120.

In 2006, Garner *et al.* reported a silver-catalysed three-component [C + NC + CC] coupling domino process, allowing a variety of highly func-tionalised chiral pyrrolidines to be synthesised at ambient temperature in good to high yields (58–94%) and diastereoselectivities ranging from 73 to 90% de.[206] The domino reaction began with the formation of a chiral imine **71** from the reaction between an aldehyde and a chiral glycyl sultam. This imine then formed a metalated azomethine ylide **72**, which underwent a 1,3-dipolar cycloaddition with an activated alkene to give the final pyrrolidine. Later, these authors have shown that, by varying the metal catalyst and the chiral auxiliary, the domino process provided direct access to four of the eight possible pyrrolidine stereoisomers.[207] For example, by mediating the reaction with a copper catalyst, such as CuOAc in THF, instead of silver catalyst such as AgOAc in dimethylsulfoxide (DMSO), the process led to diastereomers different from those obtained by using AgOAc as catalyst, in good to excellent yields (60–97%) and diastereoselectivities of up to >99% de, as shown in Scheme 1.121. In 2011, these authors illustrated the utility of this asymmetric three-component [C + NC + CC] coupling domino reaction by developing a novel entry to the naphthyridinomycin natural product family of tetrahydro-isoquinoline antibiotics.[208] Indeed, this methodology constituted the key step in the total synthesis of cyanocycline A and bioxalomycin β2.

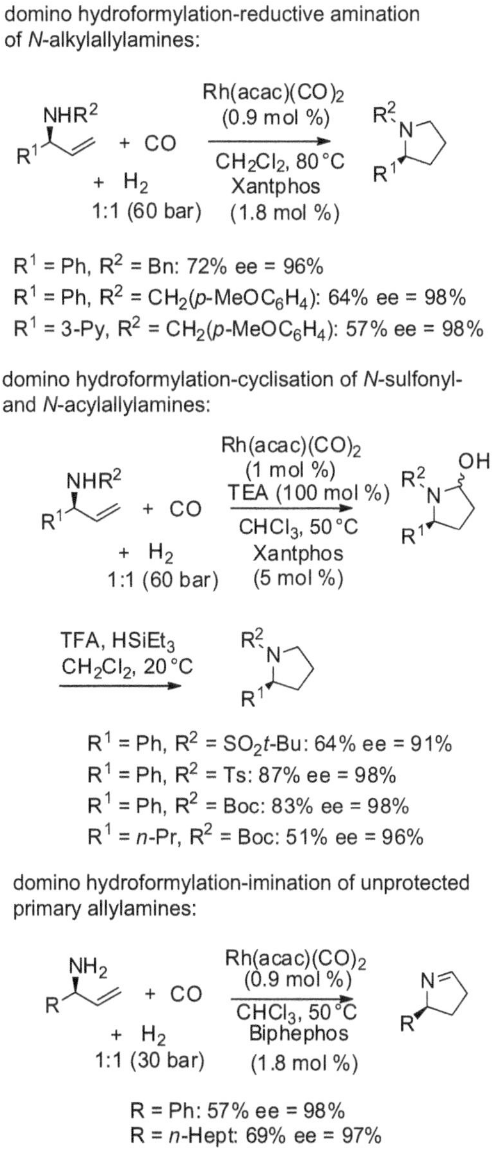
domino hydroformylation-reductive amination
of N-alkylallylamines:
NHR2
R1
+ CO
+ H2
1:1 (60 bar)
Rh(acac)(CO)2
(0.9 mol %)
CH2Cl2, 80 °C
Xantphos
(1.8 mol %)
R2
N
R1
R1 = Ph, R2 = Bn: 72% ee = 96%
R1 = Ph, R2 = CH2(p-MeOC6H4): 64% ee = 98%
R1 = 3-Py, R2 = CH2(p-MeOC6H4): 57% ee = 98%
domino hydroformylation-cyclisation of N-sulfonyl-
and N-acylallylamines:
NHR2
R1
+ CO
+ H2
1:1 (60 bar)
Rh(acac)(CO)2
(1 mol %)
TEA (100 mol %)
CHCl3, 50 °C
Xantphos
(5 mol %)
OH
R2
N
R1
TFA, HSiEt3
CH2Cl2, 20 °C
R2
N
R1
R1 = Ph, R2 = SO2t-Bu: 64% ee = 91%
R1 = Ph, R2 = Ts: 87% ee = 98%
R1 = Ph, R2 = Boc: 83% ee = 98%
R1 = n-Pr, R2 = Boc: 51% ee = 96%
domino hydroformylation-imination of unprotected
primary allylamines:
NH2
R
+ CO
+ H2
1:1 (30 bar)
Rh(acac)(CO)2
(0.9 mol %)
CHCl3, 50 °C
Biphephos
(1.8 mol %)
N
R
R = Ph: 57% ee = 98%
R = n-Hept: 69% ee = 97%

Scheme 1.120 Rh-catalysed three-component domino hydroformylation–cyclisation reactions.

In 2012, the same authors applied this methodology to an efficient synthesis of the neuramidase inhibitor A-315675.[209] As shown in Scheme 1.122, the fully functionalised core of this target was assembled in a single step through a copper-catalysed *exo*-selective asymmetric three-component domino [C + NC + CC] coupling reaction of a chiral α-acetamidoaldehyde, a chiral glycylsultam, and ethyl thioacrylate. This process afforded the corresponding highly functionalised

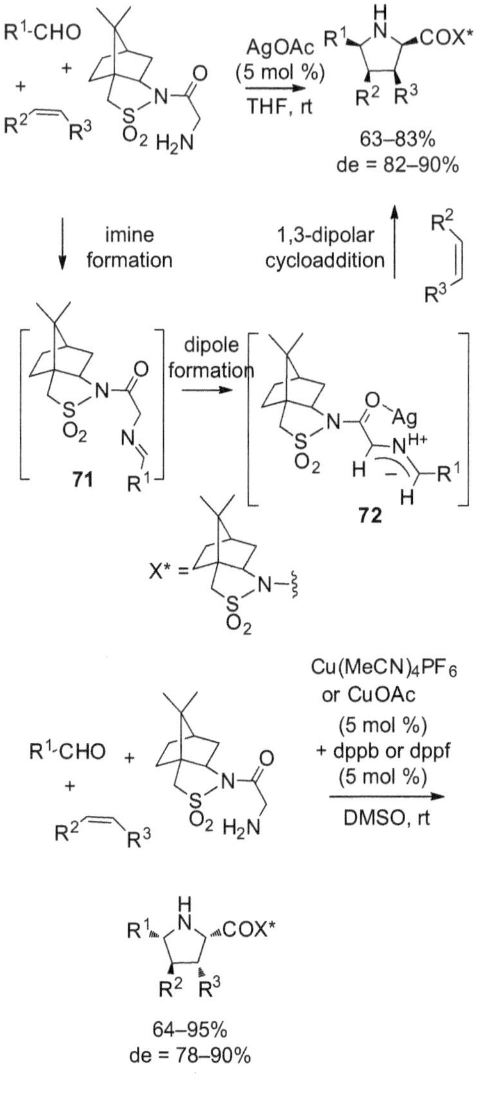

Scheme 1.121 Ag- and Cu-catalysed three-component domino imine formation–1,3-dipolar cycloaddition [C + NC + CC] processes.

pyrrolidine in 76% yield and diastereoselectivity of 90% de. This chiral product was further converted into neuramidase inhibitor A-315675 in five steps.

Another type of asymmetric three-component domino process has been reported by Katsumura *et al.*[210] It dealt with a palladium-catalysed domino aminoacetal formation–Stille–Migita coupling–6π-azaelectrocyclisation–aminoacetal formation reaction of vinylstannanes, vinyliodides, and a chiral *cis*-aminoindanol derivative. As shown in Scheme 1.123, the sequence began with the *in situ* formation of an aminoacetal from the chiral *cis*-aminoindanol

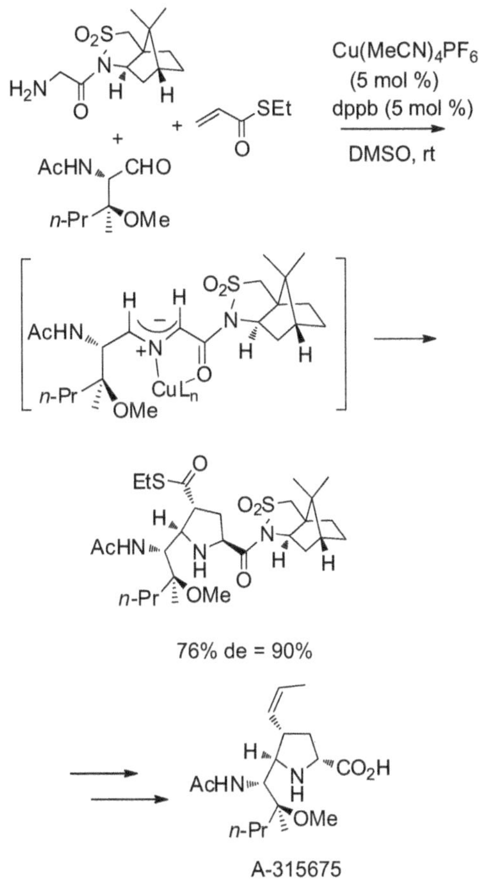

Scheme 1.122 Synthesis of A-315675 through Cu-catalysed three-component domino [C + NC + CC] coupling process.

derivative and the vinyliodide. This aminoacetal formation protected the unstable aldehyde moiety in vinyliodide, and thus successfully achieved the subsequent Stille–Migita coupling with vinyl stannane. The coupling product formed was in equilibrium with a 1-azatriene, which spontaneously cyclised into the corresponding dihydropyridine. The reactive enamine moiety in this dihydropyridine was then trapped by the proximal hydroxyl group of *cis*-aminoindanol, giving rise to the final product. The scope of this methodology was extended to a range of vinylstannanes, and tri- as well as tetra-substituted vinyliodides, providing a variety of chiral 2,4-disubstituted, 1,2,5,6-tetrahydropyridines, and 2,4,5-trisubstituted 2,5-tetrahydropyridines in good yields and diastereoselectivities of up to 100% de (Scheme 1.123).

This powerful methodology was applied to the total synthesis of three natural indole alkaloids, such as (−)-dendroprimine,[211] (−)-corynantheidine,[212] and (−)-20-epiuleine,[213] as depicted in Scheme 1.124.

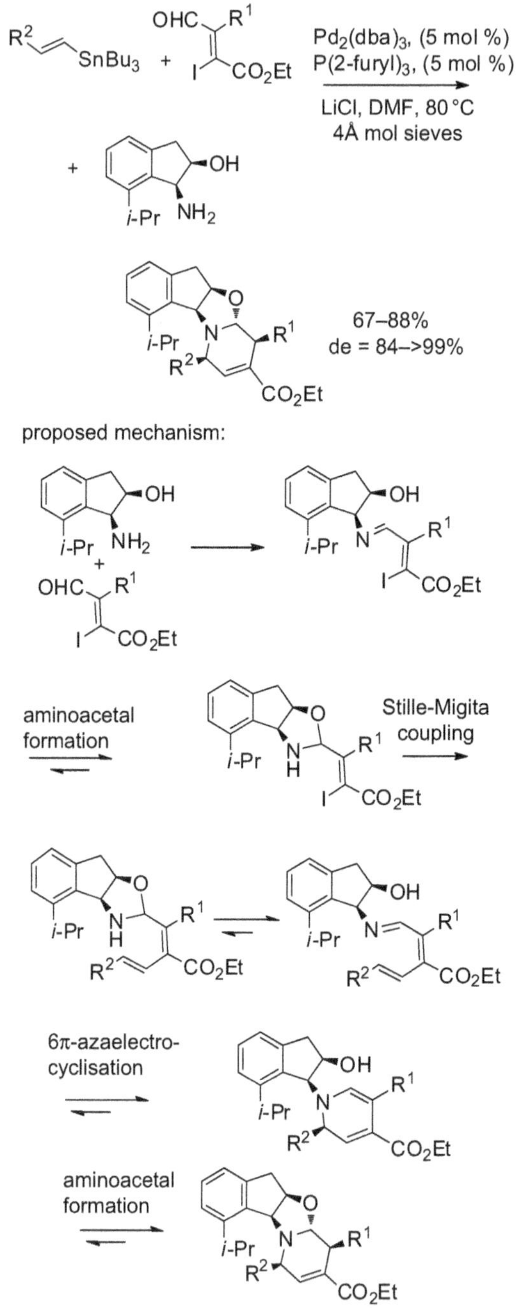

Scheme 1.123 Pd-catalysed three-component domino aminoacetal formation–Stille–
Migita coupling–6π-azaelectrocyclisation–aminoacetal formation
reaction.

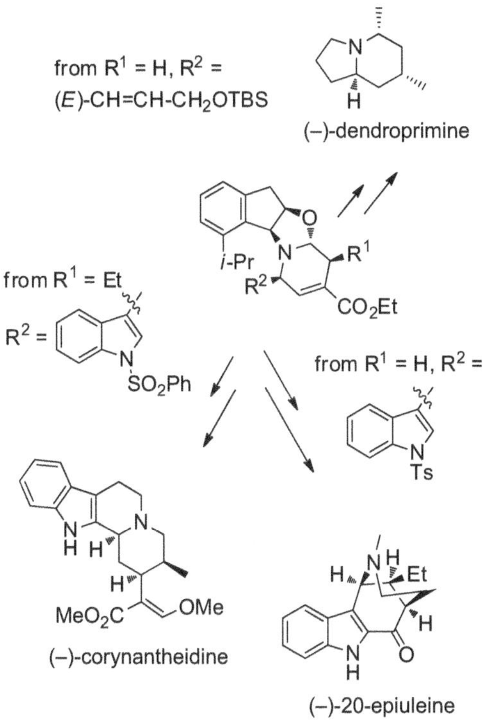

from R^1 = H, R^2 =
(*E*)-CH=CH-CH$_2$OTBS

(−)-dendroprimine

from R^1 = Et

R^2 =

SO$_2$Ph

from R^1 = H, R^2 =

Ts

i-Pr

R^2

R^1

CO$_2$Et

MeO$_2$C   OMe

(−)-corynantheidine

H   N   H   Et

H

(−)-20-epiuleine

Scheme 1.124 Synthesis of (−)-dendroprimine, (−)-corynantheidine, and (−)-epiuleine.

Diastereoselective nickel-catalysed asymmetric four-component domino Reformatsky reactions have been developed by Dondoni *et al.* with the aim of achieving chiral *C*-glycosyl β-amino esters.[138] As shown in Scheme 1.125, an *in situ* initial coupling of chiral *C*-glycosyl aldehydes with *p*-methoxy-benzylamine gave the corresponding imines. This reacted with a bromozinc enolate, which was generated *in situ* from dimethylzinc and ethyl bromoacetate, providing the corresponding chiral Reformatsky products in moderate to good yields and as single diastereomers in all cases of substrates studied.

Another type of nickel-catalysed asymmetric multicomponent domino reaction was reported by Montgomery *et al.*, in 2006.[214] In this case, the reaction consisted in a three-component coupling of a chiral α-silyloxyaldehyde with an alkynylsilane in the presence of (*i*-Pr)$_3$SiH as the reducing agent. This domino aldehyde–alkyne reductive coupling process, performed in the presence of Ni(cod)$_2$ and *N*-heterocyclic carbene generated from imidazolium, afforded the corresponding protected allylic alcohols in excellent yields and diastereo-selectivities of up to 96% de, as shown in Scheme 1.126. These chiral products could be easily converted into the corresponding *anti*-1,2-diols with conser-vation of the enantioselectivity by treatment with *n*-Bu$_4$NF.

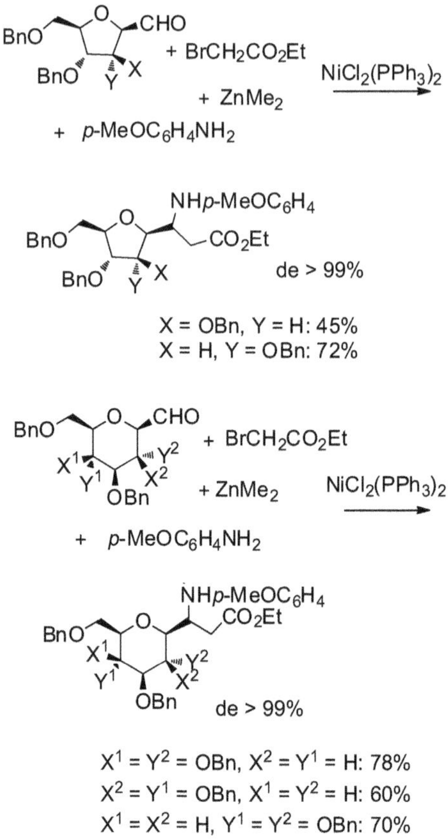

Scheme 1.125 Ni-catalysed four-component domino Reformatsky reactions.

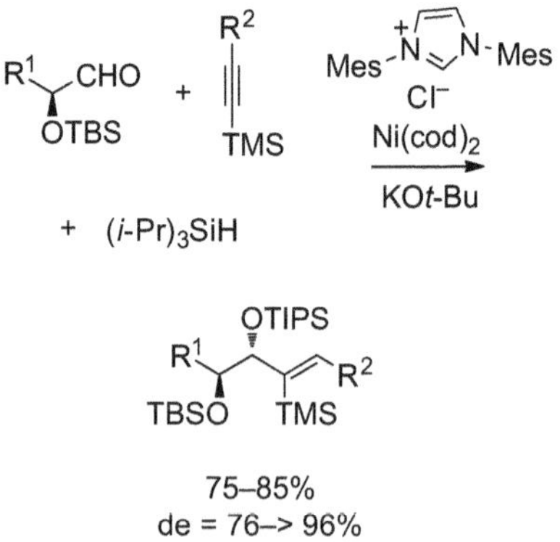

Scheme 1.126 Ni-catalysed three-component domino aldehydes–alkynes reductive coupling reaction.

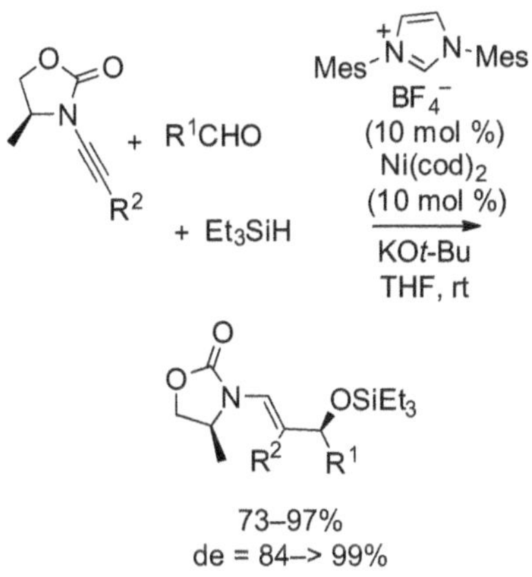

73–97%
de = 84–> 99%

Scheme 1.127　Ni-catalysed three-component domino aldehydes–ynamides reductive coupling reaction.

In 2011, Sato *et al.* reported the asymmetric synthesis of chiral γ-siloxy-enamides through diastereoselective coupling of chiral oxazolidinone-derived ynamides, aldehydes, and triethylsilane mediated by a nickel catalyst.[215] As shown in Scheme 1.127, both excellent yields and excellent diastereoselectivities were achieved in the formation of the corresponding chiral coupling products. The authors assumed that the process proceeded through the formation of oxanickelacycles to afford γ-siloxyenamide derivatives in a highly regio- and stereoselective manner.

Asymmetric titanium-catalysed domino reactions have also been described. As an example, Ghosh *et al.* have successfully developed the synthesis of chiral substituted pyrrolidine and proline derivatives on the basis of a diastereo-selective titanium-catalysed three-component domino reaction, occurring between a chiral phenyldihydrofuran, an *N*-tosylimino ester, and various silane reagents.[216] As shown in Scheme 1.128, a range of chiral functionalised pyrrolidines were stereoselectively produced in good yields and excellent diastereoselectivities of up to 98% de by using a variety of silane reagents. In order to explain the formation of these products, the authors have proposed that the domino process began with the reaction of the chiral phenyldi-hydrofuran with *N*-tosylimino ester in the presence of TiCl$_4$ to give the corresponding oxocarbenium ion **73**. This intermediate reacted with the silane reagent to provide the corresponding tetrahydrofuran **74**, which was then activated by TiCl$_4$ to produce a novel oxonium ion **75**. The S$_N$2 nucleophilic attack of sulfonamide nucleophile (*N*Ts) to oxonium ion **75** provided the final pyrrolidine, as depicted in Scheme 1.128.

In 2007, Kerr *et al.* reported the total synthesis of the natural product (+)-nakadomarin Λ, which constitutes one of the most architecturally beautiful classes of alkaloids.[217] An asymmetric ytterbium-catalysed three-component domino reaction, occurring between a chiral cyclopropane, *p*-methoxybenzyl-hydroxylamine, and a furfural derivative, was the key step of this synthesis,

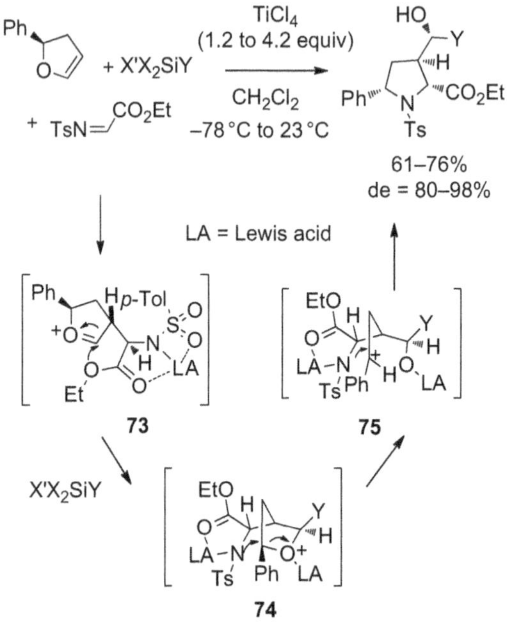

Scheme 1.128   Ti-catalysed three-component reaction of phenyldihydrofuran, *N*-tosylimino ethyl ester, and silane reagents.

allowing an enantiopure highly functionalised tetrahydro-1,2-oxazine to be achieved in both excellent yield and enantioselectivity, as shown in Scheme 1.129. This key product was further converted into expected (+)-nakadomarin A in 21 steps.

Finally, Che *et al.* have described asymmetric gold(III)-catalysed three-component coupling reactions of chiral prolinol derivatives, alkynes, and aldehydes, providing the corresponding chiral propargylamines in good to excellent yields and diastereoselectivities of up to 98% de, as shown in Scheme 1.130.[218] These processes were promoted by a salen gold(III) complex in water, which could be repeatedly used for 10 reaction cycles, leading to an overall turnover number of 812. A proposed mechanism for the three-component coupling reaction of aldehyde, amine, and alkyne is illustrated in Scheme 1.130. Aldehyde was first condensed *in situ* with the secondary amine to give an iminium ion, while the gold(III) complex activated the C–H bond of the terminal alkyne to generate a gold acetylide intermediate. The gold acetylide intermediate further underwent a nucleophilic attack on the iminium ion to give the final propargylamine.

1.3.10.2   Other Multicomponent Reactions

In 2006, Kita *et al.* described asymmetric three-component domino aminal formation–bromation reaction of cyclohexa-2,5-dienyl-1-methylaldehyde with

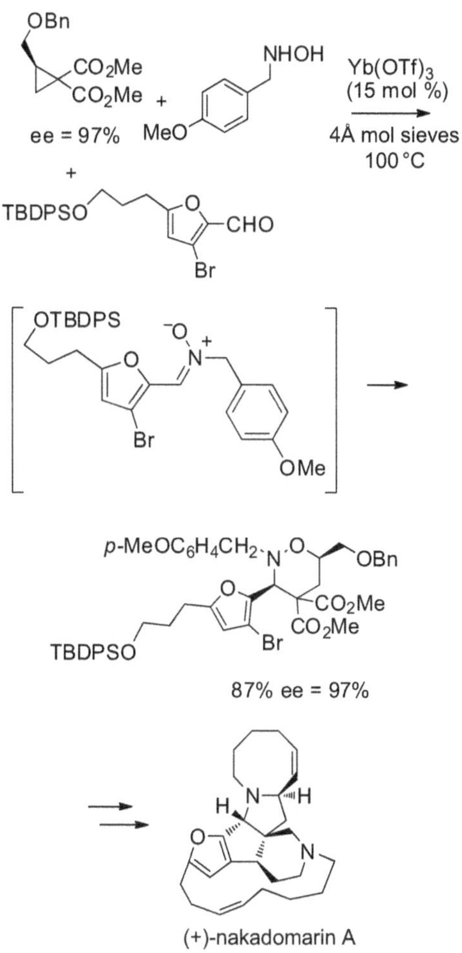

Scheme 1.129 Synthesis of (+)-nakadomarin A through Yb-catalysed three-component reaction of a chiral cyclopropane derivative, *p*-methoxy-benzylhydroxylamine, and a furfural derivative.

optically pure 1,2-di(4-methoxyphenyl)-1,2-diamine, and NBS.[219] This process provided the corresponding enantiopure tricyclic domino product in 57% yield as a single stereoisomer, through discrimination of the two olefins of cyclohexa-2,5-dienyl-1-methylaldehyde, as shown in Scheme 1.131. The utility of this methodology was illustrated by converting this enantiopure product into the natural product (−)-γ-lycorane.

The key step of a total synthesis of naturally occurring and biologically active (−)-dibromophakellstatin, which was reported by Lindel *et al.* in 2007, consisted in an asymmetric three-component imidazolidinone annulation reaction of a chiral tricycle enamide with two equivalents of $EtO_2CNHOTs$ in the presence of CaO.[220] As shown in Scheme 1.132, this domino process

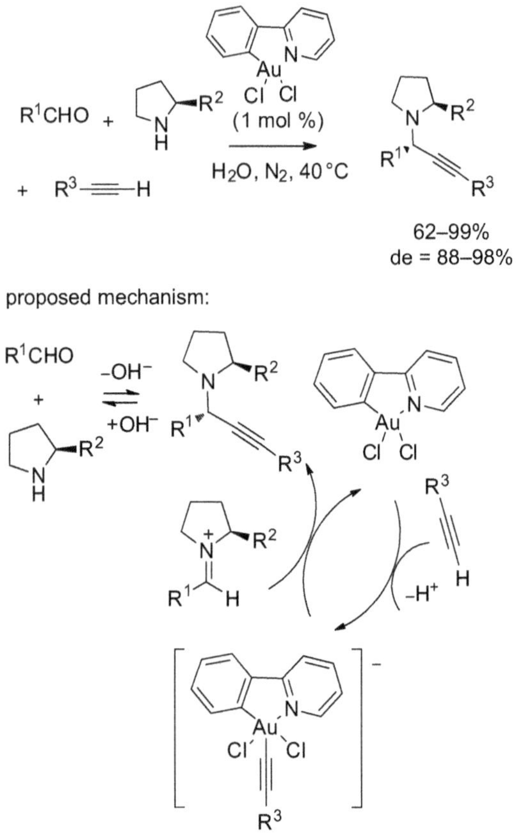

62–99%
de = 88–98%

proposed mechanism:

Scheme 1.130   Au-catalysed three-component reaction of (*S*)-prolinol derivatives, alkynes, and aldehydes.

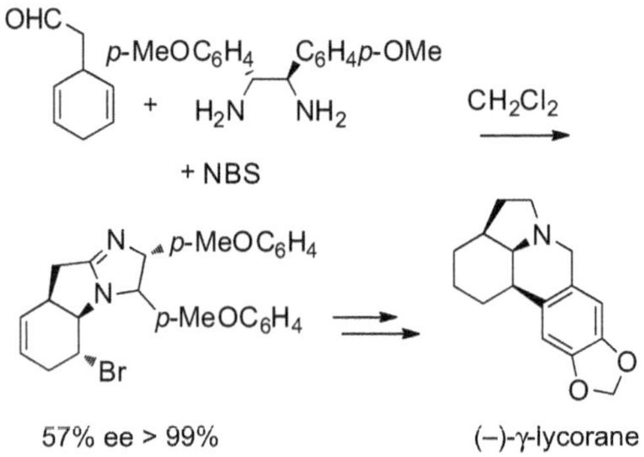

57% ee > 99%   (–)-γ-lycorane

Scheme 1.131   Synthesis of (–)-γ-lycorane through three-component domino aminal formation–bromation reaction.

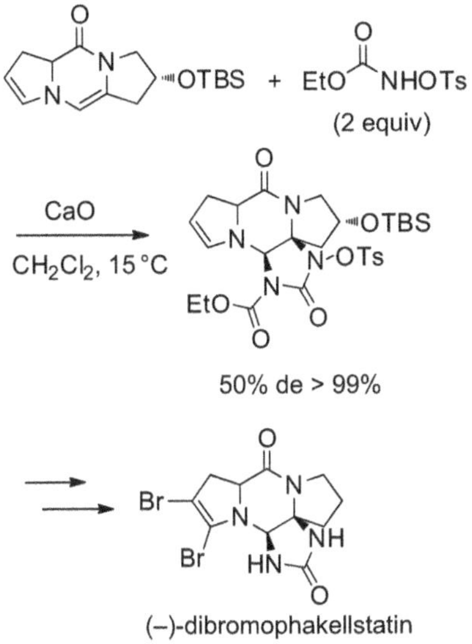

Scheme 1.132 Synthesis of (–)-dibromophakellstatin through three-component imidazolidinone annulation reaction.

provided the corresponding tetracycle in 50% yield as a single stereoisomer. This key enantiopure product was subsequently converted into the expected (–)-dibromophakellstatin in four steps.

In 2007, Beller *et al.* reported the diastereoselective three-component reaction of (*S*)-methyl pyroglutamate with aldehydes, and dienophiles such as *N*-methyl maleimide or maleic anhydride, which led to a range of chiral substituted 1-amido-2-cyclohexenes.[221] As shown in Scheme 1.133, these bicyclic products were formed in moderate to good yields and as single diastereomers. A possible mechanism of this reaction is depicted in Scheme 1.133 in which the domino reaction began with the formation of 1-(*N*-acylamino)-1,3-butadienes **76** as key intermediates, which were generated through the condensation of aldehydes to amides. These intermediates were further trapped by dienophiles through a subsequent Diels–Alder cycloaddition.

In another context, Scheidt *et al.* have developed a stereoselective synthesis of a chiral highly substituted β-hydroxy-γ-lactam by using a chiral β-silyloxy homoenolate that could be accessed from an (*L*)-phenylalaninol-derived acetamide enolate of an amide and an acylsilane. As shown in Scheme 1.134, the acylsilane acted sequentially as an electrophilic/nucleophilic moiety in this process by undergoing a 1,2-silyl group migration (1,2-Brook rearrangement). The unconventional nucleophilic species **77** subsequently underwent addition to an imine to provide the corresponding γ-amino-β-hydroxy amide in a one-pot reaction with a good yield and a high level of diastereoselectivity of up to 86% de, as shown in Scheme 1.134.[222] The authors have demonstrated that the subsequent exposure of this γ-amino-β-hydroxy amide to microwave

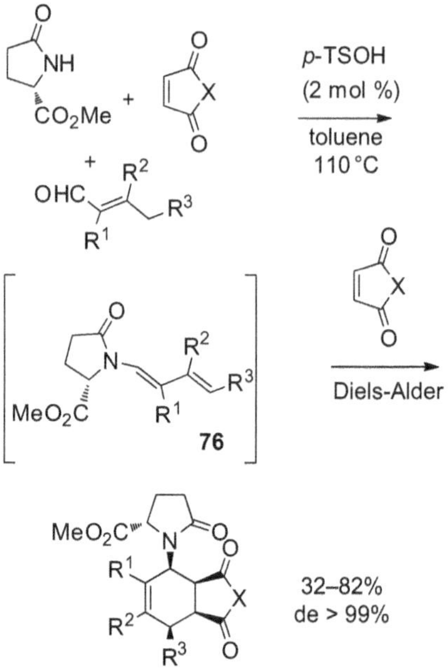

Scheme 1.133   Three-component reaction of chiral amides, aldehydes, and dienophiles.

irradiation and acidic conditions promoted a cyclisation to form the corresponding γ-lactam in high enantiomeric excesses of 87% ee (Scheme 1.134). Given that a wide number of biologically active products include γ-lactam moieties, the importance of this novel methodology, allowing chiral highly substituted γ-lactams to be achieved, was demonstrated.

In 2008, Rodriguez *et al.* reported a diastereoselective synthesis of 1,4-diazepines on the basis of a multicomponent reaction occurring among β-ketoamides, aromatic aldehydes, and cyclic or acyclic 1,2-diamines.[223] The 1,4-diazepine derivatives were achieved in low to excellent yields with a complete relative diastereoselectivity in all cases of substrates studied, as shown in Scheme 1.135. In some cases, better results were obtained by simply heating the mixture of substrates at 120 °C without solvent than by performing the reaction in toluene at 110 °C in the presence of molecular sieves. Later, the authors extended this methodology to β-ketoesters which were reacted with aromatic aldehydes and cyclic 1,2-diamines to provide the corresponding 1,4-diazepine derivatives in low to good yields (≤56%) and complete relative diastereoselectivity.[224]

In 2010, Barluenga *et al.* reported an enantioselective synthesis of 4-hydroxy-2-cyclohexenones through a four-component reaction, occurring among aryl- and heteroarylcarbene chromium complexes, *in situ* prepared lithium enolate of (*S*)-3-acetyl-4-benzyl-2-oxazolidinone, and two equivalents of propargylic organomagnesium bromides.[225] This remarkable domino

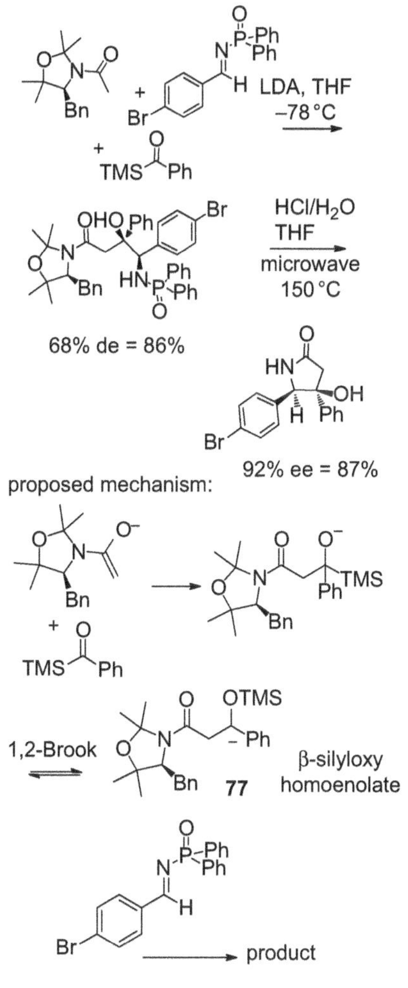

Scheme 1.134 Three-component reaction of a chiral amide, an acylsilane, and an imine.

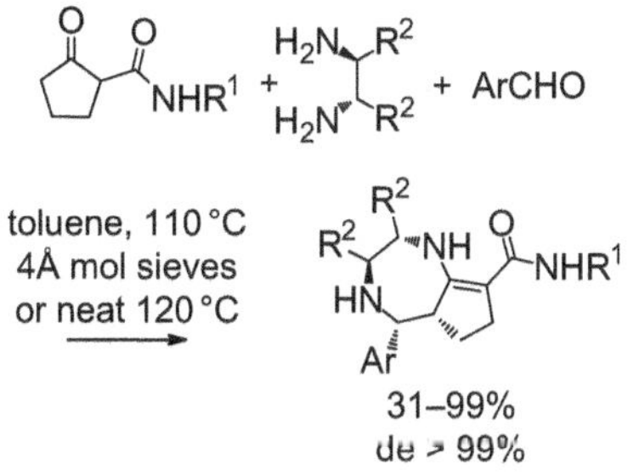

Scheme 1.135 Three-component reaction of chiral 1,2-amines, β-ketoamides, and aromatic aldehydes.

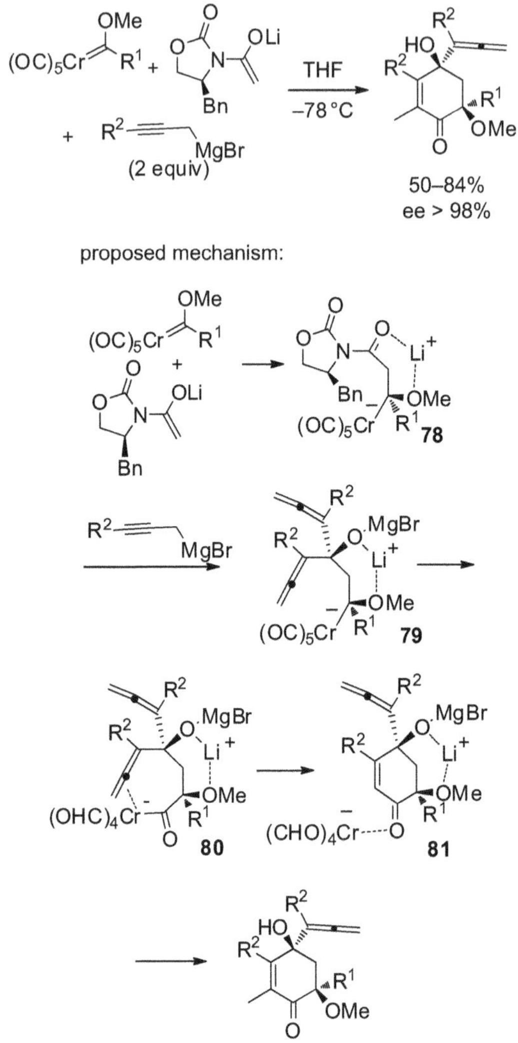

Scheme 1.136 Four-component reaction of (*S*)-3-acetyl-4-benzyl-2-oxazolidinone enolate, carbene chromium complexes, and two equivalents of propargylic organomagnesium bromides.

process afforded the corresponding almost enantiopure densely functionalised 2-cyclohexenones containing two quaternary stereogenic centres at the α and γ positions. As shown in Scheme 1.136, these products were produced in moderate to high yields as single stereoisomers in all cases of substrates studied. The authors assumed that the process began with the initial addition of the imide lithium enolate to the carbene carbon atom of the chromium complex, which led to the corresponding lithium alkylpentacarbonylchromate intermediate **78**. A subsequent double addition of the organomagnesium derivative to the exocyclic carbonyl group of the *N*-acyl-2-oxazolidinone moiety, which entailed an unprecedented removal of this chiral auxiliary group, proceeded regioselectively, incorporating two allenyl units, and provided intermediate **79**. Insertion of CO into the C(sp^3)–Cr σ bond of intermediate **79** afforded the allenyl substituted lithium acyltetracarbonylchromate complex **80**. A final intramolecular carbometalation reaction of the terminal C=C bond of one of the allene groups produced allylchromate intermediate **81**, which was protonated to give the final product.

   In addition, a three-component reaction among sulfonylimidates, silyl glyoxylates, and chiral *N*-*tert*-butanesulfinyl aldimines was developed by Lu *et al.*, in 2011.[226] This process promoted the formation of two contiguous stereogenic centres, two C–C bonds, one C–N bond, and one O–Si bond, together with the cleavage of the chiral auxiliary in a one-pot reaction. It afforded the corresponding chiral substituted cyclic sulfonylamidines in good to high yields, diastereoselectivity of >90% de in all cases of substrates studied, combined with remarkable enantioselectivities, as shown in Scheme 1.137. The domino reaction was supposed to begin with the addition of the lithium aza-enolate of sulfonylimidate to the silyl glyoxylate, which triggered a Brook

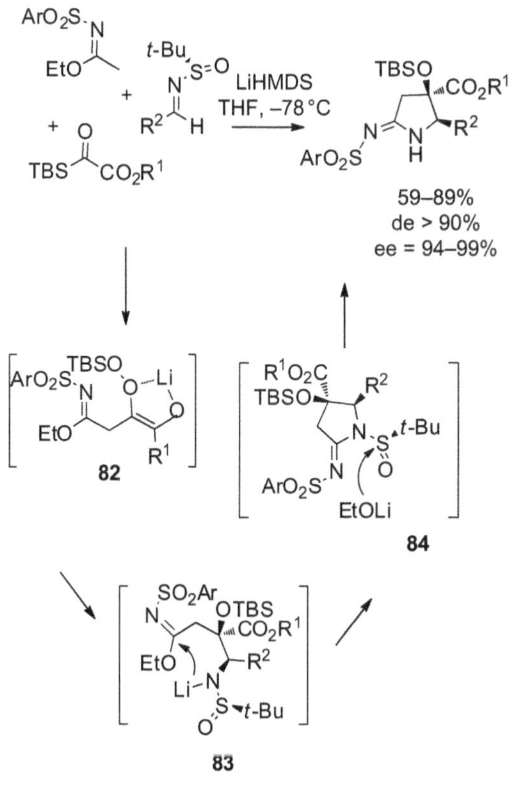

Scheme 1.137   Three-component reaction of chiral *N*-*tert*-butanesulfinyl aldimines, sulfonylimidates, and silyl glyoxylates.

rearrangement, providing a novel enolate **82**. This enolate further underwent addition to the chiral *N-tert*-butanesulfinyl aldimine, in which the imine participated as the second electrophile, providing intermediate **83**. The anionic nitrogen-induced cyclisation into **84** and subsequent desulfinylation by nucleophilic attack by the extruding ethoxide led to the final product.

A novel family of chiral fluorinated 1,4-dihydropyridines have been generated by Fustero *et al.* on the basis of an asymmetric Hantzsch-type process, occurring among alkyl propiolates, fluorinated nitriles, and (*R*)-(+)-allyl *p*-tolyl sulfoxide as chiral auxiliary.[227] The reaction began with the formation of enamino sulfoxides by reaction of (*R*)-(+)-allyl *p*-tolyl sulfoxide through the γ-position with fluorinated nitriles. These enamino sulfoxides **85** subsequently underwent an aza-Michael addition with alkyl propiolates to give the corresponding enolates **86**, which then cyclised through an intramolecular Michael addition, as shown in Scheme 1.138. The reaction took place with complete selectivity, allowing the formation of a variety of enantiopure fluorinated 1,4-dihydropyridines to be achieved as single stereoisomers in moderate to good yields.

Finally, Bella *et al.* have described a novel three-component domino reaction occurring between L-proline lithium salt, 2-cyclohexen-1-one, and aliphatic aldehydes, which afforded the corresponding chiral 4-alkylidene-2-cyclohexen-1-ones with four stereogenic centres.[228] In all cases of aldehydes studied, the reaction gave a single stereoisomer with >98% de combined with moderate yields, as shown in Scheme 1.139. In order to explain these results, the authors have proposed that the chiral lithium salt reacted with aldehyde to give the corresponding iminium salt **87**. This iminium ion gave the corresponding

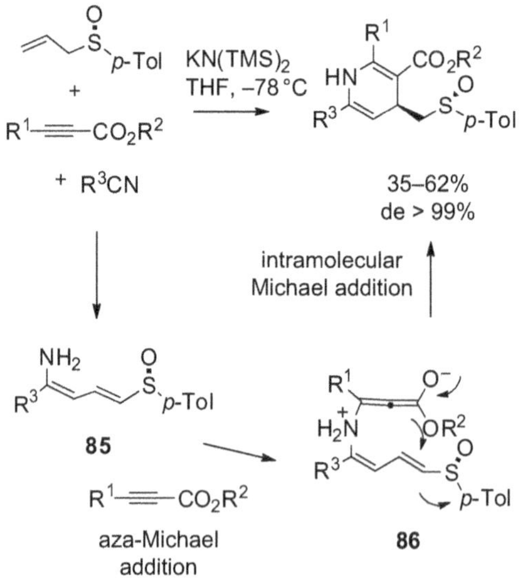

Scheme 1.138 Three-component reaction of (*R*)-(+)-allyl *p*-tolyl sulfoxyde, alkyl propiolates, and fluorinated nitriles.

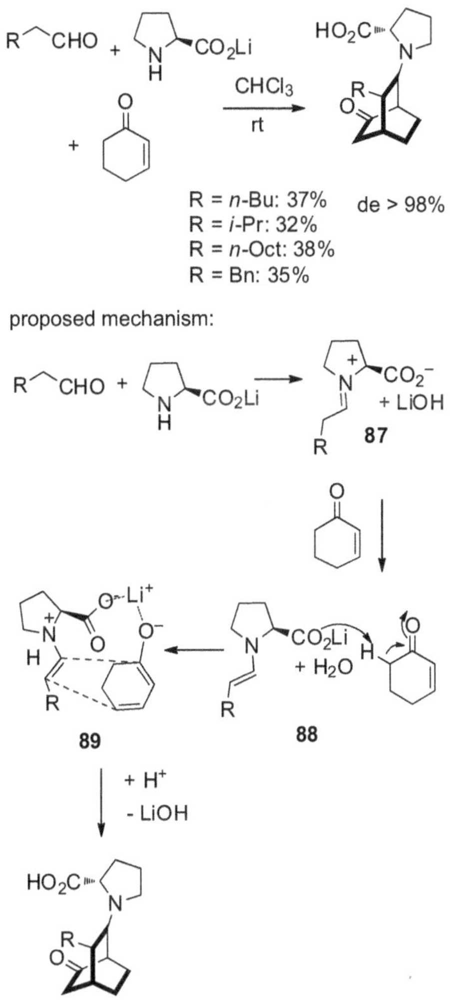
R = n-Bu: 37%
R = i-Pr: 32%
R = n-Oct: 38%
R = Bn: 35%
de > 98%
CHCl3
rt
proposed mechanism:
+ LiOH
87
+ H2O
88
89
+ H+
- LiOH

Scheme 1.139 Three-component reaction of L-proline lithium salt, aliphatic aldehydes, and 2-cyclohexen-1-one.

enamine **88** which deprotonated 2-cyclohexen-1-one, providing the formation of an activated diene **89**. Subsequently, the cycloaddition reaction of this diene with enamine occurred, affording the final bicyclic cycloadduct.

1.4 Conclusions

This chapter illustrates the power and diversity of asymmetric domino reactions based on the use of chiral substrates including chiral auxiliaries, which have quickly become a powerful, fascinating, and highly efficient tool in organic synthesis. The concept of domino sequences has allowed high molecular complexity to be attained easily, with very often high levels of stereocontrol in simple operational procedures, and advantages of savings in solvent, time, energy, and costs. The use of one- and two-component, as well as multicomponent, domino reactions in asymmetric synthesis is increasing constantly. Such single-step reactions allow the synthesis of a wide range of structurally diverse and complex chiral molecules from simple substrates in an economically favourable manner by avoiding the use of costly and time-consuming protection–deprotection processes, as well as purification procedures for intermediates. It must be noted that it was difficult to locate all the published examples of asymmetric domino reactions involving chiral substrates, because many are incorporated in total syntheses advertised under different keywords. The cases cited in this chapter have been selected to highlight the most promising applications of asymmetric domino reactions to organic synthesis.

References

1. E. N. Jacobsen, A. Pfaltz and H. Yamamoto (eds.), *Comprehensive Asymmetric Catalysis*, Springer: Berlin, 1999.
2. (a) D. Enders, C. Grondal and M. R. M. Hüttl, *Angew. Chem., Int. Ed.*, 2007, **46**, 1570–1581; (b) G. Guillena, D. J. Ramon and M. Yus, *Tetrahedron: Asymmetry*, 2007, **18**, 693–700; (c) D. M. D'Souza and T. J. J. Müller, *Chem. Soc. Rev.*, 2007, **36**, 1095–1108; (d) A.-N. Alba, X. Companyo, M. Viciano and R. Rios, *Curr. Org. Chem.*, 2009, **13**, 1432–1474; (e) H. Pellissier, *Adv. Synth. Catal.*, 2012, **354**, 237–294.
3. (a) P. I. Dalko and L. Moisan, *Angew. Chem., Int. Ed.*, 2001, **40**, 3726–3748; (b) A. Berkessel and H. Gröger, (eds.), *Asymmetric Organo-catalysis – From Biomimetic Concepts to Powerful Methods for Asymmetric Synthesis*, Wiley-VCH, Weinheim, 2005; (c) J. Seayad and B. List, *Org. Biomol. Chem.*, 2005, **3**, 719–724; (d) M. S. Taylor and E. N. Jacobsen, *Angew. Chem., Int. Ed.*, 2006, **45**, 1520–1543; (e) P. I. Dalko (eds.), *Enantioselective Organocatalysis*, Wiley-VCH, Weinheim, 2007; (f) P. I. Dalko, *Chimia*, 2007, **61**, 213–218; (g) H. Pellissier, *Tetrahedron*, 2007, **63**, 9267–9331; (h) A. G. Doyle and E. N. Jacobsen, *Chem. Rev.*, 2007, **107**, 5713–5743; (i) M. G. Gaunt, C. C. C. Johansson, A. McNally and N. C. Vo, *Drug Discov. Today*, 2007, **2**, 8–27; (j) B. List, (ed.), Special Issue on Organocatalysis, *Chem. Rev.*, 2007, **107**(12), 5413–5883; (k) D. W. C. MacMillan, *Nature*, 2008, **455**, 304–308; (l) X. Yu and W. Wang, *Chem. Asian. J.*, 2008, **3**, 516–532; (m) A. Dondoni and A. Massi, *Angew. Chem., Int. Ed.*, 2008, **47**, 4638–4660; (n) P. Melchiorre, M. Marigo, A. Carlone and G. Bartoli, *Angew. Chem., Int. Ed.*, 2008, **47**, 6138–6171; (o) F. Peng and Z. Shao, *J. Mol. Catal. A*, 2008, **285**, 1–13; (p) C. F. Barbas, *Angew. Chem., Int. Ed.*, 2008, **47**, 42–47; (q) C. Palomo, M. Oiarbide and R. Lopez, *Chem. Soc. Rev.*, 2009, **38**, 632–653; (r) S. Bertelsen and K. A. Jørgensen, *Chem. Soc. Rev.*, 2009, **38**, 2178–2189; (s) L.-W. Xu, J. Luo and Y. Lu, *Chem. Commun.*, 2009, 1807–1821; (t) M. Bella and T. Gasperi, *Synthesis*, 2009,

1583–1614; (u) Q. Rao, S. Luo and L. Gong, *Chin. Sci. Bull.*, 2010, **55**(17), 1742–1752; (v) D. Kampen, C. M. Reisinger and B. List, *Top. Curr. Chem.*, 2010, **291**, 395–456; (w) H. Pellissier (ed.), *Recent Developments in Asymmetric Organocatalysis*, Royal Society of Chemistry, Cambridge, 2010; (x) R. Mahrwald (ed.), *Enantioselective Organocatalysed Reactions*, Springer, Berlin, 2011; (y) F. Giacalone, M. Gruttadauria, P. Agrigento and R. Noto, *Chem. Soc. Rev.*, 2012, **41**, 2406–2447.

4. (a) R. M. de Figueiredo and M. Christmann, *Eur. J. Org. Chem.*, 2007, 2575–2600; (b) E. Marquès-Lopez, R. P. Herrera and M. Christmann, *Nat. Prod. Rep.*, 2010, **27**, 1138–1167.

5. L. F. Tietze, G. Brasche and K. Gericke, in *Domino Reactions in Organic Synthesis*, Wiley-VCH, Weinheim, 2006.

6. H. Pellissier, *Tetrahedron*, 2006, **62**, 1619–1665.

7. (a) L. F. Tietze and U. Beifuss, *Angew. Chem., Int. Ed. Engl.*, 1993, **32**, 131–163; (b) L. F. Tietze, *Chem. Ind. (London)*, 1995, **12**, 453–457; (c) L. F. Tietze, *Chem. Rev.*, 1996, **96**, 115–136; (d) L. F. Tietze and M. E. Lieb, *Curr. Opin. Chem. Biol.*, 1998, **2**, 363–371.

8. (a) L. F. Tietze and A. Modi, *Med. Res. Rev.*, 2000, **20**, 304–322; (b) J. Zhu and H. Bienaymé (eds.), *Multicomponent Reactions*, Wiley-VCH, Weinheim, 2005; (c) D. J. Ramon and M. Yus, *Angew. Chem., Int. Ed.*, 2005, **44**, 1602–1634; (d) R. V. A. Orru and E. Ruijter (eds.), Synthesis of heterocycles via multicomponent reactions, in *Topics in Heterocyclic Chemistry*, Springer, Berlin, 2010, vols. I and II.

9. T. Komnenos, *Justus Liebigs Ann. Chem.*, 1883, **218**, 145–169.

10. P. Perlmutter, in *Conjugate Addition Reactions in Organic Synthesis*, Pergamon Press, Oxford, 1992.

11. (a) D. Almasi, D. A. Alonso and C. Najera, *Tetrahedron: Asymmetry*, 2007, **18**, 299–365; (b) J. L. Vicario, D. Badia and L. Carrillo, *Synthesis*, 2007, 2065–2092; (c) S. B. Tsogoeva, *Eur. J. Org. Chem.*, 2007, 1701–1716; (d) L. F. Tietze and A. Düfert in *Catalytic Asymmetric Conjugate Reactions*, ed. A. Cordova, Wiley-VCH, Weinheim 2010, p. 321.

12. N. M. Garrido, D. Diez, S. H. Dominguez, M. Garcia, M. R. Sanchez and S. G. Davies, *Tetrahedron: Asymmetry*, 2006, **17**, 2183–2186.

13. J. R. Scheerer, J. F. Lawrence, G. C. Wang and D. A. Evans, *J. Am. Chem. Soc.*, 2007, **129**, 8968–8969.

14. H. Zhang, M. S. Reddy, S. Phoenix and P. Deslongchamps, *Angew. Chem., Int. Ed.*, 2008, **47**, 1272–1275.

15. H. Kinoshita, T. Osamura, K. Mizuno, S. Kinoshita, T. Iwamura, S.-i. Watanabe, T. Kataoka, O. Muraoka and G. Tanabe, *Chem. Eur. J.*, 2006, **12**, 3896–3897.

16. S. G. Davies, N. Mujtaba, P. M. Roberts, A. D. Smith and J. E. Thomson, *Org. Lett.*, 2009, **11**, 1959–1962.

17. U. Groth, C. Kesenheimer and P. Kreye, *Synlett*, 2006, 2223–2226.

18. M. Casey and R. McCarthy, *Synlett*, 2011, 801–804.

19. C. Aydillo, G. Jimenez-Oses, A. Avenoza, J. H. Busto, J. M. Peregrina and M. M. Zurbano, *J. Org. Chem.*, 2011, **76**, 6990–6996.

20. Y. Koseki, K. Fujino, A. Takeshita, H. Sato and T. Nagasaka, *Tetrahedron: Asymmetry*, 2007, **18**, 1533–1536.

21. H. Wu, X. Xiao and Y. Qin, *Synlett*, 2011, 907–910.

22. A. K. Hajare, L. S. Datrange, S. Vyas, D. Bhuniya and D. S. Reddy, *Tetrahedron Lett.*, 2010, **51**, 5291–5293.

23. A. K. Atta and T. Pathak, *Eur. J. Org. Chem.*, 2010, 6810–6819.

24. S. Roscales and A. G. Csaky, *Org. Lett.*, 2012, **14**, 1187–1193.

25. Q.-G. Wang, X.-M. Deng, B.-H. Zhu, L.-W. Ye, X.-L. Sun, C.-Y. Li, C.-Y. Zhu, Q. Shen and Y. Tang, *J. Am. Chem. Soc.*, 2008, **130**, 5408–5409.

26. B.-H. Zhu, R. Zhou, J.-C. Zheng, X.-M. Deng, X.-L. Sun, Q. Shen and Y. Tang, *J. Org. Chem.*, 2010, **75**, 3454–3457.

27. L.-W. Ye, S.-B. Wang, Q.-G. Wang, X.-L. Sun, Y. Tang and Y.-G. Zhou, *Chem. Commun.*, 2009, 3092–3094.

28. J. Weng, Y.-B. Li, R.-B. Wang, F.-Q. Li, C. Liu, A. S. C. Chan and G. Lu, *J. Org. Chem.*, 2010, **75**, 3125–3128.

29. H. Ishikawa, M. Honma and Y. Hayashi, *Angew. Chem., Int. Ed.*, 2011, **50**, 2824–2827.

30. J. D. Cuthbertson, A. A. Godfrey and R. J. K. Taylor, *Tetrahedron Lett.*, 2011, **52**, 2024–2027.

31. S. Fustero, J. Moscardo, M. Sanchez-Rosello, E. Rodriguez and P. Barrio, *Org. Lett.*, 2010, **12**, 5494–5497.

32. W. Yao, Y. Wu, G. Wang, Y. Zhang and C. Ma, *Angew. Chem., Int. Ed.*, 2009, **48**, 9713–9716.

33. Z. E. Wilson and M. A. Brimble, *Org. Biomol. Chem.*, 2010, **8**, 1284–1286.

34. (a) S. Takiguchi, T. Iizuka, Y.-s. Kumakura, K. Murasaki, N. Ban, K. Higuchi and T. Kawasaki, *J. Org. Chem.*, 2010, **75**, 1126–1131; (b) T. Kawasaki, M. Shinada, D. Kamimura, M. Ohzono and A. Ogawa, *Chem. Commun.*, 2006, 420–422.

35. M. K. Ghorai, S. Halder and R. K. Das, *J. Org. Chem.*, 2010, **75**, 7061–7072.

36. D. C. Beshore and A. B. Smith, *J. Am. Chem. Soc.*, 2008, **130**, 13778–13779.

37. L. F. Tietze, N. Tölle, D. Kratzert and D. Stalke, *Org. Lett.*, 2009, **11**, 5230–5233.

38. S. W. Cho and D. Romo, *Org. Lett.*, 2007, **9**, 1537–1540.

39. J. Ruble, J. Craig, A. R. Hurd, T. A. Johnson, D. A. Sherry, M. R. Barbachyn, P. L. Toogood, G. L. Bundy, D. R. Graber and G. M. Kamilar, *J. Am. Chem. Soc.*, 2009, **131**, 3991–3997.

40. P. Fischer, M. Gruner, A. Jäger, O. Kataeva and P. Metz, *Chem. Eur. J.*, 2011, **17**, 13334–13340.

41. T. G. Elfoprd and D. G. Hall, *J. Am. Chem. Soc.*, 2010, **132**, 1488–1489.

42. J. L. G. Ruano, J. Aleman, S. Catalan, V. Marcos, S. Monteagudo, A. Parra, C. del Pozo and S. Fustero, *Angew. Chem., Int. Ed.*, 2008, **47**, 7941–7944.

43. N. M. Garrido, M. Garcia, D. Diez, M. R. Sanchez, F. Sanz and J. G. Urones, *Org. Lett.*, 2008, **10**, 1687–1690.
44. N. M. Garrido, M. Garcia, M. R. Sanchez, D. Diez and J. G. Urones, *Synlett*, 2010, 387–390.
45. N. M. Garrido, M. R. Sanchez, D. Diez, F. Sanz and J. G. Urones, *Tetrahedron: Asymmetry*, 2011, **22**, 872–880.
46. N. Satoh, T. Akiba, S. Yokoshima and T. Fukuyama, *Tetrahedron*, 2009, **65**, 3239–3245.
47. S. J. Gharpure and S. R. B. Reddy, *Org. Lett.*, 2009, **11**, 2519–2522.
48. N. C. Giampietro, J. W. Kampf and J. P. Wolfe, *J. Am. Chem. Soc.*, 2009, **131**, 12556–12557.
49. N. C. Giampietro and J. P. Wolfe, *Angew. Chem., Int. Ed.*, 2010, **49**, 2922–2924.
50. P.-A. Nocquet, D. Hazelard and P. Compain, *Eur. J. Org. Chem.*, 2011, 6619–6635.
51. D. R. Ijzendoorn, P. N. M. Botman and R. H. Blaauw, *Org. Lett.*, 2006, **8**, 239–242.
52. N. R. Mente, J. D. Neighbors and D. F. Wiemer, *J. Org. Chem.*, 2008, **73**, 7963–7970.
53. K. Watanabe, J. Sakurai, H. Abe and T. Katoh, *Chem. Commun.*, 2010, **46**, 4055–4057.
54. E. Fenster, C. Fehl and J. Aube, *Org. Lett.*, 2011, **13**, 2614–2617.
55. M.-A. Beaulieu, K. C. Guérard, G. Maertens, C. Sabot and S. Canesi, *J. Org. Chem.*, 2011, **76**, 9460–9471.
56. J. Poulin, C. M. Grisé-Bard and L. Barriault, *Chem. Soc. Rev.*, 2009, **38**, 3092–3101.
57. Y. Sasaki, D. Kato and D. L. Boger, *J. Am. Chem. Soc.*, 2010, **132**, 13533–13544.
58. X. Y. Mak, A. L. Crombie and R. L. Danheiser, *J. Org. Chem.*, 2011, **76**, 1852–1873.
59. (a) E. L. O. Sauer, J. H. Hooper, T. Woo and L. Barriault, *J. Am. Chem. Soc.*, 2007, **129**, 2112–2119; (b) S. Arns and L. Barriault, *Chem. Commun.*, 2007, 2211–2221.
60. S. Dong, E. Hamel, R. Bai, D. G. Covell, J. A. Beutler and J. A. Porco, *Angew. Chem., Int. Ed.*, 2009, **48**, 1494–1497.
61. S. Dong, K. J. Cahill, M.-II. Kang, N. H. Colburn, C. J. Henrich, J. A. Wilson, J. A. Beutler, R. P. Johnson and J. A. Porco, *J. Org. Chem.*, 2011, **76**, 8944–8954.
62. M. C. Carreno, A. Somoza, M. Ribagorda and A. Urbano, *Chem. Eur. J.*, 2007, **13**, 879–890.
63. (a) M. P. Doyle and D. C. Forbes, *Chem. Rev.*, 1998, **98**, 911–935; (b) K. V. Gothelf and K. A. Jørgensen, *Chem. Rev.*, 1998, **98**, 863–909; (c) A. Padwa and M. D. Weingarten, *Chem. Rev.*, 1996, **96**, 223–269; (d) S. Karlsson and H.-E. Högberg, *Org. Prep. Proc. Int.*, 2001, **33**, 103–172; (e) I. N. N. Namboothiri and A. Hassner, *Top. Curr. Chem.*, 2001, **216**, 1–49; (f) S. Kanemasa, *Synlett*, 2002,

1371–1387; (g) K. V. Gothelf, *Synthesis*, 2002, 211–247; (h) G. Broggini, G. Molteni, A. Terraneo and G. Zecchi, *Heterocycles*, 2003, **59**, 823–858; (i) I. Coldham and R. Hufton, *Chem. Rev.*, 2005, **105**, 2765–2810; (j) H. Pellissier, *Tetrahedron*, 2007, **63**, 3235–3285; (k) S. Kanemasa, *Heterocycles*, 2010, **82**, 87–200.

64. A. Padwa, *Acc. Chem. Res.*, 1991, **24**, 22–28.

65. (a) A. Padwa and S. F. Hornbuckle, *Chem. Rev.*, 1991, **91**, 263–309; (b) A. Padwa, *Helv. Chim. Acta*, 2005, **88**, 1357–1374; (c) A. Padwa, *J. Organomet. Chem.*, 2005, **690**, 5533–5540; (d) A. Padwa, *Chem. Soc. Rev.*, 2009, **38**, 3072–3081.

66. Y. Hirata, S. Nakamura, N. Watanabe, O. Kataoka, A. M. Kurrosaki, M. Anada, S. Kitagaki, M. Shiro and S. Hashimoto, *Chem. Eur. J.*, 2006, **12**, 8898–8925.

67. S. Nakamura, Y. Sugano, F. Kikuchi and S. Hashimoto, *Angew. Chem., Int. Ed.*, 2006, **45**, 6532–6535.

68. S. Lou, P. Dai and S. E. Schaus, *J. Org. Chem.*, 2007, **72**, 9998–10008.

69. Z. Geng, B. Chen and P. Chiu, *Angew. Chem., Int. Ed.*, 2006, **45**, 6197–6201.

70. S. K. Lam and P. Chiu, *Chem. Eur. J.*, 2007, **13**, 9589–9599.

71. D. Muroni, M. Mucedda and A. Saba, *Tetrahedron: Asymmetry*, 2009, **20**, 1154–1159.

72. J. Li, J. M. Suh and E. Chin, *Org. Lett.*, 2010, **12**, 4712–4715.

73. (a) G. Poli, G. Giambastiani and A. Heumann, *Tetrahedron*, 2000, **56**, 5959–6150; (b) I. Nakamura and Y. Yamamoto, *Chem. Rev.*, 2004, **104**, 2127–2198; (c) L. F. Tietze, H. Ila and H. P. Bell, *Chem. Rev.*, 2004, **104**, 3453–3516; (d) G. Zeni and R. Larock, *Chem. Rev.*, 2004, **104**, 2285–2309; (e) J.-C. Wasilke, S. J. Obrey, R. T. Baker and G. C. Bazan, *Chem. Rev.*, 2005, **105**, 1001–1020; (f) S. Cacchi and G. Fabrizi, *Chem. Rev.*, 2005, **105**, 2873–2920.

74. A. De Meijere and S. Bräse, *J. Organomet. Chem.*, 1999, **576**, 88–110.

75. (a) R. F. Heck, *J. Am. Chem. Soc.*, 1968, **90**, 5518–5526; (b) T. Mizoroki, *Bull. Chem. Soc. Jpn.*, 1971, **44**, 581–581; (c) R. F. Heck and J. P. Nolley, *J. Org. Chem.*, 1972, **37**, 2320–2322; (d) R. F. Heck, *Org. React.*, 1982, **27**, 345–390; (e) A. de Meijere and F. E. Meyer, *Angew. Chem., Int. Ed. Engl.*, 1994, **33**, 2379–2411; (f) M. Shibasaki, C. D. J. Boden and A. Kojima, *Tetrahedron*, 1997, **53**, 7371–7393; (g) P. J. Guiry, A. Hennessy and J. Cahill, *Top. Catal.*, 1997, **4**, 311–326; (h) I. P. Beletskaya and A. V. Cheprakov, *Chem. Rev.*, 2000, **100**, 3009–3066; (i) C. Amatore and A. Jutard, *Acc. Chem. Res.*, 2000, **33**, 314–321; (j) A. B. Dounay and L. E. Overman, *Chem. Rev.*, 2003, **103**, 2945–2963; (k) M. Shibasaki, E. M. Vogl and T. Ohshima, *Adv. Synth. Catal.*, 2004, **346**, 1533–1552; (l) A. G. Coyne, M. O. Fitzpatrick and P. J. Guiry in *The Mizoroki–Heck Reaction*, ed. M. Oestreich, Wiley, Chichester, 2009; (m) D. Mc Cartney and P. J. Guiry, *Chem. Soc. Rev.*, 2011, **40**, 5122–5150.

76. (a) S. Bräse and A. de Meijere, in *Metal-Catalysed Cross Coupling Reactions*, eds. A. de Meijere and F. Diederich, Wiley-VCH, Weinheim,

2004; (b) A. B. Dounay and L. E. Overman, in *The Mizoroki – Heck Reaction*, ed. M. Oestreich, Wiley, Chichester, 2009; (c) L. F. Tietze and A. Düfert, *Pure Appl. Chem.*, 2010, **82**, 1375–1392.

77. (a) O. Reiser, M. Weber and A. De Meijere, *Angew. Chem., Int. Ed. Engl.*, 1989, **59**, 74–91; (b) K. Albrecht, O. Reiser, M. Weber and A. De Meijere, *Angew. Chem., Int. Ed. Engl.*, 1989, **28**, 1037–1038; (c) S. P. Watson, G. R. Knox and N. M. Heron, *Tetrahedron Lett.*, 1994, **35**, 9763–9766; (d) R. Grigg, P. Kennewell, A. Teasdale and V. Sridharan, *Tetrahedron Lett.*, 1993, **34**, 153–156; (e) R. Grigg and V. Sridharan, *Tetrahedron Lett.*, 1992, **33**, 7965–7968; (f) F. E. Meyer, K. H. Ang, A. G. Steinig and A. de Meijere, *Synlett*, 1994, 191–193; (g) G. D. Harris, R. J. Herr and S. M. Weinreb, *J. Org. Chem.*, 1993, **58**, 5452–5464.

78. L. F. Tietze, A. Düfert, F. Lotz, L. Sölter, K. Oum, T. Lenzer, T. Beck and R. Herbst-Immer, *J. Am. Chem. Soc.*, 2009, **131**, 17879–17884.

79. D. Giguère, P. Cloutier and R. Roy, *J. Org. Chem.*, 2009, **74**, 8480–8483.

80. D. L. Priebbenow, S. G. Stewart and F. M. Pfeffer, *Tetrahedron Lett.*, 2012, **53**, 1468–1471.

81. F. Silva, M. Reiter, R. Mills-Webb, M. Sawicki, D. Klär, N. Bensel, A. Wagner and V. Gouverneur, *J. Org. Chem.*, 2006, **71**, 8390–8394.

82. J. Boukouvalas, M. Pouliot, J. Robichaud, S. MacNeil and V. Snieckus, *Org. Lett.*, 2006, **8**, 3597–3599.

83. S. Inuki, A. Iwata, S. Oishi, N. Fujii and H. Ohno, *J. Org. Chem.*, 2011, **76**, 2072–2083.

84. P. Gogoi, R. Sigüeiro, S. Eduardo and A. Mourino, *Chem. Eur. J.*, 2010, **16**, 1432–1439.

85. L. Wang, P. Li and D. Menche, *Angew. Chem., Int. Ed.*, 2010, **49**, 9270–9273.

86. (a) S. P. Nolan and H. Clavier, *Chem. Soc. Rev.*, 2010, **39**, 3305–3101; (b) J. Cossy, S. Arseniyadis and C. Meyer (eds.), *Metathesis in Natural Product Synthesis*, Wiley-VCH, Weinheim, 2010; (c) D. Astruc, *New J. Chem.*, 2005, **29**, 42–56.

87. S. Fustero, D. Jimenez, C. Sanchez-Rosello and C. del Pozo, *J. Am. Chem. Soc.*, 2007, **129**, 6700–6701.

88. S. Fustero, S. Monteagudo, F. Sanchez-Rosello, Flores, P. Barrio and C. del Pozo, *Chem. Eur. J.*, 2010, **16**, 9835–9845.

89. S.-H. Kwon, H.-J. Lee and C.-W. Cho, *Bull. Korean Chem. Soc.*, 2011, **32**, 315–318.

90. H. Fuwa, K. Noto and M. Sasaki, *Org. Lett.*, 2010, **12**, 1636–1639.

91. H. Ark, H. Kim and J. Hong, *Org. Lett.*, 2011, **13**, 3742–3745.

92. D. A. Kummer, J. B. Brenneman and S. F. Martin, *Tetrahedron*, 2006, **62**, 11437–11449.

93. J. Li and D. Lee, *Chem. Asian J.*, 2010, **5**, 1298–1302.

94. J. A. Henderson and A. J. Phillips, *Angew. Chem., Int. Ed.*, 2008, **47**, 8499–8501.

95. A. C. Hart and A. J. Phillips, *J. Am. Chem. Soc.*, 2006, **128**, 1094–1095.

96. K. L. Jackson, J. A. Henderson, H. Motoyoshi and A. J. Phillips, *Angew. Chem., Int. Ed.*, 2009, **48**, 2346–2450.

97. S. Mondal, C. K. Malik and S. Ghosh, *Tetrahedron Lett.*, 2008, **49**, 5649–5651.

98. K. Matcha, S. Maity, C. K. Malik and S. Ghosh, *Tetrahedron Lett.*, 2010, **51**, 2754–2757.

99. M. Schubert and P. Metz, *Angew. Chem., Int. Ed.*, 2011, **50**, 2954–2956.

100. B. M. Trost, H. Yang, C. S. Brindle and G. Dong, *Chem. Eur. J.*, 2011, **17**, 9777–9788.

101. D. T. Ngoc, M. Albicker, L Schneider and N. Cramer, *Org. Biomol. Chem.*, 2010, **8**, 1781–1784.

102. (a) G. C. Lloyd-Jones, *Org. Biomol. Chem.*, 2003, **1**, 215–236; (b) A. R. Chianese, S. J. Lee and M. R. Gagné, *Angew. Chem., Int. Ed.*, 2007, **46**, 4042–4046; (c) A. Fürstner and P. W. Davies, *Angew. Chem., Int. Ed.*, 2007, **46**, 3410–3449; (d) A. S. K. Hashmi, *Chem. Rev.*, 2007, **107**, 3180–3211; (e) S. I. Lee and N. Chatani, *Chem. Commun.*, 2009, 371–384; (f) P. Y. Toullec and V. Michelet, *Top. Curr. Chem.*, 2011, **302**, 31–80.

103. Z. Chen, Y.-X. Zhang, Y.-H. Wang, L.-L. Zhu, H. Liu, X.-X. Li and L. Guo, *Org. Lett.*, 2010, **12**, 3468–3471.

104. W. Rao, D. Susanti and P. W. H. Chan, *J. Am. Chem. Soc.*, 2011, **133**, 15248–15249.

105. H. Shi, L. Fang, C. Tan, L. Shi, W. Zhang, C.-c. Li, T. Luo and Z. Yang, *J. Am. Chem. Soc.*, 2011, **133**, 14944–14947.

106. K. Lee, H. Kim and J. Hong, *Org. Lett.*, 2011, **13**, 2722–2725.

107. J. C. Green and T. R. R. Pettus, *J. Am. Chem. Soc.*, 2011, **133**, 1603–1608.

108. J. Peed, I. R. Davies, L. R. Peacock, J. E. Taylor, G. Kociok-Köhn and S. D. Bull, *J. Org. Chem.*, 2012, **77**, 543–555.

109. S. Chandrasekhar, B. Saritha, V. Jagadeshwar and S. J. Prakash, Tetrahedron, *Asymmetry*, 2006, **17**, 1380–1386.

110. H. Fujioka, S. Matsuda, M. Horai, E. Fujii, M. Morishita, N. Nishiguchi, K. Hata and Y. Kita, *Chem. Eur. J.*, 2007, **13**, 5238–5243.

111. A. Chanu, I. Safir, R. Basak, A. Chiaroni and S. Arseniyadis, *Org. Lett.*, 2007, **9**, 1351–1354.

112. A. Corbu, G. Gauron, J. M. Castro, M. Dakir and S. Arseniyadis, Tetrahedron, *Asymmetry*, 2008, **19**, 1730–1743.

113. J. A. Gonzalez-Vera, T. Garcia-Lopez and R. Herranz, *J. Org. Chem.*, 2007, **72**, 5395–5398.

114. R. Hayashi, J. B. Feltenberger and R. P. Hsung, *Org. Lett.*, 2010, **12**, 1152–1155.

115. K. C. Nicolaou, D. Sarlah, T. R. Wu and W. Zhan, *J. Org. Chem.*, 2009, **48**, 6870–6874.

116. M. K. Ghorai and D. P. Tiwari, *J. Org. Chem.*, 2010, **75**, 6173–6181.

117. J. D. White, C. M. Lincoln, J. Yang, W. H. C. Martin and D. B. Chan, *J. Org. Chem.*, 2008, **73**, 4139–4150.

118. M. Schelwies, A. Farwick, F. Rominger and G. Helmchen, *J. Org. Chem.*, 2010, **75**, 7917–7919.

119. H. T. Cao, T. Roisnel and R. Grée, *Eur. J. Org. Chem.*, 2011, 6405–6408.

120. S.-i. Ikeda, *Acc. Chem. Res.*, 2000, **33**, 511–519.

121. J. Petrignet, T. Roisnel and R. Grée, *Tetrahedron Lett.*, 2006, **47**, 7745–7748.

122. A. Mames, S. Stecko, P. Mikolajczyk, M. Soluch, B. Furmann and M. Chmielewski, *J. Org. Chem.*, 2010, **75**, 7580–7587.

123. Y. Liu, W. Xu and X. Wang, *Org. Lett.*, 2010, **12**, 1448–1451.

124. (a) D. E. Fogg and E. N. dos Santos, *Coord. Chem. Rev.*, 2004, **248**, 2365–2379; (b) C. J. Chapman and C. G. Frost, *Synthesis*, 2007, 1–21.

125. (a) N. Hall, *Science*, 1994, **266**, 32–34; (b) P. A. Wender, S. T. Handy and D. L. Wright, *Chem. Ind., (London)*, 1997, 765–769.

126. (a) I. Ugi, R. Meyr, U. Fetzer and C. Steinbrückner, *Angew. Chem.*, 1959, **71**, 386–388; (b) R. V. A. Orru and M. de Greef, *Synthesis*, 2003, 1471–1499.

127. S. G. Davies, P. M. Roberts and A. D. Smith, *Org. Biomol. Chem.*, 2007, **5**, 1405–1415.

128. M. Koutsaplis, P. C. Andrews, S. D. Bull, P. J. Duggan, B. H. Fraser and P. Jensen, *Chem. Commun.*, 2007, 3580–3582.

129. Y. Coquerel, M.-H. Filippini, D. Bensa and J. Rodriguez, *Chem. Eur. J.*, 2008, **14**, 3078–3092.

130. Y. Chen, C. Zhong, J. L. Petersen, N. G. Akhmedov and X. Shi, *Org. Lett.*, 2009, **11**, 2333–2336.

131. M. Ozeki, S. Ochi, N. Hayama, S. Hosoi, T. Kajimoto and M. Node, *J. Org. Chem.*, 2010, **75**, 4201–4211.

132. A. Kamimura, H. Okawa, Y. Morisaki, S. Ishikawa and H. Uno, *J. Org. Chem.*, 2007, **72**, 3569–3572.

133. R. Noël, M.-C. Fareau-Bellassoued, C. Vanucci-Bacqué and G. Lhommet, *Synthesis*, 2008, 1948–1954.

134. J. C. Gonzalez-Gomez, F. Foubelo and M. Yus, *Tetrahedron Lett.*, 2008, **49**, 2343–2361.

135. Z. Xu, T. Buechler, K. Wheeler and H. Wang, *Chem. Eur., J.*, 2010, **16**, 2972–2980.

136. (a) G. Schneiber and U. Fechner, *Nat. Rev. Discovery*, 2005, **4**, 649–663; (b) S. B. Mhaske and N. P. Argade, *Tetrahedron*, 2006, **62**, 9787–9826; (c) C. C. Hughes and D. Trauner, *Angew. Chem., Int. Ed.*, 2002, **41**, 4556–4559.

137. (a) A. Hantzsch, *Justus Liebigs Ann. Chem.*, 1882, **215**, 1–82; (b) G. Jones, in *Comprehensive Heterocyclic Chemistry II*, ed. A. R. Katritzky, C. W. Rees and E. F. V. Scriven, Pergamon, Oxford, 1996; vol. V, p. 167.

138. A. Dondoni and A. Massi, *Acc. Chem. Res.*, 2006, **39**, 451–463.

139. A. Dondoni, A. Massi and M. Aldhoun, *J. Org. Chem.*, 2007, **72**, 7677–7687.

140. D. R. B. Ducatti, A. Massi, M. D. Noseda, M. E. R. Duarte and A. Dondoni, *Org. Biomol. Chem.*, 2009, **7**, 1980–1986.

141. A. Dömling and I. Ugi, *Angew. Chem., Int. Ed.*, 2000, **39**, 3168–3210.
142. (a) I. Ugi, *Angew. Chem., Int. Ed. Engl.*, 1982, **21**, 810–819; (b) I. Ugi, *J. Prakt. Chem.*, 1997, **339**, 499–516.
143. F. Sigmuller, R. Herrmann and I. Ugi, *Tetrahedron*, 1986, **42**, 5931–5941.
144. (a) H. Kunz and W. Pfrengle, *Tetrahedron*, 1988, **44**, 5487–5494; (b) H. Kunz, W. Pfrengle and W. Sager, *Tetrahedron Lett.*, 1989, **30**, 4109–4110; (c) H. Kunz and W. Pfrengle, *J. Am. Chem. Soc.*, 1988, **110**, 651–652; (d) H. Kunz, W. Pfrengle, K. Rück and W. Sager, *Synthesis*, 1991, 1039–1042.
145. (a) M. Goebel and I. Ugi, *Synthesis*, 1991, 1095–1098; (b) S. Lehnhoff, M. Goebel, R. M. Karl, R. Klösel and I. Ugi, *Angew. Chem., Int. Ed. Engl.*, 1995, **34**, 1104–1107.
146. D. Kadzimirsz, D. Hildebrandt, K. Merz and G. Dyker, *Chem. Commun.*, 2006, 661–662.
147. A. Basso, L. Banfi, R. Riva and G. Guanti, *Tetrahedron*, 2006, **62**, 8830–8837.
148. A. Basso, L. Banfi, G. Guanti and R. Riva, *Tetrahedron*, 2010, **66**, 2390–2397.
149. S. Di Micco, R. Vitale, M. Pellecchia, M. F. Rega, R. Riva, A. Basso and G. Bifulco, *J. Med. Chem.*, 2009, **52**, 7856–7867.
150. (a) I. Kanizsai, Z. Szakonyi, R. Sillanpää and F. Fülöp, *Tetrahedron Lett.*, 2006, **47**, 9113–9116; (b) I. Kanizsai, S. Gyonfalvi, Z. Szakonyi, R. Sillanpää and F. Fülöp, *Green Chem.*, 2007, **9**, 357–360.
151. A. V. Gulevich, E. S. Balenkova and V. G. Nenajdenko, *J. Org. Chem.*, 2007, **72**, 7878–7885.
152. V. G. Nenajdenko, A. L. Reznichenko and E. S. Balenkova, *Tetrahedron*, 2007, **63**, 3031–3041.
153. L. Banfi, A. Basso, G. Guanti, S. Merlo, C. Repetto and R. Riva, *Tetrahedron*, 2008, **64**, 1114–1134.
154. (a) K. M. Bonger, T. Wennekes, D. V. Filippov, G. Lodder, G. A. van der Marel and H. S. Overkleeft, *Eur. J. Org. Chem.*, 2008, 3678–3688; (b) K. M. Bonger, T. Wennekes, S. V. P. de Lavoir, D. Esposito, R. J. B. H. N. van der Berg, R. E. J. N. Litjens, G. A. van der Marel and H. S. Overkleeft, *QSAR Com. Sci.*, 2006, **5-6**, 491–503.
155. A. Basso, L. Banfi, G. Guanti and R. Riva, *Org. Biomol. Chem.*, 2009, **7**, 253–258.
156. A. Trabocchi, D. Scarpi and A. Guarna, *Amino Acids*, 2008, **34**, 1–24.
157. A. M. Belostotskii, H. E. Gottlieb and M. Shokhen, *J. Org. Chem.*, 2002, **67**, 9257–9266.
158. J. M. Lehn, *Fortschr. Chem. Forsch.*, 1970, **15**, 311–377.
159. A. Znabet, E. Ruijter, F. J. J. De Kanter, V. Köhler, M. Helliwell, N. J. Turner and R. V. A. Orru, *Angew. Chem., Int. Ed.*, 2010, **49**, 5289–5292.
160. A. Znabet, J. Zonneveld, E. Janssen, F. J. J. De Kanter, M. Helliwell, N. J. Turner, E. Ruijter and R. V. A. Orru, *Chem. Commun.*, 2010, **46**, 7706–7708.

161. B. Nicholson, G. K. Lloyd, B. R. Miller, M. A. Palladino, Y. Kiso, Y. Hayashi and S. T. C. Neuteboom, *Anti-Cancer Drugs*, 2006, **17**, 25–31.
162. V. Köhler, K. R. Bailey, A. Znabet, J. Raftery, M. Helliwell and N. J. Turner, *Angew. Chem., Int. Ed.*, 2010, **49**, 2182–2184.
163. A. Znabet, M. M. Polak, E. Janssen, F. J. J. de Kanter, N. J. Turner, R. V. A. Orru and E. Ruijter, *Chem. Commun.*, 2010, **46**, 7918–7920.
164. W. Wang, S. Joyner, K. A. S. Khoury and A. Dömling, *Org. Biomol. Chem.*, 2010, **8**, 529–532.
165. G. M. Chinigo, A. Breder and E. M. Carreira, *Org. Lett.*, 2011, **13**, 78–81.
166. A. Strecker, *Justus Liebigs Ann. Chem*, 1850, **75**, 25–51.
167. (a) J. Wang, X. Liu and X. Feng, *Chem. Rev.*, 2011, **111**, 6947–6983; (b) P. Merino, E. Marques-Lopez, T. Tejero and R. P. Herrera, *Tetrahedron*, 2009, **65**, 1219–1234; (c) J. Gawronski, N. Wascinska and J. Gajewy, *Chem. Rev.*, 2008, **108**, 5227–5252; (d) S. J. Connon, *Angew. Chem., Int. Ed.*, 2008, **47**, 1176–1178; (e) H. Gröger, *Chem. Rev.*, 2003, **103**, 2795–2827.
168. Y. Pérez-Fuertes, J. E. Taylor, D. A. Tickell, M. F. Mahon, S. D. Bull and T. D. James, *J. Org. Chem.*, 2011, **76**, 6038–6047.
169. C. Mannich and W. Krosche, *Arch. Pharm.*, 1912, **250**, 647–667.
170. (a) M. Arend, B. Westermann and N. Risch, *Angew. Chem., Int. Ed.*, 1998, **37**, 1044–1070; (b) A. Cordova, *Acc. Chem. Res.*, 2004, **37**, 102–112; (c) J. M. M. Verkade, L. J. C. Van Hemert, P. J. L. M. Quaedflieg and F. P. J. T. Rutjes, *Chem. Soc. Rev.*, 2008, **37**, 29–41.
171. R. G. Arrayas and J. C. Carretero, *Chem. Soc. Rev.*, 2009, **38**, 1940–1948.
172. C.-H. Wong (ed.), *Carbohydrate-based Drug Discovery*, Wiley-VCH, Weinheim, 2003.
173. L. Cappannini, C. Cimarelli, S. Giuli, G. Palmieri and M. Petrini, *Tetrahedron: Asymmetry*, 2007, **18**, 1022–1029.
174. Y. Yang, D. P. Phillips and S. Pan, *Tetrahedron Lett.*, 2011, **52**, 1549–1552.
175. (a) L. F. Tietze, C. C. Brazel, S. Hölsken, J. Magull and A. Ringe, *Angew. Chem., Int. Ed.*, 2008, **47**, 5246–5249; (b) L. F. Tietze, T. Kinzel and C. C. Brazel, *Acc. Chem. Res.*, 2009, **42**, 367–378; (c) L. F. Tietze, T. Kinzel and T. Wolfram, *Chem. Eur. J.*, 2009, **15**, 6199–6210; (d) L. F. Tietze, T. Kinzel and S. Schmatz, *Chem. Eur. J.*, 2009, **15**, 1706–1712.
176. L. F. Tietze, T. Kinzel and S. Schmatz, *J. Am. Chem. Soc.*, 2008, **130**, 4386–4395.
177. L. F. Tietze, S. Biller and T. Wolfram, *Synlett*, 2010, 2130–2132.
178. J. Pospisil and I. E. Marko, *J. Org. Chem.*, 2007, **129**, 3516–3517.
179. M. Passerini, *Gazz. Chim. Ital.*, 1921, **51**, 126–129.
180. T. Yue, D. Wang, G. Masson and J. Zhu, *Angew. Chem., Int. Ed.*, 2009, **48**, 6717–6721.
181. S. Maeda, S. Komagwa, M. Uchiyama and K. Morokuma, *Angew. Chem., Int. Ed.*, 2011, **50**, 644–649.

182. (a) H. Bock and I. Ugi, *J. Prakt. Chem.*, 1997, **339**, 385–389; (b) D. Gryko, J. Chalko and J. Jurczak, *Chirality*, 2003, **15**, 514–541; (c) R. Frey, S. G. Galbraith, S. Guelfi, C. Lamberth and M. Zeller, *Synlett*, 2003, 1536–1538.

183. P. R. Krishna, G. Dayaker and P. V. N. Reddy, *Tetrahedron Lett.*, 2006, **47**, 5977–5980.

184. P. R. Krishna and K. Lopinti, *Synlett*, 2007, 83–86.

185. P. Biginelli, *Gazz. Chim. Ital.*, 1893, **23**, 360–416.

186. (a) C. O. Kappe, *Acc. Chem. Res.*, 2000, **33**, 879–888; (b) C. O. Kappe, *QSAR Comb. Sci.*, 2003, **22**, 630–645; (c) Z. D. Aron and L. E. Overman, *Chem. Commun.*, **2004**, 253–265.

187. (a) F. J. Lopez Aparacio, J. A. Lopez Sastre, J. Molina Molina and F. J. Lopez Herrera, *An. Quim.*, 1981, **77**, 147–149; (b) A. Dondoni, A. Massi and S. Sabbatini, *Tetrahedron Lett.*, 2001, **42**, 4495–4497; (c) C. Y. Hong and Y. Kishi, *J. Am. Chem. Soc.*, 1992, **114**, 7001–7006.

188. L. D. S. Yadav, C. Awasthi, V. K. Rai and A. Rai, *Tetrahedron Lett.*, 2007, **48**, 4899–4902.

189. N. A. Petasis and I. Akritopoulou, *Tetrahedron Lett.*, 1993, **34**, 583–586.

190. N. R. Candeias, P. M. Montalbano, S. D. Cal and P. M. P. Gois, *Chem. Rev.*, 2010, **110**, 6169–6193.

191. N. A. Petasis and I. A. Zavialov, *J. Am. Chem. Soc.*, 1997, **119**, 445–446.

192. T. J. Southwood, M. C. Curry and C. A. Hutton, *Tetrahedron*, 2006, **62**, 236–242.

193. N. Kumagai, G. Muncipinto and S. L. Schreiber, *Angew. Chem., Int. Ed.*, 2006, **45**, 3635–3638.

194. M. V. Shevehuk, A. E. Sorochinsky, V. P. Khilya and V. D. Romanenko, *Synlett*, 2010, 73–76.

195. T. Ritthiwigrom and S. G. Pyne, *Org. Lett.*, 2008, **10**, 2769–2771.

196. T. Ritthiwigrom, A. C. Willis and S. G. Pyne, *J. Org. Chem.*, 2010, **75**, 815–824.

197. D. Garcia-Cuadrado, S. Barluenga and N. Winssinger, *Chem. Commun.*, 2008, 4619–4621.

198. A. H. L. Liu, M. Sugiyama, Y. Fu and C.-H. Wong, *J. Am. Chem. Soc.*, 2009, **131**, 8352–8353.

199. L. J. Whalen and C.-H. Wong, *Aldrichimica Acta*, 2006, **39**, 63–71.

200. (a) B. Winchester and G. W. J. Fleet, *Glycobiology*, 1992, **2**, 199–210; (b) P. Sears and C.-H. Wong, *Angew. Chem., Int. Ed.*, 1999, **38**, 2301–2324; (c) H. Ouchi, Y. Mihara and H. Takahata, *J. Org. Chem.*, 2005, **70**, 5207–5214.

201. S. Torssell and P. Somfai, *Adv. Synth. Catal.*, 2006, **348**, 2421–2430.

202. (a) S. Kobayashi, H. Ishitani and M. Ueno, *J. Am. Chem. Soc.*, 1998, **2**, 431–432; (b) J. Kobayashi, M. Nakamura, Y. Mori, Y. Yamashita and S. Kobayashi, *J. Am. Chem. Soc.*, 2004, **30**, 9192–9193.

203. D. J. Ager and D. R. Prakash, *Chem. Rev.*, 1996, **96**, 835–876.

204. H. Huang, X. Guo and W. Hu, *Angew. Chem., Int. Ed.*, 2007, **46**, 1337–1339.

205. P. Dübon, A. Farwick and G. Helmchen, *Synlett*, 2009, 1413–1416.
206. P. Garner, H. U. Kaniskan, J. Hu, W. J. Youngs and M. Panzner, *Org. Lett.*, 2006, **8**, 3647–3650.
207. P. Garner and H. U. Kaniskan, *Curr. Org. Synth.*, 2010, **7**, 348–357.
208. P. Garner, H. U. Kaniskan, C. M. Keyari and L. Weerasinghe, *J. Org. Chem.*, 2011, **76**, 5283–5294.
209. P. Garner, L. Weerasinghe, W. J. Youngs, B. Wright, D. Wilson and D. Jacobs, *Org. Lett.*, 2012, **14**, 1326–1329.
210. (a) T. Kobayashi, M. Nakashima, T. Hakogi, K. Tanaka and S. Katsumara, *Org. Lett.*, 2006, **8**, 3809–3812; (b) T. Kobayashi, K. Takeuchi, J. Miwa, H. Tsuchikawa and S. Katsumara, *Chem. Commun.*, 2009, 3363–3365.
211. T. Kobayashi, F. Hasegawa, K. Tanaka and S. Katsumara, *Org. Lett.*, 2006, **8**, 3813–3816.
212. Y. Li, T. Kobayashi and S. Katsumara, *Tetrahedron Lett.*, 2009, **50**, 4482–4484.
213. T. Sakaguchi, S. Kobayashi and S. Katsumura, *Org. Biomol. Chem.*, 2011, **9**, 257–264.
214. K. Sa-ei and J. Montgomery, *Org. Lett.*, 2006, **8**, 4441–4443.
215. N. Saito, T. Katayama and Y. Sato, *Heterocycles*, 2011, **82**, 1181–1187.
216. A. K. Ghosh, S. Kulkarni, C.-X. Xu and P. E. Fanwick, *Org. Lett.*, 2006, **8**, 4509–4511.
217. I. S. Young and M. A. Kerr, *J. Am. Chem. Soc.*, 2007, **129**, 1465–1469.
218. (a) V. K.-Y. Lo, Y. Liu, M.-K. Wong and C.-M. Che, *Org. Lett.*, 2006, **8**, 1529–1532; (b) V. K.-Y. Lo, K. K.-Y. Kung, M.-K. Wong and C.-M. Che, *J. Organomet. Chem.*, 2009, **694**, 583–591.
219. H. Fujioka, K. Murai, Y. Ohba, H. Hirose and Y. Kita, *Chem. Commun.*, 2006, 832–836.
220. M. Zöllinger, P. Mayer and T. Lindel, *Synlett*, 2007, 2756–2758.
221. (a) S. Hübner, D. Michalik, H. Jiao, H. Neumann, S. Klaus, D. Strübing, A. Spannenberg and M. Beller, *Chem. Asian J.*, 2007, **2**, 734–746; (b) S. Hübner, H. Jiao, D. Michalik, H. Neumann, S. Klaus, D. Strübing, A. Spanneberg and M. Beller, *Chem. Asian J.*, 2007, **2**, 720–733.
222. R. B. Lettan, C. C. Woodward and K. A. Scheidt, *Angew. Chem., Int. Ed.*, 2008, **47**, 2294–2297.
223. E. Sotoca, T. Constantieux and R. Rodriguez, *Synlett*, 2008, 1313–1316.
224. E. Sotoca, C. Allais, T. Constantieux and R. Rodriguez, *Org. Biomol. Chem.*, 2009, **7**, 1911–1920.
225. J. Barluenga, M. G. Suero, R. De la Campa and J. Florez, *Angew. Chem., Int. Ed.*, 2010, **49**, 9720–9724.
226. M. Yao and C.-D. Lu, *Org. Lett.*, 2011, **13**, 2782–2785.
227. S. Fustero, S. Catalan, M. Sanchez-Rosello, A. Simon-Fuentes and C. del Pozo, *Org. Lett.*, 2010, **12**, 3484–3487.
228. P. Renzi, P. Overgaard and M. Bella, *Org. Biomol. Chem.*, 2010, **8**, 980–983.

Asymmetric Domino Reactions Based on the Use of Chiral Metal Catalysts

2.1 Introduction

During the second half of the 20th century, transition metals have come to play a major role in synthetic organic chemistry and this has led to the development of a large number of transition metal-catalysed reactions for creating organic molecules. Transition metals have a unique ability to activate various organic compounds and through this activation they can catalyse the formation of new bonds.[1] In particular, asymmetric transition metal catalysis has emerged over the past few decades as a powerful tool to perform reactions in a highly enantioselective fashion. The economic interest in combinations of asymmetric metal catalytic processes with the concept of domino and multicomponent reactions is obvious, and it has allowed easy attainment of high molecular complexity with often excellent levels of stereocontrol in simple one-pot operational procedures. The reactions have the advantages of savings in solvent, time, energy, and costs by avoiding costly protecting groups and time-consuming purification procedures after each step. During the last six years, an explosive increase has occurred in the number of novel powerful asymmetric domino and multicomponent processes being developed on the basis of enantioselective metal catalysis, which is fast becoming fundamental in many manufacturing processes. This chapter illustrates how much asymmetric organometallic catalysis has contributed to the development of enantioselective domino and multicomponent reactions. Its goal is to cover the recent efforts of the chemical community in the development of novel enantioselective domino reactions catalysed by chiral metals, published since the beginning of 2006,

RSC Catalysis Series No. 10
Asymmetric Domino Reactions
By Hélène Pellissier
© The Royal Society of Chemistry 2013
Published by the Royal Society of Chemistry, www.rsc.org

since this field was most recently reviewed in 2006 by Tietze in a book,[2] and by this author in two reviews.[3] The chapter is divided into two parts, which deal with one- and two-component domino reactions, and multicomponent reactions, respectively. The first part is subdivided into 12 sections, dealing successively with domino reactions initiated by the Michael reaction, domino reactions initiated by an aldol reaction, domino reductive aldol reactions, domino reactions initiated by a [2 + 2 + 2] cycloaddition, domino reactions initiated by an allylic alkylation, domino Passerini-type reactions, domino carbonyl ylide-formation–1,3-dipolar cycloaddition reactions, domino reactions initiated by the Heck reaction, domino reactions initiated by the Wacker reaction, domino reactions based on cyclisations, domino radical reactions, and finally miscellaneous domino reactions which cannot be included in the other sections, because of their different mechanisms. The second part of the chapter, which concerns the multicomponent reactions, is subdivided into 10 sections, dealing respectively with multicomponent reactions initiated by the Michael reaction, multicomponent reactions based on the Mannich reaction, multicomponent reactions initiated by a pericyclic reaction, multicomponent reactions based on the Passerini reaction, multicomponent reactions initiated by the Friedel–Crafts reaction, multicomponent reactions of alkynes, aldehydes and amines, multicomponent reactions of 1,3-dienes, aldehydes and reducing agents, multicomponent reductive amination reactions of ketones, multicomponent Kabachnik–Fields reactions, and finally miscellaneous multicomponent reactions. It must be noted that only reactions catalysed by chiral metals are considered in this chapter.

2.2 One- and Two-Component Domino Reactions

2.2.1 Domino Reactions Initiated by the Michael Reaction

The nucleophilic 1,4-addition of stabilised carbon nucleophiles to electron-poor olefins, generally α,β-unsaturated carbonyl compounds, is known as the Michael addition, although it was first reported by Komnenos in 1883.[4] Michael-type reactions can be considered as one of the most powerful and reliable tools for the stereocontrolled formation of carbon–carbon and carbon–heteroatom bonds,[5] as has been demonstrated by the huge number of examples in which they have been applied as a key strategic transformation in total synthesis. Since the first catalytic domino Michael–aldol reaction reported by Noyori *et al.* in 1996,[6] there have been numerous examples of domino reactions using this methodology. Among them, a number of enantioselective domino reactions initiated by a Michael addition have been promoted by chiral palladium catalysts. It is well known that one of the most heavily studied families of transition metal catalysts comprises those based on palladium. Over the past 40 years, many new reactions have been uncovered, and activity in this area has remained high. Among the increasing number of enantioselective domino processes starting with a transition metal-catalysed reaction, the enantioselective palladium-catalysed domino transformations have seen an astounding development over the past few years.[7] Indeed, given that palladium has the advantage of being compatible with many functional groups, it constitutes an ideal catalyst for domino reactions.[1d,8] Among the recent examples of palladium-catalysed enantioselective domino reactions initiated by a Michael addition was that reported by Nishikata, which consisted in an asymmetric conjugate addition of arylboronic acids to enones, bearing a 2-acylaryl group at the β-carbon, followed by an aldol condensation, which finally afforded the corresponding chiral 1-aryl-1*H*-indenes.[9] As shown in Scheme 2.1, the domino Michael–aldol reaction was performed in aqueous isopropanol in the presence of a cationic Pd(II)–Chiraphos catalyst, providing *via* sequential carbopalladation of C–C and C–O– double bonds a series of chiral indenes in high yields of up to 99% and excellent enantioselectivities of up to 97% ee in almost all the cases of substrates tested. It must be noted that there was no clear propensity shown between the enantioselectivities and functionalities on the aromatic boronic acids or their positions, but the enantioselectivities generally improved with increasing substituent bulk. Furthermore, the *para*-substituted arylboronic acids (90–97% ee) resulted in higher enantioselectivities than those of *meta*-substituted acids (91–94% ee), whereas the presence of an *ortho*-substituent retarded the reaction. As for the substrates, aliphatic and aromatic ketones and aldehydes could be used as

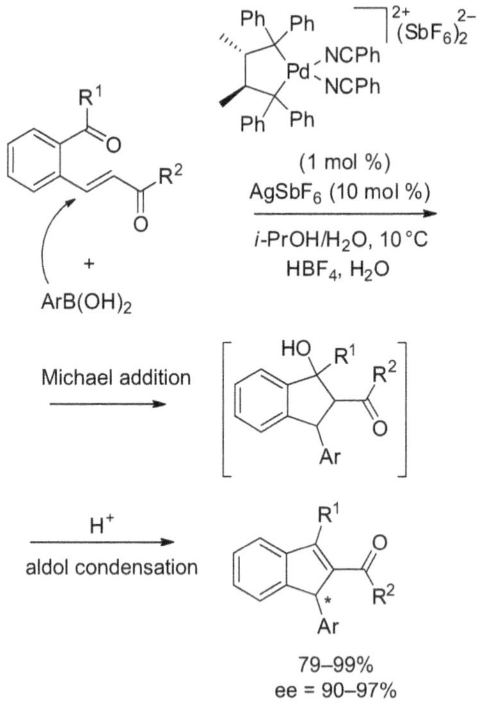

Scheme 2.1   Pd-catalysed domino Michael–aldol reaction of β-(2-acylaryl)enones with arylboronic acids.

trapping functionalities for cyclisation with the palladium(II) enolates generated from enones and arylboronic acids.

Another cationic palladium complex, prepared *in situ* from $Pd(MeCN)_4(BF_4)_2$ and axially chiral biphosphine ligands, such as (*R*)-Tol-BINAP [(*R*)-tolyl-2,2'-bis(diphenylphosphino)-1,1'-binaphthyl], was used as a catalyst by Lu and Yu to promote the enantioselective annulation of 2-formylarylboronic acids with allenoates.[10] In these conditions, the corresponding *cis*-indenol derivatives were diastereoselectively generated in generally excellent yields (92–100%) and good enantioselectivities (77–84% ee). It must be noted that, however, the scope of this process was limited to electron-deficient allenes. Moreover, the reaction was totally inhibited when the allenoate bore a methyl group at the α-position. In the same context, these authors have developed a cationic palladium-catalysed enantioselective domino annulation of *ortho*-boronate-substituted cinnamic ketones with alkynes, affording the corresponding chiral indenes in excellent yields and high enantioselectivities of up to 93% ee.[11] The cationic palladium species was prepared *in situ* by adding the chiral ligand directly to the $Pd(OTf)_2 \cdot 2H_2O$ without isolation of palladium complexes. The authors have proposed a plausible mechanism for the domino reaction, which is depicted in Scheme 2.2. First, the monohydroxo cationic palladium species **1** with vacant sites could be formed from the *in situ* reaction of $Pd(OTf)_2 \cdot 2H_2O$ and the chiral ligand. Intermediate **1** was believed to be the active catalyst and enabled smooth transmetalation with the substrate to yield a cationic arylpalladium(II) species **2**. Next, π-coordination of the carbon–carbon triple bond to the cationic palladium centre occurred to form **3**, followed by the regioselective insertion of the alkyne into the carbon–palladium bond, affording the vinylpalladium intermediate **4** (carbopalladation). Subsequently, conjugate addition of this vinylpalladium species to the cinnamic ketone occurred to yield the final product. It was also proposed that the coordinated intermediate **4** was helpful to the enantioface discrimination of α,β-unsaturated ketone, resulting in high enantioselectivities. It must be noted that this domino process has provided an efficient method for the construction of optically active, multiply substituted indene derivatives.

Moreover, the same authors have recently reported the synthesis of chiral aminoindene derivatives on the basis of a diastereo- and enantioselective domino annulation reaction of (*S*)-2-(*N-tert*-butanesulfinylimino)arylboronic acids as chiral auxiliaries with a variety of alkynes.[12] The reaction was catalysed by a chiral cationic palladium complex, $[Pd[(S,S)\text{-bdpp}](H_2O)_2]]^{2+}$ $(BF_4^-)_2$ [where bdpp is 2,4-bis(diphenylphosphino)pentane], affording through double asymmetric induction the corresponding sulfoxides as mixtures of two diastereomers. These sulfoxides were further converted through oxidation by treatment with MCPBA into the corresponding more stable sulfones with moderate to high yields and excellent enantioselectivities of >99% ee, as shown in Scheme 2.3. It must be noted that the double asymmetric induction was necessary to obtain high enantioselectivities. Indeed, the authors have performed the reaction of a corresponding racemic 2-(*N-tert*-butanesulfinylimino)arylboronic acid and an alkyne under comparable conditions and observed a very low

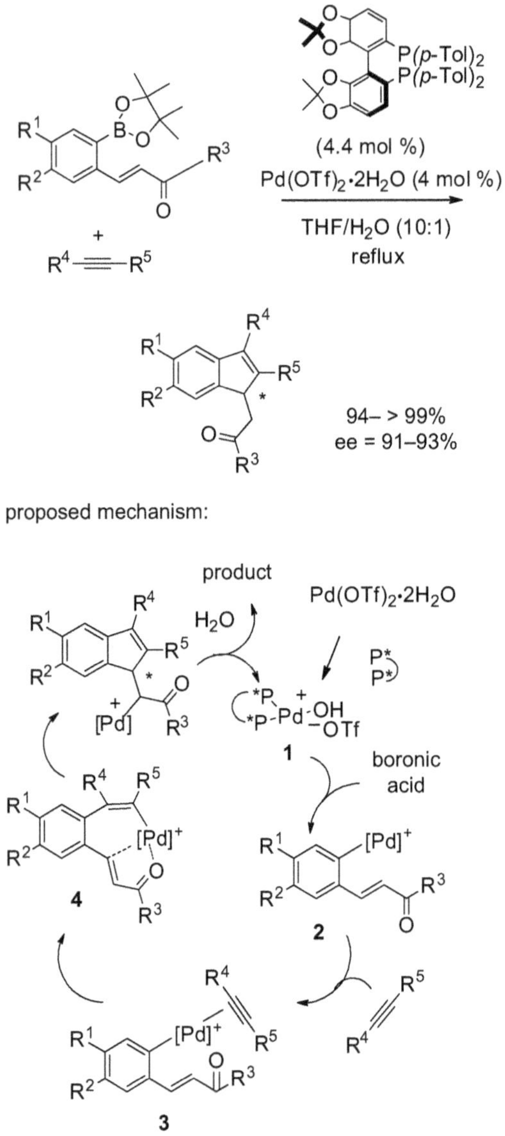

Scheme 2.2 Pd-catalysed domino annulation of *ortho*-boronate-substituted cinnamic
ketones with alkynes.

diastereoselectivity (27% de) for the formation of the corresponding domino
product. Furthermore, the authors have shown that the diastereoselectivity of the
reaction was highly dependent on the configuration of the substrate. Thus, under
comparable conditions, the reaction of a (*R*)-2-(*N*-*tert*-butanesulfinylimino)-
arylboronic acid with an alkyne was not diastereoselective. According to these
results, it was presumed that the chiral catalyst interfered with the chiral

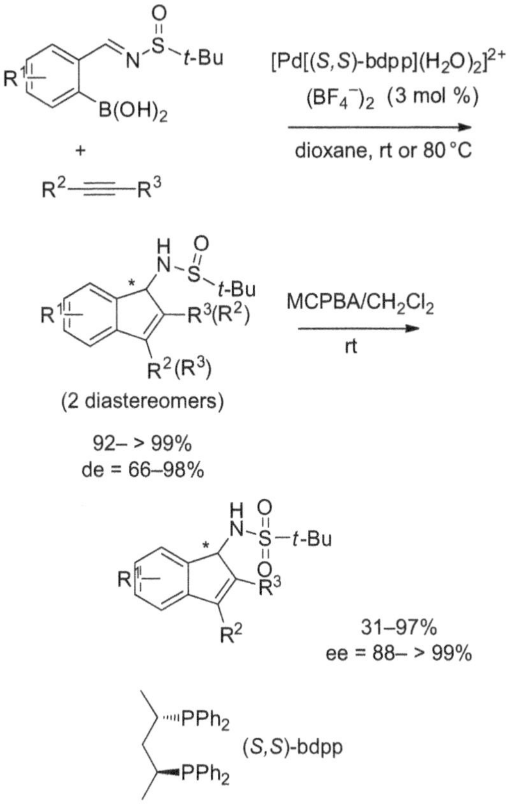

Scheme 2.3 Pd-catalysed domino annulation of (S)-2-(N-tert-butanesulfinylimino)-arylboronic acids with alkynes.

substrates, resulting in different stereocontrol; (S)-2-(N-tert-butanesulfinylimino)-arylboronic acid and the cationic palladium complex constituted a matched pair, leading to the enhancement of diastereoselectivity. In addition, a moderate enantioselectivity of 50% ee was reported by the same authors in the domino annulation of racemic 2-aroylmethoxyarylboronic acids with alkynes to give the corresponding 1-benzoxepines upon catalysis with either [Pd[(S,S)-bdpp]-(H$_2$O)$_2$]]$^{2+}$ (OTf$^-$)$_2$ or [Pd(R)-BINAP(H$_2$O)$_2$]$^{2+}$ · (OTf$^-$)$_2$].[13]

Copper-catalysed domino reactions have been used for the synthesis of carbocycles, as well as for heterocycles such as indoles, benzoxazoles, and quinoxalines. Among these reactions are enantioselective copper-catalysed domino reactions initiated by conjugate additions. A recent example was reported by Alexakis and Li, involving the Cu-catalysed enantioselective conjugate addition of dialkylzinc to bis-α,β-unsaturated carbonyl compounds, followed by the intramolecular trapping of the intermediate zinc enolate through a second intramolecular conjugate addition in the presence of chiral phosphoramidite ligands.[14] This domino Michael–Michael process afforded the corresponding chiral cyclic and heterocyclic products having three stereocentres as mixtures of two diastereomers with moderate to high diastereoselectivities of up to >98% de, and enantioselectivities of up to 94% ee, as shown in Scheme 2.4. The stereochemistry was determined to be *trans,trans* for the major products and *trans,cis* for the minor products.

Another enantioselective domino reaction initiated by a Michael addition was reported by Feringa *et al.*, in 2010.[15] In this work, the authors developed Cu-catalysed conjugate addition of Grignard reagents to 4-chloro-α,β-unsaturated esters, thioesters, and ketones, which was followed by enolate trapping through intramolecular alkylation reaction to provide the corresponding *trans*-1-alkyl-2-substituted cyclopropane esters, thioesters, and ketones, respectively, in good to high yields and enantioselectivities of up to

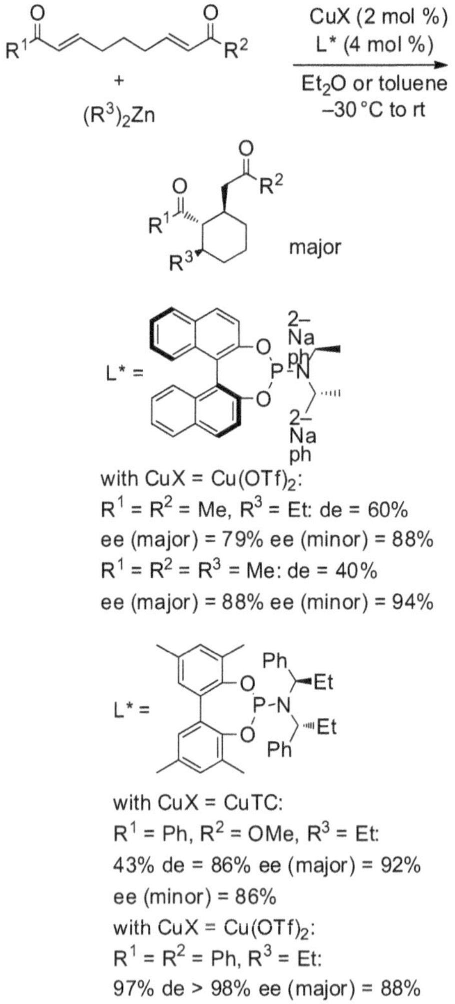

CuX (2 mol %)
L* (4 mol %)
Et2O or toluene
–30 °C to rt
(R3)2Zn
major
L* =
with CuX = Cu(OTf)2:
R1 = R2 = Me, R3 = Et: de = 60%
ee (major) = 79% ee (minor) = 88%
R1 = R2 = R3 = Me: de = 40%
ee (major) = 88% ee (minor) = 94%
L* =
with CuX = CuTC:
R1 = Ph, R2 = OMe, R3 = Et:
43% de = 86% ee (major) = 92%
ee (minor) = 86%
with CuX = Cu(OTf)2:
R1 = R2 = Ph, R3 = Et:
97% de > 98% ee (major) = 88%

Scheme 2.4 Cu-catalysed domino intramolecular Michael–Michael reaction.

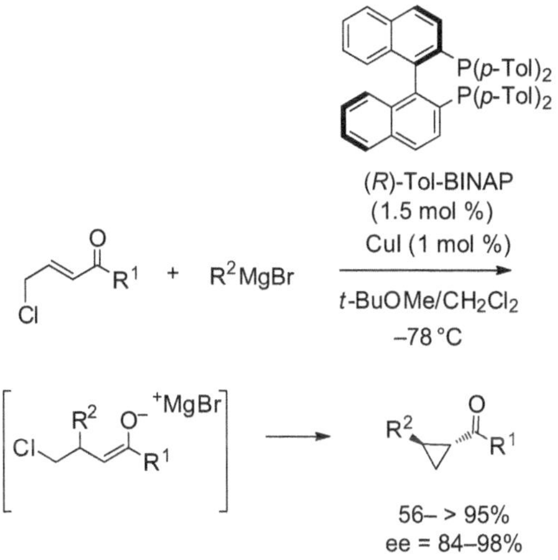

Scheme 2.5 Cu-catalysed domino Michael–intramolecular alkylation reaction.

98% ee, as shown in Scheme 2.5. These results were obtained by using (R)-Tol-BINAP as a chiral ligand. This novel and versatile methodology, employing an extremely simple catalytic system based on a commercial chiral ligand, CuI and Grignard reagents, was applied to the synthesis of key intermediates for the total syntheses of cascarillic acid and grenadamide.

A number of asymmetric domino processes involve 1,3-dicarbonyl compounds as nucleophilic partners in Michael reactions.[16] In this area, Feng *et al.* have described an efficient asymmetric synthesis of 4*H*-chromene derivatives through a domino Michael addition–cyclisation reaction catalysed by a salen–cobalt(II) complex.[17] Indeed, the asymmetric synthesis of a range of 2-amino-5-oxo-4-aryl-5,6,7,8-tetrahydro-4*H*-chromene-3-carboxylates was achieved on the basis of the reaction occurring between easily available cyclohexane-1,3-dione and ethyl 2-cyano-3-phenylacrylates in the presence of a salen–cobalt(II) complex generated *in situ* from $Co(OAc)_2 \cdot 4H_2O$ and a chiral salen derived from (R,R)-1,2-diphenylenethane-1,2-diamine and 3,5-di-*tert*-butylsalicylaldehyde. The domino process provided moderate to good yields of up to 81% combined with high enantioselectivities of up to 89% ee, as summarised in Scheme 2.6.

In 2010, Arai *et al.* investigated a catalytic asymmetric *exo'*-selective virtual [3 + 2] cycloaddition of iminoesters with *trans* nitroalkenes.[18] Usually, when a *trans* nitroalkene is used in a [3 + 2] cycloaddition, the stereoconjunction between the 3- and 4-positions is fixed in a *trans* conformation, and four diastereomers are possible, classified as *endo, exo, endo'*, and *exo'* isomers. Screening of the metal salts to study the *exo'* adduct ratio has found that nickel salts facilitated the selective production of the *exo'* products. Thus, these authors have performed the *exo'*-selective reaction of iminoesters and *trans* nitroalkenes by using a combination of $Ni(OAc)_2$ with a chiral imidazoline-aminophenol as catalyst, which provided the corresponding

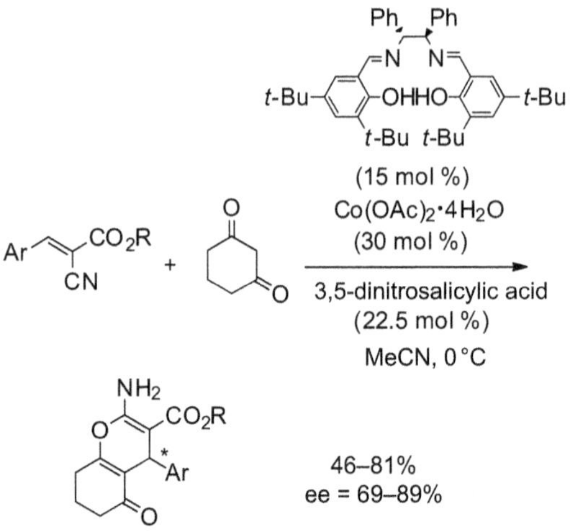

Scheme 2.6 Co-catalysed domino Michael–cyclisation reaction.

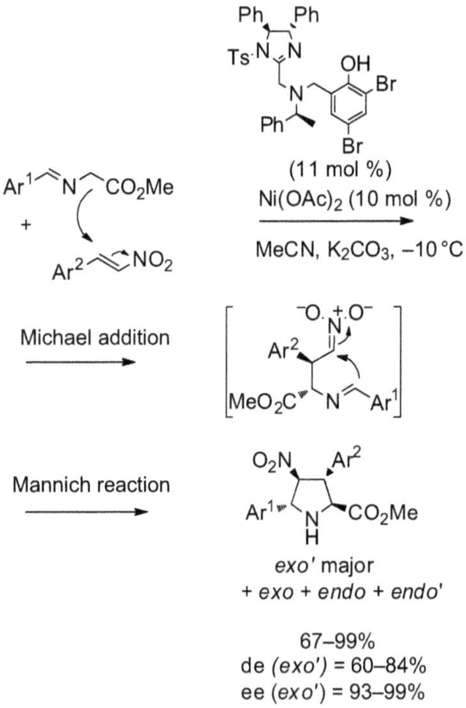

Scheme 2.7 Ni-catalysed domino Michael–Mannich reaction.

pyrrolidines in good yields and diastereoselectivities of up to 84% de combined with high enantioselectivities of up to 99% ee for the major isomers, as shown in Scheme 2.7. This novel methodology represented the first general success in the catalytic asymmetric *exo'*-selective reaction of iminoesters and nitroalkenes.

In order to explain the results, the authors have proposed that the products were generated from a domino Michael addition of the iminoesters onto the *trans* nitroalkenes, which was followed by a Mannich reaction, as depicted in Scheme 2.7.

2.2.2 Domino Reactions Initiated by an Aldol Reaction

The direct catalytic asymmetric aldol reaction is a powerful and atom-economical method for synthesising chiral β-hydroxy carbonyl compounds. Many metals and organocatalysts for reactions of aldehyde electrophiles have been developed in the past decade.[19] The use of ketone electrophiles in direct aldol reactions for the construction of a tetrasubstituted carbon stereocentre, however, is limited to either activated ketones or intramolecular reactions. Furthermore, catalytic asymmetric construction of two contiguous tetra-substituted chiral carbon stereocentres in C–C bond-forming reactions is rare, probably because of severe steric hindrance. In order to address these issues, Shibasaki *et al.* have employed a chiral magnesium catalyst to promote the domino aldol condensation–cyclisation reaction of αsubstituted α-isothio-cyanato esters with aryl, heteroaryl, alkyl, and alkenyl methyl ketones in addition to a cyclic ketone.[20] The process was catalysed by a combination of $(n\text{-Bu})_2\text{Mg}$ with a chiral Schiff base, affording at room temperature the corresponding protected α-amino-β-hydroxy esters bearing two contiguous tetrasubstituted carbon stereocentres in good to excellent yields, diastereo-selectivities of up to 94% de, and enantioselectivities of up to 98% ee, as shown in Scheme 2.8.

A chiral tin complex, $[\text{Sn}(R)\text{-Ph-PyBox}](\text{OTf})_2$ (where Ph represents phenyl, Py pyridine and Box bisoxazoline), has been shown by Zhu *et al.* to be

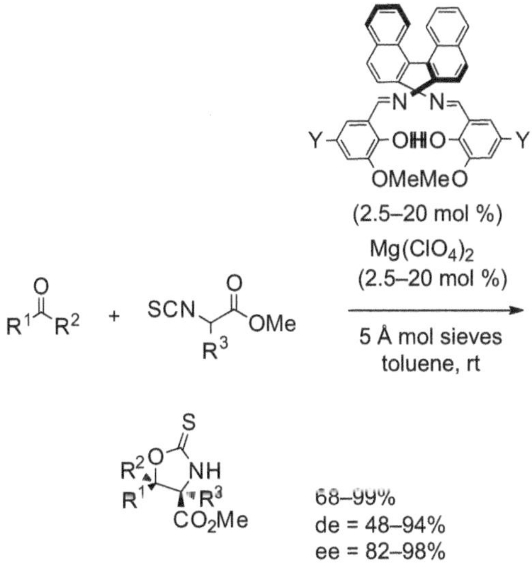

Scheme 2.8 Mg-catalysed domino aldol condensation–cyclisation reaction.

capable of catalysing the enantioselective domino condensation–cyclisation reaction of α-isocyanoacetamides with aldehydes, which afforded the corresponding 5-amino-2-(1-hydroxyalkyl)-oxazoles in low to good yields of up to 72% and low to good enantioselectivities of up to 80% ee.[21] More recently, Yanagisawa *et al.* have employed a chiral tin dibromide (Scheme 2.9) as a chiral precatalyst and sodium methoxide as a base precatalyst to induce the enantioselective domino aldol condensation–cyclisation reaction of γ-substituted

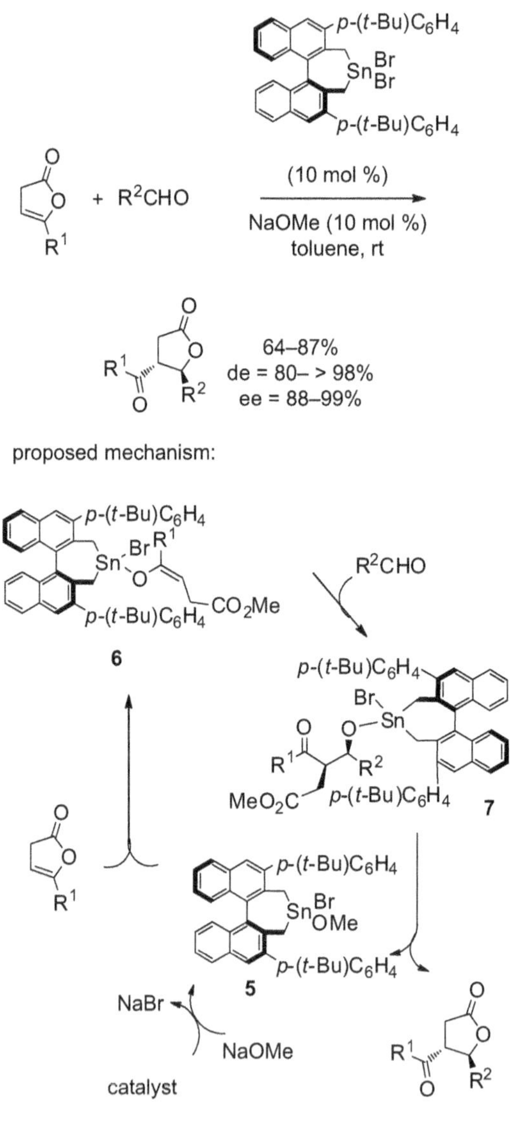

Scheme 2.9 Sn-catalysed domino aldol condensation–cyclisation reaction.

β,γ-didehydro-γ-lactones with aldehydes.[22] The corresponding chiral *trans*-β,γ-disubstituted γ-butyrolactones were selectively produced in moderate to high yields and enantioselectivities of up to 99% ee, as shown in Scheme 2.9. The authors have proposed the mechanism depicted in Scheme 2.9, which started with the reaction of the chiral tin bromidewith an equimolecular amount of sodium methoxide to give the corresponding chiral tin bromide methoxide **5**, which was the key catalyst of the process. The latter was then added to γ-substituted β,γ-didehydro-γ-lactone to yield a chiral tin enolate **6**. Subsequent aldol reaction between chiral tin enolate **6** and aldehyde provided the corresponding tin alkoxide of β-hydroxy ketone **7**. Finally, tin alkoxide **7** underwent cyclisation *via* participation of its ester moiety to afford the chiral final product with regeneration of the chiral tin bromide methoxide. The methoxycarbonyl group of intermediates **6** and **7** played a crucial role in the catalytic cycle.

A number of palladium-catalysed enantioselective domino reactions incorporate an aldol reaction in their sequence of transformations. As a recent example, Klein Gebbink *et al.* have developed an asymmetric domino aldol condensation–cyclisation reaction that occurs between α-methyl isocyanate and various aromatic aldehydes, which afforded the corresponding enantio-enriched oxazolines.[23] The process was catalysed by chiral five-coordinated NCN-pincer cationic palladium(II) complexes, providing moderate diastereo- and enantioselectivities of ≤48% de and ≤42% ee, respectively. Recently, the combination of organocatalysts and transition metal catalysts has evolved as a new strategy to carry out enantioselective transformations that could not be performed in a traditional way by simply employing one of the two catalysts.[8c,24] These transformations not only demonstrate the potential of this merged catalytic approach, but they also show that there are more options available to render a reaction highly enantioselective than testing different chiral metal-ligand complexes, organocatalysts, or additives. By using appropriate combinations of a chiral or achiral organocatalyst and an achiral or chiral transition metal catalyst, simple methods of reaction optimisation can be developed by varying one of the two existing catalysts. Furthermore, such a combination of multiple catalyst systems has opened up new avenues for developing cooperative catalyst systems where a single catalyst system alone fails to deliver sufficient catalyst reactivity and selectivity. In 2011, Oh and Kim developed a cooperative catalyst system for the highly diastereo- and enantioselective catalytic aldol reaction of methyl α-isocyanoacetate with aldehydes.[25] This domino aldol condensation–cyclisation reaction was induced by a combination of an achiral thiourea (Scheme 2.10) and a chiral cobalt complex derived from CoI$_2$, and a brucine amino diol (Scheme 2.10) as a chiral ligand in the presence of a base, such as DBU. As shown in Scheme 2.10, the corresponding chiral *trans*-oxazolines were produced in good yields and high diastereo- and enantioselectivities of up to 90% de and 98% ee, respectively. The authors assumed that the key to the successful stereocontrol of the domino process lay in the strong anion-binding interaction between the isocyanide and

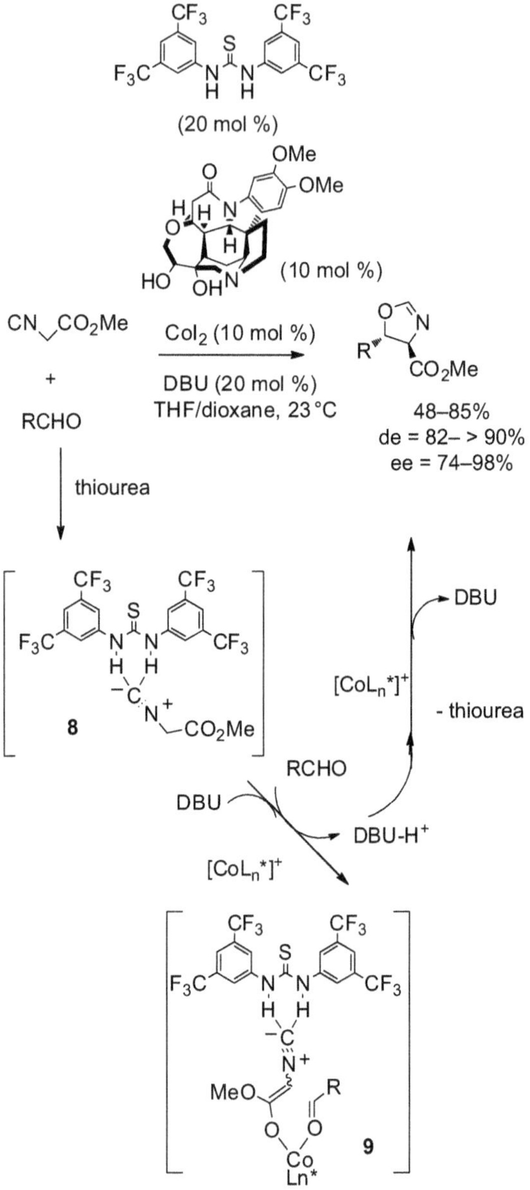

Scheme 2.10   Co-catalysed domino aldol condensation–cyclisation reaction.

thiourea catalyst, which potentially disturbed the intrinsic metal–isocyanide complexation. Indeed, anion-binding interactions between the thiourea and methyl α-isocyanoacetate in intermediate **8** provided a thiourea-assisted enolate, which was capable of coordinating to the chiral metal centre in a more organised fashion, **9**. Then, intermediate **9** cyclised to give the final product.

2.2.3 Domino Reductive Aldol Reactions

Recently, excellent progress has been achieved in the area of domino reductive aldol reactions for the construction of several contiguous stereocentres in a one-pot reaction.[26] In 2006, Shibasaki *et al.* described the first example of a catalytic enantioselective intermolecular reductive aldol reaction.[27] While copper-catalysed reductive aldol reactions between acetophenone and methyl acrylate, performed in the presence of pinacolborane as the reducing agent, provided the corresponding tertiary alcohol in only moderate enantios-electivities ($\leq 30\%$ ee) in all cases of the chiral ligands tested, reductive aldol reactions between symmetric ketones and β-disubstituted α,β-unsaturated esters afforded selectively the corresponding α-products in moderate to good enantio-induction (29–80% ee) when catalysed by a copper complex of (*R*)-Tol-BINAP. Furthermore, high to excellent enantioselectivities of up to 99% ee

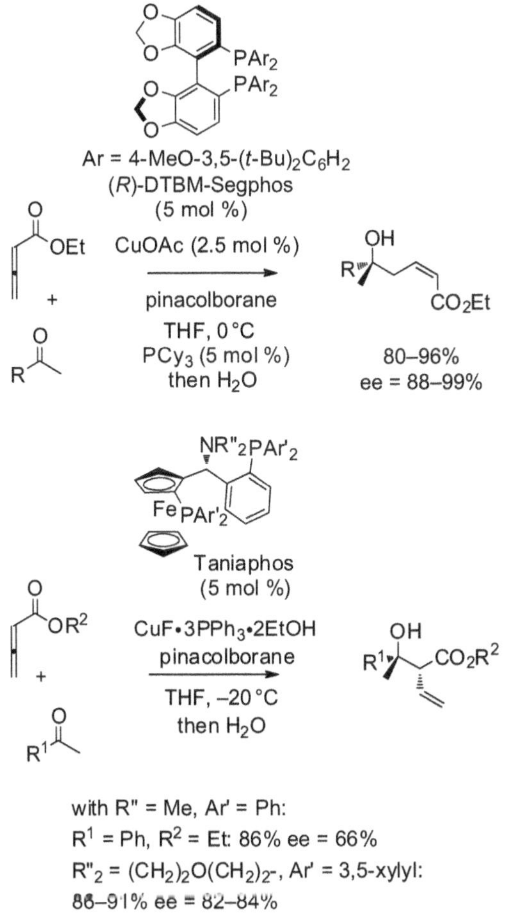

Scheme 2.11 Cu-catalysed domino reductive aldol reactions of ketones and allenic esters.

were achieved in the copper-catalysed reductive aldol reaction of allenic ethyl ester to ketones. When the reaction was catalysed by a copper(I) complex of (*R*)-DTBM-Segphos (where DTBM is ditertbutylmethoxy and Segphos is 5,5'-Bis(diphenylphosphino)-4,4'-bi-1,3-benzodioxole), it was γ-*cis*-selective because the corresponding tertiary alcohols were formed as single products in both excellent yields and enantioselectivities, as shown in Scheme 2.11. Surprisingly, the authors have found that the α- or γ-selectivity of the reaction could be switched depending on the structure of chiral diphosphine ligands used. Thus, when employing a copper(I) complex of two chiral ligands, Taniaphos (Scheme 2.11), the reaction became α-selective and provided the corresponding tertiary alcohols in high yields and enantioselectivities of up to 84% ee, as shown in Scheme 2.11.

In 2008, Lipshutz *et al.* described the first hydrometallative intramolecular cycloreduction leading to three new contiguous stereocentres in a one-pot

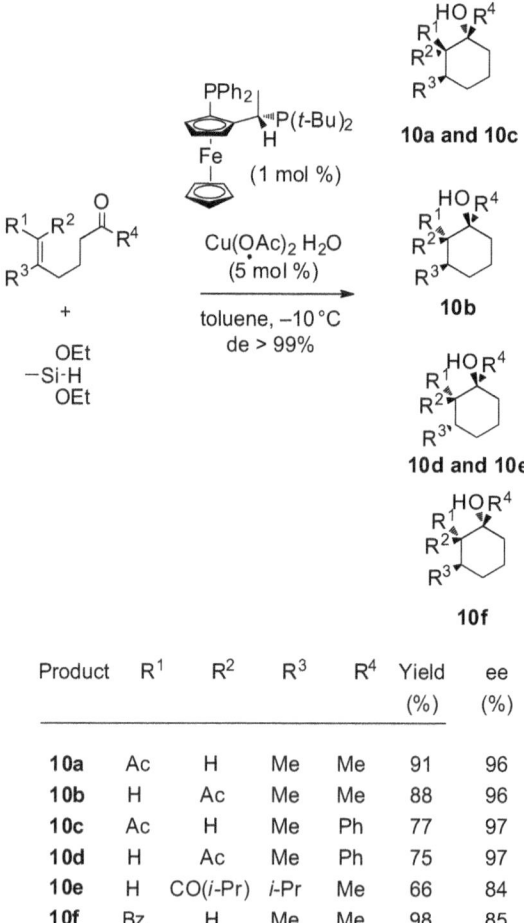

Product	R^1	R^2	R^3	R^4	Yield (%)	ee (%)
10a	Ac	H	Me	Me	91	96
10b	H	Ac	Me	Me	88	96
10c	Ac	H	Me	Ph	77	97
10d	H	Ac	Me	Ph	75	97
10e	H	CO(*i*-Pr)	*i*-Pr	Me	66	84
10f	Bz	H	Me	Me	98	85

Scheme 2.12   Cu-catalysed intramolecular domino reductive aldol–cyclisation reaction.

process, based on a chiral ligated catalytic source.[28] Thus, the enantioselective domino conjugate reduction–intramolecular aldol reaction of acyclic β,β-disubstituted keto enones provided the corresponding functionalised cyclohexanols **10a–f** as single diastereomers in both high yields and enantio-selectivities of up to 97% ee, as shown in Scheme 2.12. The generation of the three contiguous stereocentres was achieved by using a combination of Cu(OAc)$_2$ with a chiral bis-phosphine (Scheme 2.12) in the presence of diethoxymethylsilane as the reductant. In this process, the initial conjugate hydride addition generated an intermediate chiral (presumably copper) enolate through facial discrimination, which subsequently participated in an intramolecular aldol addition to ketones. Transmetalation of the resulting copper alkoxide with a particular stoichiometric silane regenerated ligated CuH.

Only few methodologies of domino reductive–aldol reactions have described the synthesis of bi- and tricyclic compounds. One example was recently reported by Riant and Deschamp, who developed a versatile methodology for the diastereo- and enantioselective domino reductive aldol–cyclisation reaction of functionalised α,β-unsaturated esters into the corresponding bicyclic domino products in moderate to high yields and enantioselectivities of up to 97% ee, as shown in Scheme 2.13.[29] The reaction was catalysed by a copper(I) complex of

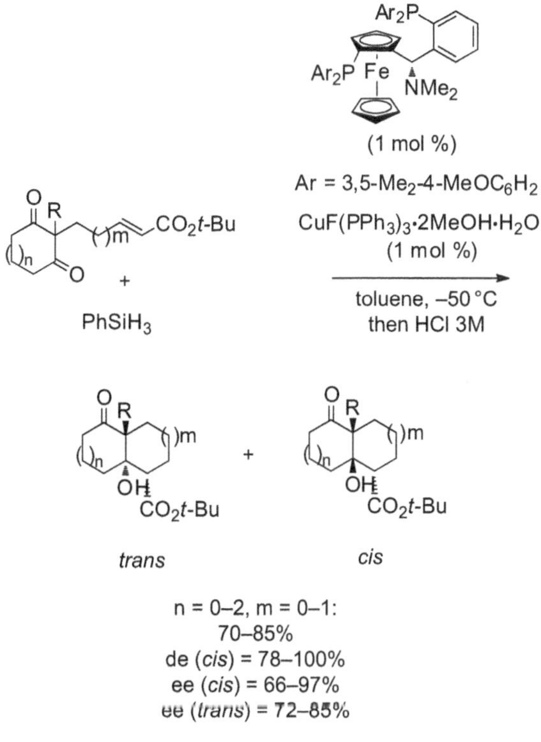

n = 0–2, m = 0–1:
70–85%
de (*cis*) = 78–100%
ee (*cis*) = 66–97%
ee (*trans*) = 72–85%

Scheme 2.13 Cu-catalysed intramolecular domino reductive aldol–cyclisation reaction.

chiral bis-phosphine (Scheme 2.13) in the presence of phenylsilane as the reductive agent. The *cis* product was produced as the major diastereomer with moderate to complete diastereosectivities. In this study, the authors demonstrated that the stereoselectivity of the process was increased with the steric hindrance of the ester moiety, because the best results were obtained with *t*-butyl esters. Furthermore, it was shown that increasing the steric bulkiness around the phosphorus atoms of the ligand gave further improvement of both *cis* : *trans* ratio and enantioselectivity.

These authors have also employed other chiral Taniaphos-based ligands to induce chirality in copper(I)-catalysed reductive aldol reactions of methyl acrylate with aldehydes, using phenylsilane as the reducing agent.[30] Although

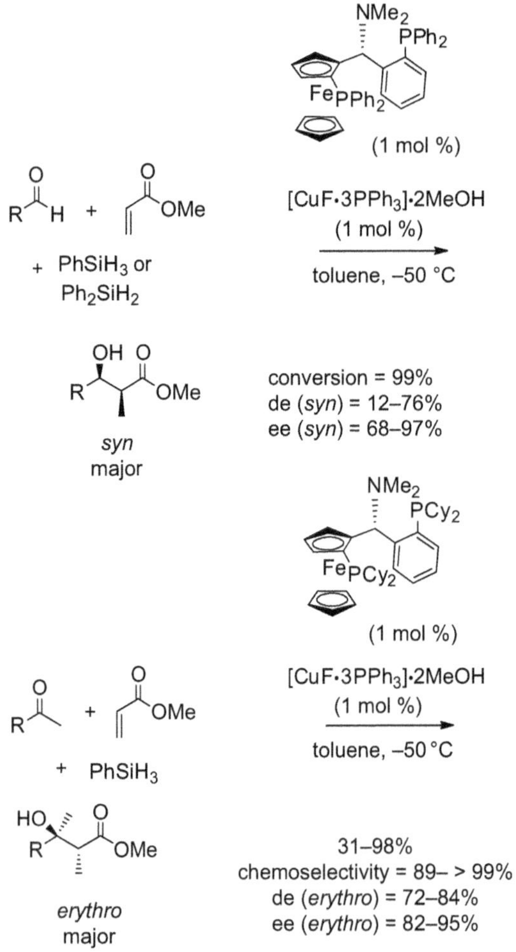

Scheme 2.14 Cu-catalysed three-component reductive aldol reactions of aldehydes and ketones.

the reaction was highly chemoselective, it gave moderate *syn*-diastereo-selectivities ($\leq 76\%$ de). However, good to excellent enantioselectivities of up to 97% ee were obtained for domino products that arose from a wide range of cyclic aliphatic, aromatic, and heteroaromatic aldehydes when a Taniaphos ligand was employed, as shown in Scheme 2.14. The scope of this methodology was extended by these authors to the reductive aldol reactions of methyl acrylate with ketones under similar conditions.[31] Thus, the multicomponent reaction of ketones with methyl acrylate and phenylsilane gave rise chemo-selectively, by using a closely related chiral Taniaphos ligand, to the corresponding tertiary alcohols as single products, which were mixtures of *erythro*- and *threo*-diastereomers. The major *erythro*-isomers were produced in moderate diastereoselectivities, albeit with good to excellent enantioselectivities of up to 97% ee, as shown in Scheme 2.14.

In 2008, Krische *et al.* reported the first enantioselective reductive aldol couplings of vinyl ketones, which were achieved through the design of a new class of TADDOL-like phosphonite ligands.[32] As shown in Scheme 2.15, the Rh-catalysed hydrogenative aldol coupling of methyl vinyl ketone or ethyl vinyl ketone with aldehyde, performed in the presence of a chiral ligand, allowed the corresponding linear aldol adducts to be obtained with high diastereo- and enantioselectivities of up to 96% de and 96% ee, respectively. More recently, copper(I)–ClickFerrophos complexes were demonstrated by Fukuzawa *et al.* to catalyse the same reactions.[33] In the case of the reductive aldol reaction of ketones, the products were produced in generally excellent yields of up to 99% and moderate to high enantioselectivities (37–95% ee) for the major *erythro* products. While yields and enantioselectivities were comparable, the diastereomeric ratio of *erythro : threo* was improved to up to 92% de when compared with that obtained when using Taniaphos ligands.

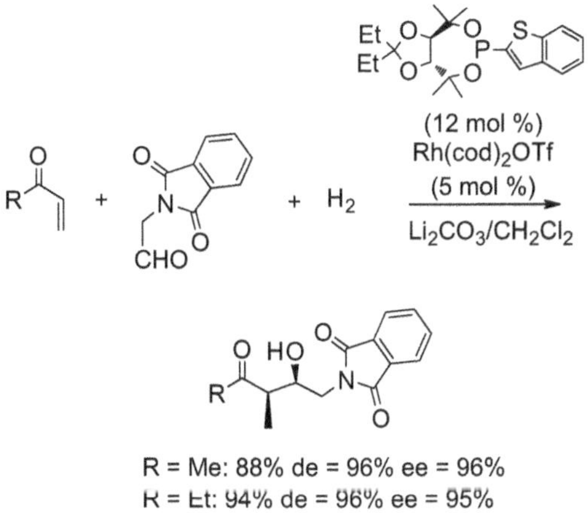

R = Me: 88% de = 96% ee = 96%
R = Et: 94% de = 96% ee = 95%

Scheme 2.15 Rh-catalysed domino reductive aldol reaction.

2.2.4 Domino Reactions Initiated by a [2 + 2 + 2] Cycloaddition

Transition metal-catalysed [2 + 2 + 2] cycloaddition of unsaturated motifs, such as alkyne and alkene, constitutes the most atom-economical and facile protocol for the construction of a six-membered ring system.[34] In particular, enantio-selective [2 + 2 + 2] cycloaddition is a fascinating protocol for the construction of chiral cyclic skeletons.[35] In recent years, this cycloaddition has been included in several enantioselective domino processes promoted by chiral catalysts of transition metals, such as rhodium and iridium. As an example, Tanaka *et al.* have developed a cationic rhodium(I)/(*R*)–Solphos complex-catalysed en-antioselective domino transesterification–[2 + 2 + 2] cycloaddition reaction of 1,6-diyne esters with tertiary propargylic alcohols, leading to the corresponding tricyclic 3,3-disubstituted phthalides.[36] As shown in Scheme 2.16, the cyclo-adducts were achieved in both high yields and enantioselectivities of up to 94% ee. Although the [2 + 2 + 2] cycloaddition constitutes the second step of this domino sequence, because this cycloaddition is the key step of the process it was decided to place it in this section dealing with domino reactions initiated by a [2 + 2 + 2] cycloaddition.

Enantioselective rhodium-catalysed domino double [2 + 2 + 2] cycloaddition reactions have been independently developed by the groups of Tanaka and Shibata. In 2006, Tanaka *et al.* established an enantioselective synthesis of tetra-*ortho*-substituted axially chiral biaryls through rhodium-catalysed domino double [2 + 2 + 2] cycloaddition reaction.[37] By using a chiral Rh(I)[+]/modified-(*S*)-Segphos complex as catalyst, the reaction of electron-deficient malonate-derived 1,6-diynes and 1,3-diynes afforded the corre-sponding *C₂*-symmetric tetra-*ortho*-substituted biaryls in moderate yields and excellent enantioselectivities of up to >99% ee, as shown in Scheme 2.17.

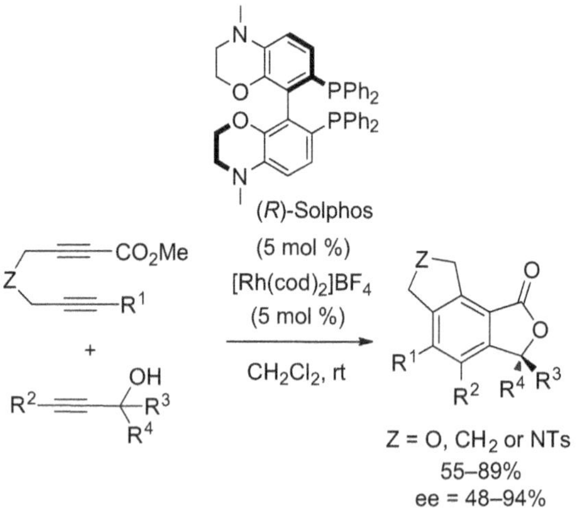

Scheme 2.16 Rh-catalysed domino transesterification–[2 + 2 + 2] cycloaddition reaction.

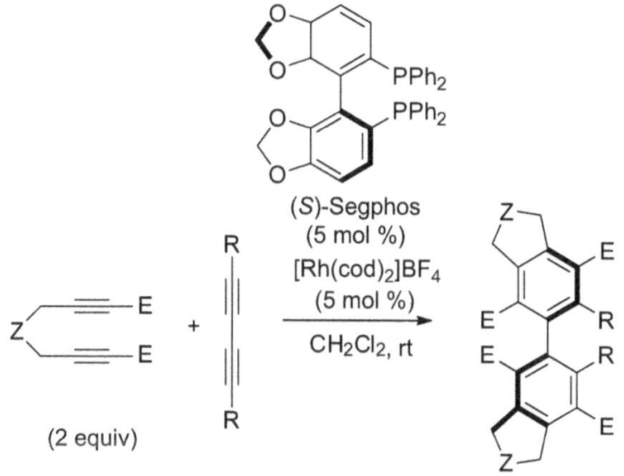

R	Z	E	Yield (%)	ee (%)
CH₂OAc	C(CO₂Me)₂	CO₂Et	59	> 99
CH₂OMe	C(CO₂Me)₂	CO₂Et	48	98
CH₂OAc	CH₂	CO₂Me	30	> 99

Scheme 2.17 Synthesis of chiral biaryls through Rh-catalysed domino double [2 + 2 + 2] cycloaddition reaction.

Furthermore, comparable reaction conditions were applied to the reaction of bis-diynenitriles, providing through intramolecular enantioselective double [2 + 2 + 2] cycloaddition the corresponding C_2-symmetric spirobipyridines in high yields and moderate enantioselectivities (≤71% ee). As an extension of this methodology, the same authors have developed the domino double [2 + 2 + 2] cycloaddition reaction of terminal tetraynes with electron-deficient monoynes to give the corresponding biaryls in moderate yields (24–44%) and good to excellent enantioselectivities (70–98% ee). Later, an intramolecular version of this methodology was also developed by these authors, allowing chiral C_2-symmetric spirobipyridine ligands to be achieved, starting from the corresponding bisdiynenitriles, in good to excellent yields (70–99%) and moderate enantioselectivities of up to 71% ee.[38]

Later, Shibata *et al.* described an enantioselective synthesis of chiral tetraphenylenes on the basis of consecutive inter- and intramolecular rhodium-catalysed [2 + 2 + 2] cycloadditions of two triynes.[39] When triynes having a phenylene-bridged 1,5-diyne moiety were submitted to a cationic rhodium catalyst of (*R*)-Cy-BINAP (where Cy is cyclohexyl), they afforded the corresponding dimerised products in good yields and high enantioselectivities of up to >99% ee, as shown in Scheme 2.18. The authors assumed that these products were formed through a mechanism, depicted in Scheme 2.18, involving successive inter- and intramolecular [2 + 2 + 2] cycloadditions. For the first time, oxidative coupling of the 1,6-diyne moiety of the first triyne gave

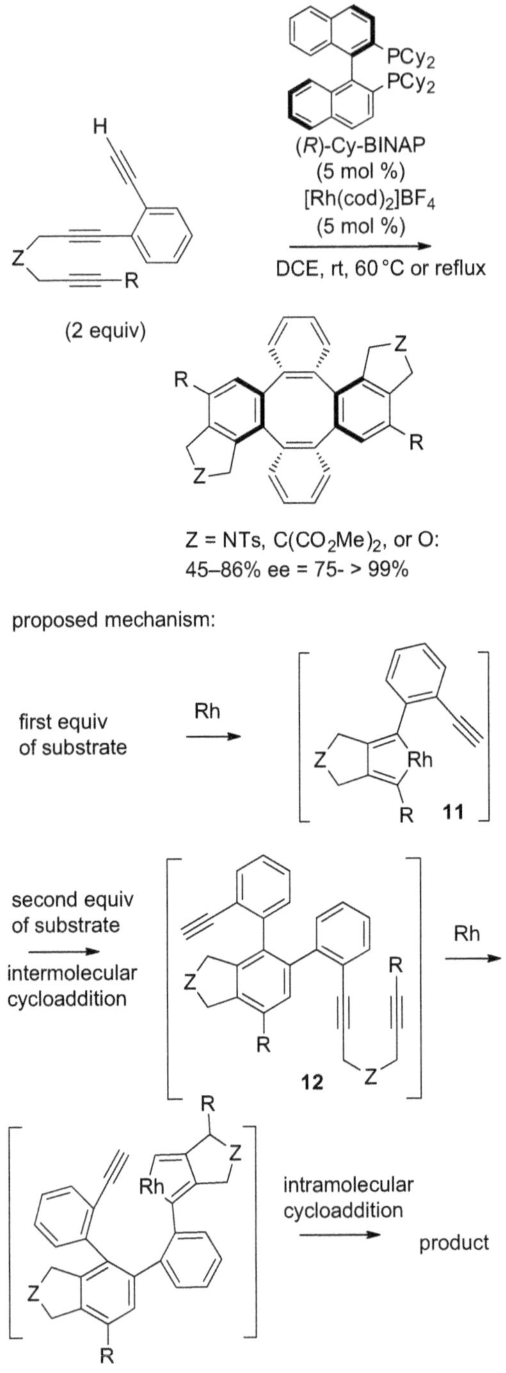

Z = NTs, C(CO$_2$Me)$_2$, or O:
45–86% ee = 75- > 99%

proposed mechanism:

Scheme 2.18 Synthesis of chiral tetraphenylenes through Rh-catalysed domino double [2 + 2 + 2] cycloaddition reaction.

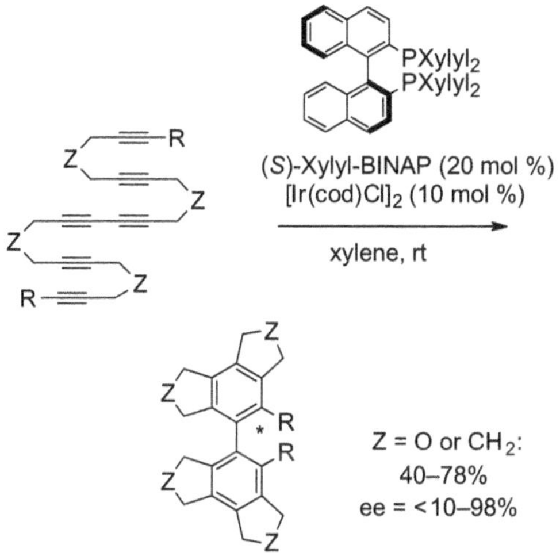

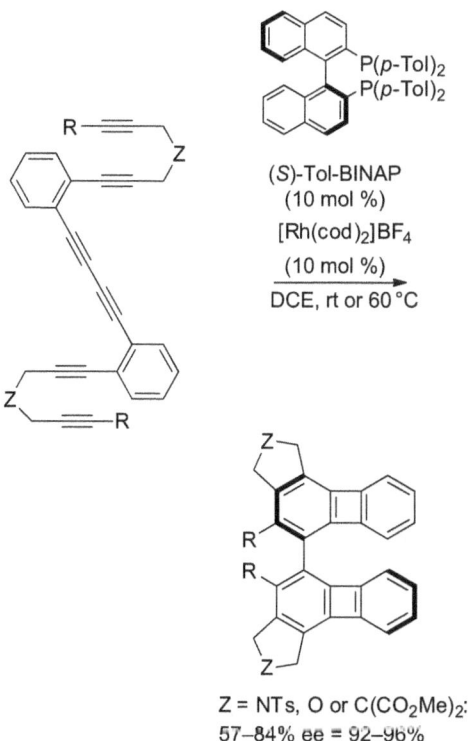

Scheme 2.19   Ir-catalysed domino intramolecular double [2+2+2] cycloaddition reaction.

Scheme 2.20   Synthesis of chiral bis(biphenylenyl) derivatives through Rh-catalysed domino double [2+2+2] cycloaddition reaction.

metalacyclopentadiene, **11**. Chemo- and regioselective intermolecular coupling with terminal alkyne moiety of the second triyne gave the primary cycloadduct **12**. Oxidative coupling of the 1,6-diyne moiety of the second triyne and intramolecular coupling with the remaining terminal alkyne moiety of the first triyne gave the final tetraphenylenes.

In addition, these authors have developed iridium-catalysed consecutive intramolecular and enantioselective $[2+2+2]$ cycloaddition domino reactions of hexaynes possessing a 1,3-diyne moiety and oxygen or nitrogen tether.[40] As shown in Scheme 2.19, the reaction of these substrates, in the presence of a combination of $[Ir(cod)Cl]_2$ and (S)-Xylyl-**BINAP**, led to the formation of the corresponding C_2-symmetrical biaryl products in good yields and low to excellent enantioselectivities of up to 98% ee.

More recently, the same authors have investigated the rhodium-catalysed enantioselective intramolecular reaction of hexaynes bearing a 1,3-diyne moiety.[41] When promoted by a chiral cationic rhodium complex of (S)-Tol-**BINAP**, the domino double $[2+2+2]$ cycloaddition reaction afforded the corresponding axially chiral bis(biphenylenyl) derivatives in good yields and good to high enantioselectivities of up to 96% ee, as shown in Scheme 2.20.

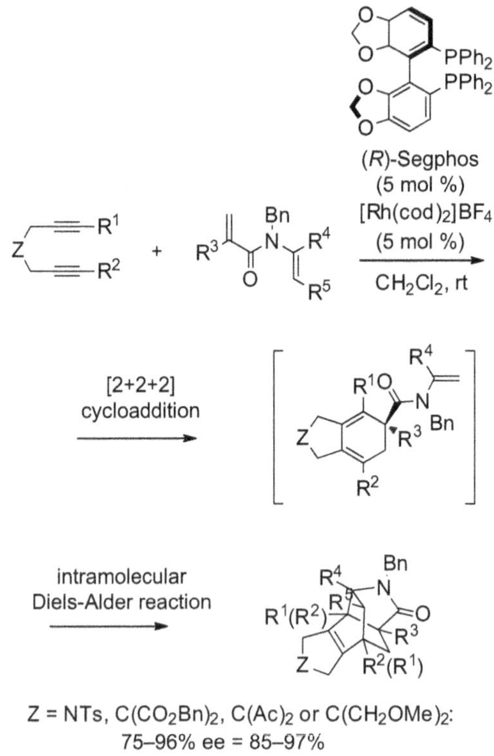

Scheme 2.21 Rh-catalysed domino intermolecular $[2+2+2]$ cycloaddition–intramolecular Diels–Alder cycloaddition reaction.

The authors explained the formation of these cycloadducts through two consecutive $[2+2+2]$ cycloadditions.

In addition, the $[2+2+2]$ cycloaddition has been associated with other reactions, such as the Diels–Alder cycloaddition, in enantioselective domino metal-catalysed reactions. As an example, Tanaka *et al.* have recently reported an efficient enantioselective domino intermolecular $[2+2+2]$ cyclo-addition–intramolecular Diels–Alder cycloaddition reaction occurring between 1,6-diynes and amide-linked 1,5-dienes bearing two sterically and/or electronically different alkene units.[42] This domino reaction was induced by a cationic rhodium(I)/(R)-Segphos complex and provided the corresponding amides in moderate to high yields and excellent enantioselectivities of up to 97% ee, as shown in Scheme 2.21. The mechanism of the reaction involved a $[2+2+2]$ cycloaddition between the two substrates, which was induced by the chiral cationic rhodium(I) catalyst, to form the corresponding cyclohexadiene containing a pendant alkene unit. A subsequent intramolecular Diels–Alder reaction of this cyclohexadiene furnished the final chiral bridged multicyclic product, as depicted in Scheme 2.21.

2.2.5 Domino Reactions Initiated by an Allylic Alkylation

Nitrogen-containing heterocycles, present in a wide number of pharmaceutically interesting compounds, have been usually prepared by palladium-catalysed asymmetric tandem allylic substitution reaction of allylic acetates, carbonates, or halides. Indeed, this reaction, also known as the Tsuji–Trost reaction, constitutes a powerful procedure for the formation of C–C, C–O, and C–N bonds. One of the early impressive examples, where this reaction had been combined with a pallada-ene reaction, was developed by Oppolzer and Gaudin, in 1987.[43] In 1993, Hayashi *et al.* reported another early work in this area with the palladium-BINAP-catalysed synthesis of optically active morpholines and piperazines in enantioselectivities of up to 61% ee.[44] The asymmetric induction of this type of process is controlled during the π-allyl–palladium intermediate prior to the second nucleophilic attack by the nucleophile. In general, the Tsuji–Trost reaction can be combined with other Pd-catalysed transformations, such as a Heck reaction or a second Tsuji–Trost reaction. As a recent excellent example, Trost and Dong have developed an enantioselective palladium-catalysed domino double allylic alkylation reaction, which enabled efficient access to chiral tricyclic pyrrolopiperazinones.[45] As shown in Scheme 2.22, the reaction of a bisallylic carbonate with a nucleophile containing both a pyrrole and a *N*-methoxyamide group, afforded the corresponding piperazinone in 82% yield and excellent enantioselectivity of >97% ee through a domino double allylic alkylation reaction, which was catalysed by a combination of $[Pd_2(dba)_3] \cdot CHCl_3$ (where dba represents (E,E)-dibenzylideneacetone) and the standard Trost ligand (R,R)-L$_{ST}$ in the presence of acetic acid as an additive. This elegant procedure was applied to a total synthesis of the natural product, (+)-agelastatin A, in five steps starting from the piperazinone.

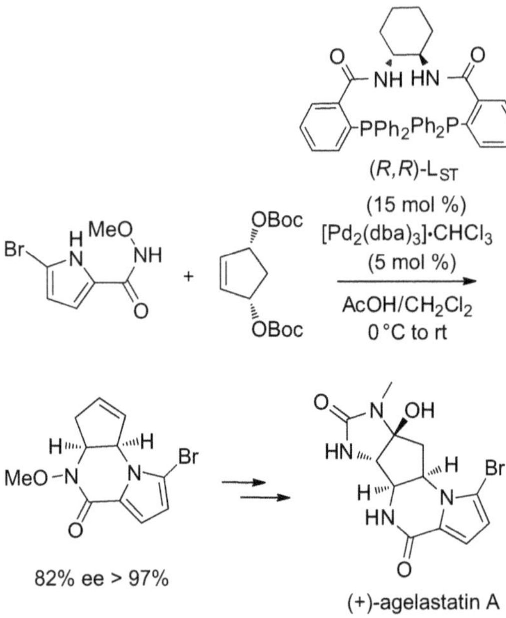

Scheme 2.22 Pd-catalysed domino double allylic alkylation reaction and synthesis of (+)-agelastatin A.

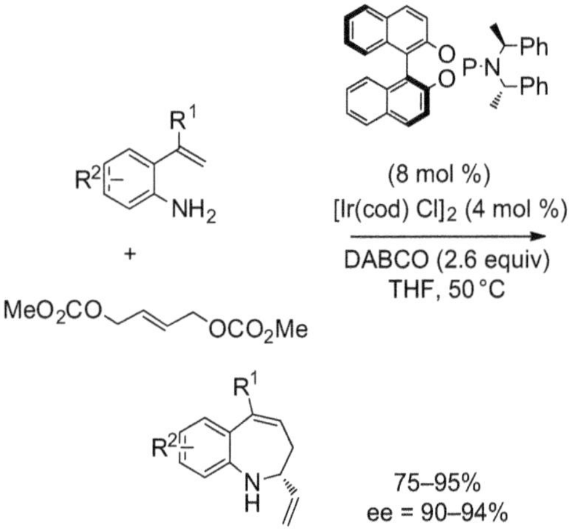

Scheme 2.23 Ir-catalysed domino allylic vinylation–amination reaction.

Later, You *et al.* reported an enantioselective synthesis of 2,3-dihydro-1*H*-benzo[*b*]azepines on the basis of an iridium-catalysed domino allylic vinylation–amination reaction using a chiral phosphoramidite ligand (Scheme 2.23).[46] Indeed, these authors have shown that a combination of this chiral ligand with [Ir(cod)Cl]₂ (where cod is cyclooctadiene) could efficiently catalyse the domino allylic vinylation and amination reaction of (*E*)-but-2-ene-1,4-diyl dimethyl dicarbonate with *ortho*-amino styrene derivatives, affording in the presence of DABCO (1,4-diazabicyclo[2.2.2]octane) as a base the corresponding 2,3-dihydro-1-*H*-benzo[*b*]azepines with high yields and enantioselectivities of up to 94% ee, as shown in Scheme 2.23. In another context, an enantioselective domino reaction including an allylstannylation followed by a Heck reaction was recently reported by Schmalz *et al.*[47] In this study, these authors have developed an enantioselective access to 3-methylene-1*H*-indanol through domino allylstannylation–Heck reaction occurring between *o*-iodobenzaldehyde and allyl tributylstannane in the presence of a palladium complex generated *in situ* from [Pd$_2$dba$_3$] and a Taniaphos chiral ligand. The domino product was produced in 52% yield and 96% ee.

2.2.6 Domino Passerini-type Reactions

The modern concept of domino and multicomponent reactions is intimately related to the reactions developed with isocyanide reagents.[48] The unique ability of isocyanides to serve as *C*-nucleophiles towards imines and aldehydes, resulting in nitrilium ion electrophiles, enables the four-component Ugi reaction and the three-component Passerini reaction, respectively.[49] Despite the strong relationship between domino and multicomponent reactions and isocyanides, there are still few reactions in which the use of isocyanide reagents leads to chiral compounds, and more particularly those which are induced by chiral catalysts. As a recent example, Zhu *et al.* reported the use of a chiral salen–aluminum(III) complex to induce chirality in a Passerini-type reaction between α-isocyanoacetamides and aldehydes to provide the corresponding 2-(1-hydroxyalkyl)-5-aminooxazoles in moderate to good yields (35–82%) and enantioselectivities (54–80% ee).[50] The conditions were applicable to both aliphatic and aromatic aldehydes and, in general, the enantioselectivity was higher for aliphatic than for aromatic ones. Two years later, the same authors have developed similar reactions in the presence of a mixture of Et$_2$AlCl in combination with a chiral phosphoric acid. A series of 2-(1-hydroxyalkyl)-5-aminooxazoles was produced from the reaction of the corresponding α-isocyanoacetamides and aldehydes in comparable yields and enantios-electivities of up to 91% ee.[51] In addition, Shibasaki *et al.* have selected the α-addition of α-isocyanoacetamides to aldehydes as a model reaction to investigate the efficiency of bimetallic Schiff base catalysts.[52] When the new heterobimetallic, Ga(O*i*-Pr)$_3$/Yb(OTf)$_3$/*o*-vanillin-derived Schiff base complex (Scheme 2.24), was used for catalytic asymmetric addition of α-isocyanoacet-amides to a range of aryl, heteroaryl, alkenyl, as well as alkyl aldehydes, the corresponding domino products were achieved in generally high yields and excellent enantioselectivities of up to 98% ee, as shown in Scheme 2.24. In this study, the authors demonstrated that the chiral diamine backbone affected both the reactivity and enantioselectivity of the reaction. For example, Schiff bases bearing a binaphthyl diamine unit gave unsatisfactory results.

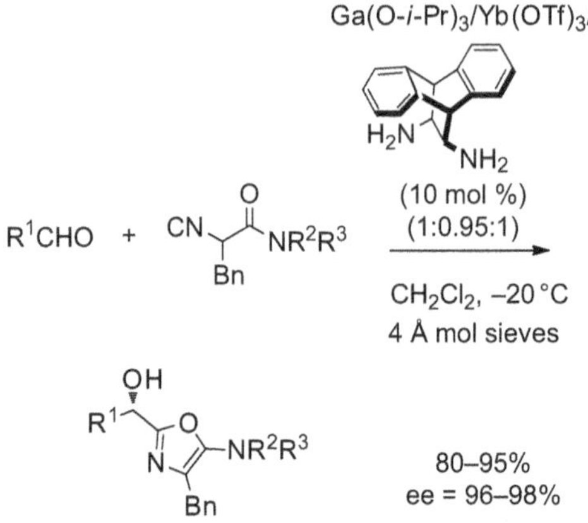

Ga(O-*i*-Pr)$_3$/Yb(OTf)$_3$/

H$_2$N   NH$_2$
(10 mol %)
(1:0.95:1)

R^{1}CHO   +   CN $\overset{\text{O}}{\underset{\text{Bn}}{\diagdown}}$ NR^{2}R^3

CH$_2$Cl$_2$, −20 °C
4 Å mol sieves

OH
R^1 $\overset{\text{O}}{\diagdown}$
N $-$NR^{2}R^3
Bn

80–95%
ee = 96–98%

Scheme 2.24   Ga- and Yb-catalysed-domino Passerini-type reaction.

2.2.7   Domino Carbonyl Ylide Formation–1,3-Dipolar Cycloaddition Reactions

Catalytic methods encompassing metal carbene intermediates constitute a vast array of transformations that offer the synthetic chemist great scope in the synthesis of many complex molecules.[53] Of these processes, the catalytic domino carbonyl ylide formation–1,3-dipolar cycloaddition offers an elegant route to highly substituted oxygen-containing heterocycles.[53b,54] This powerful methodology has been extensively advanced by the Padwa group, in particular.[53c,55] The development of a catalytic enantioselective version of this domino reaction has become a challenging objective. The primary work in this area was reported by Hodgson *et al.* in 1997, dealing with intramolecular enantioselective catalytic domino carbonyl-ylide formation–cyclisation reactions of α-diazo-β-ketoesters in enantioselectivities of up to 53% ee by using Davies' prolinate catalyst, Rh$_2$((S)-DOSP)$_4$, where DOSP represents *N*-*p*-dodecylbenzenesulfonylprolinate.[56] Ever since, the formation of keto carbenoids by treatment of diazo keto compounds with rhodium(II) salts has been broadly employed in enantioselective domino processes as the primary step. This is followed by the generation of a 1,3-dipole through an intramolecular cyclisation of the keto carbenoid onto an oxygen atom of a neighbouring keto group and an inter- or intramolecular 1,3-dipolar cycloaddition. More recently, Hashimoto *et al.* have reported the use of chiral dirhodium(II) tetrakis[*N*-benzene-fused-phthaloyl-(S)-valinate], Rh$_2$(S-BPTV)$_4$, to promote the enantioselective domino carbonyl ylide formation–1,3-dipolar cycloaddition reaction of a range of α-diazo ketones with various aromatic aldehydes, affording the corresponding bicyclic cycloadducts in good yields and enantioselectivities of up to 92% ee for the *exo*-diastereomer, which was diastereoselectively generated as a single product in almost all the cases of substrates studied, as shown in Scheme 2.25.[57] On the other hand, it must be

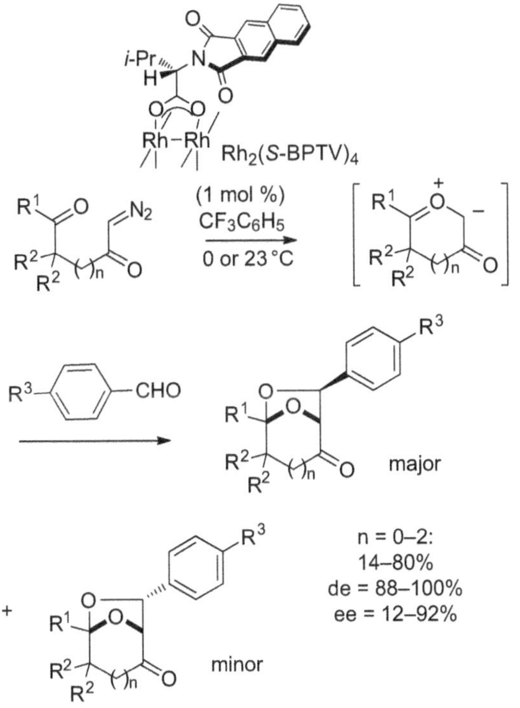

Scheme 2.25 Rh-catalysed domino carbonyl-ylide formation–1,3-dipolar cyclo-addition reaction of α-diazo ketones with aromatic aldehydes.

noted that the use of aliphatic aldehydes gave disappointingly complex mixtures of products. This methodology was applied to the construction of the *exo*-7-aryl-6,8-dioxabicyclo[3.2.1]octane framework ($R^1 = i$-Pr, $R^2 = H$, $R^3 = OH$, $n = 1$) of natural products, psoracorylifols B and C, that exhibit significant inhibitor activity against *Helicobacter pylori*.

In addition, these authors have found that dirhodium(II) tetrakis[*N*-tetrachlorophthaloyl-(*S*)-*tert*-leucinate], Rh₂(*S*-TCPTTL)₄, was an exceptionally effective catalyst for the enantioselective domino carbonyl ylide formation–1,3-dipolar cycloaddition reaction of 2-diazo-3,6-diketoesters with arylacetylenes and alkoxyacetylenes, providing the corresponding cycloadducts in good to high yields and enantioselectivities of up to 98% ee, as shown in Scheme 2.26.[58] Furthermore, the scope of this methodology could be extended to other dipolarophiles, such as styrenes, which provided by reaction with 2-diazo-3,6-diketoesters the corresponding cycloadducts with complete *exo/endo* diastereocontrol combined with high yields and excellent enantioselectivities of up to 99% ee, as shown in Scheme 2.26.

Another extension of this methodology was the enantioselective reaction of the formyl-derived carbonyl ylide, generated from the diazo compound depicted in Scheme 2.27, with 4-hydroxy-3-methoxyphenylacetylene. This provided, upon catalysis with Rh₂(*S*-TCPTTL)₄ (where TCPTTL represents

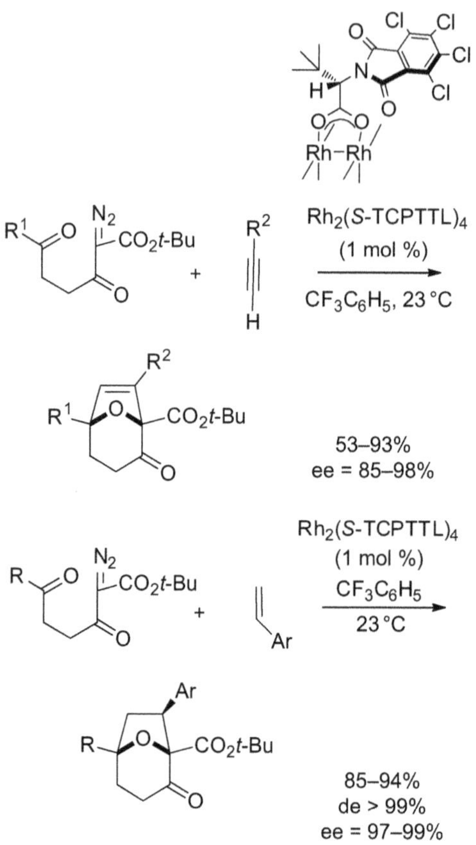

Scheme 2.26 Rh-catalysed domino carbonyl-ylide formation–1,3-dipolar cycloaddition reactions of 2-diazo-3,6-diketoesters with aryl- and alkoxyacetylenes, and styrenes.

N-tetrachlorophthaloyl-*tert*-leucinate), the corresponding domino cycloadduct in 73% yield and 95% ee.[59] This product could be converted into a
natural product, *endo*-6-aryl-8-oxabicyclo[3.2.1]oct-3-en-2-one, as shown in
Scheme 2.27. In addition, another natural and biologically active product,
descurainin, was achieved on the basis of the same domino reaction, albeit
catalysed by $Rh_2(R$-TCPTTL$)_4$.[60] Indeed, by using the (*R*)-enantiomer of the
same dirhodium(II) catalyst, the corresponding enantiomeric cycloadduct was
obtained in 77% yield and 95% ee, as shown in Scheme 2.27. This product was
further converted into the expected descurainin in 10 steps.

In 2011, the catalyst $Rh_2(S$-TCPTTL$)_4$ was also employed by the same
authors to induce the first enantioselective intermolecular cycloaddition of
carbonyl ylides with another type of dipolarophiles, such as indoles.[61] Thus, the
domino reaction of 2-diazo-3,6-diketoesters with a series of *N*-methylindoles
led to the corresponding tetracyclic cycloadducts in both high yields and
enantioselectivities of up to 99% ee, as well as excellent *exo*-diastereoselectivity

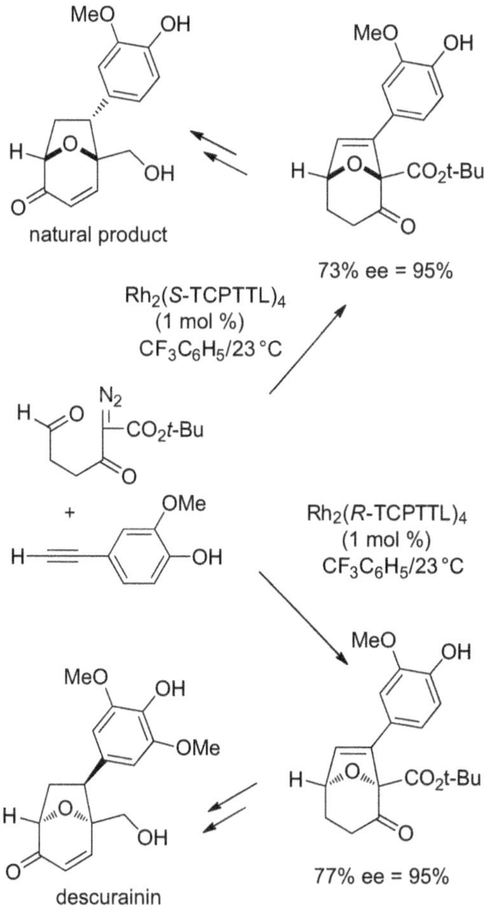

Scheme 2.27 Rh-catalysed domino carbonyl-ylide formation–1,3-dipolar cyclo-addition reaction of 2-diazo-5-formyl-3-ketoester with 4-hydroxy-3-methoxyphenylacetylene, and synthesis of natural products.

of >98% de, as shown in Scheme 2.28. In the same context, these authors have described the first example of asymmetric induction (up to 66% ee) in an intramolecular 1,3-dipolar cycloaddition of carbonyl ylides generated *in situ* from the diazo decomposition of indolyl-substituted 2-diazo-5-imido-3-ketoesters under the influence of Rh₂(S-TCPTTL)₄. This reaction constituted an elegant approach to the pentacyclic skeleton of *Aspidosperma* alkaloids.[62] In addition, Hodgson *et al.* investigated the enantioselective intramolecular domino carbonyl ylide formation–1,3-dipolar cycloaddition reaction employing sulfonyl functionality as a different type of electron-withdrawing group at the ylidic carbon, specifically using unsaturated α-diazo-β,ε-diketo sulfones.[63] Among a series of chiral catalysts tested, chiral dirhodium(II) tetrakis[*N*-benzene-fused-phthaloyl-(S)-valinate], Rh₂(S-BPTV)₄, provided the

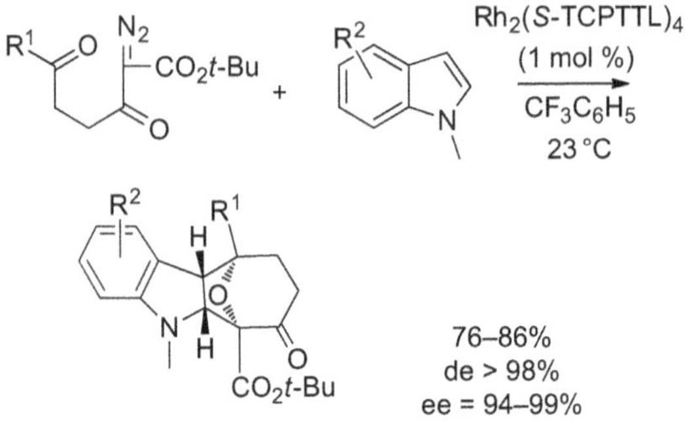

Scheme 2.28   Rh-catalysed domino carbonyl-ylide formation–1,3-dipolar cyclo-addition reaction of 2-diazo-3,6-diketoesters and indoles.

best, albeit moderate, enantioselectivity of 43% ee for the cycloadduct derived from phenyl sulfone, along with 75% yield.

Chiral catalysts of metals other than rhodium, such as nickel and silver, and chiral catalysts of ytterbium have also been employed to induce enantioselective domino carbonyl-ylide formation–1,3-dipolar cycloaddition reactions. For example, Suga *et al.* have reported 1,3-dipolar cycloaddition reactions of carbonyl ylides generated *in situ* from *o*-(*p*-bromobenzyloxy)carbonyl-α-diazoacetophenones and 3-crotonoyl-2-oxazolidinones by using a combination of a chiral Pybox ligand with Yb(OTf)$_3$ as the chiral Lewis acid catalyst.[64] Indeed, the reaction of the α-diazoacetophenone derivative with a 3-crotonoyl-2-oxazolidinone derivative afforded the corresponding *endo*-cycloadduct as a sole product with a diastereoselectivity >98% de and with extremely high enantioselectivity of 96% ee. In contrast, *o*-(methylbenzyloxy)carbonyl-α-diazoacetophenone reacted more slowly with 3-crotonoyl-2-oxazolidinones to give the corresponding domino products with moderate enantioselectivities (≤78% ee). On the other hand, these authors have developed the first successful example of reverse-electron-demand dipole-LUMO (lowest occupied molecular orbital)/dipolarophile-HOMO (highest occupied molecular orbital) controlled cycloaddition reactions between carbonyl ylides, which were generated *in situ* from *o*-methoxycarbonyl-α-diazoacetophenone and their acyl derivatives in the presence of Rh$_2$(OAc)$_4$, and vinyl ether derivatives activated by chiral Lewis acids.[65] As shown in Scheme 2.29, when cyclohexyl vinyl ether was activated by (*R*)-BINIM-4Me-2QN−Ni(II) complex chiral Lewis acids, it reacted with α,α′-dicarbonyl diazo compounds to yield the corresponding *endo*-cycloadducts in high yields, complete diastereoselectivity and high enantioselectivities of up to 97% ee.

Finally, Dias *et al.* have reported a highly efficient silver-catalysed enantioselective domino carbonyl-ylide formation–1,3-dipolar cycloaddition reaction of arylideneiminoglycinates with maleimides.[66] The process, performed in toluene in the presence of a catalytic amount of TEA, and employing 5 mol% of an equimolecular mixture of (*S*)-BINAP and AgSbF$_6$, led highly stereoselectively to the corresponding *endo* cycloadducts in high yields, excellent

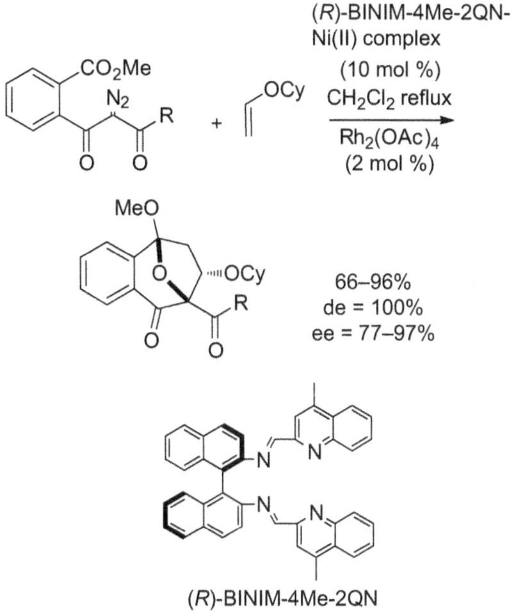

Scheme 2.29 Ni-catalysed domino carbonyl-ylide formation–1,3-dipolar cyclo-addition reaction of α,α′-dicarbonyl diazo compounds and cyclohexyl vinyl ether.

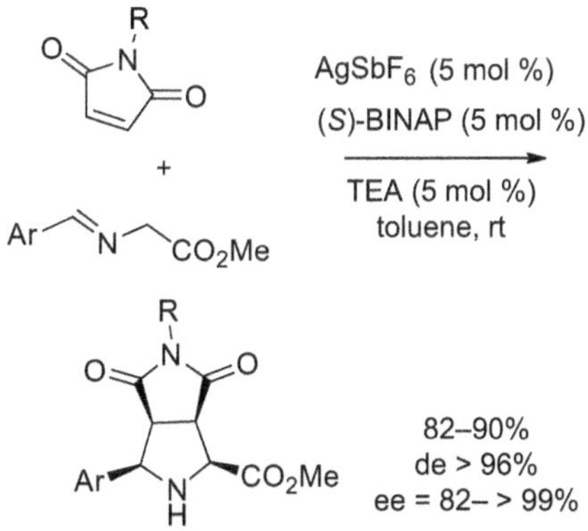

Scheme 2.30 Ag-catalysed domino carbonyl-ylide formation–1,3-dipolar cyclo-addition reaction of arylideneiminoglycinates and maleimides.

general diastereoselectivity of >96% de, and enantioselectivity of >99% ee in almost all the cases of substrates studied, as shown in Scheme 2.30. It must be noted, however, that acrylates, maleates, and fumarates were found to be unsuitable as dipolarophiles for the process, whereas good results were observed by using (*E*)-1,2-bis(phenylsulfonyl)ethylene as dipolarophile. Indeed, the corresponding *endo* cycloadducts were generated under the same conditions in high yields (80–91%), general diastereoselectivity of >96% de, and good to high enantioselectivities (88–92% ee).

2.2.8   Domino Reactions Initiated by the Heck Reaction

The Heck reaction is an important way to couple aryl and vinyl systems in the presence of palladium,[67] and has been recently exploited as a key step of many total syntheses of natural products.[68] Furthermore, it forms the keystone of a number of domino reactions.[69] In 1989, Overman *et al.* highlighted the feasibility of applying the asymmetric intramolecular Heck reaction to domino cyclisation of polyenes as a method to fashion quaternary chiral spirocyclic centres from trienyl triflates.[70] Following this seminal work, there have been many publications dealing with the application of the tandem Heck reaction, and the inclusion of the Heck reaction as a single step in asymmetric palladium-catalysed domino processes.[2] In 1996, Keay *et al.* reported the synthesis of (+)-xestoquinone by the palladium-catalysed asymmetric domino Heck coupling.[71] On the other hand, sequences of intramolecular Heck reaction–anion capture have met with great success in the synthesis of structurally diverse heterocycles and spirocycles.[72] In addition to a variety of nucleophiles employed to trap the stable σ-alkylpalladium complex, such as organoboronic acids, organotin derivatives, enolates, and π nucleophiles, cyanide has rarely been used as a terminating agent in such processes. As a recent example, Zhu *et al.* have reported an efficient synthesis of enantiomerically enriched 3-substituted-3-cyanomethyl-2-oxindoles based on an enantioselective palladium-catalysed domino intramolecular Heck-cyanation employing potassium ferro(II)cyanide, $K_4[Fe(CN)_6]$, as a trapping agent for the σ-alkylpalladium intermediate.[73] As shown in Scheme 2.31, reaction of *ortho*-iodoanilidines with potassium ferro(II)cyanide as a cyanide donor in DMF in the presence of potassium carbonate as a base, silver phosphate (Ag_3PO_4) as an additive, and a combination of bis(dibenzylideneacetone)palladium, [Pd(dba)₂], and (*S*)-DIFLUORPHOS [(*S*)-5,5′-bis(diphenylphosphino)-2,2,2′,2′-tetrafluoro-4,4′-bi-1,3-benzodioxole] as a chiral catalyst, afforded the corresponding chiral oxindoles in moderate to good yields and enantioselectivities of up to 79% ee. The authors demonstrated that the enantioselectivity of the domino process was very sensitive to the ligand structure by investigating a series of other chiral ligands, such as (*R*)-BINAP, (*R*)-Tol-BINAP, (*R*)-SYNPHOS, (*R*)-DIOP, (*R,S*)-JOSIPHOS, (*S*)-Cl-MeO-BIPHEP and (*R*)-C₃-TUNEPHOS. Among all the tested ligands, (*S*)-DIFLUORPHOS proved to be the most effective. The authors have proposed a possible reaction path which is depicted in Scheme 2.31. The oxidative addition of aryl iodide to a Pd(0) species generated *in situ* led to intermediate **13**, which underwent a 5-*exo*-trig cyclisation to afford the σ-alkylpalladium complex **14**. Displacement of the iodide by cyanide afforded palladium complex **15**, which upon reductive elimination produced the final oxindole with the concurrent regeneration of the Pd⁰ species. This novel methodology was applied to a concise total synthesis of esermethole and physostigmine, which are powerful inhibitors of acetyl- and butyryl-cholinesterase.

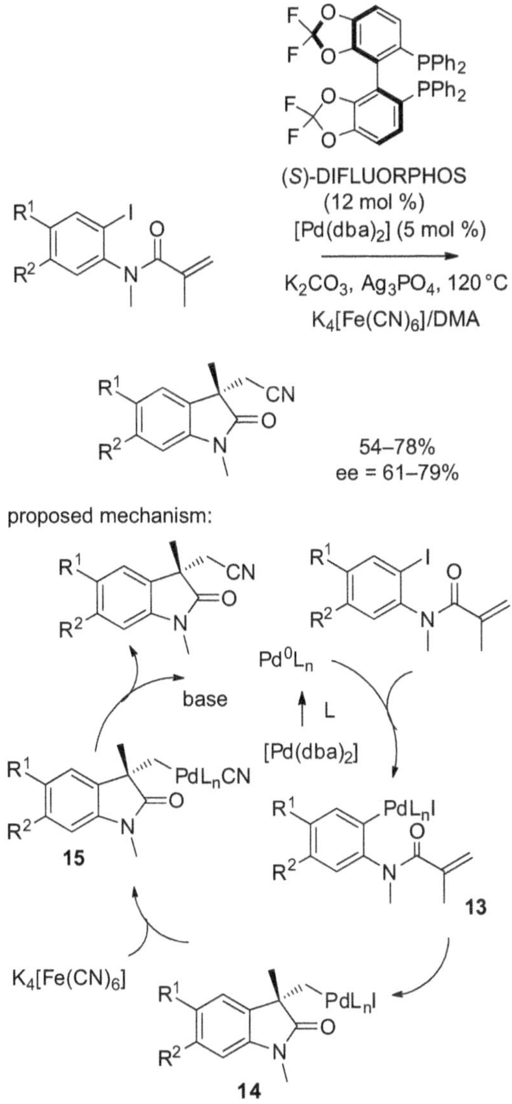

Scheme 2.31   Domino Heck–cyanation reaction.

In 2006, Keath *et al.* investigated a series of chiral ligands derived from BINAP to induce chirality in the palladium-catalysed domino Heck–cyclisation reaction of an aryl triflate, which provided the corresponding tetracyclic product (Scheme 2.32).[74] When the reaction was carried out in toluene at 105 °C upon catalysis with a combination of Pd$_2$(dba)$_3$ and (*S*)-3,3′-di-O*i*-Pr-BINAP as a catalyst, and in the presence of 1,2,2,6,6-penta-methylpiperidine (PMP) as a base, the corresponding cyclised (*R*)-product was generated as major enantiomer with 93% yield and moderate enantioselectivity of 74% ee. Surprisingly, the employment of (*S*)-BINAP as a ligand provided approximatively the same enantioselectivity and yield (ee = 72%, yield = 88%) for the same product, albeit with the opposite configuration (*S*). More recently, the same authors have demonstrated that the enantioselectivity of this reaction could be dramatically enhanced by using an electron-deficient ligand, such as (*R*)-DIFLUORPHOS, and by using ethanol rather than toluene as the solvent.[75] As shown in Scheme 2.32, the use of either an oil bath or a microwave for the reaction performed in ethanol gave rise to the expected (*R*)-product with excellent enantioselectivity ranging from 94 to >99% ee. The authors have proposed possible reasons for the observed results that use of both electron-deficient ligands and ethanol gave higher enantioselectivities, due to the ligand and ethanol stabilising the cationic intermediate **16** formed after oxidative addition, thereby slowing the rate of cyclisation onto the first olefin to form intermediate **17**. Furthermore, it must be noted that the time to completion was reduced significantly using microwave irradiation.

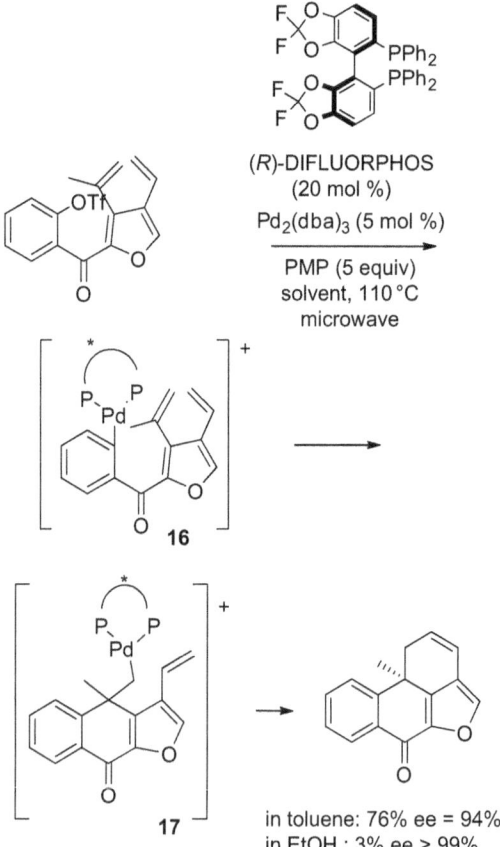

Scheme 2.32 Domino Heck–cyclisation reaction.

2.2.9 Domino Reactions Initiated by the Wacker Reaction

The most important reaction based on Pd(II)-catalysis is the Wacker reaction,[76] which is employed industrially for the synthesis of acetaldehyde starting from ethene. In 2005, the Wacker reaction was combined by Tietze *et al.* with a Heck reaction in an enantioselective domino process.[77] This highly efficient methodology was later used by the same authors as a key step in a novel total synthesis of vitamin E.[78] Indeed, this process allowed the formation of part of the chiral chroman framework, and the concurrent introduction of part of the side chain of this natural product. The reaction comprised the enantioselective Wacker cyclisation of a benzyl-protected phenol (Scheme 2.33) and methyl vinyl ketone performed in dichloromethane in the presence of catalytic amounts of Pd(TFA)$_2$, the chiral ligand (*S,S*)-Bn-BOXAX [2,2'-bis(oxazolyl)-1,1'-binaphthyl], and *p*-benzoquinone as a reoxidant, affording the corresponding chiral chroman in 84% yield and an excellent enantioselectivity of 97% ee, as shown in Scheme 2.33. The authors assumed that the first step of the domino reaction, the generation of the chiral catalyst from Pd(TFA)$_2$ and the chiral ligand, coordinated enantiofacially to the aliphatic double bond in benzyl-protected phenol (Scheme 2.33). Oxy-palladation provided the enantioselective formation of palladated chroman **18** with the correct absolute configuration at C-2 relative to α-tocopherol. Because β-hydride elimination was not possible, an intermolecular reaction with methyl vinyl ketone to form intermediate **19** occurred, and this could now undergo β-hydride elimination to yield the final chroman. This domino product was further submitted to an aldol condensation with (3*R*)-3,7-dimethyloctanal, followed successively by 1,2-addition of methyllithium, elimination of water and then hydrogenation to yield the desired vitamin E. Furthermore, a similar methodology was applied by the same authors to the enantioselective domino Wacker–Heck reaction of an alkenyl phenol, depicted in Scheme 2.33, and methyl acrylate, which provided under comparable conditions the corresponding chroman in 55% yield and 88% ee, albeit after a rather long reaction time of seven days (Scheme 2.33).[79] Attempts to increase the reaction rate by performing the process at elevated temperature resulted, however, in a significant reduction of the enantioselectivity. In addition, another asymmetric domino reaction including a Heck reaction was recently reported by Schmalz *et al.*[47] Indeed, these authors have developed an enantioselective access to 3-methylene-1*H*-indanol through a domino allylstannylation–Heck reaction, occurring between *o*-iodobenzaldehyde and allyl tributylstannane in the presence of a palladium complex generated *in situ* from [Pd$_2$dba$_3$] and a Taniaphos chiral ligand. The domino product was produced in 52% yield and 96% ee.

In 2009, Sasai *et al.* focused on the synthesis of novel chiral spiro ligands bearing *N*-heterocycles to be applied to induce palladium-catalysed asymmetric reactions, including domino Wacker-type cyclisation reactions.[80] Various types of chiral spiro ligand were prepared, such as spiro bis(isoxazole), spiro (isoxazole–isoxazoline), spiro bis(oxazoline), spiro bis(pyrazole) ligands, and further investigated as palladium ligands to promote the enantioselective domino Wacker-type cyclisation of a dialkenyl alcohol into its corresponding bicyclic product (Scheme 2.34). Among these ligands, spiro bis(isoxazoline)

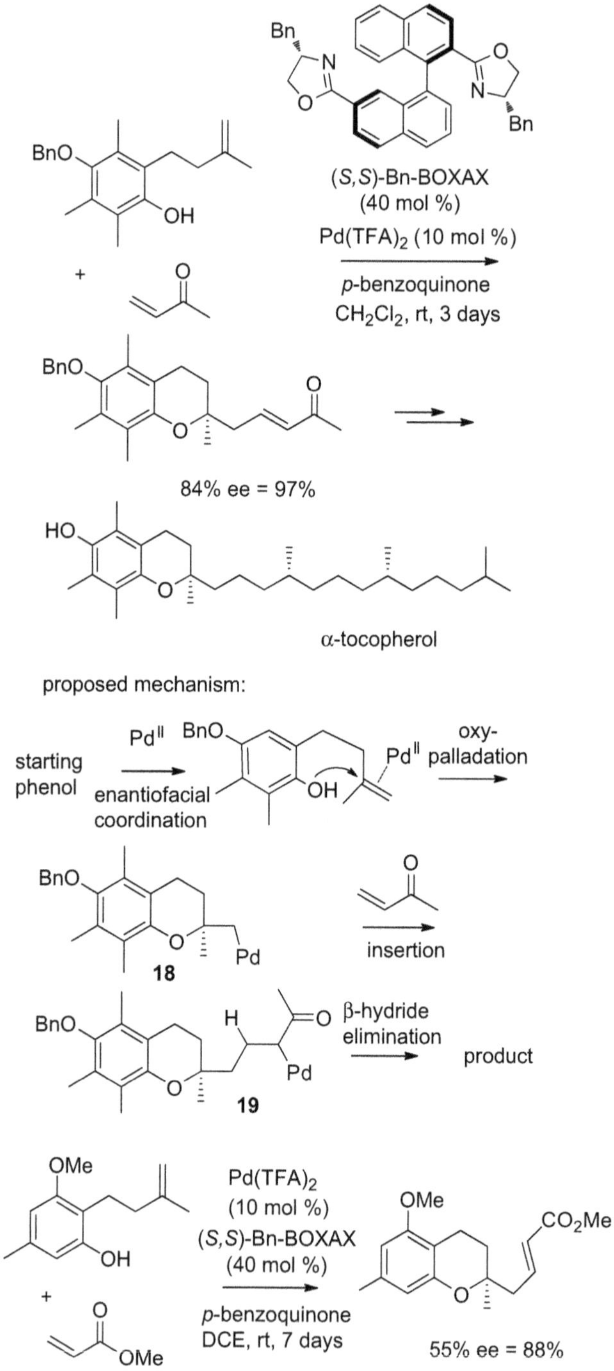

84% ee = 97%

α-tocopherol

proposed mechanism:

Scheme 2.33 Domino Wacker–Heck reactions, and synthesis of vitamin E.

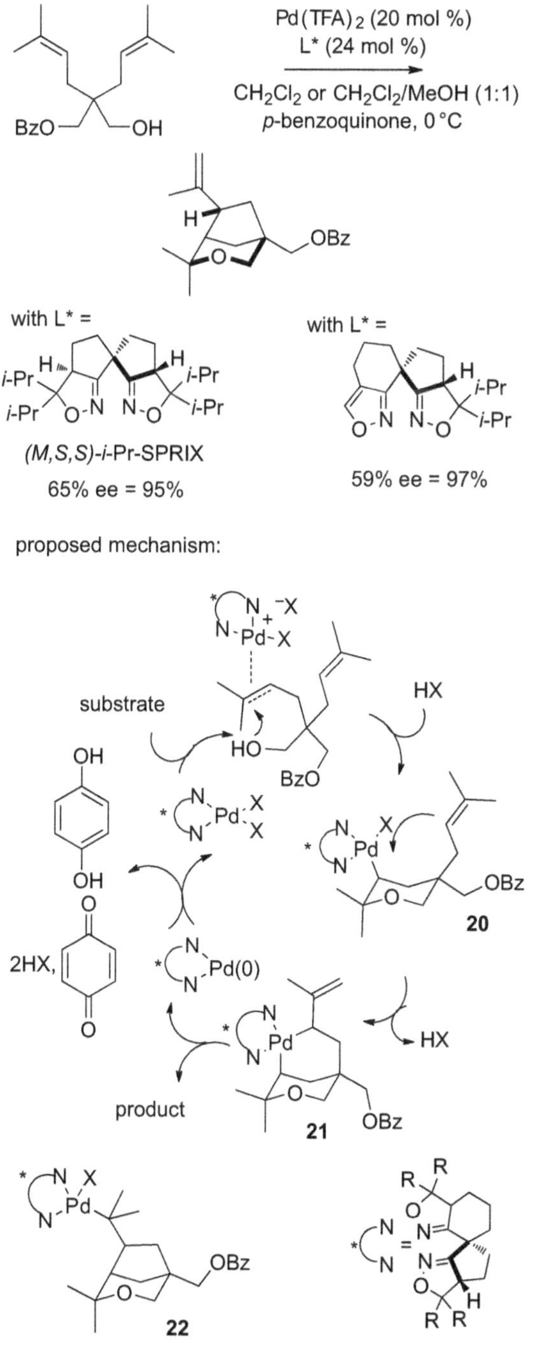

Scheme 2.34 Domino Wacker-type cyclisation reaction of dialkenyl alcohol.

ligand, *i*-Pr-SPRIX, and hybrid spiro (isoxazole–isoxazoline) ligand combined with Pd(TFA)₂ provided the best results both in terms of yields and enantio-selectivities, with values of up to 89% and 97% ee, respectively. While the domino reaction based on the use of the *i*-Pr-SPRIX ligand was performed in CH₂Cl₂ as the solvent, that employing the hybrid spiro (isoxazole–isoxazoline) ligand was carried out in a mixed solvent CH₂Cl₂/MeOH (1 : 1), as shown in Scheme 2.34. The authors have proposed the mechanism depicted in Scheme 2.34 to explain the results. Intramolecular nucleophilic attack of the hydroxyl group at the activated C–C double bond produced alkyl Pd(II) intermediate **20**. Formation of a subsequent C–C bond led to the final domino bicyclic product through the formation of palladacycle **21** by intramolecular carbopalladation or from a direct insertion of intermediate **22**.

2.2.10   Domino Reactions Based on Cyclisations

In the last few years, a number of novel enantioselective domino processes catalysed by various chiral metal complexes have been developed on the basis of cyclisation reactions. In 2006, Yang *et al.* developed a Pd(II)-catalysed enantioselective oxidative domino cyclisation reaction using readily available (–)-sparteine as the chiral ligand and molecular oxygen as a green oxidant, providing direct access to enantioenriched and structurally versatile indolines.[81] As shown in Scheme 2.35, treatment of a series of mono-substituted unsaturated anilides by molecular oxygen upon catalysis with a combination of Pd(TFA)₂ and (–)-sparteine in the presence of 3 Å molecular sieves and tertiary amine diisopropylethylamine (DIPEA) at 80 °C produced the corresponding

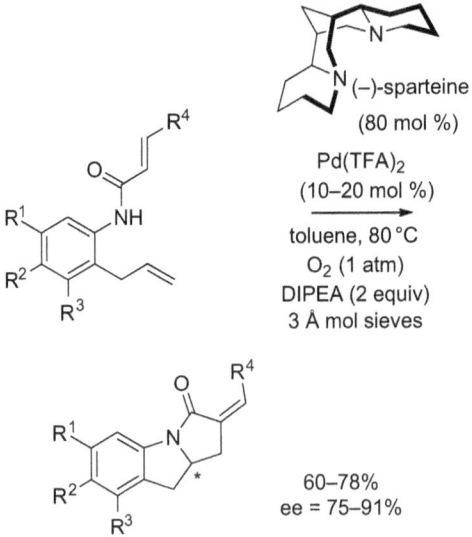

Scheme 2.35   Pd-catalysed oxidative domino cyclisation reaction of monosubstituted unsaturated anilides.

tricyclic indolines in moderate to good yields and enantioselectivities of up to 91% ee. The investigation of a series of other chiral ligands, such as (DHQ)₂PHAL, (DHQ)₂AQN, (DHQD)₂PYR, (–)-cinchonidine, hydroquinine, and (S,S)-Ph-BOX, has shown the emergence of (–)-sparteine as the most effective ligand to induce chirality in this process.

Later, these authors developed other new chiral palladium catalyst systems, comprising quinoline and chiral oxazoline units, which had the advantages of being air stable, structurally tunable, and highly diastereo- and enantioselective for a variety of disubstituted olefinic substrates. While the use of (–)-sparteine as a ligand provided good enantioselectivities in the case of oxidative domino cyclisation of monosubstituted olefinic substrates, its use for the reaction of disubstituted olefinic substrates gave only low enantioselectivities (≤14% ee). On the other hand, the use of novel catalyst systems, such as the combination of Pd(OAc)₂ with *t*-Bu-QUOX [quinoline-oxazolineSegphos: 5,5′-Bis(diphenyl-phosphino)-4,4′-bi-1,3-benzodioxole] as the ligand, allowed the formation of the corresponding tricyclic products in good yields and high enantioselectivities of up to 98% ee, as shown in Scheme 2.36.[82] Moreover, it was noteworthy that the processing of all substrates essentially yielded only one diastereomer. The authors assumed that the olefin geometry of the substrate completely controlled the relative stereochemistry of the product.

Metal-catalysed cyclisation reactions of 1,*n*-enynes have appeared as conceptually and chemically highly attractive processes because they contribute to the search for highly desirable atom economy and allow the discovery of new reactions.[83] In particular, the use of gold catalysts for organic transformations has become increasingly common over the past decade, leading to the development of a number of useful carbon–carbon and carbon–heteroatom bond-forming processes, allowing the synthesis of a wide variety of carbo- and

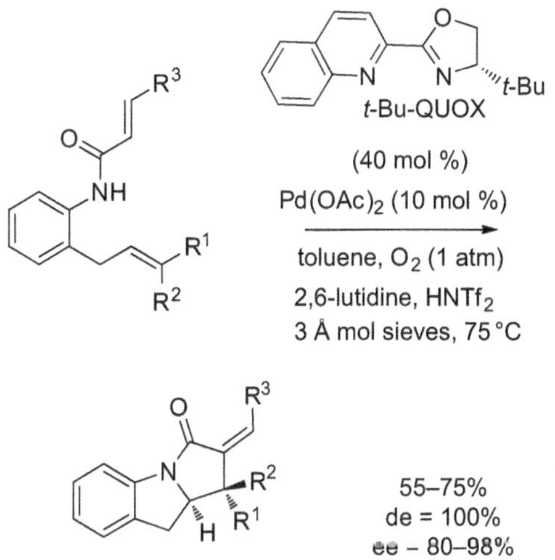

Scheme 2.36 Pd-catalysed oxidative domino cyclisation reaction of disubstituted unsaturated anilides.

heterocycles presenting a high degree of structural complexity.[84] The development of enantioselective variants of these transformations is, however, still rare,[85] probably because of the high substrate-dependency encountered with these systems. Among the numerous 1,n-enyne cyclisation reactions, Michelet *et al.* have recently developed domino processes based on cycloisomerisations performed in the presence of external nucleophiles, such as alcohols, electron-rich aromatic rings, amines, carboxylic acids, and 1,3-dicarbonyl compounds, which led to functionalised cyclic cyclopentenes.[86] In this context, these authors have reported enantioselective gold-catalysed domino hydro-arylation–cyclisation reactions of 1,6-enynes and various aromatic nucleophiles induced by a chiral Au(I) catalyst, such as (R)-MeO-dtbm-BIPHEP(AuCl)$_2$ [where BIPHEP is 2,2'-bis(diphenylphosphino)-1,1'-biphenyl], associated with AgOTf. As shown in Scheme 2.37, the corresponding arylated cyclic products were produced in high yields as single diastereomers and with good to excellent enantioselectivities of up to 98% ee by using a variety of aromatic nucleophiles, such as pyrrole, indole derivatives, and di- or trimethoxybenzene derivatives (Scheme 2.37). In this study, the authors have highlighted a general strong dependency of the intrinsic hindrance generated by both the enyne and the nucleophile. The mechanism of this domino hydroarylation–cyclisation reaction has been widely argued and is still under debate (Scheme 2.37). Based on a combined theoretical and experimental study, the nucleophilic attack–cycloisomerisation reaction has been proposed, by the group of Echavarren, to operate *via* a transient unstable cyclopropylcarbene **24**.[91g] More recently, Fürstner and Morency proposed an alternative rationale relying on the mesomeric "nonclassical carbocation", **25**.[87] The present results of asymmetric reactions showed a variation of enantioselectivity observed for a given enyne substrate in the presence of different carbon nucleophiles. In this regard, intermediates **24** and **25** did not adequately account for the formation of cyclic alkene **27**. Indeed, considering the *anti* nucleophilic attack on the cyclo-propylcarbene intermediate **24** as a stereospecific event, the external carbon nucleophile did not take part in the cyclopropylcarbene formation which represented the enantiodetermining step of the transformation. Considering the variation of enantioselectivity observed for a given enyne substrate and different carbon nucleophiles, the major contribution of intermediate **24**, and of its diastereomeric counterpart **24′**, was bound to the existence of an equilibrium between forms **23** and **24/24′**. The intervention of **25** would theoretically lead to a diastereomeric mixture of *syn/anti* derivatives if bond rotation of the carbocationic intermediates was faster than the addition of the carbon nucleophile, which had never been observed by the authors. According to these authors, the addition of the nucleophile could be concerted with the cyclisation. In this possible scenario, the intervention of a conformationally favoured intermediate **26**, as proposed by Fürstner,[87] could better account for the experimental results. It was indeed observed that the ee values increased with the size of the enyne tether. Considering a "chair-like" η^2-complex **26**, the addition of the nucleophile could occur in complete analogy with the Stork–Eschenmoser hypothesis,[88] introduced to account for the selectivity of

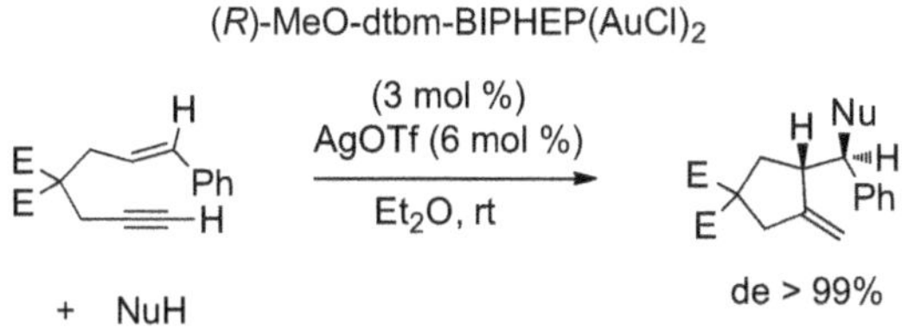

(R)-MeO-dtbm-BIPHEP(AuCl)$_2$

(3 mol %)
AgOTf (6 mol %)
Et$_2$O, rt

+ NuH

de > 99%

E	NuH	Yield (%)	ee (%)
CO$_2$Me	pyrrole	86	80
CO$_2$i-Pr	1-Me-indole	94	95
CO$_2$i-Pr	1-Me-2-Ph-indole	99	95
CO$_2$Bn	1,3,5-(MeO)$_3$C$_6$H$_2$	99	82
SO$_2$Ph	1,3-(MeO)$_2$C$_6$H$_3$	86	98
SO$_2$Ph	1,3,5-(MeO)$_3$C$_6$H$_2$	99	98
SO$_2$Ph	1,3,5-(MeO)$_3$-2-BrC$_6$H	85	94

proposed mechanism:

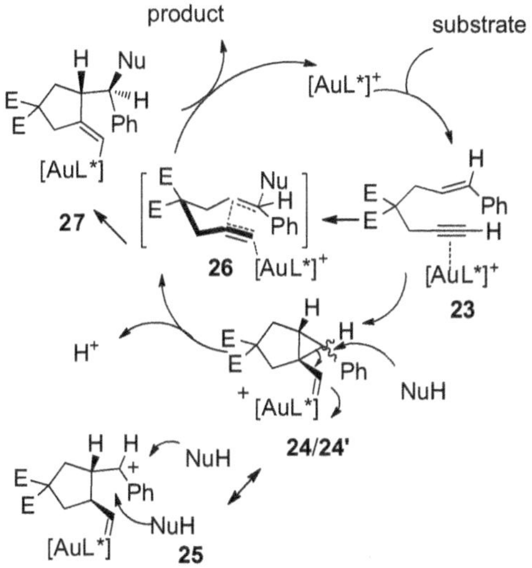

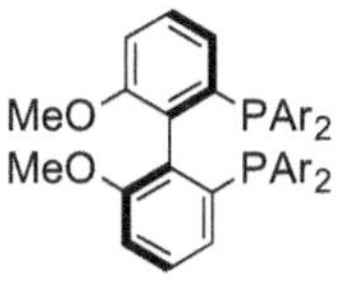

Ar = 4-MeO-3,5-(t-Bu)$_2$-C$_6$H$_2$
(R)-MeO-dtbm-BIPHEP

Scheme 2.37 Au-catalysed domino hydroarylation–cyclisation reaction.

polyene cyclisation reactions. In 2011, these authors extended the scope of this methodology to oxygen nucleophiles, such as water and methanol, affording the corresponding heterocycles in moderate to excellent yields (46–99%), complete diastereoselectivity and moderate enantioselectivities of up to 68% ee.[89] According to these results, it seemed that the nature of the nucleophile had a strong influence on the enantioselectivity of the reaction.

In 2006, Genêt *et al.* developed enantioselective platinum-catalysed domino hydroxyl- and alkoxycyclisation reactions of 1,6-enynes.[90] As shown in Scheme 2.38, when the reaction of 1,6-enynes with water or methanol was catalysed by a chiral platinum complex of (*R*)-Ph-BINEPINE (phenylbinaph-thophosphepine), a monophosphane atropisomeric ligand, in the presence of silver salts as additives, it afforded the corresponding functionalised carbo- and heterocycles in moderate to high yields and good enantioselectivities of up to 85% ee. A reasonable hypothesis for the mechanism of the process based on the Lewis acid character of the Pt catalyst is depicted in Scheme 2.38. The reaction could be initiated by the formation of π-alkynyl complex **28** through the complexation of the unsaturated triple bond to the metal catalyst. The π-alkynyl complex **28** could then evolve to give a cyclopropyl metal carbene complex **30**, which could be opened by an external nucleophile, such as methanol or water, to give rise to vinylmetallate **29**. Further protonolysis of this intermediate formed the final cycloadduct and regenerated the catalyst. It must be noted that the Pt carbene has not been isolated; consequently, the concerted addition of the nucleophile on intermediate **28** and formation of the C–C bond is also possible.

As an extension of this methodology, these authors have applied similar conditions to the enantioselective domino hydroarylation–cycloisomerisation reaction of 1,6-enynes with a variety of electron-rich aromatic and hetero-aromatic nucleophiles to give a range of functionalised cyclic derivatives in good yields and enantioselectivities, as shown in Scheme 2.39.[91] It must be noted that for all the substrates investigated, the corresponding domino products were formed in > 19 : 1 *syn : anti* diastereoselectivity. As depicted in Scheme 2.39, the reaction could evolve through initial η^2-coordination of the metal to the alkyne to give intermediate **31**, which was subsequently submitted to an intramolecular addition of the alkene, leading to cyclopropylcarbene **32**. In the presence of the external nucleophile Ar^{2}H, ring opening of the cyclo-propylcarbene and rearrangement to cyclised vinylmetal complex **33** occurred. Then, protodemetalation completed the catalytic cycle, providing the final product and regenerated catalyst.

The utility of *N*-heterocyclic carbenes as catalysts in domino reactions has received growing attention in the past few years.[92] In this area, asymmetric domino methoxycyclisations of 1,6-enynes and methanol have also been investigated by Tomioka *et al.* by using gold(I) complexes of chiral C_2-symmetric *N*-heterocyclic carbenes.[93] The corresponding functionalised cyclopentanes were achieved in high yields (93–95%), albeit with low to moderate enantioselectivities (4–56% ee). On the other hand, Sanz *et al.* have reported enantioselective gold-catalysed alkoxy- and hydroxycyclisation of

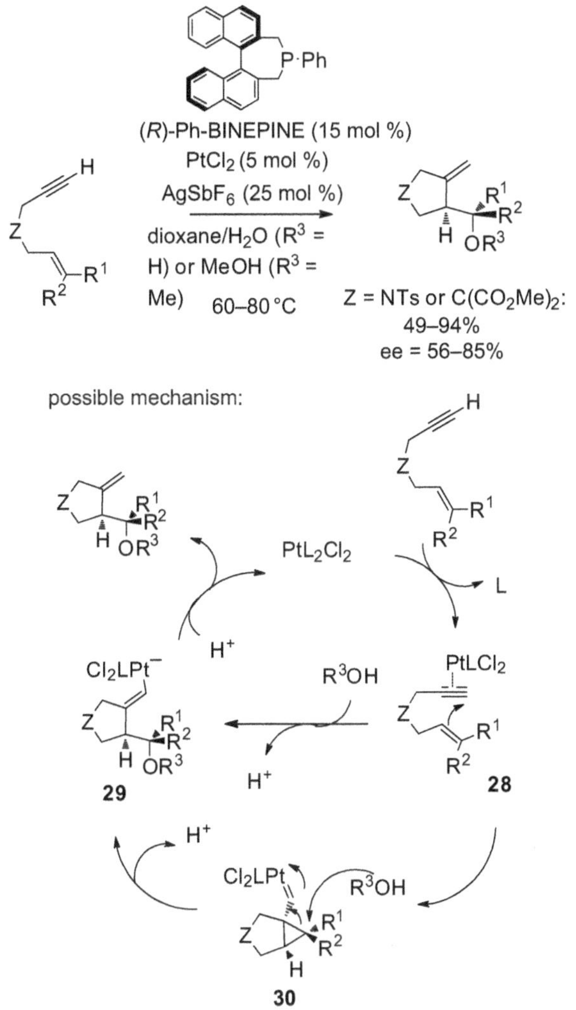

Scheme 2.38 Pt-catalysed domino alkoxy- and hydroxycyclisation reaction.

o-(alkylnyl)styrenes and alcohols and water, respectively, by employing a gold complex of (S)-3,5-Xylyl-MeO-BIPHEP in the presence of silver salts.[94] The process provided a range of chiral functionalised 1H-indene derivatives in high yields and moderate to high enantioselectivities of up to 92% ee, as shown in Scheme 2.40. The catalytic cycle depicted in Scheme 2.40 explained the formation of indenes. The reaction was initiated by coordination of the cationic gold complex to the triple bond of the o-(alkynyl)styrene to give intermediate **34**. Intramolecular addition of the alkene moiety selectively led to the cationic intermediate **35** through a 5-endo-dig-cyclisation. In the presence of alcohol R⁵OH, trapping of the carbocation **35** led to formation of vinyl gold

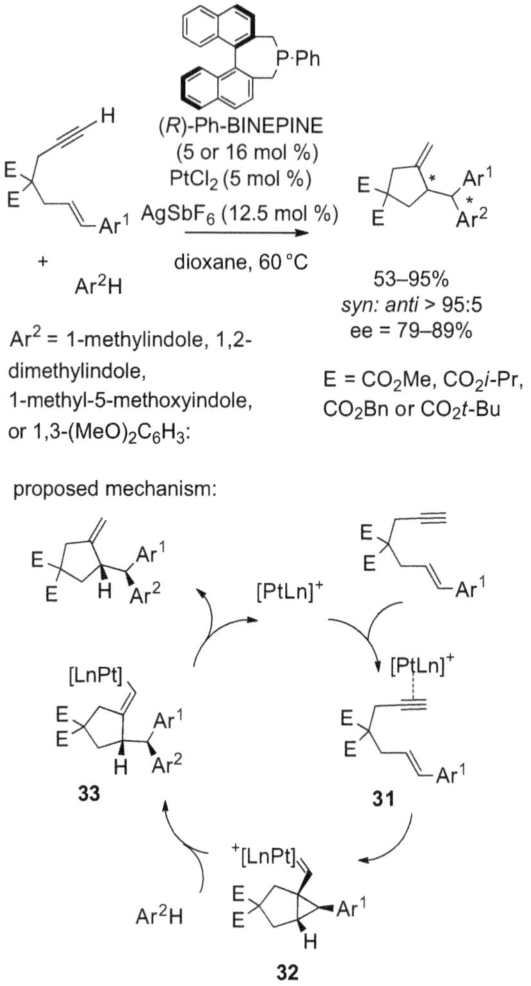

Scheme 2.39 Pt-catalysed domino hydroarylation–cycloisomerisation reaction.

intermediate **36**. A further protodemetalation furnished the final product, and regenerated catalyst.

Acyclic diaminocarbene ligands, also known as nitrogen acyclic carbenes, are potentially advantageous over the more familiar N-heterocyclic carbenes for enantioselective catalysis, because their wide N–C–N angles (116–121°) can place chiral substituents closer to the metal. In this context, Slaughter and Handa have very recently reported the use of this type of ligand to mediate enantioselective domino acetalisation–cycloisomerisation reaction of alkynylbenzaldehydes in the presence of alcohols.[95] As shown in Scheme 2.41, a series of chiral bicyclic products could be achieved in moderate to good yields and enantioselectivities of up to >99% ee by employing this new class of chiral gold(I) complex of nitrogen acyclic carbenes, such as catalysts depicted in

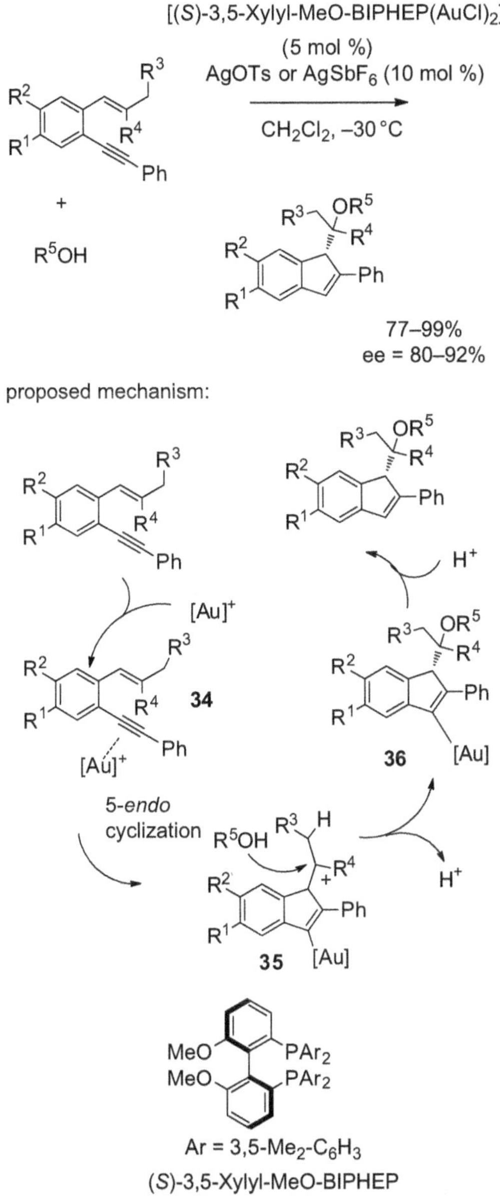

$[(S)$-3,5-Xylyl-MeO-BIPHEP(AuCl)$_2$]

(5 mol %)

AgOTs or AgSbF$_6$ (10 mol %)

CH$_2$Cl$_2$, −30 °C

77–99%
ee = 80–92%

proposed mechanism:

34

5-*endo*
cyclization

35 [Au]

36 [Au]

Ar = 3,5-Me$_2$-C$_6$H$_3$
(*S*)-3,5-Xylyl-MeO-BIPHEP

Scheme 2.40 Au-catalysed domino alkoxy- and hydroxycyclisation reaction.

Scheme 2.41 which bear chiral groups near the catalytic site. These complexes were generated through addition of bulky amines to Au(I) complexes of suitably substituted chiral biaryl isocyanides. In order to explain the results, the authors have proposed that enantioselectivity in alkynylbenzaldehyde cyclisation correlated with the presence of Au–arene interactions with an

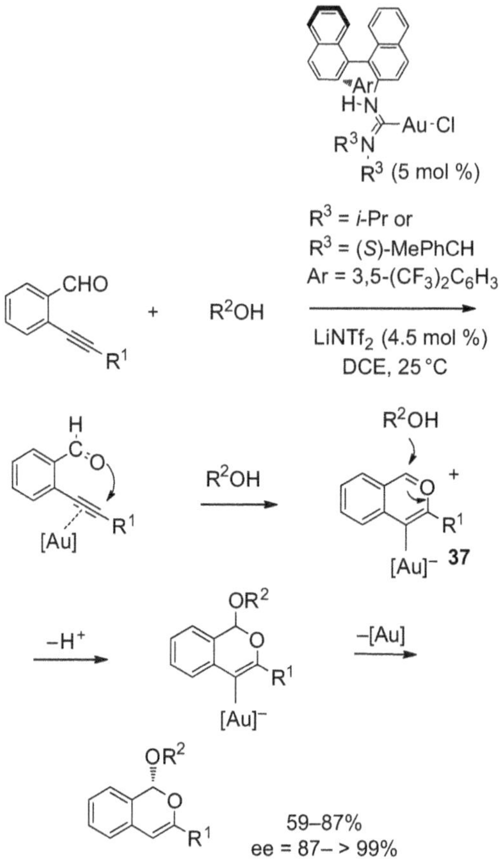

Scheme 2.41   Au-catalysed domino acetalisation–cycloisomerisation reaction.

electron-deficient aryl group. The combined structural, catalytic, and DFT results have suggested a dynamic chiral pocket in which reversible Au–aryl association, increased by intra-ligand sterics in the case of the bulkier catalyst ($R^3 = (S)$-MePhCH, where Me represents methyl), was sufficiently favourable to influence the enantiodetermining step. The reaction is supposed to evolve through the mechanism depicted in Scheme 2.41. The metal-activated alkyne formed a π-complex that was liable to nucleophilic attack directly from the oxygen of the carbonyl to give a highly reactive benzopyrylium intermediate **37** through a 6-*endo-dig* cyclisation reaction. The nucleophile could then attack this benzopyrylium intermediate and this led to catalyst recycling by protodemetallation with formation of the final isochromene.

In 2010, Zhang *et al.* reported a novel gold(I)-catalysed diastereo- and enantioselective domino cyclisation–[3 + 3]cycloaddition reaction of 2-(1-alkynyl)-2-alken-1-ones with nitrones, affording the corresponding chiral hetero-bicyclic furo[3,4-d][1,2]oxazines.[96] In this study, several chiral ligands derived from BIPHEP were investigated, such as (*R*)-C$_1$-TUNEPHOS and (*R*)-MeO dtbm BIPHEP, bearing bulky substituents on the phosphine aryl rings,

In general, the use of (*R*)-MeO-dtbm-BIPHEP as the ligand provided higher enantioselectivities of up to 98% ee than those obtained by using simple (*R*)-MeO-BIPHEP as the ligand. Furthermore, the authors have shown that placing aliphatic substituents on the olefin or the alkyne moiety of the substrates resulted in dramatic decreases in enantio- and diastereoselectivity. These results indicated that there was a steric demand to obtain excellent enantioselectivity. Moreover, in the case of some substrates, the ligand (*R*)-C_1-TUNEPHOS provided better enantioselectivities than the ligand (*R*)-MeO-dtbm-BIPHEP, which demonstrated that two strategies of modification of MeO-BIPHEH were effective, with the former being more efficient, owing to the Au–Au interaction which could make the structure more rigid. The best results are collected in Scheme 2.42, in which a plausible mechanism for this reaction is depicted.

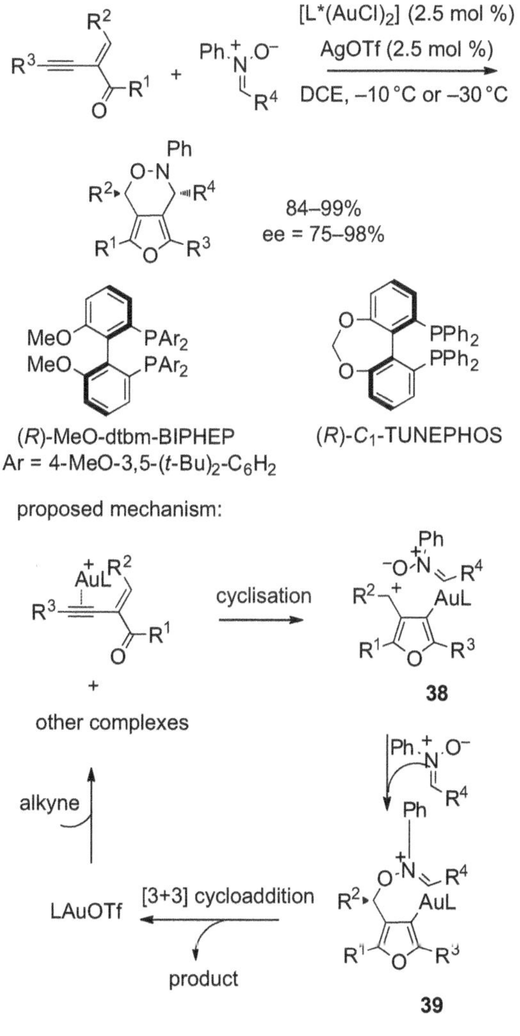

Scheme 2.42   Au-catalysed domino cyclisation–[3 + 3]cycloaddition reaction.

The cationic gold(I) species first coordinated the furanyl gold complex **38** arising from cyclisation, which was rapidly trapped by the nucleophilic oxygen atom of the nitrone to afford intermediate **39**. Subsequent intramolecular [3 + 3] cycloaddition produced the final product as well as regenerating the gold catalyst.

These authors have recently applied ligand (*R*)-MeO-dtbm-BIPHEP to induce another type of gold-catalysed domino reaction based on an intramolecular cyclisation–1,5-hydride transfer–cyclisation reaction of 2-(1-alkynyl)-2-alken-1-ones.[97] This process allowed a range of chiral furan-fused azepine derivatives to be achieved, generally in both high yields and high enantioselectivities, as shown in Scheme 2.43. These results represented the first example of an enantioselective redox-neutral domino reaction catalysed by gold(I) that resulted in the direct functionalization of unreactive sp^3 C–H bonds. A plausible mechanism for this reaction is depicted in Scheme 2.43. Activation of the alkyne moiety of the substrate by the gold catalyst caused a heterocyclisation (first cyclisation) to give furanyl intermediate **40**, possessing a reactive carbocation. After 1,5-hydride shift to produce intermediate **41**,

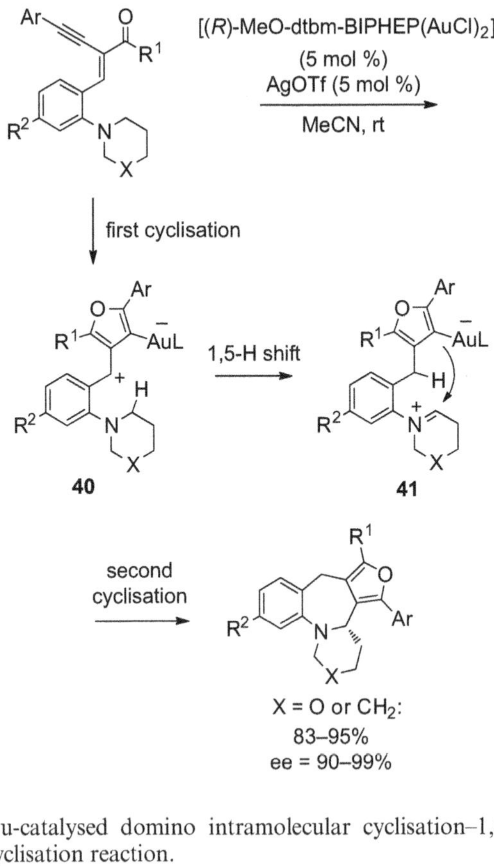

$X = O$ or CH_2:
83–95%
ee = 90–99%

Scheme 2.43 Au-catalysed domino intramolecular cyclisation–1,5-hydride transfer–cyclisation reaction.

followed by subsequent ring closure (second cyclisation), the final furan-fused azepine derivative was achieved.

In another context, Toste *et al.* have reported the use of the same chiral ligand (*R*)-MeO-dtbm-BIPHEP and its closely related analogue (*R*)-MeO-dtb-BIPHEP to induce chirality in gold(I)-catalysed enantioselective domino polycyclisation reactions of a range of functionalised 1,6-enynes.[98] As shown in Scheme 2.44, the 1,6-enynes were bicyclised under gold catalysis into the corresponding bicyclic lactones in good yields and high enantioselectivities of 92% ee through a highly enantioselective polyene cyclisation reaction in which transition metal-promoted alkyne activation served as the cyclisation initiating event. The scope of this methodology was extended to phenoxy-substituted phenyl alkynes, which provided the corresponding chiral hexahy-droxanthene derivatives in excellent yields and enantioselectivities of up to 94% ee, as shown in Scheme 2.44. Additionally, the use of an electron-rich aryl group, such as 3,5-dimethoxyphenyl, as a nucleophile on the substrate allowed for the enantioselective formation of the corresponding domino product, which contained a benzylic quaternary centre in 98% yield and 94% ee (Scheme 2.44). These results constituted the first example of a highly enantioselective polyene cyclisation reaction in which transition metal-promoted alkyne activation served as the cyclisation initiating event. In another context, Liu *et al.* have recently reported the stereoselective synthesis of azacyclic products through a gold-catalysed domino redox–[2 + 2 + 2]cycloaddition reaction of 1-alkynyl-2-nitrobenzenes and a range of alkenes.[99] The authors have developed an enantioselective version of this methodology by employing a gold complex of (*R*)-DM-Segphos. When this ligand was used to induce the domino reaction between 1-ethynyl-2-nitrobenzene and 2,3-dihydrofuran, the corresponding azacyclic product was obtained as a single diastereomer with a moderate enantioselectivity of 73% ee.

In 2008, Glueck *et al.* showed that enantioselective domino alky-lation–arylation reacton of primary phosphines with 1-bromo-8-chloromethylnaphthalene could be catalysed by Pt(DuPhos) complexes to provide the corresponding P-stereogenic 1-phosphaacenaphthenes in moderate enantioselectivities of up to 74% ee.[100] In the same year, Gagné *et al.* reported the enantioselective platinum-catalysed cascade cyclisation of 3,7-dienol substrates to give the corresponding bicyclic products.[101] Among a wide variety of chiral readily available diphosphine ligands investigated in this process, a platinum complex of (*S*)-Xylylphanephos was shown to be the most effective, providing low to good enantioselectivities of up to 87% ee. The reaction was compatible with monosubstitution and 1,2-disubstitution at the terminal alkene; however, *Z* alkenes were better behaved than the *E* alkenes. In each case of the substrates investigated, a single stereo- and regioisomer of the product was obtained. Although perfect stereospecificity was observed in the *E* and *Z* substrates, they differed markedly in their enantioselectivities. Indeed, the terminal *Z* alkene cyclised with the highest enantioselectivities (87% ee), whereas non-substituted terminal alkenes provided moderate to good enanti-oselectivities (79% ee), and terminal *E* alkenes poor enantioselectivities

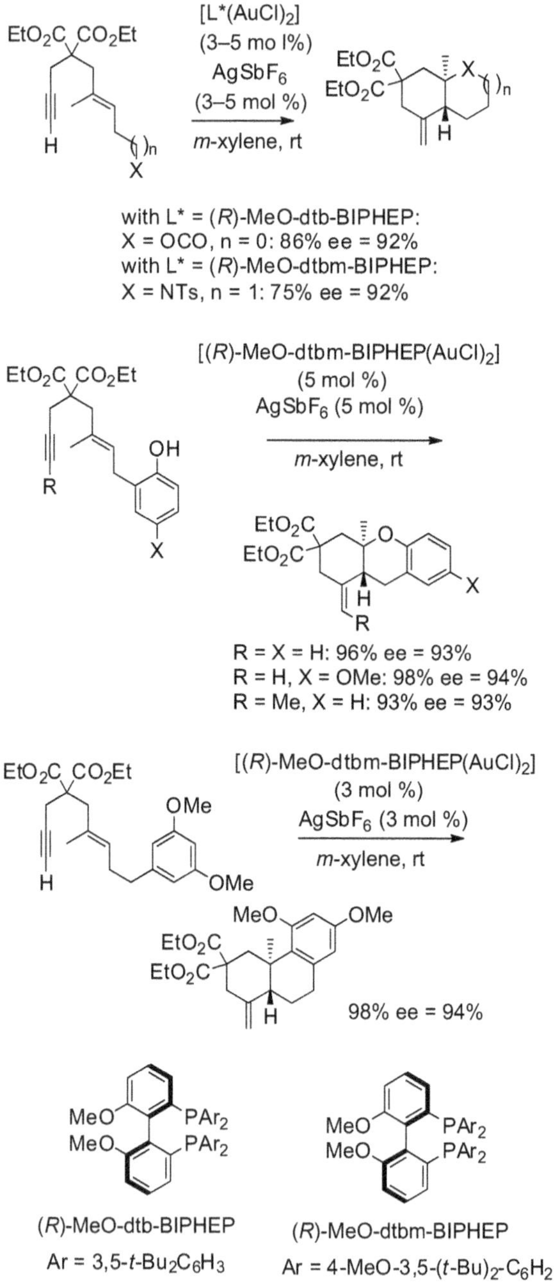

EtO₂C CO₂Et

[L*(AuCl)₂]
(3–5 mo l%)

AgSbF₆
(3–5 mol %)

m-xylene, rt

with L* = (*R*)-MeO-dtb-BIPHEP:
X = OCO, n = 0: 86% ee = 92%
with L* = (*R*)-MeO-dtbm-BIPHEP:
X = NTs, n = 1: 75% ee = 92%

EtO₂C CO₂Et

[(*R*)-MeO-dtbm-BIPHEP(AuCl)₂]
(5 mol %)

AgSbF₆ (5 mol %)

m-xylene, rt

R = X = H: 96% ee = 93%
R = H, X = OMe: 98% ee = 94%
R = Me, X = H: 93% ee = 93%

EtO₂C CO₂Et

[(*R*)-MeO-dtbm-BIPHEP(AuCl)₂]
(3 mol %)

AgSbF₆ (3 mol %)

m-xylene, rt

98% ee = 94%

(*R*)-MeO-dtb-BIPHEP

Ar = 3,5-*t*-Bu₂C₆H₃

(*R*)-MeO-dtbm-BIPHEP

Ar = 4-MeO-3,5-(*t*-Bu)₂-C₆H₂

Scheme 2.44 Au-catalysed domino polycyclisation reactions.

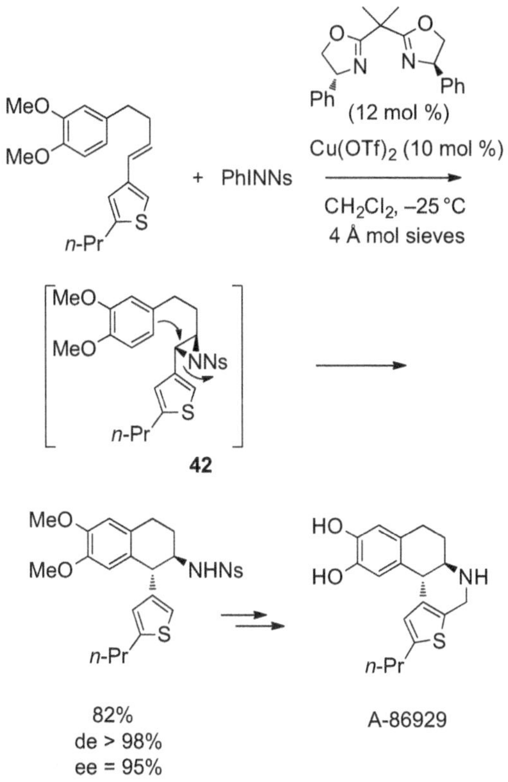

Scheme 2.45 Cu-catalysed domino aziridination–Friedel–Crafts cyclisation reaction, and synthesis of dopamine D1 agonist, A-86929.

(10–12% ee). On the other hand, a copper-catalysed enantioselective domino reaction based on an aziridination reaction followed by a Friedel–Crafts cyclisation was recently developed by Hajra and Bar.[102] As shown in Scheme 2.45, a styrene was reacted with PhINNs (where INNs is N-(p-nitro-benzenesulfonyl)iminophenyliodinane) in the presence of Cu(OTf)$_2$ and (R)-phenyl glycinol-derived (R)-Box ligand, to give the corresponding chiral bicyclic domino product in 82% yield, almost complete *trans*-diastereoselectivity, and enantioselectivity of 95% ee. As depicted in Scheme 2.45, the styrene was converted in a first step into the corresponding chiral aziridine **42** by reaction with PhINNs. Subsequently, aziridine **42** was submitted to ring-opening through a Friedel–Crafts reaction type cyclisation to lead to the final domino product. The utility of this methodology was illustrated in the total synthesis of a dopamine D1 agonist, A-86929.

In a recent work, Toste *et al.* have reported the enantioselective synthesis of highly substituted furans on the basis of a copper(II)-catalysed domino cycloisomerisation–indole addition reaction.[103] As shown in Scheme 2.46, copper(II) catalysis of a chiral phosphine promoted the intramolecular

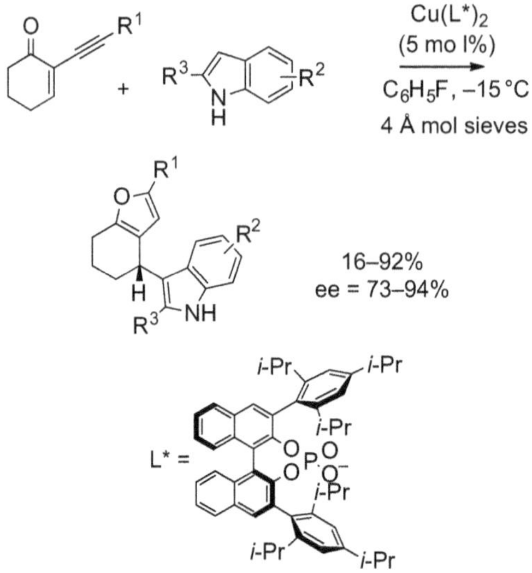

Scheme 2.46 Cu-catalysed domino cycloisomerisation–nucleophile addition reaction.

heterocyclisation of a range of 2-(1-alkynyl)-2-alkene-1-ones and facilitated high levels of enantioselectivity in the subsequent nucleophilic attack of indoles, providing the corresponding tetracyclic domino products. The reaction showed a wide substrate scope because both aromatic and aliphatic alkynes gave high yields and enantioselectivities of up to 94% ee. Similarly, electronic variations in the aryl ring of the indole scaffold were also tolerated, as both electron-donating and -withdrawing substituents gave the desired products in high yields and enantioselectivities. However, 2-methyl indole ($R^2 = H$, $R^3 = Me$) proved to be a difficult substrate for the reaction with phenyl alkyne ($R^1 = Ph$) because only 16% yield was reached in this case. The authors have demonstrated that formation of a copper(II)–indole species was important for the catalysis of the process.

2.2.11 Domino Radical Reactions

In recent years, studies on enantioselective radical reactions have achieved some remarkable success, particularly in intermolecular reactions.[104] In contrast, only a few reports have described enantioselective radical cyclisations. As a recent example, Takemoto *et al.* have developed enantioselective domino radical addition–cyclisation reactions of hydroxamate esters with alkyl iodides.[105] These reactions were induced by one equivalent of a combination of $Zn(OTf)_2$ with the chiral ligand depicted in Scheme 2.47, in the presence of 1 equivalent of triethylborane, which played the role of radical initiator. The process afforded diastereoselectively the corresponding *cis*-lactams in moderate yields and enantioselectivities of up to 88% ee. As shown in Scheme 2.47, the domino reaction began with the addition of a radical arising from alkyl iodide to hydroxamate esters, which generated radical intermediates **43**. The latter subsequently cyclised to give novel intermediates **44**. Then, an iodide atom-transfer process from secondary or tertiary alkyl iodide (R^2I) to unstable primary intermediate radicals occurred, providing the final lactams. It must be noted that the reaction of a non-substituted hydroxamate ester with alkyl iodides led to the formation of the corresponding *trans*-lactams in higher enantioselectivities of up to 92% ee, albeit with lower yields and diastereo-selectivities (Scheme 2.47). In spite of the good results obtained by using this methodology, it suffered from the disadvantage of requiring 1 equivalent of catalyst.

The scope of this methodology was extended to the domino radical addition–cyclisation reaction of closely related hydroxamate esters bearing an alkyne moiety instead of the alkene moiety, which afforded the corresponding

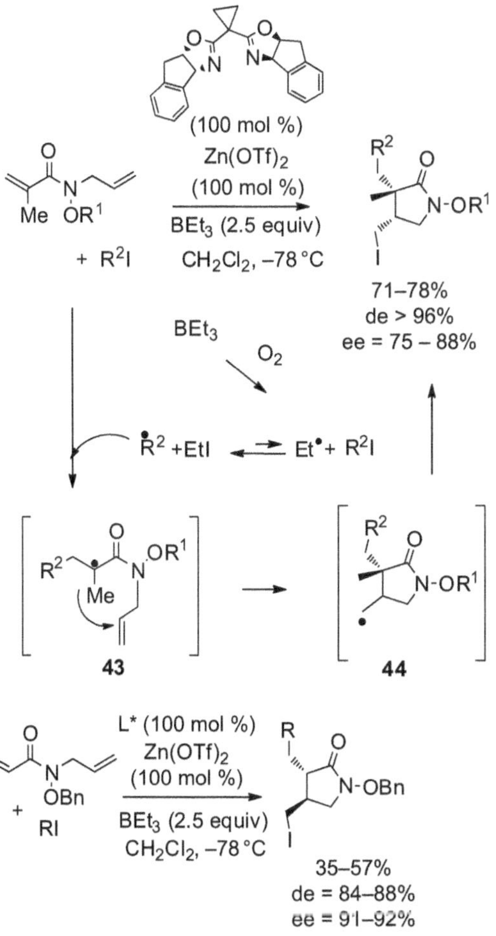

Scheme 2.47   Zn-catalysed domino radical addition–cyclisation reactions of hydro-xamate esters with alkyl iodides.

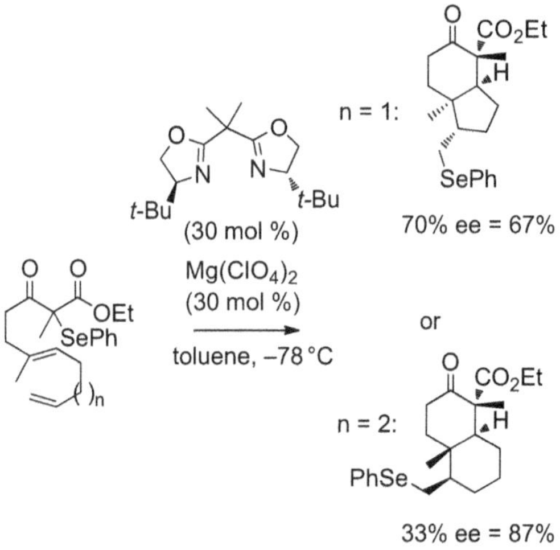

Scheme 2.48   Mg-catalysed domino group-transfer–radical cyclisation reaction.

domino products in both high yields and high diastereoselectivities, combined with good enantioselectivities of up to 92% ee.[105] The authors have shown that reduction of the catalyst loading to 10 mol% resulted in a decrease of the chemical yield and enantioselectivity. Furthermore, these conditions were also applied to the enantioselective zinc-catalysed domino radical addition–cyclisation reaction of oxime ethers, which produced the corresponding *trans*-lactams in moderate yields, and diastereo- and enantioselectivities of up to 86% de and 86% ee, respectively.[106] Finally, Yang *et al.* reported the first enantioselective catalytic domino group-transfer–radical cyclisation reaction, in 2006.[107] This work delt with the enantioselective PhSe-group-transfer radical cyclisation reaction of α(phenylseleno)-β-ketoesters which was catalysed by a complex generated from Mg(ClO4)2 and a chiral bisoxazoline (Scheme 2.48) to give the corresponding bicyclic domino products. Substrate with n = 1 led to the corresponding *cis*-6,5-fused ring product in 70% yield and 67% ee, whereas substrate with n = 2 underwent 6-*endo*/6-*exo* cyclisation to give the corresponding 6,6-*trans*-fused ring product in 33% yield and 87% ee, as shown in Scheme 2.48.

2.2.12   Miscellaneous Domino Reactions

In 2007, Willis *et al.* described the enantioselective addition of imide depicted in Scheme 2.49 to *N*-tosylimines, providing the corresponding cyclised Mannich adducts in both generally excellent yields and enantioselectivities of up to 99% ee, albeit with moderate to good *anti* diastereoselectivities, as shown in Scheme 2.49.[108] The reaction was catalysed by a chiral magnesium complex generated *in situ* from Mg(ClO4)2 and a chiral bisoxazoline ligand, which was selected from a series of chiral ligands tested, including other bisoxazoline

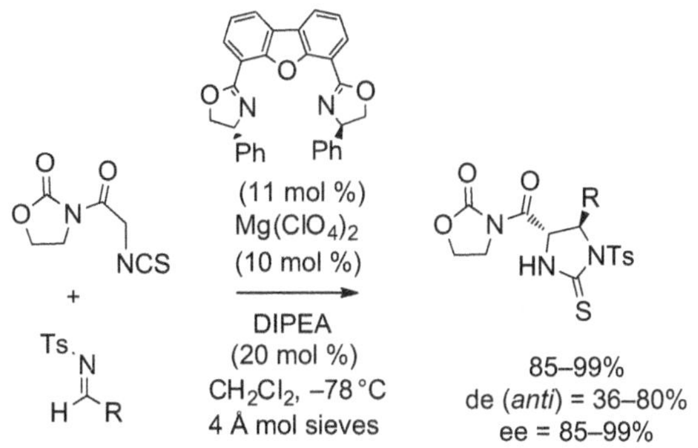

Scheme 2.49 Mg-catalysed Mannich reaction.

ligands such as PyBox ligands. As shown in Scheme 2.49, a variety of aryl-, alkenyl-, and alkyl-derived imines could be employed, allowing a novel enantioselective route to *anti* configured protected α,β-diamino acids on the basis of a direct enantioselective Mannich reaction.

In another context, Fu and Son have established that, through the use of a planar chiral bipyridine ligand, enantioselective copper-catalysed [4 + 1] cyclo-additions of α,β-unsaturated ketones with diazoacetates could produce highly substituted 2,3-dihydrofurans in good yields, and high diastereo- and enanti-oselectivities of up to >90% de and 92% ee, respectively.[109] The best results were reached with enones having unsaturated substituents. The transition metal-mediated conversion of alkynes, alkenes, and carbon monoxide in a formal [2 + 2 + 1] cycloaddition manner, commonly known as the Pauson–Khand reaction, is an elegant protocol for the construction of cyclopentenones.[110] Asymmetric versions of this reaction have been included in various domino reactions. In this context, Kwong *et al.* have reported tandem rhodium-catalysed cooperative decarbonylation and asymmetric Pauson–Khand-type cyclisation reactions of 1,6-enynes in which a formate or an aldehyde was a CO surrogate.[111] In the presence of a chiral rhodium complex, generated from [Rh(cod)Cl]₂ and (*S*)-BisbenzodioxanPhos as chiral ligand, a series of oxygen-, amine-, and carbon-tethered 1,6-enynes reacted under microwave conditions with cinnam-aldehyde as a source of CO to give the corresponding cyclopentenones in moderate to good yields (40–73%), and moderate to good enantioselectivities (49–90% ee). In another context, Nishibayashi *et al.* have reported the ruthenium-catalysed enantioselective formal [3 + 3] cycloaddition of propargylic alcohols with 2-naphthols, which afforded the corresponding naphthopyran derivatives in moderate to good yields and high enantioselectivities of up to 99% ee, as shown in Scheme 2.50.[112] The process was induced by a chiral thiolate-bridged diruthenium complex in combination with NaBH₄BF₄. The authors assumed that the process proceeded *via* stepwise reactions of propargylation and intramolecular cyclisation, in which ruthenium–allenylidene and vinylidene complexes worked as key reactive intermediates, respectively.

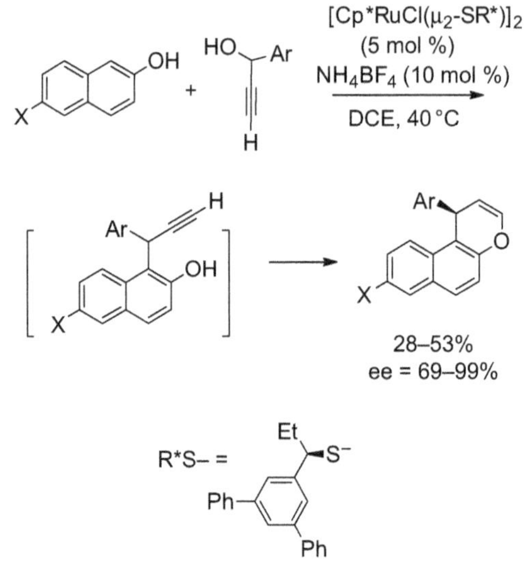

Scheme 2.50   Ru-catalysed domino propargylation–intramolecular cyclisation reaction.

The enantioselective reductive amination of ketones in a one-pot reaction without isolation of the intermediates, by using multifunctional catalysts, remains a challenge of high importance. In this context, Bruneau and Bondarev have developed an enantioselective domino reductive amination of 2-tetralone into the corresponding chiral amine.[113] By parallel screening of a small library of ligands incorporated with three different metals, such as ruthenium, rhodium, and iridium, the catalytic system {Ru(cod)(O$_2$CCF$_3$)$_2$ + (S)-MeOBIPHEP} was found to be superior. When this system was used in the presence of hydrogen, it allowed the chiral amine to be produced with enantioselectivity of up to 47% ee. On the other hand, Londregan *et al.* have developed an enantioselective synthesis of (R)-3-amino-3,4-dihydro-1H-[1,8]naphthyridin-2-one based on a novel asymmetric domino reduction–lactamisation reaction of the corresponding enone.[114] Upon hydrogenation in the presence of Ru((R)-BINAP)OAc$_2$ as catalyst, the bicyclic (R)-product was achieved in both excellent yield and enantioselectivity of 97% and 98% ee, respectively, as shown in Scheme 2.51. Furthermore, the authors have demonstrated that by simply switching the chirality of the ruthenium catalyst to Ru((S)-BINAP)OAc$_2$ under identical reaction conditions, the corresponding enantiomeric (S)-product was obtained in 97% yield and 97% ee.

In 2010, Peters *et al.* applied the cooperative activation by a soft bimetallic catalyst, a hard Brønsted acid, and a hard Brønsted base to the formation of highly enantioenriched, diastereomerically pure masked α-amino acids bearing adjacent quaternary and tertiary stereocentres on the basis of a domino azlactone formation–Michael addition reaction starting from N-benzoylated amino acids and α,β-unsaturated ketones.[115] As shown in Scheme 2.52, the best results for the domino reaction were achieved when bispalladacycle ferrocene

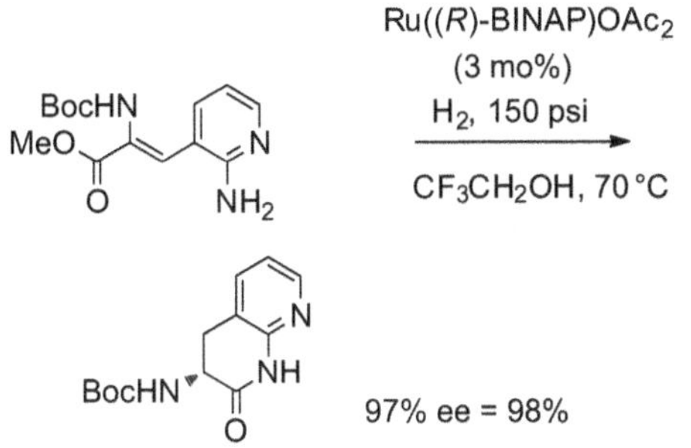

Scheme 2.51 Ru-catalysed domino hydrogenation–lactamisation reaction.

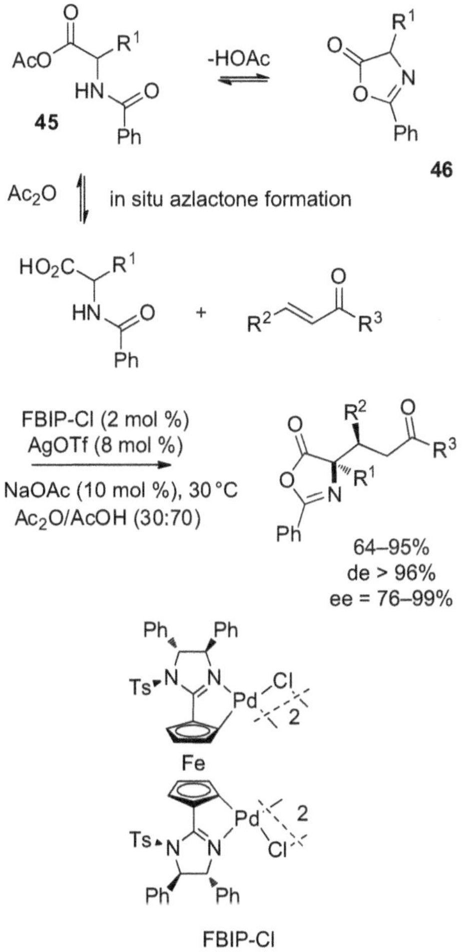

Scheme 2.52 Ag-catalysed domino azlactone formation–Michael addition reaction.

bis-imidazoline bis-palladacycle (FBIP)-Cl was activated by AgOTf in the presence of catalytic amounts of NaOAc. Given that the activated catalyst was stable towards acetic anhydride, the *in situ* formation of azlactones could be achieved through *O*-acylation with acetic anhydride of *N*-benzoylated amino acids, generating mixed anhydrides **45**. For a second time, the azlactones **46** were added to α,β-unsaturated ketones through a Michael addition to give the final products in high yields combined with excellent diastereo- and enantios-electivities of up to >96% de and 99% ee, respectively. The formed chiral products could be applied to prepare bicyclic dipeptides.

In 2011, chiral 2-substituted-1,5-benzodiazepine derivatives were synthesised for the first time from an enantioselective domino reaction involving *o*-phenylenediamine and 2'-hydroxychalcones.[116] The process was induced by a titanium complex formed from a chiral ligand derived from (*S*)-BINOL (1,1'-bi-2-naphthol) and L-prolineamide, and Ti(O*i*-Pr)$_4$. It led to the formation of a range of 2-aryl-1,5-benzodiazepine derivatives in good yields and enantioselectivities of up to 82% ee, as shown in Scheme 2.53. The 2'-hydroxy group of the α,β-unsaturated ketones was demonstrated to be critical for both the reactivity and stereoinduction of the process. In order to explain these results, the authors have proposed that the α,β-unsaturated ketimine intermediate **47** formed for the first time might be stabilised by an intramolecular hydrogen bond of the hydroxyl group. In the presence of the chiral titanium complex, intermediate **47** could coordinate to the metal with the oxygen atom of the hydroxyl group and the nitrogen atom of the imine moiety to generate intermediate **48**. In intermediate **48**, the amino group preferred to attack the β-*Si* face of the C = C bond because there was less steric hindrance between the aniline moieties of the ligand. Thus, the (*S*)-products were afforded through a final aza-Michael addition and the chiral titanium complex was released.

The forerunner in the cobalt-catalysed domino processes was that developed by Vollhardt *et al.* with their excellent synthesis of steroids initiated by a [2 + 2 + 2] cycloaddition.[117] Since then, a number of other cobalt-catalysed domino reactions have been developed. Among them, Feng *et al.* have reported a highly enantioselective synthesis of tetrahydroquinolines *via* cobalt(II)-catalysed domino 1,5-hydride transfer–cyclisation reaction.[118] As shown in Scheme 2.54, a chiral catalyst generated from a L-proline-derived *N,N'*-dioxide and Co(BF$_4$)$_2$ · 6H$_2$O was applied to the asymmetric intramolecular hydride transfer initiated cyclisation reaction of a series of *o*-dialkylamino-substituted alkylidene malonate derivatives to provide the corresponding biologically interesting tetrahydroquinolines in high to excellent yields and high enantios-electivities of up to 90% ee. The mechanism of the process involved the formation of the zwitterionic intermediate **49** through intramolecular hydride transfer, which subsequently cyclised to give the final product (Scheme 2.54).

In another context, Davies *et al.* have recently developed a novel domino reaction initiated by the formation of an oxygen ylide and which involved vinyldiazoacetates and allylic alcohols as the substrates. The enantioselective domino reaction generated highly functionalised chiral cyclopentanes

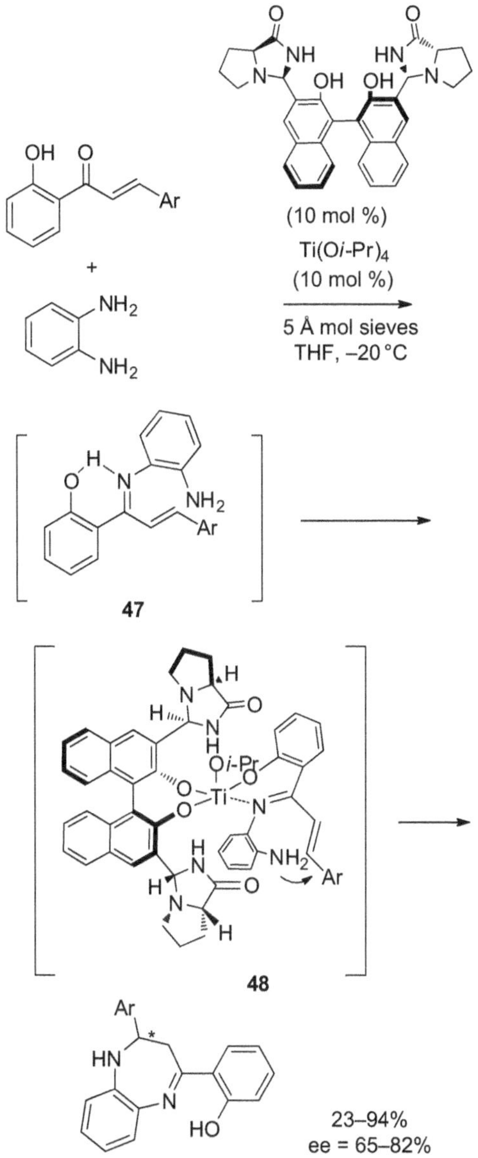

Scheme 2.53   Ti-catalysed domino imine formation–aza-Michael addition reaction.

bearing four stereogenic centres as almost single diastereomers (de > 90%) and moderate to high enantioselectivities of up to 92% ee, as shown in Scheme 2.55.[119] The process was catalysed by a combination of $Rh_2(S\text{-}DOSP)_4$ (*N*-*p*-dodecylbenzenesulfonylprolinate) and $Sc(OTf)_3$ and was supposed to evolve through five distinct steps. The authors assumed that the first step of the sequence was a rhodium-catalysed oxygen ylide formation, which was followed

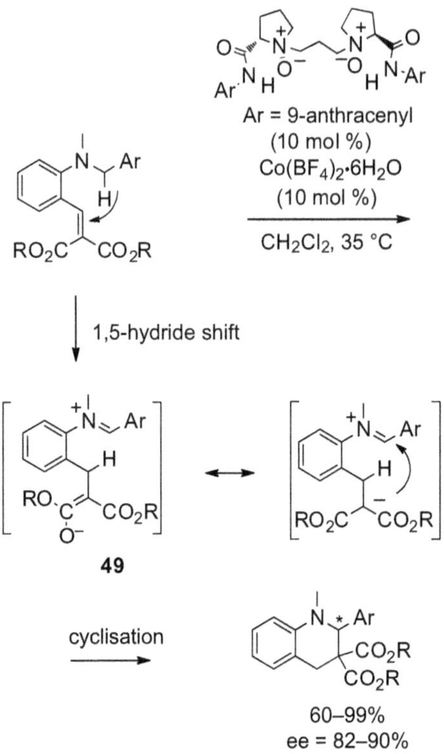

Scheme 2.54   Co-catalysed domino 1,5-hydride transfer–cyclisation reaction.

by a [2,3]-sigmatropic rearrangement, an oxy-Cope rearrangement, a keto/enol tautomerisation, and then finally a carbonyl ene reaction, as summarised in Scheme 2.55.

Finally, Lautens and Boyer have reported a novel enantioselective domino process in which rhodium was demonstrated to promote four reactions successively: an asymmetric ring-opening reaction, allylic alcohol isomerisation, acetalisation, and oxidation.[120] In this work, a series of chiral bicyclo[2.2.2]lactone products were achieved in good yields and excellent enantioselectivities of up to >98% ee by a reaction occurring between the corresponding doubly bridgehead substituted oxabicyclic alkene and a range of dialkyl amines, as shown in Scheme 2.56. The asymmetric induction of the domino process was provided by using a cationic RhOTf/(R,Sp)-Josiphos system as chiral catalyst. This result constituted the first intermolecular rhodium-catalysed asymmetric ring-opening reaction of this type of substrate. In order to explain the results, the authors have proposed the mechanism depicted in Scheme 2.56. It was reasoned that the asymmetric ring-opening reaction was the first step of the process, leading to intermediate **50** which contained an alkene function that could isomerise in a second step to the

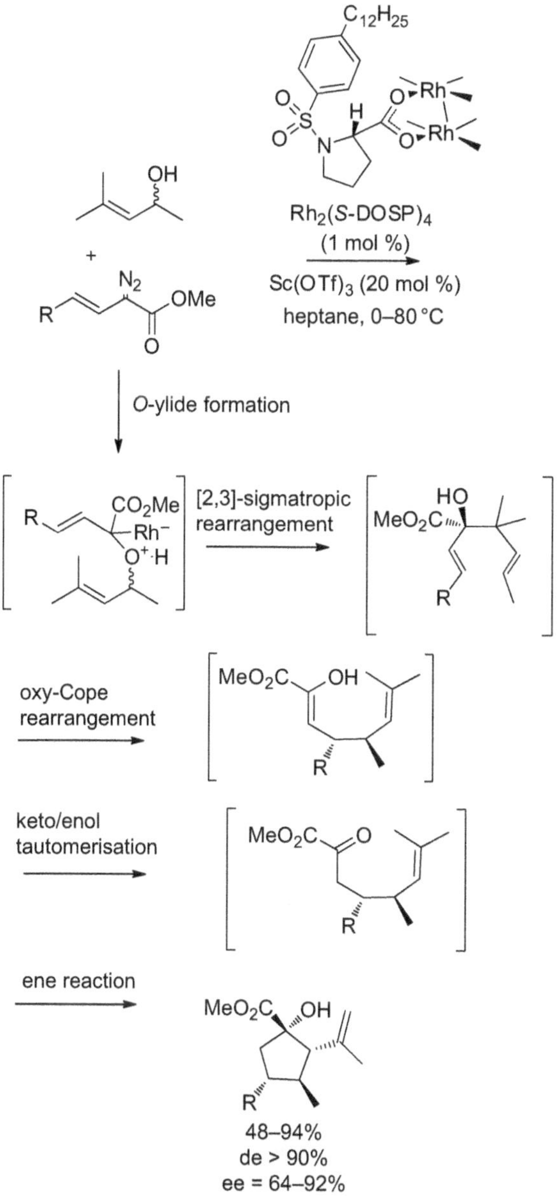

Scheme 2.55 Rh- and Sc-catalysed domino oxygen ylide formation–[2,3]-sigmatropic rearrangement–oxy-Cope rearrangement–keto/enol tautomerisation–carbonyl ene reaction of vinyldiazoacetates and an allylic alcohol.

corresponding intermediate aldehyde **51**. The latter, thus formed in close proximity to the tertiary alcohol function, led to the formation of intermediate hemiacetal **52**, which was finally oxidised under the reaction conditions to give the observed product.

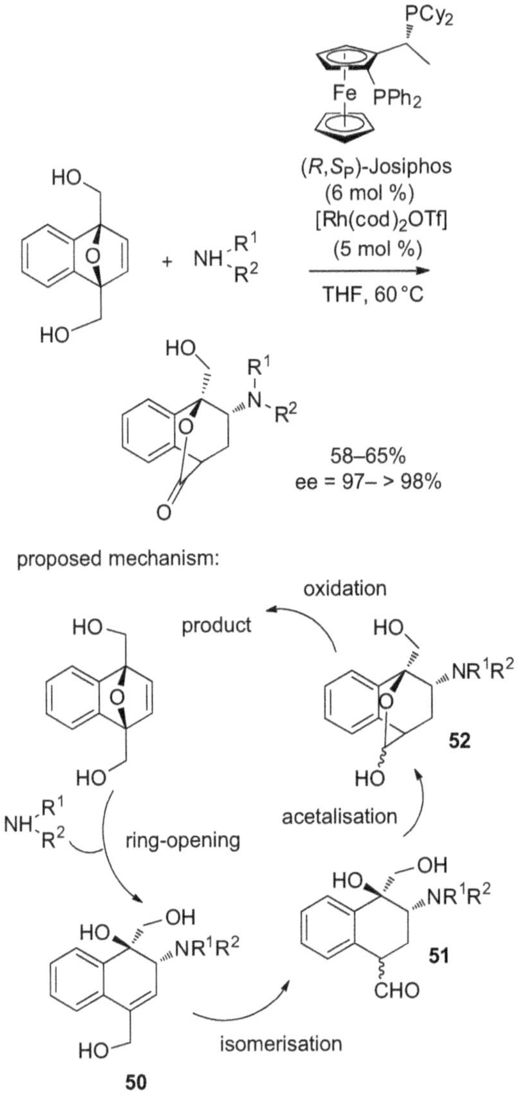

Scheme 2.56   Rh-catalysed domino ring-opening–isomerisation–acetalisation–oxidation
reaction.

2.3   Multicomponent Reactions

In spite of the recent emergence of organocatalysis, enantioselective metal-catalysed multicomponent reactions represent the majority of catalytic enantioselective multicomponent reactions. Asymmetric organometallic catalysis has contributed considerably to the development of novel enantioselective metal-catalysed multicomponent reactions. A range of metal catalysts, such as copper, rhodium, gold, platinum, palladium, cobalt, ruthenium, scandium, magnesium, titanium, zinc, aluminium, nickel, silver, and iridium, as well as tin, have been demonstrated to induce a large number of types of these powerful one-pot reactions, which allow easy attainment of high molecular complexity in an economically favourable way. They have the advantage of saving solvent, time, energy, and costs by avoiding costly protecting groups and time-consuming purification procedures after each step. Even though the history of multicomponent reactions dates back to the second half of the 19th century, with the reactions of Strecker, Hantzsch, and Biginelli, it was only in recent decades with the work of Ugi that the concept of the multicomponent reaction has emerged as a powerful tool in synthetic chemistry.[121]

2.3.1 Multicomponent Reactions Initiated by the Michael Reaction

Reactions which evolved through nucleophilic 1,4-addition of stabilised carbon nucleophiles to electron-poor olefins, generally α,β-unsaturated carbonyl compounds, are considered to be Michael additions. Michael-type reaction can be considered as one of the most powerful and reliable tools for the stereo-controlled formation of carbon–carbon and carbon–heteroatom bonds, as has been demonstrated by the huge number of examples in which it has been applied as a key strategic transformation in total synthesis. Since the first catalytic domino Michael–aldol reaction, reported by Noyori *et al.* in 1996,[6] there have been numerous examples of domino and multicomponent domino reactions initiated by the Michael reaction. Among them, a number of enantioselective metal-catalysed Michael-initiated multicomponent reactions have been successfully developed. As a recent example, Yus *et al.* have reported convenient access to enantiopure β-amino ketones on the basis of a multicomponent reaction of dialkyl zinc reagents, cyclic enones, and chiral *N-tert*-butanesulfinimines catalysed by a copper complex of a chiral phosphoramidite ligand (Scheme 2.57).[122] Three contiguous stereocentres and two carbon–carbon bonds could be generated with excellent stereocontrol through a double asymmetric induction, arising from the chiral ligand and the chiral *N-tert*-butanesulfinimine used, as shown in Scheme 2.57. The authors assumed that, whereas the enantioselection at the cycle stereocentres was governed by the phosphoramidite auxiliary, in the case of the aminic α-C-stereocentre the asymmetric induction came from the *tert*-butylsulfinyl moiety.

In the same area, Huang *et al.* have reported a highly diastereo- and enantioselective construction of three contiguous acyclic stereogenic centres through a copper-catalysed multicomponent domino Michael–Mannich reaction of organozinc reagents, such as $ZnEt_2$, and acyclic α,β-unsaturated ketones in the presence of imines to afford the corresponding β-aminocarbonyl derivatives.[123] High yields and enantioselectivities of up to 95% ee were reached along with moderate to good diastereoselectivities of up to 86% de, when a combination of CuBr and a chiral phosphite was employed as catalyst.

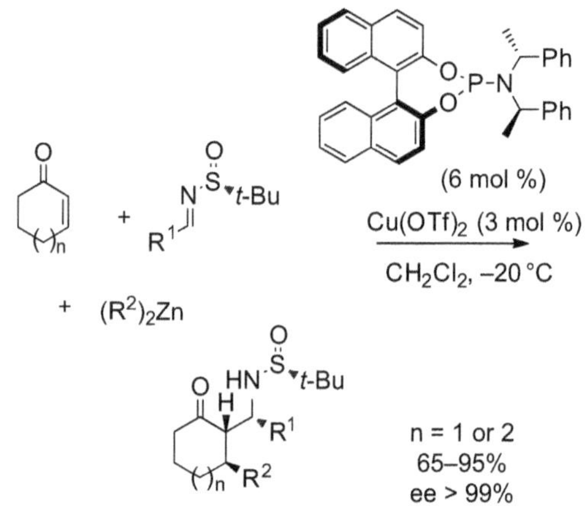

Scheme 2.57 Cu-catalysed three-component domino Michael–Mannich reaction.

The scope of the process was broad, because both aromatic and aliphatic imines provided high enantioselectivities. Since the first catalytic domino Michael–aldol reaction was reported by Noyori in 1996, there have been numerous examples of domino and multicomponent reactions of this type, including enantioselective versions, based on this powerful methodology.6 In 2009, Shibasaki *et al.* reported an enantioselective multicomponent domino Michael–aldol reaction occurring between 3-phenyl-2-cyclohexen-1-one, benzaldehyde, and pinacolborane.[124] In the presence of a copper complex of a chiral diphosphine, conjugate boration of the enone, followed by aldol condensation onto benzaldehyde, led to the corresponding chiral tertiary organoboric ester, which was directly oxidised into the corresponding diol in 71% yield, good diastereoselectivity of 74% de, and enantioselectivity of 91% ee. More recently, Huang *et al.* have developed highly diastereo- and enantioselective multicomponent double Michael reactions, allowing a range of chiral functionalised pyrrolidines bearing multiple stereocentres to be achieved.[125] The process involved diethylzinc, α,β-unsaturated ketones, and nitroalkenes as the three components, and was catalysed by a combination of CuCl and a chiral ligand depicted in Scheme 2.58. As shown in Scheme 2.58, the double Michael adducts possessing three contiguous stereocentres were generally obtained in high yields, with remarkable diastereo- and enantioselectivities of up to >98% de, and 97% ee, respectively. The dramatic effect of neutral copper in the domino reaction was disclosed, which provided evidence that the electrophilic catalyst precursor played a crucial role in the control of stereoselectivity. This exceptional methodology cumulates several advantages, such as the employment of a low catalyst loading of 1 mol%, a broad scope, and generally excellent levels of diastereo- and enantioselectivities reached in the control of three contiguous stereocentres.

The Henry reaction has often been associated with the Michael reaction in successful asymmetric domino sequences.[126] In this context, a novel highly

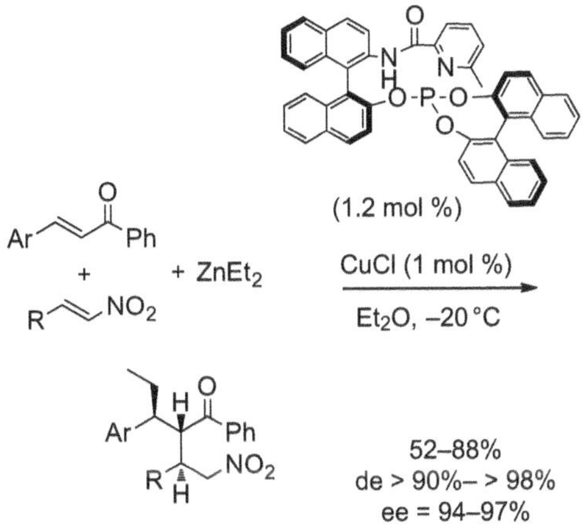

Scheme 2.58 Cu-catalysed three-component domino Michael–Michael reaction.

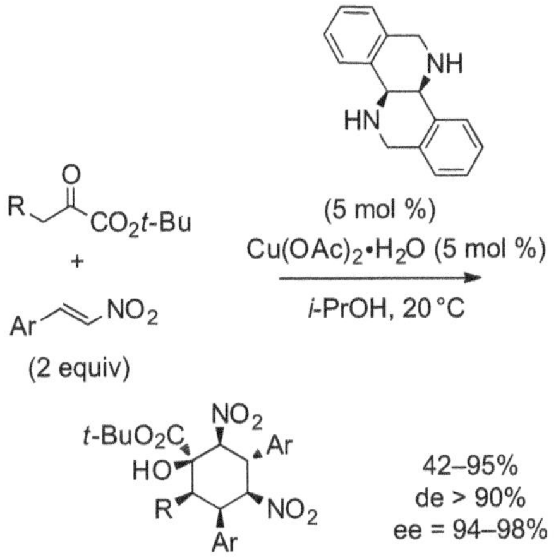

Scheme 2.59 Cu-catalysed three-component domino Michael–Michael–Henry reaction.

enantio- and diastereoselective copper-catalysed domino Michael–Michael–Henry reaction was reported by Huang *et al.*, in 2012.[127] This remarkable three-component reaction involved two equivalents of nitroalkenes and α-ketoesters, which produced the corresponding highly functionalised cyclohexane carboxylates, having six stereogenic centres including one quaternary, in excellent yields, diastereo-, and enantioselectivities of up to >90% de, and 98% ee, respectively (Scheme 2.59). This highly efficient formal [2+2+2] annulation provided the best results when catalysed by a combination of $Cu(OAc)_2 \cdot H_2O$ and the chiral ligand depicted in Scheme 2.59. The performance of this exceptional process is situated in the generally excellent levels of diastereo- and enantioselectivities obtained to control six contiguous stereocentres under mild conditions.

2.3.2   Multicomponent Reactions Based on the Mannich Reaction

The classic direct Mannich reaction discovered in 1912,[128] is an aminoalkylation of carbonylic compounds involving ammonia (or a primary or secondary amine derivative), a non-enolisable aldehyde (usually formaldehyde) or a ketone, and an enolisable carbonyl compound, leading to β-aminocarbonyl derivatives.[129] In the last few years, several groups have developed efficient enantioselective Mannich-type reactions by using chiral catalysts of various metals, such as scandium, silver, tin, zirconium, and copper. For example, Feng *et al.* have applied a scandium complex of chiral ligand to promote a three-component Mannich-type reaction occurring between aldehydes, *o*-anisidine, and a ketene silyl acetal, as shown in Scheme 2.60.[130] Indeed, a series of chiral

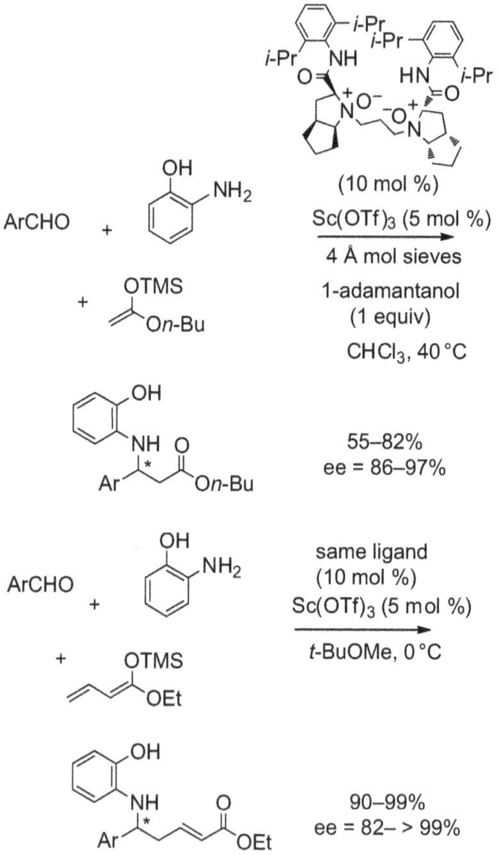

Scheme 2.60   Sc-catalysed three-component Mannich-type reactions.

β-amino esters were achieved through the reaction of *in situ* generated aldimines and ketene silyl acetal in the presence of a combination of a chiral ligand with $Sc(OTf)_3$ and 1-adamantol. As shown in Scheme 2.60, a number of products were produced in moderate to good yields and high enantios-electivities of up to 97% ee. Later, these authors applied the same complex derived from the same chiral ligand and $Sc(OTf)_3$ to the enantioselective catalysis of a three-component vinylogous Mannich reaction of various aldehydes, *o*-anisidine, and an acyclic silyl dienol ester.[131] The process afforded the corresponding highly functionalised δ-amino-α,β-unsaturated esters in excellent yields, complete regioselectivity and enantioselectivities of up to >99% ee, as shown in Scheme 2.60.

In 2008, Hoveyda *et al.* described a remarkable protocol for three-component silver(I)-catalysed enantioselective vinylogous Mannich reactions of *in situ* generated alkyl-substituted aldimines, including those bearing heteroatom-containing substituents, and readily available siloxyfurans.[132] Particularly high efficiency and exceptional diastereo- and enantioselectivities were achieved through the use of *o*-thiomethyl-*p*-methoxyaniline-derived aldimines, as shown in Scheme 2.61. The processes were promoted by a combination of AgOAc and as little as 1 mol% of an easily accessible amino acid-based chiral ligand (Scheme 2.61). The corresponding unsaturated δ-lactones were obtained in 44–92% yield, and in general with diastereo- and enantioselectivities of up to >96% de and >98% ee, respectively. Later, Zanardi *et al.* applied similar conditions to achieve corresponding silver-catalysed diastereo- and enantioselective vinylogous Mannich reactions of pyrrole-based silyl dienolates with *in situ* generated alkyl-substituted aldimines.[133] The reactions provided the corresponding unsaturated lactams in good to high yields, complete diastereoselectivity in favour of the

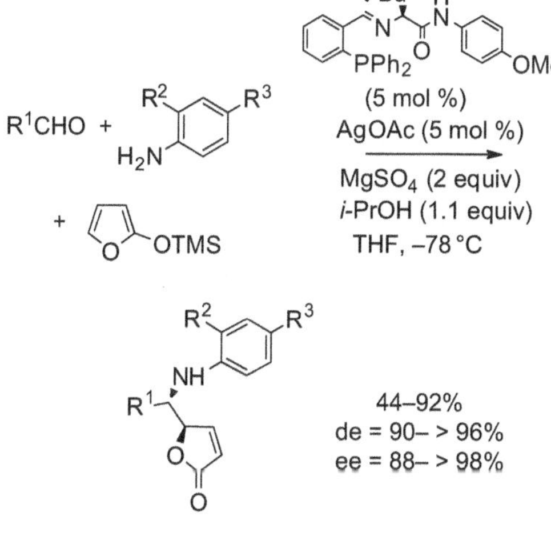

Scheme 2.61   Ag-catalysed three-component vinylogous Mannich reaction.

anti-configured isomer, and high enantioselectivities of up to 96% ee. The utility of the Mannich products was demonstrated in the synthesis of an unprecedented perhydrofuro[3,2-*b*]pyrrolone product, which is an aza-analogue of naturally occurring (+)-goniofufurone.

The three-component Mannich reaction presents a disavantage in that a moisture-sensitive Lewis acid cannot be used to catalyse the reaction because a stoichiometric amount of water is generated when an imine is formed from an amine and an aldehyde. In 2009, Yanagisawa *et al.* found that an *in situ* generated chiral tin bromide alkoxide possessing a binaphthyl structure catalysed the enantioselective Mannich-type reaction of alkenyl trichloroacetates, ethyl glyoxalate, and aniline derivatives smoothly even in the presence of water or alcohol as well as a polar substituent, such as an amino group or a phenolic hydroxyl group.[134] Indeed, a series of chiral β-amino ketones were *syn*-selectively achieved in high yields and enantioselectivities of up to 98% ee on the basis of this process, performed in the presence of an (*S*)-BINOL-derived tin dibromide, possessing a 4-trifluoromethylphenyl group at the 3- and 3′-positions, as chiral precatalyst in the presence of sodium ethoxide, sodium iodide, and ethanol. It must be noted that even *in situ* generated imines possessing a polar amino group gave satisfactory results. In addition, Kobayashi *et al.* have developed enantioselective Mannich-type reaction of 5-hexynal, 2-amino-*m*-cresol and a ketene silyl acetal derived from phenyl propionate by using an *in situ* generated chiral zirconium catalyst of (*S*)-6,6′-(C$_2$F$_5$)$_2$-BINOL.[135] As shown in Scheme 2.62, the process led to the formation of the corresponding Mannich adduct in 85% yield, good *syn*-stereoselectivity (*syn : anti* = 91 : 9) in combination with excellent enantioselectivity of 96% ee for the major *syn*-product. The product was further converted into onchidin, which is a natural cytotoxic, C_2-symmetric cyclic decadepsipeptide. In addition, palladium(II)-catalysed enantioselective three-component aminomethylation of β-ketoesters using formalin and benzylamine trifluoromethanesulfonic acid salt have been developed by Sodeoka *et al.*[136] In the presence of a chiral palladium complex of biphosphine, the domino reaction afforded the corresponding aminomethylated β-ketoesters in almost quantitative yields. Among the ligands tested, (*R*)-Segphos gave the best enantioselectivity of 68% ee. Unfortunately, distinct improvement of the enantiomeric excess was not achieved, although the effects of solvent, reaction temperature, and substituents on the nitrogen atom were examined.

Finally, a chiral copper(I) complex of (*R*)-DIFLUORPHOS was used by Shibasaki *et al.* to induce the first catalytic enantioselective reductive Mannich reaction of ketimines.[137] Indeed, the multicomponent reaction of ketimines, α,β-unsaturated esters and (EtO)$_3$SiH as the reducing agent provided the corresponding amines, containing contiguous tetra- and trisubstituted carbons, in good yields, moderate to good diastereoselectivities, and enantioselectivities of up to 91% ee. It must be noted that this methodology constituted the first entry to the catalytic asymmetric synthesis of β^{2,3,3}-amino acid derivatives.

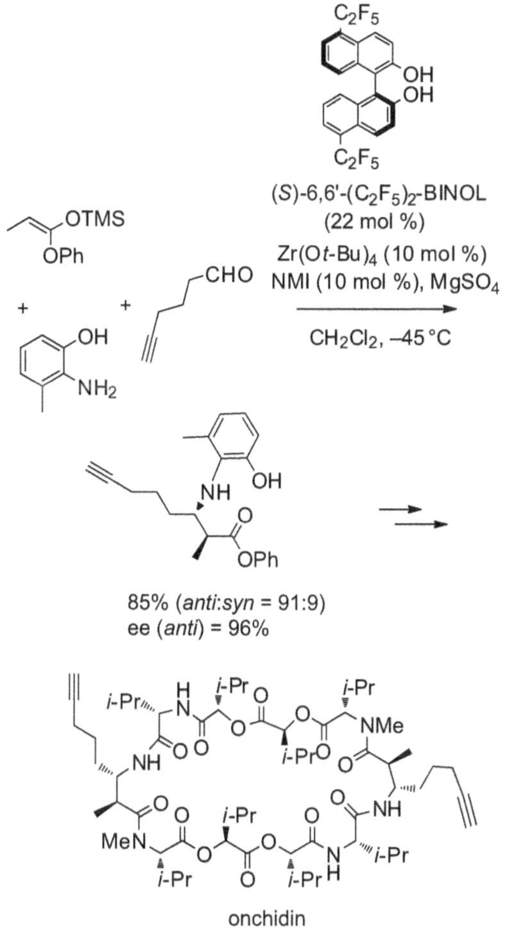

Scheme 2.62 Zr-catalysed three-component Mannich-type reaction, and synthesis of onchidin.

2.3.3 Multicomponent Reactions Initiated by a Pericyclic Reaction

A number of asymmetric domino reactions initiated by pericyclic reactions have been reported.[138] In particular, multicomponent reactions including a pericyclic reaction, such as Diels–Alder-, Knoevenagel-, 1,3-dipolar cyclo-addition, and [2+2+2] cycloaddition have been reported. In recent years, several examples of enantioselective metal-catalysed multicomponent reactions initiated by a (hetero)Diels–Alder or a [2+2+2] cycloaddition have been developed. For example, a cobalt(I)-catalysed Diels–Alder reaction of a 1-boron-functionalised 1,3-diene with an alkyne was the first key step in a domino Diels–Alder–allylboration reaction interconnecting three components to provide the corresponding multifunctionalised domino product in good yield (87%) and moderate enantioselectivity of 71% ee.[139] The second step of the sequence was the allylation of the aldehyde. The product was regio- and diastereoselectively produced by using a combination of $CoBr_2$ with chiral ligand (*S,S*)-Norphos (2,3-bis(diphenylphosphino)-bicyclo[2.2.1]hept-5-ene). Later, Feng *et al.* reported the synthesis of a range of chiral ring-fused tetrahydroquinolines containing three contiguous stereocentres on the basis of a remarkable enantioselective three-component inverse electron-demand aza-Diels–Alder reaction catalysed by a chiral scandium complex.[140] As shown in Scheme 2.63, the reaction of various aldehydes with substituted *o*-anisidines and cyclopentadiene gave rise to the corresponding tricyclic products in high yields, and high to excellent diastereo- and enantioselectivities of up to >98% de and >99% ee, respectively. This novel and efficient process opened a novel route to this important class of products which often exhibit potential biological activities.

Transition metal-catalysed [2+2+2] cycloaddition of unsaturated motifs, such as alkyne and alkene, constitutes the most atom-economical and facile protocol for the construction of a six-membered ring system.[34] In particular, the enantioselective [2+2+2] cycloaddition is a fascinating protocol for the construction of chiral cyclic skeletons.[35a–b] In recent years, this cycloaddition has been included in several enantioselective domino processes, including multicomponent reactions. As an example, Shibata *et al.* have employed an *in situ* prepared chiral iridium complex of (*S,S*)-Chiraphos [2,3-bis(diphe-nylphosphine)butane] to induce a pseudo-three-component consecutive and enantioselective [2+2+2] cycloaddition of tetraynes with two molecules of protected but-2-yne-1,4-diols to give the corresponding C_2-symmetrical

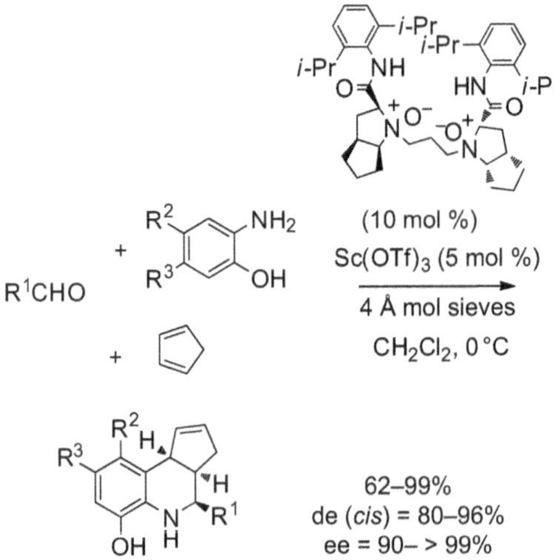

Scheme 2.63   Sc-catalysed three-component inverse electron-demand aza-Diels–Alder reaction.

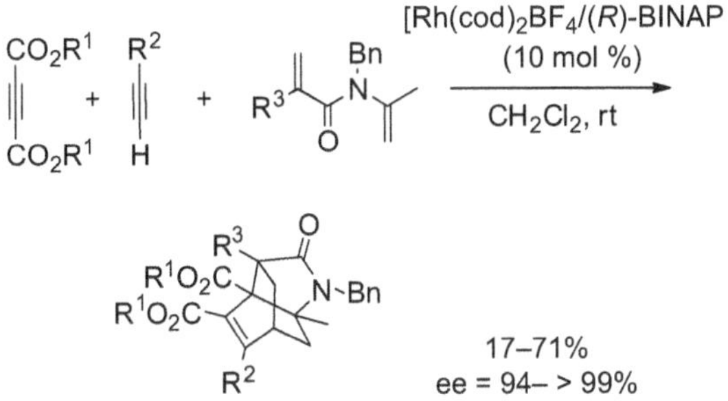

Scheme 2.64 Rh-catalysed three-component domino intermolecular [2+2+2] cycloaddition–intramolecular Diels–Alder reaction.

quateraryl products in moderate to good yields and enantioselectivities of up to 86% ee.[40] In 2011, Tanaka *et al.* reported an efficient enantioselective domino intermolecular [2+2+2]–intramolecular Diels–Alder reaction occurring between 1,6-diynes and amide-linked 1,5-dienes.[42] The authors have extended the scope of this methodology to the corresponding three-component version of the process. Indeed, the reaction of dialkyl acetylenedicarboxylates, terminal alkynes, and amide-linked 1,5-dienes, performed in the presence of a cationic rhodium(I)/(*R*)-BINAP complex, provided the corresponding amides in moderate to high yields and excellent enantioselectivities of up to 99% ee, as shown in Scheme 2.64. In all cases of the substrates studied, the bridged multicyclic products were achieved as single and almost enantiopure regioisomers. In addition, Somfai and Torrsell have described three-component 1,3-dipolar cycloadditions of *in situ* generated carbonyl ylides to aldimines induced by chiral rhodium-(II) carboxylate catalysts.[141] For example, the reaction of benzylidene-benzylamine, benzaldehyde and ethyl diazoacetate gave rise to the corresponding *syn*-β-amino alcohol in low to moderate yield (14–62%), modest diastereoselectivity (42–88% de) combined with moderate enantioselectivity of up to 64% ee. In spite of its moderate enantioselectivities, this process allowed interesting building blocks for the synthesis of natural products and other biologically important products that are *syn*-β-amino alcohols to be achieved.

2.3.4 Multicomponent Reactions Based on the Passerini Reaction

The modern concept of multicomponent reactions is intimately related to the reactions developed with isocyanide reagents.[48] The unique ability of isocyanides to serve as *C*-nucleophiles towards imines and aldehydes, resulting in nitrilium ion electrophiles, enables the four-component Ugi reaction and the three-component Passerini reaction, respectively.[49] Despite the strong relationship between multicomponent reactions and isocyanides, there are still few reactions in which the use of isocyanide reagents leads to chiral compounds, and more particularly those which are induced by chiral catalysts. In 2008, Zhu *et al.* developed an efficient catalytic enantioselective Passerini three-component reaction, which involved an aldehyde, a carboxylic acid, and an isocyanide with the generation of a stereogenic centre to afford a α-acylo-xyamide.[142] When a chiral *N,N'*-bis(3,5-di-*tert*-butylsalicylidene)-(*R,R*)-cyclohexane-1,2-diamine was used as the supporting ligand in association with Et₂AlCl, the authors showed that the reaction was applicable to a wide range of nonchelating aldehydes, carboxylic acids, and isocyanides. It allowed a range of α-acyloxyamides to be achieved in good yields and high to excellent enanti-oselectivities of up to >99% ee. By replacing aliphatic isocyanides with less reactive aromatic isocyanides, increased selectivity was observed for this process. Furthermore, a variety of carboxylic acids were compatible with the reaction, including 2-chloroacetic acid. The presence of chloroacetyl func-tionality in the Passerini products provided an interesting handle for subsequent functionalisation reactions. A deacetylated derivative was assigned the (*S*)-configuration, indicating that the isocyanide attacked preferentially on the *Re*-face of the aldehyde. Passerini-type reactions have the general char-acteristics of the Passerini reaction, but one of the inputs is replaced by a different compound with similar electronic properties. On the basis that the PK_a of hydrazoic acid was rather similar to that of carboxylic acids and the azide ion was expected to be a better nucleophile than a carboxylate ion,[143] an asymmetric Passerini-type three-component reaction for the synthesis of 5-(1-hydroxyalkyl)-tetrazoles, which used hydrazoic acid instead of the carboxylic acid component, was developed by Zhu *et al.*, in 2008.[144] This process constituted the first enantioselective Passerini-type reaction of aldehydes, isocyanides, and hydrazoic acid to give the corresponding 5-(1-hydroxyalkyl)tetrazoles. This three-component reaction was induced by the [(salen)Al(III)Me] complex depicted in Scheme 2.65, and was applied to a

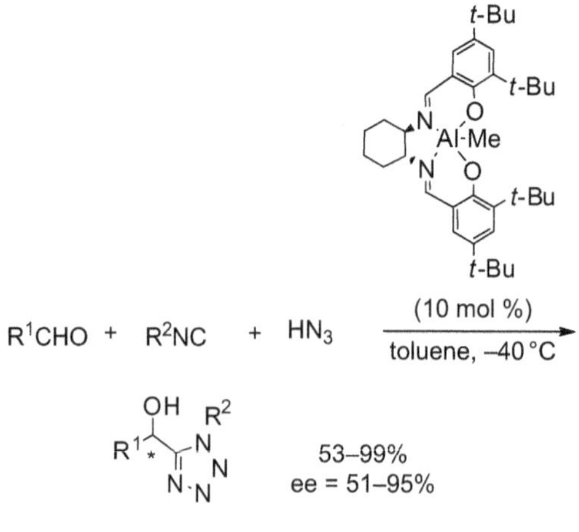

Scheme 2.65   Al-catalysed three-component Passerini-type reaction.

wide range of linear as well as α-branched aliphatic aldehydes and to both aromatic and aliphatic isocyanides, affording the corresponding tetrazoles in good to excellent yields and enantioselectivities, as shown in Sheme 2.63. It must be noted that the three-component Passerini reaction has seldom been used to prepare tetrazoles, which constitute important heterocycles in medicinal chemistry, owing to their increased stability towards metabolic degradation pathways.

2.3.5 Multicomponent Reactions Initiated by the Friedel–Crafts Reaction

The asymmetric Friedel–Crafts reaction is one of the most powerful methods to synthesise optically active aromatic compounds and has been included in various enantioselective domino reactions. As a recent example, Arai *et al.* have reported the enantioselective multicomponent domino Friedel–Crafts–Henry reaction of indoles, nitroalkenes, and aldehydes catalysed by a combination of $Cu(OTf)_2$ with a chiral imidazoline–aminophenol ligand.[145] The corresponding domino products bearing three contiguous stereocentres were obtained in high yields, remarkable enantioselectivities of up to 99% ee, and moderate to good diastereoselectivities of up to 90% de (Scheme 2.66). These products could be envisaged readily as key starting materials in the synthesis of hydroxytryptamines.

It is well known that one of the most heavily studied families of transition metal catalysts comprises those based on palladium. Over the past 40 years, many new reactions have been uncovered, and activity in this area has remained high. Among the increasing number of enantioselective multi-component processes starting with a transition metal-catalysed reaction, the

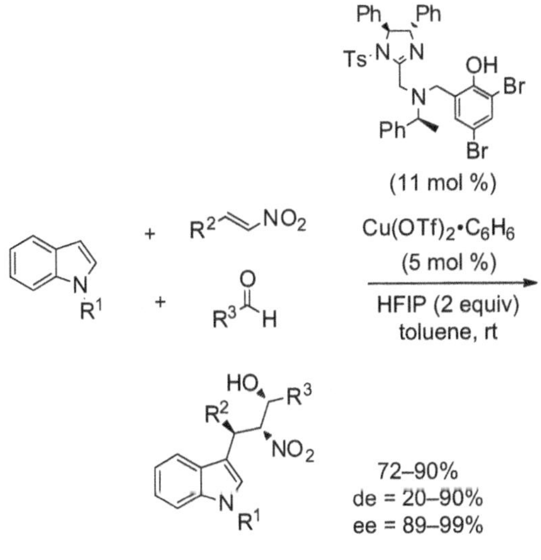

Scheme 2.66 Cu-catalysed three-component domino Friedel–Crafts–Henry reaction.

enantioselective palladium-catalysed multicomponent transformations have seen an astounding development over the past few years.[7] Indeed, since palladium has the advantage of being compatible with many functional groups, it constitutes an ideal catalyst for multicomponent reactions.[1d,8] In 2011, Zhou *et al.* developed an enantioselective approach to 2,3-disubstituted indolines through consecutive Bronsted acid/palladium complex-promoted domino reactions.[146] Thus, the reaction of 2-substituted indoles with aldehydes performed in the presence of a combination of Pd(TFA)$_2$, (*R*)-H8-BINAP as the ligand, and TsOH under hydrogen atmosphere led to the corresponding 2,3-disubstituted indolines in good yields and remarkable enantioselectivities of up to 98% ee, as shown in Scheme 2.67. The authors assumed that the process began with a Bronsted acid-promoted Friedel–Crafts reaction of 2-substituted indoles and aldehydes to give the intermediate vinylogous iminium **53**, followed by two asymmetric hydrogenation steps to accomplish this process (Scheme 2.67).

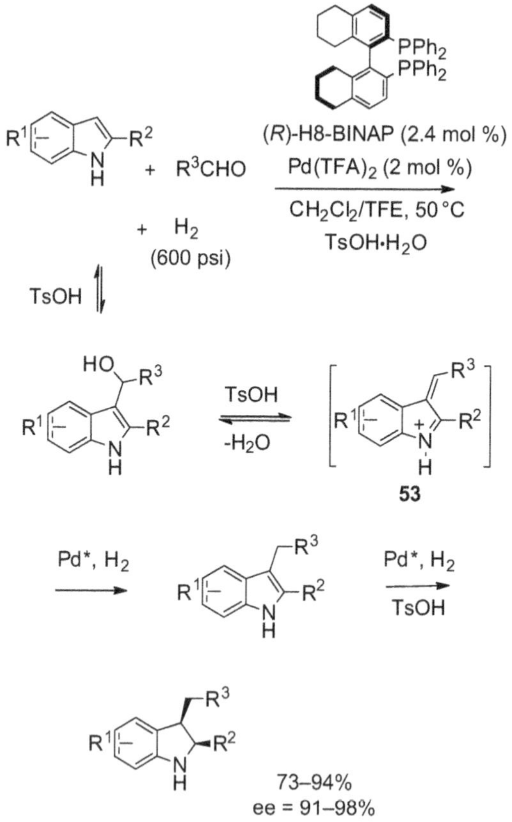

Scheme 2.67   Pd-catalysed three-component domino Friedel–Crafts–hydrogenation reaction.

2.3.6 Multicomponent Reactions of Alkynes, Aldehydes and Amines

. The copper-catalysed three-component reaction of terminal alkynes, aldehydes, and amines, providing the corresponding chiral propargylamines, has been investigated recently by several groups. In 2006, Knochel and Gommermann employed CuBr combined with (R)-QUINAP [1-(2-diphenylphosphino-1-naphthyl)isoquinoline] as the ligand to induce the reaction with secondary amines, which furnished the corresponding propargylamines in both excellent yields and enantioselectivities of up to 98% ee.[147] As shown in Scheme 2.68, the best results were reached in the case of silylated propargylamines. The applicability of these products in natural product synthesis was successfully demonstrated in the total synthesis of the alkaloid (S)-(+)-coniine.

In the same context, Carreira *et al.* have developed the enantioselective three-component reaction of aldehydes and alkynes with 4-piperidone hydrochloride hydrate, which afforded the corresponding tertiary propargylamines in good yields and enantioselectivities of up to 96% ee.[148] These results were obtained by using (R,R)-PINAP {4-[2-(diphenylphosphino)-1-naphthalenyl]-N-[1-phenylethyl]-1-phthalazinamine} as the copper chiral ligand. The use of 4-piperidone as the amine component not only provided access to a useful building block but also highlighted the exquisite chemoselectivity of the process. In the same area, Singh and Bisai have reported the same type of three-component reaction with anilines and catalysed by a combination of Cu(I)PF$_6$ and a chiral C_2-symmetric PYBOX [2,6-bis(2-oxazolyl)pyridine] ligand.[149] This nice process could be applied to a wide variety of aromatic aldehydes, leading to the corresponding aromatic alkynylamines with good to excellent yields (61–99%) and enantioselectivities (85–99% ee). The previously mentioned works deal with three-component reactions of aldehydes and amines with aryl- and silyl-substituted alkynes. In 2010, Nakamura *et al.* extended the scope of this type of three-component reaction to the use of aliphatic terminal alkynes.[150] The reaction of these alkynes with a range of aldehydes and *p*-anisidine afforded, in the

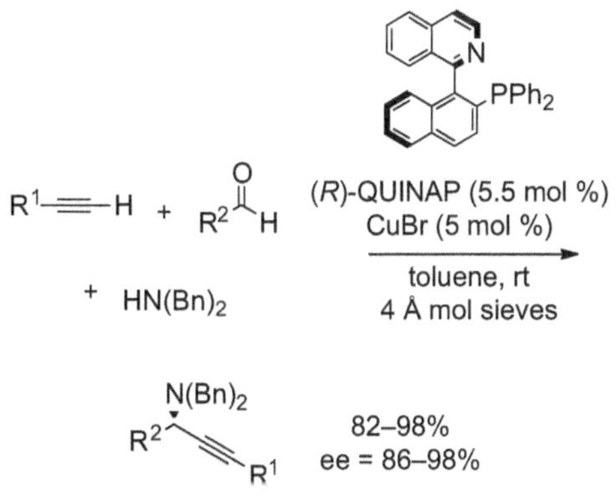

$$R^1 \!\!=\!\!-H \;+\; R^2 \overset{O}{\underset{}{\overset{\|}{C}}} H$$

$$+\; HN(Bn)_2$$

(R)-QUINAP (5.5 mol %)
CuBr (5 mol %)
———————————
toluene, rt
4 Å mol sieves

$$R^2 \overset{N(Bn)_2}{\diagup} \!\!=\!\! R^1$$

82–98%
ee = 86–98%

Scheme 2.68 Cu-catalysed three-component reaction of alkynes, aldehydes, and secondary amines.

presence of a catalytic amount of a combination of Cu(OTf)$_2$ with a closely related chiral C_2-symmetric PYBOX ligand, the corresponding chiral amines in good yields and remarkable enantioselectivities of up to 98% ee. It must be noted that Benaglia *et al.* have also studied the three-component reaction of phenylacetylene with amines and aldehydes upon catalysis by copper complexes of chiral bis-imines.[151] The best enantioselectivities of ≤75% ee were obtained by using a chiral bis-imine readily prepared from commercially available binaphthyl diamine. In addition, comparable levels of enantios-electivity (≤74% ee) were reported by Chan *et al.* for the three-component reaction of ethyl glyoxylate, *p*-anisidine and aliphatic, aromatic alkynes catalysed by a combination of CuOTf·0.5C$_6$H$_6$ with another chiral PYBOX ligand.[152] Later, Gevorgyan *et al.* developed an efficient synthesis of chiral 3-aminoindolines, the key step of which was the enantioselective copper-catalysed three-component reaction of piperidine, arylacetylenes, and *N*-(2-formylphenyl)-4-methylbenzenesulfonamide.[153] This key step was induced by a combination of CuCl with Trost's C_2-symmetric biphosphine ligand, providing the corresponding key propargylamines in good yields and moderate to high enantioselectivities (52–93% ee). These products were subsequently converted into expected chiral indolines through desilylation followed by copper-catalysed cyclisation. Zinc-catalysed enantioselective three-component reaction of terminal alkynes, aldehydes and *o*-anisidine, which provided the corresponding chiral propargylamines, has been investigated by Bolm *et al.*[154] In this work, a combination of ZnMe$_2$ with a chiral β-amino alcohol derived from norephedrine was employed to promote the domino reaction, providing propargylamines. The scope of this process could be extended to include alkynes other than phenylacetylene. It must be noted that the use of trimethylsilylethyne and various alkyl-substituted alkynes led to the corresponding products in moderate yields and enantioselectivities, and, moreover, the methodology found a limitation in the inability to use electron-poor acetylenes such as methyl propiolate. On the other hand, a range of amines, stemming from aromatic, heteroaromatic, and aliphatic α-branched aldehydes, in combination with *o*-anisidine and phenylacetylene, could be prepared in moderate to high yields with enantiomeric excesses ranging from 68–97% ee. In the same area, an enantioselective synthesis of *N*-tosyl-(*E*)-(2-en-3-ynyl)-amines has been developed by Xu *et al.* on the basis of a zinc-mediated addition of two alkynes to *N*-tosylaldimines.[155] As shown in Scheme 2.69, the three-component reaction allowed a range of chiral *N*-tosylamines to be afforded in high yields and enantioselectivities of up to 99% ee when a BINOL-derived ligand was employed. In order to explain the formation of these products, the authors have proposed the mechanism depicted in Scheme 2.69, in which an alkynylzinc was formed *in situ*. First, the 1,2-addition of this alkynylzinc to *N*-tosylaldimine afforded a π-complex intermediate **54**, which underwent an attack by another alkynylzinc to form a cyclic intermediate **55** under reflux conditions. The carbon–zinc bond could be cleaved by an excess of alkyne to give the intermediate **56**. Subsequently, the reaction was quenched with HCl to give the final product.

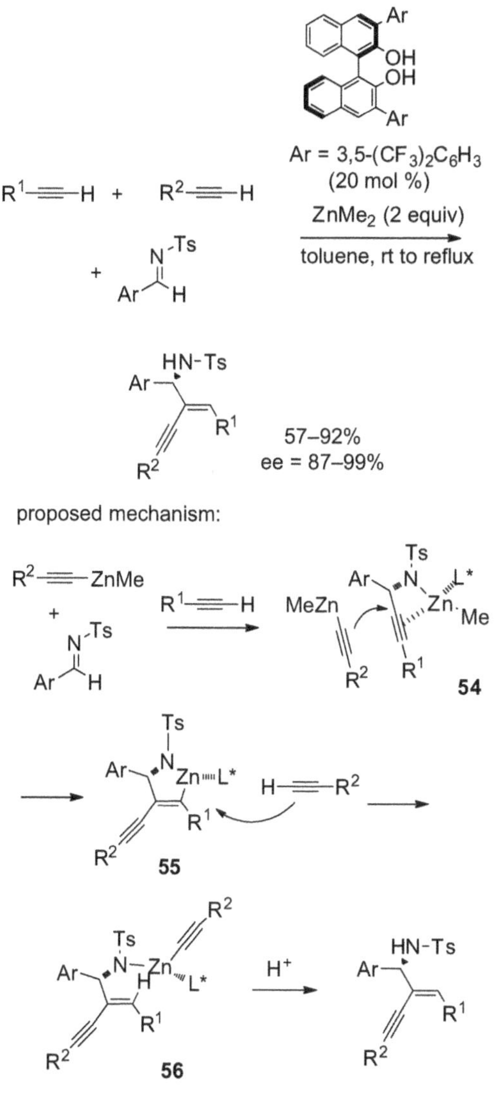

Scheme 2.69 Zn-catalysed three-component reaction of *N*-tosylaldimines and alkynes.

2.3.7 Multicomponent Reactions of 1,3-Dienes, Aldehydes and Reducing Agents

In 2007, Sato *et al.* developed nickel(0)-catalysed three-component coupling of 1,3-dienes, aldehydes, and triethylsilane as the reducing agent to give regio-, diastereo-, and enantioselectively the corresponding β-triethylsilyloxy (*Z*)-alkenes.[156] On the basis of the screening of various chiral N-heterocyclic carbene precursors, a chiral imidazolium salt having 1-(mesitylphenyl)propyl

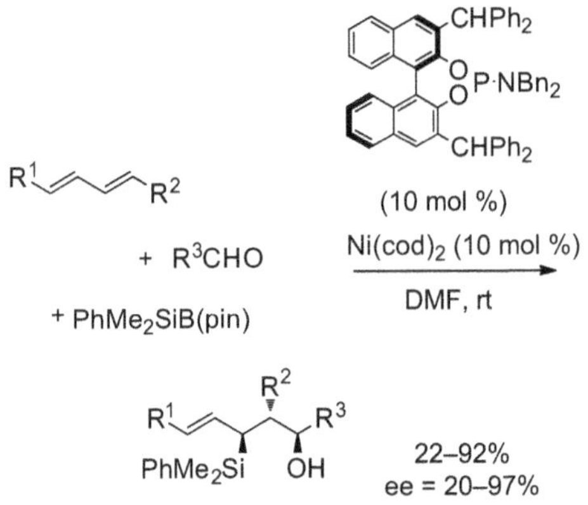

CHPh$_2$

O

P·NBn$_2$

O

CHPh$_2$

R^1 ⟍⟋⟍ R^2

(10 mol %)

+ R^{3}CHO   Ni(cod)$_2$ (10 mol %)

+ PhMe$_2$SiB(pin)   DMF, rt

R^2

R^1 ⟍⟋ R^3

PhMe$_2$Si   OH   22–92%
                         ee = 20–97%

Scheme 2.70   Ni-catalysed three-component reaction of 1,3-dienes, aldehydes, and silylboranes.

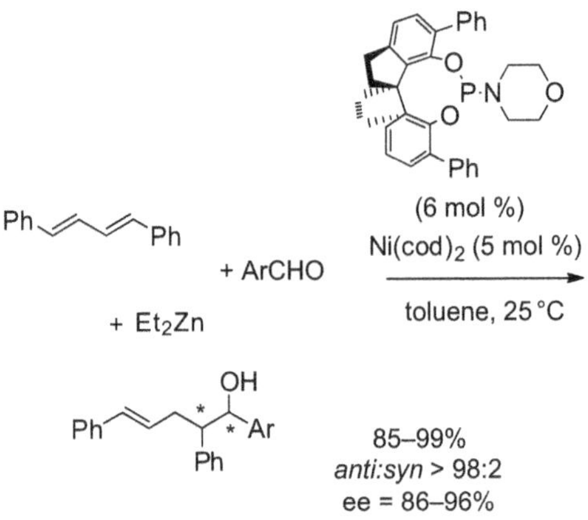

Ph

O

P·N   O

O

Ph

Ph ⟍⟋⟍ Ph   (6 mol %)

Ni(cod)$_2$ (5 mol %)

+ ArCHO

+ Et$_2$Zn   toluene, 25 °C

OH

Ph ⟍⟋ Ar

Ph   85–99%
           anti:syn > 98:2
           ee = 86–96%

Scheme 2.71   Ni-catalysed three-component reaction of 1,3-dienes, aldehydes, and ZnEt$_2$.

groups on the nitrogen was selected as the most efficient ligand for nickel to induce chirality in the process, allowing various coupling products to be synthesised in high yields and good to high enantioselectivities of up to 97% ee. The catalyst was generated *in situ* from Ni(cod)$_2$, the chiral imidazolium salt ligand, with Cs$_2$CO$_3$ as a base. Very recently, the same authors have investigated a closely related coupling reaction of 1,3-dienes with aldehydes, using (dimethylphenylsilyl)pinacolborane as the reductant instead of triethylsilylane as in the previous work.[157] In this case, a chiral phosphoramidite was selected as the most efficient ligand for nickel, allowing the corresponding (*E*)-silanes to be produced in moderate to high yields and enantioselectivities of up to 97% ee, as shown in Scheme 2.70. It must be noted that the coupling of internal

1,3-dienes gave generally lower yields (22–51%). In the case of each substrate studied, a single diastereomer was isolated.

In addition, Zhou *et al.* have developed enantioselective reductive coupling of 1,3-dienes and aldehydes by using ZnEt$_2$ as the reducing agent and nickel complexes of chiral spiro phosphoramidites, such as that depicted in Scheme 2.71.[158] In this case, the three-component reaction provided the corresponding chiral bishomoallylic alcohols in excellent yields and diastereoselectivities (*anti : syn* >99 : 1) in almost all the cases of substrates studied, combined with high enantioselectivities of up to 96% ee, as shown in Scheme 2.71.

2.3.8 Multicomponent Reductive Amination Reactions of Ketones

In 2009, chiral (diphosphine) palladium(II) dibromide catalysts were successfully employed by Rubio-Pérez *et al.* to develop enantioselective domino reductive amination of ketones.[159] As shown in Scheme 2.72, various chiral amines could be synthesised from the corresponding ketones by three-component reaction with anilines under 800 psi of hydrogen pressure with 2.5 mol% of catalyst loading. The best results were reached by using (*R*)-BINAP as the chiral ligand, which provided good yields and good to excellent enantioselectivities of up to 99% ee.

In the same year, Xiao *et al.* investigated three-component iridium-catalysed reductive amination of a wide variety of ketones.[160] The reaction of a range of aliphatic ketones with aniline derivatives under hydrogen atmosphere was catalysed by an iridium complex bearing a chiral phosphate, providing the corresponding chiral amines in high yields and enantioselectivities of up to 95% ee, as shown in Scheme 2.73. When the authors tried to extend the scope of the reaction to aromatic ketones, they found that no reaction occurred. On the other hand, using a closely related chiral iridium catalyst at a catalyst loading

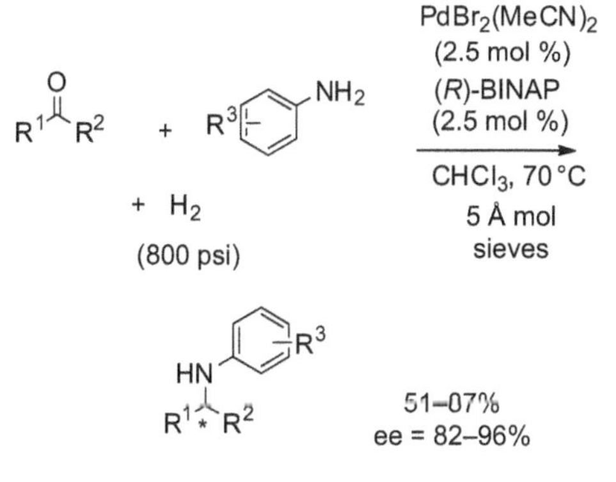

Scheme 2.72 Pd-catalysed three-component reductive amination reaction of ketones.

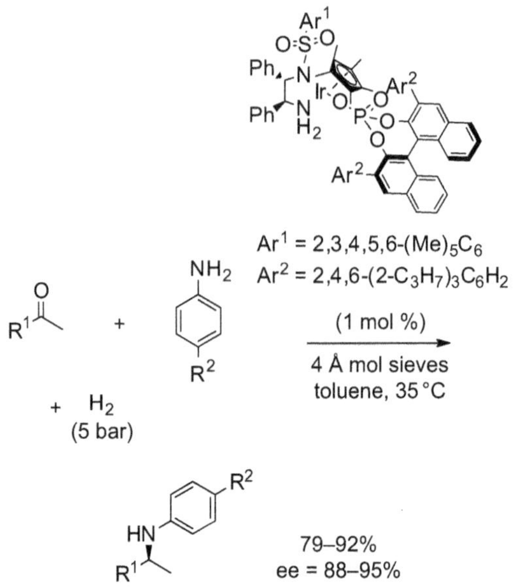

Scheme 2.73 Ir-catalysed three-component reductive amination of ketones.

as low as 0.005 mol% allowed the process to be achieved, providing the corresponding chiral amines in generally excellent yields (91–94%) and enantioselectivities (88–97% ee).

2.3.9 Multicomponent Kabachnik–Fields Reactions

Optically active α-aminophosphonates are important synthetic intermediates for the preparation of various natural products. In this context, Feng *et al.* have reported an enantioselective three-component Kabachnik–Fields reaction induced by a chiral *N,N'*-dioxide catalyst.[161] A series of chiral α-amino phosphonates were produced in moderate to good yields (76–96%) and enantio-selectivities of up to 87% ee through the domino reaction occurring among aldehydes, *o*-aminophenol, and diphenyl phosphite. In the same year, Shibata *et al.* developed a direct enantioselective three-component Kabachnik–Fields reaction of aldehydes, *p*-aminophenol, and diarylphosphites catalysed by a zinc(II) complex of a chiral C_2-symmetric pybim ligand.[162] As shown in Scheme 2.74, a range of α-aminophosphonates could be achieved in almost quantitative yields in all cases of the substrates studied, with high enantio-selectivities of up to 93% ee. Although the scandium-catalysed process developed by Feng *et al.*, described above, employed a lower catalyst loading (5 mol% instead of 10 mol%) and a milder reaction temperature (−20 °C instead of −50 °C) in comparison with this zinc-catalysed process, the latter provided both higher yields and higher enantioselectivities. Moreover, the scope of the

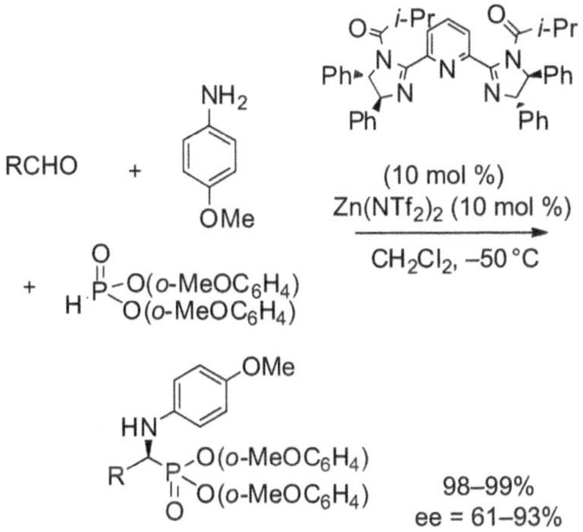

Scheme 2.74 Zn-catalysed three-component Kabachnik–Fields reaction.

zinc-catalysed process was broader because both aromatic and aliphatic aldehydes provided good results.

2.3.10 Miscellaneous Multicomponent Reactions

Gold(I) and gold(III) complexes have recently been utilized to catalyse a wide variety of transformations, including three-component reactions.[83e,83g,163] As a recent example, Toste and Campbell have reported an Au(I)-catalysed three-component reaction of imines, terminal alkynes, and *p*-toluenesulfonyliso-cyanate to give cyclic carbamimidates.[164] The authors obtained the highest efficiency with enantioselectivities of up to 91% ee by catalysing the reaction with a monophosphine gold(I) catalyst bearing a chiral sulfonyl urea as ligand, which was selected from among a range of other chiral ligands, including other sulfonyl ureas, carbamates, alkyl/aryl ureas, acyl ureas, amides, and sulfonamides. The process led regioselectively to five-membered carbami-midates as major products, along with corresponding six-membered carbami-midates as minor regioisomers. The authors assumed that the reaction proceeded through alkynylation of aryl–aryl imines generating the corre-sponding propargylamines **57**, which were trapped with *p*-TsNCO to produce acyclic ureas **58**. The latter were subsequently submitted to 5-*exo-dig* cycli-sation to provide the final cyclic five-membered carbamimidates as major products in moderate to high yields, regio- and enantioselectivities, as shown in Scheme 2.75.

On the other hand, the combination of Rh(OAc)$_4$ with a chiral zirconium complex of a BINOL derivative was demonstrated by Hu *et al.* to catalyse an efficient three-component enantioselective aldol-type reaction of aryl diazoacetates, aldehydes, and benzyl alcohol.[165] The authors assumed that the process occurred by trapping a reactive alcoholic oxonium ylide, which was

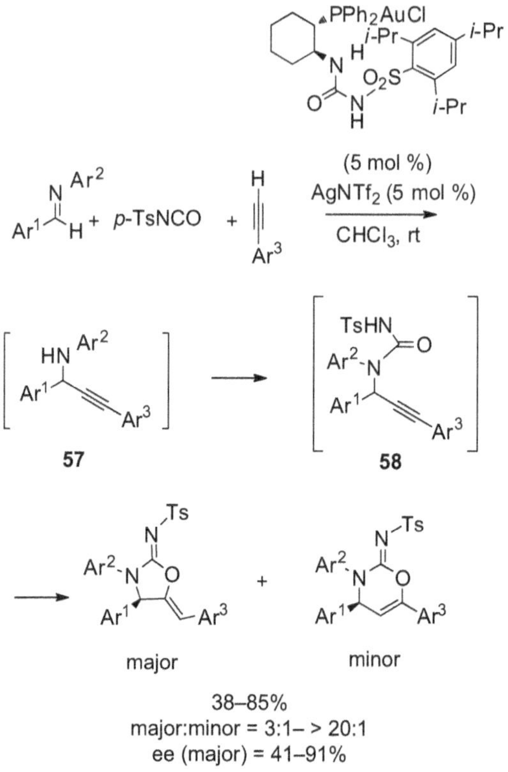

Scheme 2.75 Au-catalysed three-component reaction of imines, terminal alkynes, and *p*-toluenesulfonylisocyanate.

formed *in situ* from the diazoacetate and benzyl alcohol, with aldehyde activated by the zirconium complex to give, through an aldol-type addition, the final products. This three-component reaction provided a convenient and highly enantioselective route to the construction of an important class of compounds for both organic and medicinal chemistry, which are α,β-dihydroxy acid derivatives containing two contiguous tetrasubstituted carbon centres. The products were obtained in moderate to good yields (40–82%) and *erythro : threo* ratios, combined with high enantioselectivities of up to 98% ee for the major *erythro* product. A highly efficient nickel-catalysed reductive coupling of alkynes and imines using Et$_2$Zn as a reductant was achieved by the same authors, affording a range of chiral allylic amines with high yields and chemoselectivities.[166] Chiral induction was achieved by employing a nickel catalyst containing a chiral spiro phosphine ligand, which allowed enantio-selectivities of up to 97% ee to be reached. It must be noted that the best results were obtained in the case of aromatic alkynes. In addition, Montgomery *et al.* have developed asymmetric nickel-catalysed reductive coupling of aldehydes and alkynes, using triethylsilane as the reductant agent.[167] When the process was catalysed by nickel complexes of chiral *N*-heterocyclic carbene ligands

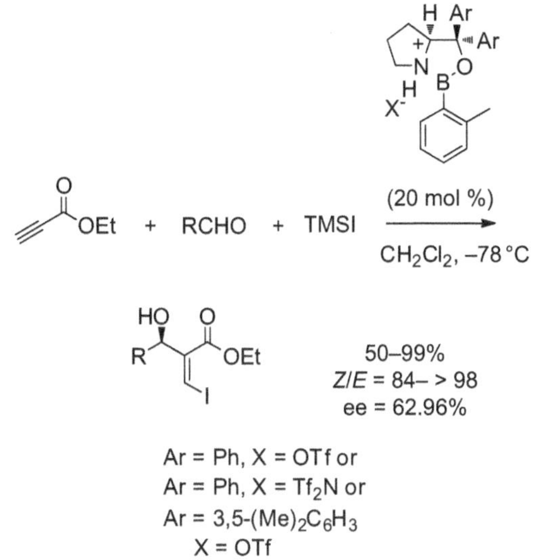

Scheme 2.76   B-catalysed three-component reaction of aldehydes, ethyl propiolate, and TMSI.

derived from C_2-symmetric diamines in the presence of KO-t-Bu as a base, it provided the corresponding chiral silyl ethers in moderate to high yields (47–98%) associated with moderate to good enantioselectivities (65–85% ee). In another context, Ryu *et al.* have developed a highly enantioselective catalytic three-component coupling reaction among an aldehyde, ethyl propiolate, and TMSI to give the corresponding (R)-(Z)-β-iodo Morita–Baylis–Hillman esters.[168] The process was induced by (S)-oxazaborolidinium catalysts, providing (R)-products in moderate to excellent yields of up to 99%, excellent Z-diastereoselectivity, along with high enantioselectivities of up to 96% ee, as shown in Scheme 2.76. Furthermore, the authors have shown that it was possible to prepare the other (S)-enantiomers by performing the reaction upon catalysis with (R)-oxazaborolidinium catalysts. In addition, the (R)-domino products could be further converted through a single step into corresponding (Z)-β-branched derivatives with retention of configuration.

In addition, remarkable levels of regio- and enantioselectivities of up to >90 : 10 and 99% ee, respectively, were reported by Murakami *et al.* in an intermolecular formal [2 + 2 + 2] cycloaddition reaction of two molecules of isocyanates with allenes.[169] The three-component reaction was catalysed by a combination of Ni(cod)$_2$ with unsymmetrical phosphino-oxazoline chiral ligand (S,S)-i-Pr-FOXAP (ferrocenyloxazolinylphosphine). The latter was selected from among a range of various chiral ligands, such as C_2-symmetric biphosphine ligands (S,S)-Chiraphos, (S,S)-Norphos, and (S)-BINAP, which gave lower regioselectivities. This process provided an efficient access to chiral dihydropyrimidine-2,4-diones in moderate to good yields, as shown in Scheme 2.77.

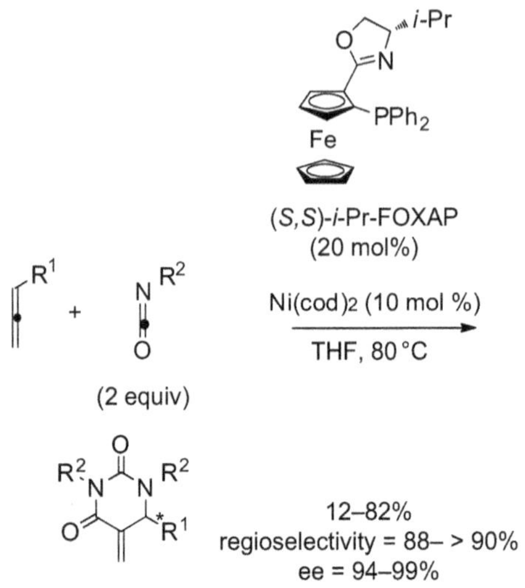

(S,S)-i-Pr-FOXAP
(20 mol%)

Ni(cod)₂ (10 mol %)
──────────────────
THF, 80 °C

(2 equiv)

12–82%
regioselectivity = 88– > 90%
ee = 94–99%

Scheme 2.77   Ni-catalysed three-component reaction of two equivalents of isocyanates and allenes.

On the other hand, Zhao *et al.* have recently developed an enantioselective pseudo-three-component reaction based on double allylations of sodium sulfide.[170] The reaction was catalysed by a chiral iridium complex, generated *in situ* from [Ir(cod)Cl]₂ and a chiral phosphoramidite ligand, in the presence of cesium fluoride and water, yielding the corresponding enantiopure chiral C_2-symmetric bis(1-substituted-allyl)sulfanes in good to excellent yields and diastereoselectivities of up to >98% de, as shown in Scheme 2.78. This work constituted the first example in which sodium sulfide was employed as an atom-economical inorganic nucleophile in transition metal-catalysed allylation substitutions. It is interesting to note that the double allylation reaction efficiently generated two C–S chiral centres in one pot.

Highly functionalised δ-lactones have been produced by Shibasaki *et al.* in the presence of a catalytic amount of a combination of Cu(OAc)₂ with (*R*)-DIFLUORPHOS through an enantioselective three-component assembly of dialkylzincs, allenic ethyl ester, and unactivated ketones.[171] This process allowed the construction of two C–C bonds and one tetrasubstituted chiral centre simultaneously. Conjugate addition of alkyl-copper species to an allenic ester produced highly active copper enolate *in situ*, and the successive asymmetric aldol addition to ketones followed by lactonisation afforded the final lactones in high yields and enantioselectivities of up to 98% ee, as shown in Scheme 2.79. It must be noted that the use of an additive, such as DMSO, HMPA (hexamethylphosphoramide) or Ph₂S = O, was important for obtaining high yields, with suppression of the undesired α-addition pathway.

Feng *et al.* have demonstrated that chiral *N*,*N*'-dioxide–scandium(III) complexes exhibited an excellent ability to activate various electrophiles and showed strong asymmetry-induction capability for many reactions. In this context, these authors have developed modular and tuneable C_2-symmetric

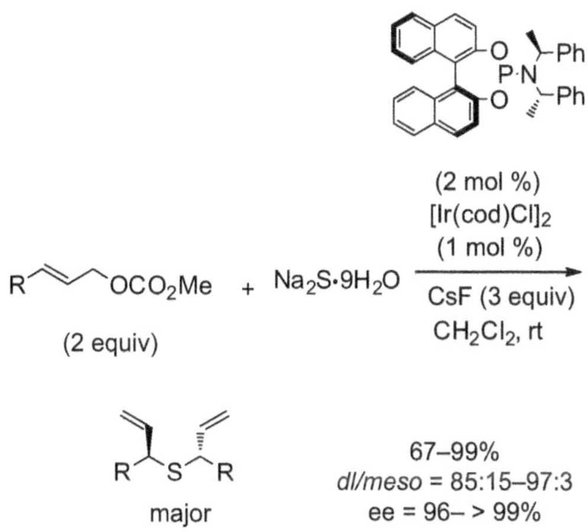

(2 mol %)
[Ir(cod)Cl]₂
(1 mol %)

R⌢OCO₂Me + Na₂S·9H₂O $\xrightarrow{\text{CsF (3 equiv)}}$
(2 equiv)   CH₂Cl₂, rt

major

67–99%
dl/meso = 85:15–97:3
ee = 96– > 99%

Scheme 2.78 Ir-catalysed pseudo-three-component double allylation of sodium sulfide.

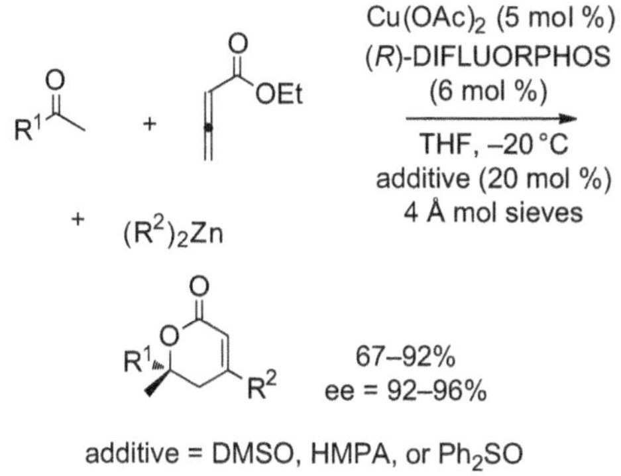

Cu(OAc)₂ (5 mol %)
(R)-DIFLUORPHOS
(6 mol %)

$\xrightarrow{\text{THF, }-20\,^{\circ}\text{C}}$

additive (20 mol %)
4 Å mol sieves

67–92%
ee = 92–96%

additive = DMSO, HMPA, or Ph₂SO

Scheme 2.79 Cu-catalysed three-component alkylative aldol reaction.

N,N'-dioxide ligands, easily synthesised from readily accessible chiral amino acids and amines. These ligands were further applied as scandium ligands to promote the enantioselective direct allylation of aldimines to give the corresponding homoallylic amines.[172] Among these ligands, a chiral chiral *N,N'*-dioxide was selected as the most efficient to perform the three-component reaction of a wide variety of aldimines generated *in situ* from aldehydes and *o*-anisidine, providing the domino products in high yields and remarkable enantioselectivities of up to 97% ee, as shown in Scheme 2.80.

In 2011, Wulff *et al.* described the first multicomponent catalytic asymmetric aziridination reaction, which incorporated a very simple protocol.[173] This process dealt with a three-component reaction of aldehydes, bis(dimethyl-anisyl)methyl amine (MEDAM-NH₂), and ethyl diazo acetate to provide the corresponding chiral aziridine-2-carboxylic esters. When promoted by a chiral boroxinate catalyst generated *in situ* from B(OPh)₃ and chiral

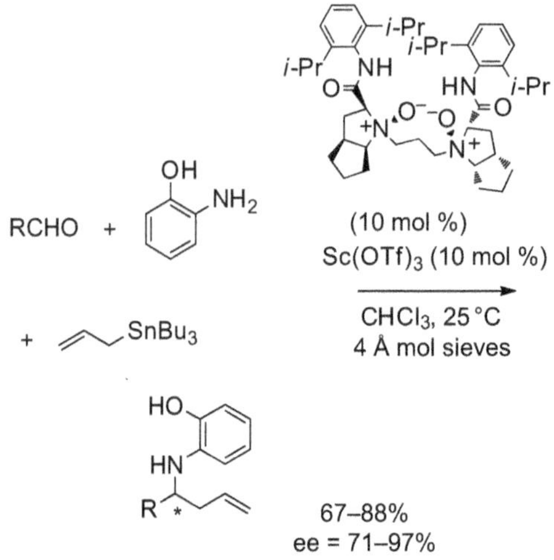

RCHO +

+ [allyl]SnBu₃

(10 mol %)
Sc(OTf)₃ (10 mol %)

CHCl₃, 25 °C
4 Å mol sieves

67–88%
ee = 71–97%

Scheme 2.80 Sc-catalysed three-component allylation of *in situ* generated aldimines.

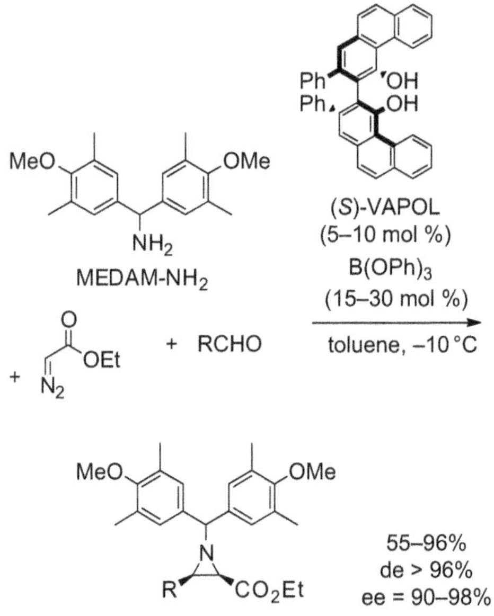

MEDAM-NH₂

+ RCHO

(S)-VAPOL
(5–10 mol %)

B(OPh)₃
(15–30 mol %)

toluene, –10 °C

55–96%
de > 96%
ee = 90–98%

Scheme 2.81 B-catalysed three-component aziridination of aldehydes.

ligand (S)-VAPOL (2,2′-diphenyl-[3,3′-biphenanthrene]-4,4′-diol), the reaction afforded products with good to high yields, excellent diastereoselectivity of >96% de in all cases of the substrates studied, and remarkable enantios-electivities ranging from 90 to 98% ee, as shown in Scheme 2.81. This novel methodology furnished an effective solution to the long-standing problem of imines derived from unbranched aliphatic aldehydes which cannot be purified.

In 2007, Tietze *et al.* reported a new palladium-catalysed three-component domino Wacker–carbonylation reaction, which allowed the synthesis of a wide range of chiral chromans and benzodioxins to be achieved.[174] The combination of a CO insertion with an enantioselective Wacker process was described for the first time. Indeed, a series of chiral chromans were produced through the three-component reaction of unsaturated phenols with alcohols under CO-atmosphere in moderate to good yields and high enantioselectivities of up to 99% ee, as shown in Scheme 2.82. The scope of the methodology was extended to monoallyl ethers, providing under the same conditions the corresponding 2,3-dihydrobenzo[1,4]dioxins in high yields and generally excellent enantio-selectivities ranging from 95 to 99% ee (Scheme 2.82). These processes were catalysed by palladium(II)trifluoroacetate as the palladium source, *p*-benzoquinone as reoxidant, and (*S*,*S*)-Bn-BOXAX as chiral ligand, under a CO-atmosphere as a source of CO at ambient pressure. Later, the same authors employed a more economical methodology employing the same catalyst, albeit used in lower catalyst loadings, such as 12 mol% instead of 40 mol% of (*S*,*S*)-Bn-BOXAX, combined with 3 mol% instead of 10 mol% of

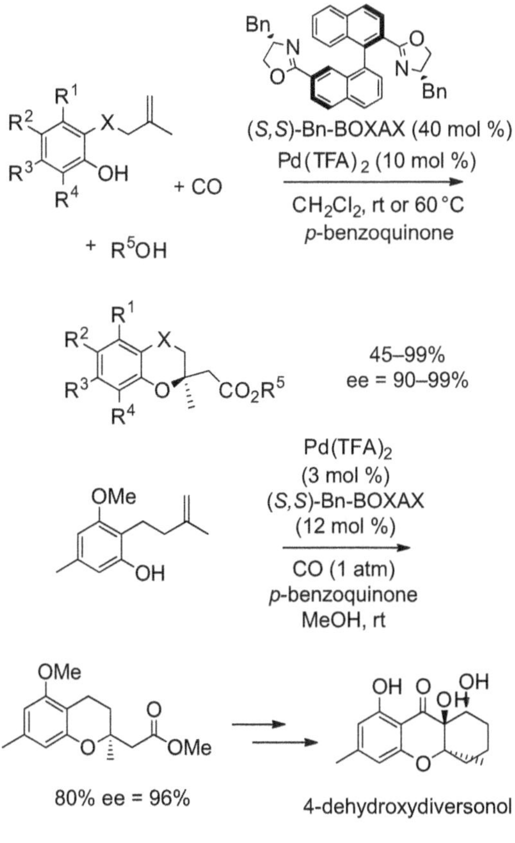

Scheme 2.82   Pd-catalysed three-component Wacker–carbonylation reactions.

palladium(II)trifluoroacetate. Moreover, the reaction was performed in methanol at room temperature instead of dichloromethane at 60 °C. In these conditions, the alkenyl phenol depicted in Scheme 2.82 provided the corresponding ester in 80% yield and 96% ee. This domino product could be subsequently converted into e natural product 4-dehydroxydiversonol.[79a]

Lanthanides have also been employed to catalyse enantioselective domino reactions. As a recent example, Mlynarski *et al.* have developed a ytterbium complex-catalysed direct asymmetric domino aldol–Tishchenko reaction of two molecules of aromatic aldehydes with aliphatic and aromatic ketones.[175] This three-component reaction was promoted by a combination of ytterbium triflate and (1*R*,2*S*)-1-phenyl-2-(1-pyrrolidynyl)-1-propanol derived from ephedrine. It afforded the corresponding domino products, which were subsequently converted into corresponding 1,3-diols through saponification, because the assigned 1,2-*anti*-1,3-*anti* stereochemistry of the domino products could be readily supported by nuclear magnetic resonance (NMR) spectroscopic analysis of separately derived diols. The yields and enantioselectivities in 1,3-diols were moderate to good (25–92% yield, 53–86% ee), while *anti* diastereocontrol was high in all cases of substrates studied. This methodology offerred a simple yet powerful way to prepare enantiomerically enriched *anti*-1,3-diols with high diastereocontrol from unmodified substrates. It must be noted that there are a few examples of enantioselective C–C bond formation leading to quaternary centres which involve free radical intermediates. In this context, Cobb *et al.* have developed a novel asymmetric domino addition–trapping radical reaction, establishing a carbon chiral quaternary centre.[176] The multicomponent reaction occurred between an α,β-unsatured-amide, allyltributylstannane, and an alkyl iodide in the presence of a combination of Yb(OTf)$_3$ and a chiral pybox ligand as the catalyst system. Moderate to good yields associated with low to moderate enantioselectivities (15–73% ee) were obtained for the corresponding α-allyl amides bearing a quaternary stereogenic centre in the α-position. A good enantioselectivity of 73% ee could be reached in only one case of a substrate arising from the use of *tert*-butyl iodide. In addition, a rare example of dual-metal enantiocatalysis was reported by Hu *et al.*, in 2010.[177] These authors have shown that the cooperative catalysis of [Rh(OAc)$_4$]$_2$ with a chiral zinc complex of a (*S*)-*t*-Bu-box ligand (Scheme 2.83) applied to the three-component reaction of diazo compounds with H$_2$O and α,β-unsaturated 2-acyl imidazoles afforded the corresponding γ-hydroxyketones which bore a stereogenic quaternary carbon centre. As shown in Scheme 2.83, these highly functionalised products were achieved in good yields and high diastereo- and enantioselectivities of up to 98% de and 99% ee, respectively. The authors have proposed that the rhodium catalyst promoted the formation of an oxonium ylide generated *in situ* from the aryl diazoacetate and water, which was further trapped by α,β-unsaturated 2-acyl imidazole activated by the zinc catalyst through a Michael-type addition to give the final product. Electronic variations on the aromatic substituents of diazoacetates as well as imidazoles were tolerated in this process. The products were readily converted into the corresponding γ-hydroxy carboxylic acid

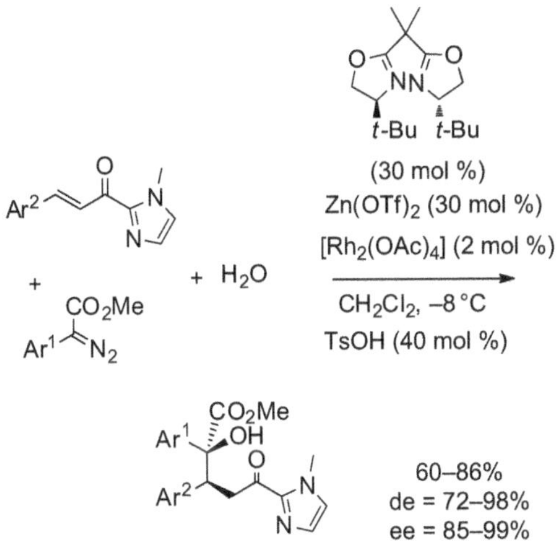

Scheme 2.83 Rh- and Zn-catalysed three-component reaction of diazo compounds, α,β-unsaturated 2-acyl imidazoles, and water.

derivatives, which constitute key intermediates in the construction of natural products and pharmaceuticals.

The Reformatsky reaction is the well-recognised carbon–carbon bond-forming reaction of α-halo esters with aldehydes or ketones in the presence of Zn metal to give β-hydroxy esters. Recently, a rhodium- and nickel-catalysed Reformatsky reaction was described in which dialkylzinc (methyl or ethyl), acting as the Zn source, reacted smoothly with carbonyl compounds and imines. Taking advantage of (1S,2R)-N-methylephedrine as a cheap and recoverable chiral ligand, Cozzi and Rivalta have discovered the first homogeneous enantioselective nickel-catalysed imino Reformatsky reaction.[178] The process was a three-component reaction occurring between aromatic aldehydes, α-bromo esters and o-anisidine, in which ZnMe$_2$ played multiple roles as a dehydrating agent for the formation of the imine, reductant of the Ni(II) salt to Ni(0), and coordinating metal. In this context, a range of chiral β-amino esters could be synthesised in moderate to good yields (40–67%) and enantioselectivities of up to 92% ee on the basis of this domino reaction. In 2010, Dias *et al.* reported a small but remarkable study on a silver-catalysed enantioselective three-component reaction of glycine methyl ester hydrochloride with N-methylmaleimide, and benzaldehyde, which afforded, through domino carbonyl-ylide formation–1,3-dipolar cycloaddition reaction, the corresponding *endo* adduct in 88% yield, and both almost complete diastereo-, and enantioselectivities, as shown in Scheme 2.84.[66] The process was performed in toluene in the presence of a catalytic amount of TEA, and employed 5 mol% of an equimolecular mixture of (S)-BINAP and AgSbF$_6$ as the catalyst system. Furthermore, this methodology could be extended to the three-component reaction of glycine methyl ester hydrochloride, 3-pyridine carbaldehyde, and a disulfone, which afforded the corresponding *endo* product (Scheme 2.84).

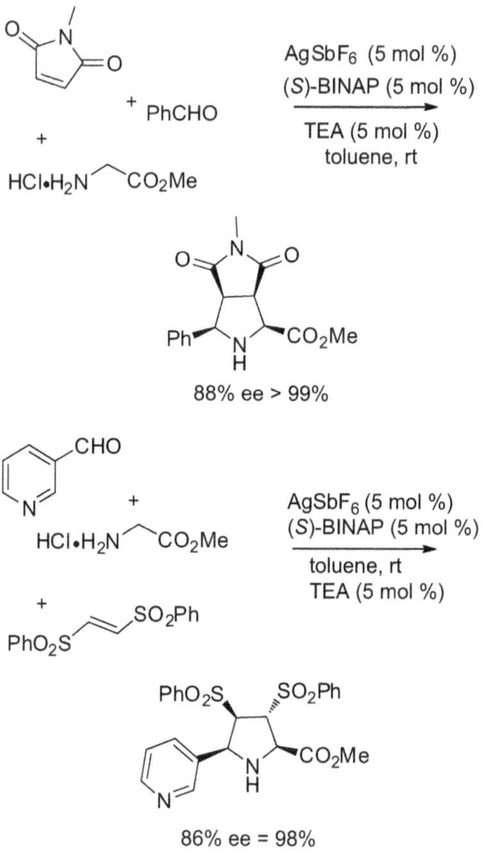

88% ee > 99%

86% ee = 98%

Scheme 2.84 Ag-catalysed three-component domino carbonyl-ylide formation–1,3-dipolar cycloaddition reactions.

Once again, impressive results were obtained, and this product was formed in 86% yieldand complete diastereoselectivity, combined with enantioselectivity of 98% ee, as shown in Scheme 2.84. The selection of the substrates resulted from a domino two-component study involving preformed imines, reported in the same work. The counterion of Ag(I) played an important role in this sensitive reaction. Indeed, it was found that the reactions proceeded smoothly in the presence of the weakly coordinating SbF_6^- counterion, providing high yields and enantioselectivities, whereas no conversion was observed when ClO_4^- was used as the counterion. In another context, the Pauson–Khand reaction is a formal $[2+2+1]$ cycloaddition in which an alkyne, an alkene, and CO furnish a cyclopentenone derivative.[179] A recent example of a multi-component asymmetric version of this reaction was described by Verdaguer *et al.* with the first asymmetric intermolecular cobalt-catalysed Pauson–Khand reaction, in 2007.[180] In this work, the use of camphor-derived chiral ligands provided low to high yields (12–98%) and modest enantioselectivities of $\leq 40\%$ ee.

2.4 Conclusions

This chapter illustrates how much asymmetric organometallic catalysis has contributed to the development of enantioselective domino and multi-component reactions. It updates the major progress in the field of enantio-selective one-, two-, and multicomponent domino reactions promoted by chiral metal catalysts, covering the literature since the beginning of 2006. It demon-strates the power of these elegant one-pot processes of two or more bond-forming reactions, evolving under identical conditions in which the subsequent transformation takes place at the functionalities obtained in the former transformation, following the same principles that are found in biosynthesis in Nature. These fascinating reactions have rapidly become one of the most current fields in organic chemistry. During the last 6 years, an explosive number of novel powerful asymmetric domino and multicomponent processes have been developed on the basis of asymmetric metal catalysis, which is fast becoming fundamental in many manufacturing processes. In particular, a number of novel enantioselective domino reactions have been promoted by chiral rhodium and gold catalysts. In some cases, up to eight stereogenic centres could be generated in excellent stereoselectivities. The wide variety of these novel highly efficient domino processes well reflects that of the metals employed to induce them. Indeed, an increasing number of different metals such as magnesium, scandium, titanium, ruthenium, cobalt, rhodium, iridium, nickel, palladium, platinum, copper, silver, gold, zinc, and aluminium, as well as tin, have been found to be effective catalysts. The economic interest in combinations of asymmetric metal catalytic processes with the concept of domino and multi-component reactions is obvious, and has allowed high molecular complexity to be attained easily, often with excellent levels of stereocontrol, in simple operational one-pot procedures. They have the advantages of savings in solvent, time, energy, and costs by avoiding costly protecting groups and time-consuming purification procedures after each step. Undoubtedly, the future direction in this field is to continue to expand the scope of enantioselective domino and multicomponent reactions through the combination of different types of reactions and the employment of novel chiral catalysts, and to apply these powerful strategies to the synthesis of biologically interesting molecules, including natural products, novel chiral ligands, and functional materials.

References

1. (a) L. F. Tietze, I. Hiriyakkanavar and H. P. Bell, *Chem. Rev.*, 2004, **104**, 3453–3516; (b) E. Negishi, *Handbook of Organopalladium Chemistry for Organic Synthesis*, John Wiley & Sons, Inc., Hoboken NJ, 2002, **2**, pp. 1689–1705; (c) M. Beller and C. Bolm, *Transition Metals for Organic Synthesis*, Wiley-VCH, Weinheim, 1998, vols 1 and 2; (d) G. Poli, G. Giambastiani and A. Heumann, *Tetrahedron*, 2000, **56**, 5959–5989; (e) A. de Meijere, P. von Zezschwitz, H. Nüske and B. Stulgies, *J. Organomet. Chem.*, 2002, **653**, 129–140.

2. L. F. Tietze, G. Brasche and K. Gericke, *Domino Reactions in Organic Synthesis*, Wiley-VCH, Weinheim, 2006.
3. (a) H. Pellissier, *Tetrahedron*, 2006, **62**, 2143–2173; (b) H. Pellissier, *Tetrahedron*, 2006, **62**, 1619–1665.
4. T. Komnenos, *Justus Liebigs Ann. Chem.*, 1883, **218**, 145–169.
5. (a) P. Perlmutter in *Conjugate Addition Reactions in Organic Synthesis*, Pergamon Press, Oxford, 1992; (b) N. Krause and A. Hoffmann-Roder, *Synthesis*, 2001, 171–196; (c) M. P. Sibi and S. Manyem, *Tetrahedron*, 2000, **56**, 8033–8061; (d) M. Kanai and M. Shibasaki in *Catalytic Asymmetric Synthesis*, Wiley, New York, 2nd edn 2000, p. 569.
6. M. Kitamura, T. Miki, K. Nakano and R. Noyori, *Tetrahedron Lett.*, 1996, **37**, 5141–5144.
7. T. Vlaar, E. Ruijter and R. V. A. Orru, *Adv. Synth. Catal.*, 2011, **353**, 809–841.
8. (a) I. Nakamura and Y. Yamamoto, *Chem. Rev.*, 2004, **104**, 2127–2198; (b) G. Zeni and R. Larock, *Chem. Rev.*, 2004, **104**, 2285–2310; (c) J.-C. Wasilke, S. J. Obrey, R. T. Baker and G. C. Bazan, *Chem. Rev.*, 2005, **105**, 1001–1020; (d) S. Cacchi and G. Fabrizi, *Chem. Rev.*, 2005, **105**, 2873–2920.
9. T. Nishikata, *Synlett*, 2007, 3055–3057.
10. X. Yu and X. Lu, *Org. Lett.*, 2009, **11**, 4366–4369.
11. F. Zhou, M. Yang and X. Lu, *Org. Lett.*, 2009, **11**, 1405–1408.
12. X. Yu and X. Lu, *Adv. Synth. Catal.*, 2011, **353**, 2805–2813.
13. G. Liu and X. Lu, *Adv. Synth. Catal.*, 2007, **349**, 2247–2252.
14. K. Li and A. Alexakis, *Chem. Eur. J.*, 2007, **13**, 3765–3771.
15. T. Den Hartog, A. Rudolph, B. Macia, A. J. Minnaard and B. L. Feringa, *J. Am. Chem. Soc.*, 2010, **132**, 14349–14351.
16. (a) D. Bonne, Y. Coquerel, T. Constancieux and J. Rodriguez, *Tetrahedron: Asymmetry*, 2010, **21**, 1085–1109; (b) C. Simon, T. Constancieux and J. Rodriguez, *Eur. J. Org. Chem.*, 2004, 4957–4980.
17. Z. Dong, X. Liu, J. Feng, M. Wang, L. Li and X. Feng, *Eur. J. Org. Chem.*, 2011, 137–142.
18. T. Arai, N. Yokoyama, A. Mishiro and H. Sato, *Angew. Chem., Int. Ed.*, 2010, **49**, 7895–7898.
19. *Modern Aldol Reactions*, ed. R. Mahrwald, Wiley-VCH, Weinheim, 2004.
20. T. Yoshino, H. Morimoto, G. Lu, S. Matsunaga and M. Shibasaki, *J. Am. Chem. Soc.*, 2009, **131**, 17082–17083.
21. S. Wang, M.-X. Wang, D.-X. Wang and J. Zhu, *Eur. J. Org. Chem.*, 2007, 4076–4080.
22. A. Yanagisawa, N. Kushihara and K. Yoshida, *Org. Lett.*, 2011, **13**, 1576–1578.
23. S. Gosiewska, S. Martinez Herreras, M. Lutz, A. L. Spek, R. W. A. Havenith, G. P. M. van Klink, G. van Koten and R. J. M. Klein Gebbink, *Organometallics*, 2008, **27**, 2549–2559.
24. (a) J. M. Lee, Y. Na, H. Han and S. Chang, *Chem. Soc. Rev.*, 2004, **33**, 302–312; (b) D. H. Paull, C. J. Abraham, M. T. Scerba, E. Alden-Danforth and T. Lectka, *Acc. Chem. Res.*, 2008, **41**, 655–663; (c) Z. Shao and H. Zhang, *Chem. Soc. Rev.*, 2009, **38**, 2745–2755; (d) M. Rueping,

R. M. Koenigs and I. Atodiresei, *Chem. Eur. J.*, 2010, **16**, 9350–9365; (e) C. Zhong and X. Shi, *Eur. J. Org. Chem.*, 2010, 2999–3025; (f) J. Zhou, *Chem. Asian J.*, 2010, **5**, 422–434.

25. H. Young Kim and K. Oh, *Org. Lett.*, 2011, **13**, 1306–1309.
26. (a) P. Chiu, *Synthesis*, 2004, 2210–2215; (b) H. Nishiyama and T. Shiomi, *Top. Curr. Chem.*, 2007, **279**, 105–137.
27. (a) D. Zhao, K. Oisaki, M. Kanai and M. Shibasaki, *Tetrahedron Lett.*, 2006, **47**, 1403–1407; (b) D. Zhao, K. Oisaki, M. Kanai and M. Shibasaki, *J. Am. Chem. Soc.*, 2006, **126**, 14440–14441.
28. B. H. Lipshutz, B. Amorelli and J. B. Unger, *J. Am. Chem. Soc.*, 2008, **130**, 14378–14379.
29. J. Deschamp and O. Riant, *Org. Lett.*, 2009, **11**, 1217–1220.
30. O. Chuzel, J. Deschamp, C. Chauster and O. Riant, *Org. Lett.*, 2006, **8**, 5943–5946.
31. J. Deschamp, O. Chuzel, J. Hannedouche and O. Riant, *Angew. Chem., Int. Ed.*, 2006, **45**, 1292–1297.
32. (a) C. Bee, S. B. Han, A. Hassan, H. Iida and M. J. Krische, *J. Am. Chem. Soc.*, 2008, **130**, 2746–2747; (b) S. B. Han, A. Hassan and M. J. Krische, *Synthesis*, 2008, 2669–2679.
33. M. Kita, H. Oki, K. Ogata and S.-i. Fukuzawa, *Synlett*, 2009, 1299–1302.
34. (a) Y. Yamamoto, *Curr. Org. Chem.*, 2005, **9**, 503–509; (b) S. Kotha, E. Brahmachary and K. Lahiri, *Eur. J. Org. Chem.*, 2005, 4741–4761; (c) P. R. Chopade and J. Louie, *Adv. Synth. Catal.*, 2006, **348**, 2307–2327; (d) Y. Shibata and K. Tanaka, *Synthesis*, 2012, 323–350.
35. (a) T. Shibata and K. Tsuchikama, *Org. Biomol. Chem.*, 2008, 1317–1323; (b) K. Tanaka, *Chem. Asian J.*, 2009, **4**, 508–518.
36. K. Tanaka, T. Osaka, K. Noguchi and M. Hirano, *Org. Lett.*, 2007, **9**, 1307–1310.
37. G. Nishida, N. Suzuki, K. Noguchi and K. Tanaka, *Org. Lett.*, 2006, **8**, 3489–3492.
38. A. Wada, K. Noguchi, M. Hirano and K. Tanaka, *Org. Lett.*, 2007, **9**, 1295–1298.
39. T. Shibata, T. Chiba, H. Hirashima, Y. Ueno and K. Endo, *Angew. Chem., Int. Ed.*, 2009, **48**, 8066–8069.
40. T. Shibata, S. Yoshida, Y. Arai, M. Otsuka and K. Endo, *Tetrahedron*, 2008, **64**, 821–830.
41. T. Shibata, T. Chiba, H. Hirashima, Y. Ueno and K. Endo, *Heteroatom Chem.*, 2011, **22**, 363–370.
42. M. Kobayashi, T. Suda, K. Noguchi and K. Tanaka, *Angew. Chem., Int. Ed.*, 2011, **50**, 1664–1667.
43. W. Oppolzer and J.-M. Gaudin, *Helv. Chim. Acta*, 1987, **70**, 1477–1481.
44. Y. Uozumi, A. Tanahashi and T. Hayashi, *J. Org. Chem.*, 1993, **58**, 6826–6832.
45. B. M. Trost and G. Dong, *Chem. Eur. J.*, 2009, **15**, 6910–6919.
46. (a) H. He, W.-B. Liu, L.-X. Dai and S.-L. You, *Angew. Chem., Int. Ed.*, 2010, **49**, 1496–1499; (b) K.-Y. Ye, H. He, W.-B. Liu, L.-X. Dai, G. Helmchen and S.-L. You, *J. Am. Chem. Soc.*, 2011, **133**, 19006–19014.

47. J. Schütte, S. Ye and H.-G. Schmalz, *Synlett*, 2011, 2725–2729.

48. A. Doemling and I. Ugi, *Angew. Chem., Int. Ed.*, 2000, **39**, 3168–3210.

49. A. Doemling, *Chem. Rev.*, 2006, **106**, 17–89.

50. S.-X. Wang, M.-X. Wang, D.-X. Wang and J. Zhu, *Org. Lett.*, 2007, **9**, 3615–3618.

51. T. Yue, M.-X. Wang, D.-X. Wang, G. Masson and J. Zhu, *J. Org. Chem.*, 2009, **74**, 8396–8399.

52. H. Mihara, Y. Xu, N. E. Shepherd, S. Matsunaga and M. Shibasaki, *J. Am. Chem. Soc.*, 2009, **131**, 8384–8385.

53. (a) M. P. Doyle and D. C. Forbes, *Chem. Rev.*, 1998, **98**, 911–935; (b) K. V. Gothelf and K. A. Jorgensen, *Chem. Rev.*, 1998, **98**, 863–909; (c) A. Padwa and M. D. Weingarten, *Chem. Rev.*, 1996, **96**, 223–269; (d) S. Karlsson and H.-E. Högberg, *Org. Prep. Proc. Int.*, 2001, **33**, 103–172; (e) I. N. N. Namboothiri and A. Hassner, *Top. Curr. Chem.*, 2001, **216**, 1–49; (f) S. Kanemasa, *Synlett*, 2002, 1371–1387; (g) K. V. Gothelf, *Synthesis*, 2002, 211–247; (h) G. Broggini, G. Molteni, A. Terraneo and G. Zecchi, *Heterocycles*, 2003, **59**, 823–858; (i) I. Coldham and R. Hufton, *Chem. Rev.*, 2005, **105**, 2765–2809; (j) H. Pellissier, *Tetrahedron*, 2007, **63**, 3235–3285; (k) S. Kanemasa, *Heterocycles*, 2010, **82**, 87–200.

54. A. Padwa, *Acc. Chem. Res.*, 1991, **24**, 22–28.

55. (a) A. Padwa and S. F. Hornbuckle, *Chem. Rev.*, 1991, **91**, 263–309; (b) A. Padwa, *Helv. Chim. Acta*, 2005, **88**, 1357–1374; (c) A. Padwa, *J. Organomet. Chem.*, 2005, **690**, 5533–5540; (d) A. Padwa, *Chem. Soc. Rev.*, 2009, **38**, 3072–3081.

56. D. M. Hodgson, P. A. Stupple and C. Johnstone, *Tetrahedron Lett.*, 1997, **38**, 6471–6472.

57. (a) H. Tsutsui, N. Shimada, T. Abe, M. Anada, M. Nakajima, S. Nakamura, H. Nambu and S. Hashimoto, *Adv. Synth. Catal.*, 2007, **349**, 521–526; (b) Y. Kurosaki, N. Shimada, M. Anada, H. Nambu and S. Hashimoto, *Bull. Korean Chem. Soc.*, 2010, **31**, 694–696.

58. N. Shimada, M. Anada, S. Nakamura, H. Nambu, H. Tsutsui and S. Hashimoto, *Org. Lett.*, 2008, **10**, 3603–3606.

59. N. Shimada, T. Hanari, Y. Kurosaki, K. Takeda, M. Anada, H. Nambu, M. Shiro and S. Hashimoto, *J. Org. Chem.*, 2010, **75**, 6039–6042.

60. N. Shimada, T. Hanari, Y. Kurosaki, M. Anada, H. Nambu and S. Hashimoto, *Tetrahedron Lett.*, 2010, **51**, 6572–6575.

61. N. Shimada, T. Oohara, J. Krishnamurthi, H. Nambu and S. Hashimoto, *Org. Lett.*, 2011, **13**, 6284–6287.

62. H. Nambu, M. Hikime, J. Krishnamurthi, M. Kamiya, N. Shimada and S. Hashimoto, *Tetrahedron Lett.*, 2009, **50**, 3675–3678.

63. D. M. Hodgson, R. Glen and A. J. Redgrave, *Tetrahedron: Asymmetry*, 2009, **20**, 754–757.

64. H. Suga, T. Suzuki, K. Inoue and A. Kakehi, *Tetrahedron*, 2006, **62**, 9218–9225.

65. (a) H. Suga, D. Hishimoto, S. Higuchi, M. Ohtsuka, T. Arikawa, T. Tsuchida, A. Kakehi and T. Baba, *Org. Lett.*, 2007, **9**, 4359–4362;

(b) H. Suga, S. Higuchi, M. Ohtsuka, D. Ishimoto, T. Airkawa, Y. Hashimoto, S. Misawa, T. Tsuchida, A. Kakehi and T. Baba, *Tetrahedron*, 2010, **66**, 3070–3089.

66. M. Martin-Rodriguez, C. Najera, J. M. Sansano, P. R. R. Costa, E. C. de Lima and A. G. Dias, *Synlett*, 2010, 962–966.

67. (a) R. F. Heck, *J. Am. Chem. Soc.*, 1968, **90**, 5518–5526; (b) T. Mizoroki, *Bull. Chem. Soc. Jpn.*, 1971, **44**, 581–581; (c) R. F. Heck and J. P. Nolley, *J. Org. Chem.*, 1972, **37**, 2320–2322; (d) R. F. Heck, *Org. React.*, 1982, **27**, 345–390; (e) A. de Meijere and F. E. Meyer, *Angew. Chem., Int. Ed. Engl.*, 1994, **33**, 2379–2411; (f) M. Shibasaki, C. D. J. Boden and A. Kojima, *Tetrahedron*, 1997, **53**, 7371–7395; (g) P. J. Guiry, A. Hennessy and J. Cahill, *Top. Catal.*, 1997, **4**, 311–326; (h) I. P. Beletskaya and A. V. Cheprakov, *Chem. Rev.*, 2000, **100**, 3009–3066; (i) C. Amatore and A. Jutard, *Acc. Chem. Res.*, 2000, **33**, 314–321; (j) A. B. Dounay and L. E. Overman, *Chem. Rev.*, 2003, **103**, 2945–2964; (k) M. Shibasaki, E. M. Vogl and T. Ohshima, *Adv. Synth. Catal.*, 2004, **346**, 1533–1552; (l) A. G. Coyne, M. O. Fitzpatrick and P. J. Guiry in *The Mizoroki – Heck Reaction*, ed. M. Oestreich, Wiley, Chichester, 2009; (m) D. Mc Cartney and P. J. Guiry, *Chem. Soc. Rev.*, 2011, **40**, 5122–5150.

68. (a) S. Bräse and A. de Meijere, in *Metal-Catalysed Cross Coupling Reactions*, ed. A. de Meijere and F. Diederich, Wiley-VCH, Weinheim, 2004; (b) A. B. Dounay and L. E. Overman, in *The Mizoroki – Heck Reaction*, ed. M. Oestreich, Wiley, Chichester, 2009.

69. (a) O. Reiser, M. Weber and A. de Meijere, *Angew. Chem., Int. Ed. Engl.*, 1989, **59**, 7479; (b) K. Albrecht, O. Reiser, M. Weber and A. de Meijere, *Angew. Chem., Int. Ed. Engl.*, 1989, **28**, 1037–1038; (c) S. P. Watson, G. R. Knox and N. M. Heron, *Tetrahedron Lett.*, 1994, **35**, 9763–9766; (d) R. Grigg, P. Kennewell, A. Teasdale and V. Sridharan, *Tetrahedron Lett.*, 1993, **34**, 153–156; (e) R. Grigg and V. Sridharan, *Tetrahedron Lett.*, 1992, **33**, 7965–7968; (f) F. E. Meyer, K. H. Ang, A. G. Steinig and A. de Meijere, *Synlett*, 1994, 191–193; (g) G. D. Harris, R. J. Herr and S. M. Weinreb, *J. Org. Chem.*, 1993, **58**, 5452–5464.

70. N. E. Carpenter, D. J. Kucera and L. E. Overman, *J. Org. Chem.*, 1989, **54**, 5846–5848.

71. S. P. Maddaford, N. G. Andersen, W. A. Cristofoli and B. A. Keay, *J. Am. Chem. Soc.*, 1996, **118**, 10766–10773.

72. R. Crigg and V. Sridharan, *J. Organomet. Chem.*, 1999, **576**, 65–87.

73. A. Pinto, Y. Jia, L. Neuville and J. Zhu, *Chem. Eur. J.*, 2007, **13**, 961–967.

74. (a) J. M. Hopkins, E. Gorobets, B. M. M. Wheatley, M. Parvez and B. A. Keay, *Synlett*, 2006, 3120–3124; (b) D. A. Rankic, D. Lucciola and B. A. Keay, *Tetrahedron Lett.*, 2010, **51**, 5724–5727.

75. D. Lucciola and B. A. Keay, *Synlett*, 2011, 1618–1622.

76. (a) J. Reinhard, *Angew. Chem., Int. Ed.*, 2009, **48**, 9034–9037; (b) R. I. McDonald, G. Liu and S. S. Stahl, *Chem. Rev.*, 2011, **111**, 2981–3019.

77. L. F. Tietze, K. M. Sommer, J. Zinngrebe and F. Stecker, *Angew. Chem., Int. Ed.*, 2005, **44**, 257–259.

78. (a) L. F. Tietze, F. Stecker, J. Zinngrebe and K. M. Sommer, *Chem. Eur. J.*, 2006, **12**, 8770–8776; (b) L. F. Tietze and T. Kinzel, *Pure Appl. Chem.*, 2007, **79**, 629–650.

79. (a) L. F. Tietze, D. A. Spiegl, F. Stecker, J. Major, C. Raith and C. Große, *Chem. Eur. J.*, 2008, **14**, 8956–8963; (b) L. F. Tietze and A. Düfert, *Pure Appl. Chem.*, 2010, **82**, 1375–1392.

80. G. B. Bajracharya, M. A. Arai, P. S. Koranne, T. Suzuki, S. Takizawa and H. Sasai, *Bull. Chem. Soc. Jpn.*, 2009, **82**, 285–302.

81. K.-T. Yip, M. Yang, K.-L. Law, N.-Y. Zhu and D. Yang, *J. Am. Chem. Soc.*, 2006, **128**, 3130–3131.

82. W. He, K.-T. Yip, N.-Y. Zhu and D. Yang, *Org. Lett.*, 2009, **11**, 5626–5628.

83. (a) B. M. Trost and M. J. Krische, *Synlett*, 1998, 1–16; (b) C. Aubert, O. Buisine and M. Malacria, *Chem. Rev.*, 2002, **102**, 813–834; (c) I. J. S. Fairlamb, *Angew. Chem., Int. Ed.*, 2004, **43**, 1048–1052; (d) C. Nevado and A. M. Echavarren, *Synthesis*, 2005, 167–182; (e) L. Zhang, J. Sun and S. A. Kozmin, *Adv. Synth. Catal.*, 2006, **348**, 2271–2296; (f) V. Michelet, P. Y. Toullec and J.-P. Genêt, *Angew. Chem., Int. Ed.*, 2008, **47**, 4268–4315; (g) E. Jimenez-Nunez and E. Echavarren, *Chem. Rev.*, 2008, **108**, 3326–3350; (h) P. Belmont and E. Parker, *Eur. J. Org. Chem.*, 2009, 6075–6089.

84. (a) G. C. Lloyd-Jones, *Org. Biomol. Chem.*, 2003, **1**, 215–236; (b) A. R. Chianese, S. J. Lee and M. R. Gagné, *Angew. Chem., Int. Ed.*, 2007, **46**, 4042–4046; (c) A. Fürstner and P. W. Davies, *Angew. Chem., Int. Ed.*, 2007, **46**, 3410–3449; (d) A. S. K. Hashmi, *Chem. Rev.*, 2007, **107**, 3180–3211; (e) S. I. Lee and N. Chatani, *Chem. Commun.*, 2009, 371–384; (f) P. Y. Toullec and V. Michelet, *Top. Curr. Chem.*, 2011, **302**, 31–80.

85. (a) N. Bongers and N. Krause, *Angew. Chem., Int. Ed.*, 2008, **47**, 2178–2181; (b) R. A. Widenhoefer, *Chem. Eur. J.*, 2008, **14**, 5382–5391; (c) S. Sengupta and X. Shi, *ChemCatChem*, 2010, **2**, 609–619; (d) A. Pradal, P. Y. Toullec and V. Michelet, *Synthesis*, 2011, 1501–1514.

86. (a) C.-M. Chao, E. Genin, P. Y. Toullec, J.-P. Genêt and V. Michelet, *J. Organomet. Chem.*, 2009, **694**, 538–545; (b) C.-M. Chao, M. R. Vitale, P. Y. Toullec, J.-P. Genêt and V. Michelet, *Chem. Eur. J.*, 2009, **15**, 1319–1323.

87. (a) A. Fürstner and L. Morency, *Angew. Chem., Int. Ed.*, 2008, **47**, 5030–5033; (b) A. S. K. Hashmi, *Angew. Chem., Int. Ed.*, 2008, **47**, 6754–6756.

88. (a) G. Stork and A. W. Burgstahler, *J. Am. Chem. Soc.*, 1955, **77**, 5068–5077; (b) A. Eschenmoser, L. Ruzicka, O. Jeger and D. Arigoni, *Helv. Chim. Acta*, 1955, **38**, 1890–1904; (c) A. Eschenmoser and D. Arigoni, *Helv. Chim. Acta*, 2005, **88**, 3011–3025.

89. A. Pradal, C.-M. Chao, M. R. Vitale, P. Y. Toullec and V. Michelet, *Tetrahedron*, 2011, **67**, 4371–4377.

90. V. Michelet, L. Charruault, S. Gladiali and J.-P. Genêt, *Pure Appl. Chem.*, 2006, **78**, 397–407.

91. P. Y. Toullec, C.-M. Chao, Q. Chen, S. Gladiali, J.-P. Genêt and V. Michelet, *Adv. Synth. Catal.*, 2008, **350**, 2401–2408.

92. A. Grossmann and D. Enders, *Angew. Chem., Int. Ed.*, 2011, **50**, 2–14.

93. Y. Matsumoto, K. B. Selim, H. Nakanishi, K.-i. Yamada, Y. Yamamoto and K. Tomioka, *Tetrahedron Lett.*, 2010, **51**, 404–406.

94. A. Martinez, P. Garcia-Garcia, M. A. Fernandez-Rodriguez, F. Rodriguez and R. Sanz, *Angew. Chem., Int. Ed.*, 2010, **49**, 4633–4637.

95. S. Handa and L. M. Slaughter, *Angew. Chem., Int. Ed.*, 2012, **51**, 2912–2915.

96. F. Liu, D. Qian, L. Li, X. Zhao and J. Zhang, *Angew. Chem., Int. Ed.*, 2010, **49**, 6669–6672.

97. G. Zhou, F. Liu and J. Zhang, *Chem. Eur. J.*, 2011, **17**, 3101–3104.

98. S. G. Sethofer, T. Mayer and F. D. Toste, *J. Am. Chem. Soc.*, 2010, **132**, 8276–8277.

99. A. M. Jadhav, S. Bhunia, H.-Y. Liao and R.-S. Liu, *J. Am. Chem. Soc.*, 2011, **133**, 1769–1771.

100. B. J. Anderson, M. A. Guino-o, D. S. Glueck, J. A. Golen, A. G. DiPasquale, L. M. Liable-Sands and A. L. Reingold, *Org. Lett.*, 2008, **10**, 4425–4428.

101. C. A. Mullen, A. N. Campbell and M. R. Gagné, *Angew. Chem., Int. Ed.*, 2008, **47**, 6011–6014.

102. S. Hajra and S. Bar, *Chem. Commun.*, 2011, 3981–3982.

103. V. Rauniyar, Z. J. Wang, H. E. Burks and F. D. Toste, *J. Am. Chem. Soc.*, 2011, **133**, 8486–8489.

104. (a) P. Renaud and M. Gerster, *Angew. Chem., Int. Ed.*, 1998, **37**, 2562–2579; (b) M. P. Sibi and N. A. Porter, *Acc. Chem. Res.*, 1999, **32**, 163–171; (c) G. Bar and A. F. Parsons, *Chem. Soc. Rev.*, 2003, **32**, 251–263; (d) M. P. Sibi, S. Manyem and J. Zimmerman, *Chem. Rev.*, 2003, **103**, 3263–3296.

105. H. Miyabe, R. Asada, A. Toyoda and Y. Takemoto, *Angew. Chem., Int. Ed.*, 2006, **45**, 5863–5866.

106. (a) H. Miyaka, A. Toyoda and Y. Takemoto, *Synlett*, 2007, 1885–1888; (b) H. Miyabe and Y. Takemoto, *Chem. Eur. J.*, 2007, **13**, 7280–7286.

107. D. Yang, B.-F. Zheng, Q. Gao, S. Gu and N.-Y. Zhu, *Angew. Chem., Int. Ed.*, 2006, **45**, 255–258.

108. G. A. Cutting, N. E. Stainforth, M. P. John, G. Kociok-Köhn and M. C. Willis, *J. Am. Chem. Soc.*, 2007, **129**, 10632–10633.

109. S. Son and G. C. Fu, *J. Am. Chem. Soc.*, 2007, **129**, 1046–1047.

110. (a) L. V. R. Bonaga and M. E. Krafft, *Tetrahedron*, 2004, **60**, 9795–9833; (b) K. H. Park and Y. K. Chung, *Synlett*, 2005, 545–559; (c) S. E. Gibson and N. Mainolfi, *Angew. Chem., Int. Ed.*, 2005, **44**, 3022–3037; (d) T. Shibata, *Adv. Synth. Catal.*, 2006, **348**, 2328–2336.

111. (a) H. W. Lee, A. S. C. Chan and F. Y. Kwong, *Chem. Commun.*, 2007, 2633–2635; (b) H. W. Lee, L. N. Lee, A. S. C. Chan and F. Y. Kwong, *Eur. J. Org. Chem.*, 2008, 3403–3406.

112. (a) K. Kanao, Y. Miyake and Y. Nishibayashi, *Organometallics*, 2010, **29**, 2126–2131; (b) Y. Nishibayashi, *Synthesis*, 2012, 489–503.

113. O. Bondarev and C. Bruneau, *Tetrahedron: Asymmetry*, 2010, **21**, 1350–1354

114. A. T. Londregan, D. Bernhardson, J. Bradow, T. M. Makowski, G. Storer, J. Warmus, C. Wooten and X. Yang, *Tetrahedron: Asymmetry*, 2010, **21**, 2072–2075.

115. M. Weber, S. Jautze, W. Frey and R. Peters, *J. Am. Chem. Soc.*, 2010, **132**, 12222–12225.

116. X. Fu, J. Feng, Z. Dong, L. Lin, X. Liu and X. Feng, *Eur. J. Org. Chem.*, 2011, 5233–5236.

117. S. H. Lecker, N. H. Nguyen and K. P. C. Vollhardt, *J. Am. Chem. Soc.*, 1986, **108**, 856–858.

118. W. Cao, X. Liu, W. Wang, L. Lin and X. Feng, *Org. Lett.*, 2011, **13**, 600–603.

119. B. T. Parr, Z. Li and H. M. L. Davies, *Chem. Sci.*, 2011, **2**, 2378–2382.

120. A. Boyer and M. Lautens, *Angew. Chem., Int. Ed.*, 2011, **50**, 7346–7349.

121. (a) I. Ugi, R. Meyr, U. Fetzer and C. Steinbrückner, *Angew. Chem.*, 1959, **71**, 386–388; (b) R. V. A. Orru and M. de Greef, *Synthesis*, 2003, 1471–1499; (c) E. Ruijter, R. Scheffelaar and R. V. A. Orru, *Angew. Chem. Int. Ed.*, 2011, **50**, 6234–6246.

122. J. C. Gonzales-Gomez, F. Foubelo and M. Yus, *Tetrahedron Lett.*, 2008, **49**, 2343–2347.

123. S. Guo, Y. Xie, X. Hu, C. Xia and H. Huang, *Angew. Chem., Int. Ed.*, 2010, **49**, 2728–2731.

124. I.-H. Chen, L. Yin, W. Itano, M. Kanai and M. Shibasaki, *J. Am. Chem. Soc.*, 2009, **131**, 11664–11665.

125. S. Guo, Y. Xie, X. Hu and H. Huang, *Org. Lett.*, 2011, **13**, 5596–5599.

126. (a) J. Boruwa, N. Gogoi, P. P. Saikia and N. C. Barua, *Tetrahedron: Asymmetry*, 2006, **17**, 3315–3326; (b) C. Palomo, M. Oiarbide and A. Laso, *Eur. J. Org. Chem.*, 2007, 2561–2574; (c) E. Marquès-Lopez, P. Merino, T. Tejero and R. P. Herrera, *Eur. J. Org. Chem.*, 2009, 2401–2420; (d) Y. Alvarez-Casao, E. Marques-Lopez and R. P. Herrera, *Symmetry*, 2011, **3**, 220–245.

127. D. Shi, Y. Xie, H. Zhou, C. Xia and H. Huang, *Angew. Chem., Int. Ed.*, 2012, **51**, 1248–1251.

128. C. Mannich and W. Krosche, *Arch. Pharm.*, 1912, **250**, 647–667.

129. (a) M. Arend, B. Westermann and N. Risch, *Angew. Chem., Int. Ed.*, 1998, **37**, 1044–1070; (b) A. Cordova, *Acc. Chem. Res.*, 2004, **37**, 102–112; (c) J. M. M. Verkade, L. J. C. Van Hemert, P. J. L. M. Quaedflieg and F. P. J. T. Rutjes, *Chem. Soc. Rev.*, 2008, **37**, 29–41; (d) R. Gomez Arrayas and J. C. Carretero, *Chem. Soc. Rev.*, 2009, **38**, 1940–1948.

130. S. Chen, Z. Hou, Y. Zhu, J. Wang, L. Lin, X. Liu and X. Feng, *Chem. Eur. J.*, 2009, **15**, 5884–5887.

131. Q. Zhang, Y. Hui, X. Zhou, L. Lin, X. Liu and X. Feng, *Adv. Synth. Catal.*, 2010, **352**, 976–980.

132. H. Mandai, K. Mandai, M. L. Snapper and H. Hoveyda, *J. Am. Chem. Soc.*, 2008, **130**, 17961–17969

133. B. Ranieri, C. Curti, L. Battistini, A. Sartori, L. Pinna, G. Casiraghi and F. Zanardi, *J. Org. Chem.*, 2011, **76**, 10291–10298.

134. A. Izumiseki, K. Yoshida and A. Yanagisawa, *Org. Lett.*, 2009, **11**, 5310–5313.

135. S. Kobayashi, J. Kobayashi, R. Yazaki and M. Ueno, *Chem. Asian J.*, 2007, **2**, 135–144.

136. Y. Hamashima, N. Sasamoto, N. Umebayashi and M. Sodeoka, *Chem. Asian J.*, 2008, **3**, 1443–1455.

137. Y. Du, L.-W. Xu, Y. Shimizu, K. Oisaki, M. Kanai and M. Shibasaki, *J. Am. Chem. Soc.*, 2008, **130**, 16146–16147.

138. J. Poulin, C. M. Grisé-Bard and L. Barriault, *Chem. Soc. Rev.*, 2009, **38**, 3092–3101.

139. G. Hilt, W. Hess and K. Harms, *Org. Lett.*, 2006, **8**, 3287–3290.

140. M. Xie, X. Chen, Y. Zhu, B. Gao, L. Lin, X. Liu and X. Feng, *Angew. Chem., Int. Ed.*, 2010, **49**, 3799–3802.

141. S. Torssell and P. Somfai, *Adv. Synth. Catal.*, 2006, **348**, 2421–2430.

142. S.-X. Wang, M.-X. Wang, D.-X. Wang and J. Zhu, *Angew. Chem., Int. Ed.*, 2008, **47**, 388–391.

143. J. M. McManus and R. M. Herbst, *J. Org. Chem.*, 1959, **24**, 1643–1649.

144. T. Yue, M.-X. Wang, D.-X. Wang and J. Zhu, *Angew. Chem., Int. Ed.*, 2008, **47**, 9454–9457.

145. (a) T. Arai and N. Yokoyama, *Angew. Chem., Int. Ed.*, 2008, **47**, 4989–4992; (b) T. Arai, M. Wasai and N. Yokoyama, *J. Org. Chem.*, 2011, **76**, 2909–2912.

146. Y. Duan, M.-W. Chen, Z.-S. Ye, D.-S. Wang, Q.-A. Chen and Y.-G. Zhou, *Chem. Eur. J.*, 2011, **17**, 7193–7197.

147. N. Gommermann and P. Knochel, *Chem. Eur. J.*, 2006, **12**, 4380–4392.

148. P. Aschwanden, C. R. J. Stephenson and E. M. Carreira, *Org. Lett.*, 2006, **8**, 2437–2440.

149. A. Bisai and V. K. Singh, *Org. Lett.*, 2006, **8**, 2405–2408.

150. S. Nakamura, M. Ohara, Y. Nakamura, N. Shibata and T. Toru, *Chem. Eur. J.*, 2010, **16**, 2360–2362.

151. F. Colombo, M. Benaglia, S. Orlandi and F. Usuelli, *J. Mol. Catal. A*, 2006, **260**, 128–134.

152. Z. Shao, X. Pu, X. Li, B. Fan and A. S. C. Chan, *Tetrahedron: Asymmetry*, 2009, **20**, 225–229.

153. D. Chernyak, N. Chernyak and V. Gevorgyan, *Adv. Synth. Catal.*, 2010, **352**, 961–966.

154. L. Zani, T. Eichhorn and C. Bolm, *Chem. Eur. J.*, 2007, **13**, 2587–2600.

155. C. Yin, X.-Q. Hu, X.-P. Hui and P.-F. Xu, *Adv. Synth. Catal.*, 2009, **351**, 1512–1516.

156. Y. Sato, Y. Hinata, R. Seki, Y. Oonishi and N. Saito, *Org. Lett.*, 2007, **9**, 5597–5599.

157. N. Saito, A. Kobayashi and Y. Sato, *Angew. Chem., Int. Ed.*, 2012, **51**, 1228–1231.

158. Y. Yang, S.-F. Zhu, H.-F. Duan, C.-Y. Zhou, L.-X. Wang and Q.-L. Zhou, *J. Am. Chem. Soc.*, 2007, **129**, 2248–2249.

159. L. Rubio-Pérez, F. J. Pérez-Flores, P. Sharma, L. Velasco and A. Cabrera, *Org. Lett.*, 2009, **11**, 265–268.

160. C. Li, B. Villa-Marcos and J. Xiao, *J. Am. Chem. Soc.*, 2009, **131**, 6967–6969.

161. X. Zhou, D. Shang, Q. Zhang, L. Lin, X. Liu and X. Feng, *Org. Lett.*, 2009, **11**, 1401–1404.

162. M. Ohara, S. Nakamura and N. Shibata, *Adv. Synth. Catal.*, 2011, **353**, 3285–3289.

163. (a) M. Bandini, *Chem. Soc. Rev.*, 2011, **40**, 1358–1367; (b) A. Fürstner, *Chem. Soc. Rev.*, 2009, **38**, 3208–3221; (c) Z. Li, C. Brouwer and C. He, *Chem. Rev.*, 2008, **108**, 3239–3265; (d) A. Arcadi, *Chem. Rev.*, 2008, **108**, 3266–3325; (e) D. J. Gorin, B. D. Sherry and F. D. Toste, *Chem. Rev.*, 2008, **108**, 3351–3378; (f) A. S. K. Hashmi and M. Rudolph, *Chem. Soc. Rev.*, 2008, **37**, 1766–1775; (g) H. C. Shen, *Tetrahedron*, 2008, **64**, 3885–3903; (h) R. Skouta and C.-J. Li, *Tetrahedron*, 2008, **64**, 4917–4938.

164. M. J. Campbell and F. D. Toste, *Chem. Sci.*, 2011, **2**, 1369–1378.

165. X. Zhang, H. Huang, X. Guo, X. Guan, L. Yang and W. Hu, *Angew. Chem., Int. Ed.*, 2008, **47**, 6647–6649.

166. C.-Y. Zhou, S.-F. Zhu, L.-X. Wang and Q.-L. Zhou, *J. Am. Chem. Soc.*, 2010, **132**, 10955–10957.

167. M. R. Chaulagain, G. J. Sormunen and J. Montgomery, *J. Am. Chem. Soc.*, 2007, **129**, 9568–9569.

168. B. K. Senapati, G.-S. Hwang, S. Lee and D. H. Ryu, *Angew. Chem., Int. Ed.*, 2009, **48**, 4398–4401.

169. T. Miura, M. Morimoto and M. Murakami, *J. Am. Chem. Soc.*, 2010, **132**, 15836–15838.

170. S. Zheng, W. Huang, N. Gao, R. Cui, M. Zhang and X. Zhao, *Chem. Commun.*, 2011, 6969–6971.

171. K. Oisaki, D. Zhao, M. Kanai and M. Shibasaki, *J. Am. Chem. Soc.*, 2007, **129**, 7439–7443.

172. X. Li, X. Liu, Y. Fu, L. Wang, L. Zhou and X. Feng, *Chem. Eur. J.*, 2008, **14**, 4796–4798.

173. A. K. Gupta, M. Mukherjee and W. D. Wulff, *Org. Lett.*, 2011, **13**, 5866–5869.

174. L. F. Tietze, J. Zinngrebe, D. A. Spiegl and F. Stecker, *Heterocycles*, 2007, **74**, 473–489.

175. J. Mlynarski, B. Rakiel, M. Stodulski, A. Suszcynska and J. Frelek, *Chem. Eur. J.*, 2006, **12**, 8158–8167.

176. S. Rajikumar, K. Shankland, G. D. Brown and A. J. A. Cobb, *Synlett*, 2006, 689–692.

177. X.-Y. Guan, L.-P. Yang and W. Hu, *Angew. Chem., Int. Ed.*, 2010, **49**, 2190–2192.

178. P. G. Cozzi and E. Rivalta, *Pure Appl. Chem.*, 2006, **78**, 287–291.

179. (a) S. E. Gibson and A. Stevenazzi, *Angew. Chem., Int. Ed.*, 2003, **42**, 1800–1810; (b) J. Blanco-Urgoiti, L. Anorbe, L. Pérez-Serrano, G. Dominguez and J. Pérez-Castells, *Chem. Soc. Rev.*, 2004, **33**, 32–42.

180. A. Lledo, J. Solà, X. Verdaguer, A. Riera and M. A. Maestro, *Adv. Synth. Catal.*, 2007, **349**, 2121–2128.

Asymmetric Domino Reactions Based on the Use of Chiral Organocatalysts

3.1 Introduction

The economic interest in combinations of asymmetric catalytic processes with domino reactions is obvious. While the end of the 20th century was dominated by the use of metal,[1] and biocatalysis,[2] a change in perception has occurred during the last decade when several reports confirmed that relatively simple organic molecules, such as proline, could be highly effective and remarkably enantioselective catalysts of a variety of fundamentally important transformations.[3] This rediscovery has initiated an explosive growth of research activities in organocatalysis, both in industry and in academia. Organocatalysts have several important advantages, because they are usually robust, inexpensive, readily available, and non-toxic. Enantioselective organocatalytic processes have reached maturity in recent years with an impressive and steadily increasing number of publications regarding the applications of this type of reaction, which paint a comprehensive picture of their real possibilities in organic synthesis.[4] Hence, the application of chiral organocatalysts has permitted the preparation of a number of highly valuable chiral products with the exclusion of any trace of hazardous metals and with several advantages from an economic and environmental point of view.[5] The ability of organocatalysts to promote a wide range of reactions by different activation modes makes organocatalysis ideal for its application in domino reactions, which proceed in a one-pot procedure to build complex frameworks from simple starting compounds. These organocatalysed domino reactions are often highly efficient and follow, in some way, different biomimetic pathways, with the same

RSC Catalysis Series No. 10
Asymmetric Domino Reactions
By Hélène Pellissier
© The Royal Society of Chemistry 2013
Published by the Royal Society of Chemistry, www.rsc.org

principles that are found in biosynthesis in nature. In this context, chemists have devoted more and more effort to the development of new and powerful strategies in domino reactions that avoid the use of costly and time-consuming protection–deprotection processes, as well as purification procedures for intermediates. The efficiency of the enantioselective domino processes can be evaluated by the bonds formed, the stereoselectivity achieved, and the complexity of the newly formed molecules. The goal of this chapter is to cover the recent efforts of the chemical community in the development of novel enantioselective organocatalytic domino reactions, published since the beginning of 2006, as in the two other chapters of the book. The general field of asymmetric domino reactions, using chiral substrates as well as chiral catalysts, was previously collected by Tietze in a book,[6] and by this author in two reviews,[7] all published in 2006. The domino reactions are catalogued on the basis of the reaction types involved in the first (two) synthetic step(s). They have been selected according to Tietze's definition, qualifying a domino reaction as a reaction involving two or more bond-forming transformations which take place under the same reaction conditions, without adding additional reagents and catalysts, and in which the subsequent reactions are a consequence of the functionality produced by bond formation or fragmentation in the previous step.[6,8] In order to facilitate presentation, the chapter is divided into two parts. The first part deals with organocatalytic enantioselective domino reactions involving one or two components, whereas the second part describes multicomponent reactions, which are domino reactions involving at least three substrates and forming products that contain significant portions of all reactants, ideally all atoms.[9]

3.2   One- and Two-Component Domino Reactions

Asymmetric organocatalytic domino reactions are often catalysed by chiral amines, especially chiral secondary amines, because they are capable of both enamine and iminium catalysis.[10] Their ability to tolerate numerous functional groups makes them ideal for the design of new asymmetric organocatalytic domino reactions, although their scope is mainly limited to carbonyl systems. The combination of iminium and enamine activations in a single organo-catalysed operation can serve as an attractive platform in the design of new catalytic cascade processes, based on the involvement of intermediate enamine and iminium species, in which several bond-forming steps take place in a single operation. Since the first organocatalytic domino reaction, reported by Bui and Barbas in 2000,[11] the use of these two powerful methodologies based on LUMO-lowering iminium ion activation and HOMO-raising enamine activation has been studied intensively, significantly contributing to the success of the rapidly developing area of asymmetric organocatalysis. Generally, chiral secondary amines, such as diarylprolinol ethers or MacMillan's chiral imidazolidinones,[12] are most often used to activate α,β-unsaturated aldehydes (or ketones) by forming electron-deficient iminium ions, which render the β-carbons more electrophilic than their carbonyl precursors for nucleophilic attack (Scheme 3.1). All the reactions based on this strategy have in common

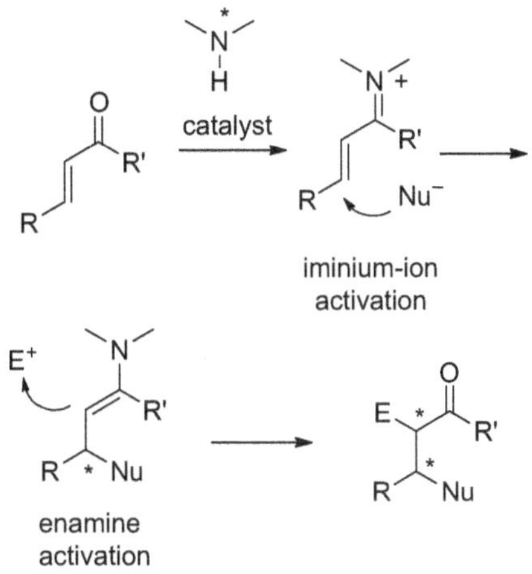

Scheme 3.1 Mechanism of iminium–enamine activation mode.

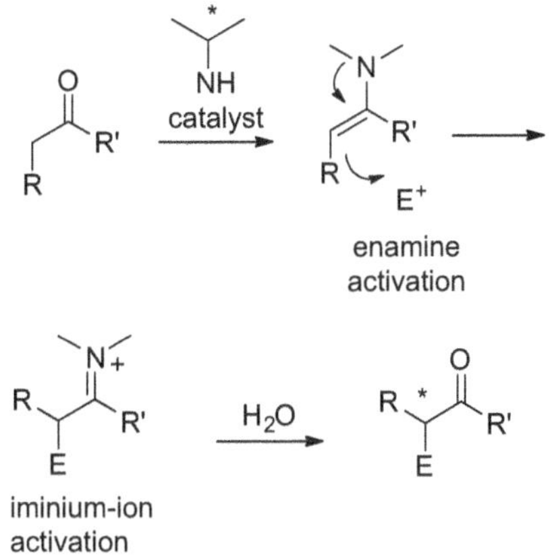

Scheme 3.2 Mechanism of enamine–iminium activation mode.

the use of an α,β-unsaturated aldehyde or an α,β-unsaturated ketone and are catalysed by a chiral secondary amine. The general mechanism of the iminium–enamine activation mode is depicted in Scheme 3.1. An α,β-unsaturated aldehyde (or ketone) is activated by a secondary chiral amine catalyst, forming an iminium-ion intermediate, which then undergoes a conjugate addition with a nucleophile. The resulting enamine can then undergo a second reaction with an electrophile to afford the double-substituted product, which generally contains two novel stereocentres.

On the other hand, chiral secondary amines are also used for promoting formation of electron-rich enamines from enolisable aldehydes or ketones, which then react with various electrophiles to afford products.[13] For example, proline is generally less effective in iminium catalysis, particularly with enals or enones. Enamine (-iminium) activation, depicted in Scheme 3.2, has gained significant attention for controlling the absolute configuration in the α-func-tionalization of aldehydes and ketones by a variety of electrophilic reagents.

The reactions will be organised by the reaction types involved in the first synthetic (two) step(s) and also by the type of activation mode of their components.

3.2.1 Domino Reactions Initiated by the Michael Reaction

3.2.1.1 Domino Michael–Michael Reactions

The nucleophilic 1,4-addition of stabilised carbon nucleophiles to electron-poor olefins, generally α,β-unsaturated carbonyl compounds, is known as the Michael addition, although it was first reported by Komnenos in 1883.[14] Michael-type reactions can be considered as one of the most powerful and reliable tools for the stereocontrolled formation of carbon–carbon and carbon–heteroatom bonds,[15] as has been demonstrated by the huge number of examples in which it has been applied as a key strategic transformation in total synthesis. As a consequence, in recent years, many different organocatalytic versions of this important transformation have been reported, using a wide variety of nucleophiles and conjugate acceptors.[16] The iminium–enamine approach is currently the strategy most employed in developing new asymmetric organocatalytic domino reactions. All examples based on this concept have in common the use of enal or enone systems. Early examples based on the iminium–enamine concept have been successively developed by Barbas and Bui with the L-proline-catalysed synthesis of the Wieland–Miescher ketone through domino Michael–aldol reaction of methyl vinyl ketone and 2-methyl 1,3-cyclohexanedione;[11] Jorgensen *et al.* with Michael–aldol reactions of β-ketoesters and enones catalysed by imidazolidine catalyst;[17] and then, at nearly the same time, Gryko with L-proline-catalysed domino Michael–aldol reactions of methyl vinyl ketone and 1,3-diketones;[18] List *et al.* with intramolecular reductive Michael cyclisations of enal enones catalysed by imidazolidinone catalysts;[19] and MacMillan *et al.* with domino Michael–intermolecular halogenation reactions catalysed by imidazolidinone catalysts.[20] A more recent work reported by Wang *et al.* described the synthesis of highly functionalised tetrasubstituted chiral five-membered carbocycles on the basis of a domino Michael–Michael reaction of γ-malonate-α,β-unsaturated esters with α,β-unsaturated aldehydes.[21] This process was catalysed by chiral diphenylprolinol trimethylsilyl ether in ethanol media, and afforded through the iminium–enamine activation mode the corresponding cyclopentanes bearing three stereogenic centres in high yields, diastereoselectivities of up to >90% de, and enantioselectivities of up to 99% ee, as shown in Scheme 3.3. Later, Cordova *et al.* developed a related process based on the use of the same organocatalyst, which allowed the construction of cyclopentanes through a domino nitro-Michael–Michael sequence.[22] Instead of malonate derivatives, these authors employed a γ-nitro-α,β-unsaturated ester as the nucleophile for the initial Michael addition, providing nitrogen-, formyl-, and ester-functionalised

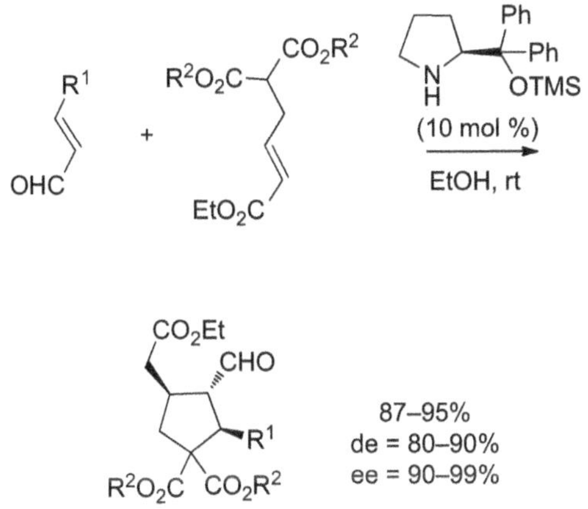

Scheme 3.3   Domino Michael–Michael reaction of α,β-unsaturated aldehydes with γ-malonate-α,β-unsaturated esters through iminium–enamine activation mode.

cyclopentane derivatives with four stereogenic centres with good to high yields (70–88%) and generally excellent enantioselectivities (97–99% ee), albeit combined with moderate diastereoselectivities (40–60% de).

In 2009, Brenner and McGarraugh reported the generation of highly substituted fused chiral carbocycles based on a new enantioselective Michael–Michael cascade reaction of conjugated β-ketoesters in which the olefin was part of a carbocycle with α,β-unsaturated aldehydes.[23] This process, catalysed by chiral diphenylprolinol trimethylsilyl ether, allowed the corresponding domino Michael–Michael products to be achieved as mixtures of two diastereomers in high yields (61–87%) and a high diastereomeric ratio of ≥91:9 in all cases of the substrates studied. The major products were obtained in excellent enantioselectivities of ≥96% ee as a mixture of two epimers (α and β), as shown in Scheme 3.4. The authors have proposed the mechanism depicted in Scheme 3.4 based on an iminium–enamine approach. First, the amine catalyst activated α,β-unsaturated aldehydes to form the corresponding iminium species, which then added conjugated β-ketoesters, leading to the corresponding Michael products. These enamines were further submitted to an intramolecular Michael addition, occurring with the Michael acceptor approaching the enamine from the face opposite, and the proton at C3 projected out towards the bulky group, which afforded the corresponding final domino Michael–Michael β-products as major products with excellent enantioselectivities of up to 99% ee. The minor formation of epimerised α-products was explained by the authors through the formation of the corresponding enamines of the first-formed epimer β-products.

In 2010, the same catalyst was employed by Ma to promote an enantioselective synthesis of polysubstituted cyclopentanones through domino Michael–Michael reactions of a range of α,β-unsaturated aldehydes with a keto ester bearing a highly electron-deficient olefin unit.[24] This process allowed the

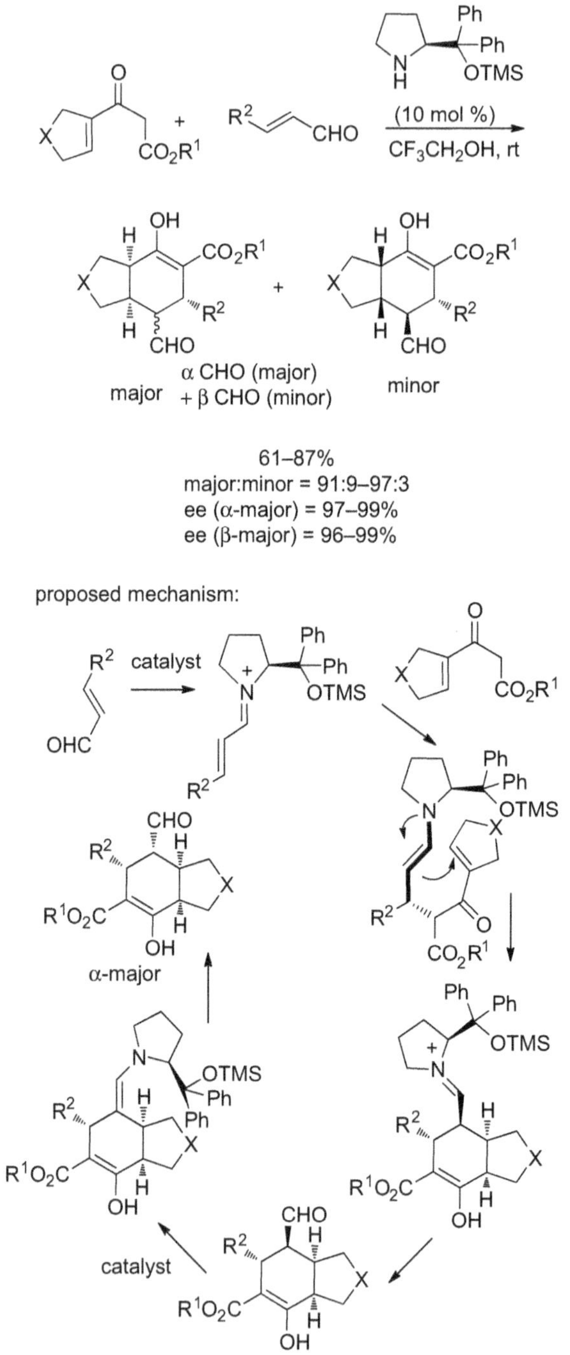

Scheme 3.4 Domino Michael–Michael reaction of α,β-unsaturated aldehydes with conjugated β-ketoesters through iminium–enamine activation mode.

formation of four contiguous stereocentres in the cyclopentanone ring in one step, with excellent enantioselectivity of up to >99% ee. As shown in Scheme 3.5, an iminium–enamine activation mode was proposed to explain the results. Thus, the authors assumed that, after the Michael addition of an anion generated from the keto ester onto iminium through the transition state depicted in Scheme 3.5, the newly formed enamine moiety in the enamine intermediate attacked the highly electron-deficient olefin part to afford the intramolecular Michael adduct. The scope of the process was extended to various β-, α,β- and β,β-substituted α,β-unsaturated aldehydes, including cyclic α,β-unsaturated aldehydes, which led to the formation of the corresponding

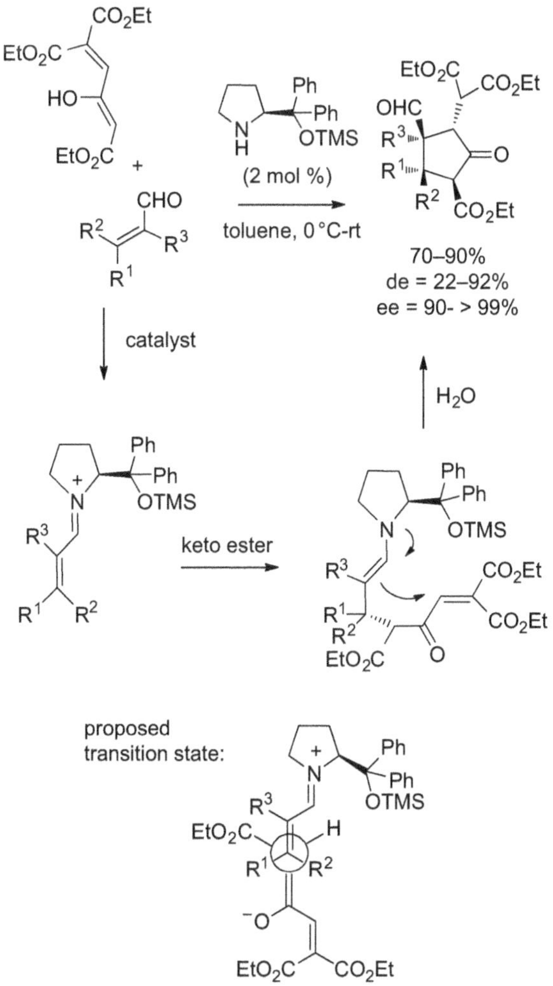

Scheme 3.5   Domino Michael–Michael reaction with β-, α,β-, and β,β-substituted α,β-unsaturated aldehydes with β-ketoesters bearing an electron-deficient olefin unit through iminium–enamine activation mode.

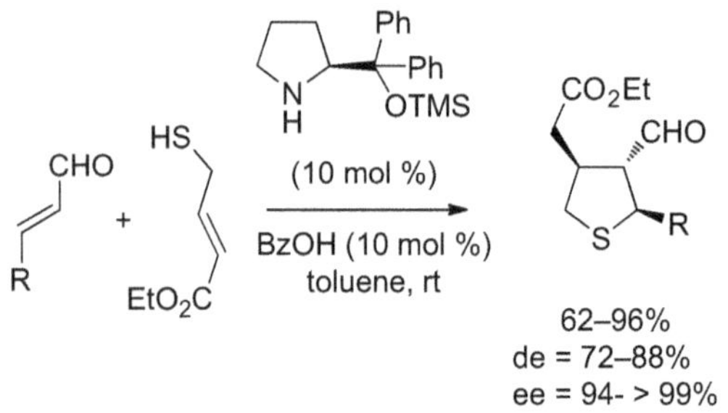

Scheme 3.6 Domino thia-Michael–Michael reaction of α,β-unsaturated aldehydes with 4-mercapto-2-butenoate through iminium–enamine activation mode.

major stereomers in generally high yields, excellent enantioselectivities, and moderate to high diastereoselectivities (22–92% de). On comparing these results with those from many Michael additions of α,β-unsaturated aldehydes catalysed with secondary amines, it must be noted that this novel double Michael addition process required less catalyst (2 mol%). In most cases, the minor diastereomer was assumed to be the corresponding 2-epimer of the domino product.

In marked contrast to the Michael addition of carbon-nucleophiles, the addition of non-carbon nucleophiles, such as amines,[16b] thiols,[25] phosphorus,[26] and alcohols[27] (hetero-Michael addition) has gained considerably less interest in recent decades. This holds especially true for the conjugate addition of thiols, the thia-Michael addition. As an example, Wang *et al.* have reported the synthesis of chiral tetrahydrothiophenes on the basis of a domino thia-Michael–Michael reaction of α,β-unsaturated aldehydes with *trans*-ethyl 4-mercapto-2-butenoate catalysed by chiral diphenylprolinol trimethylsilyl ether, which evolved through an iminium–enamine activation mode.[28] After the first thio-Michael addition, the enamine intermediate promoted a conjugated attack to the α,β-unsaturated ester, furnishing the thiophene ring. As shown in Scheme 3.6, a variety of aromatic, heteroaromatic, and aliphatic enals could be employed, providing the corresponding highly functionalised chiral tetrahydrothiophenes with generation of three new stereogenic centres in good to high yields and diastereoselectivities of up to 88% de, combined with excellent enantioselectivities of up to >99% ee.

The iminium–enamine activation mode can also be envisaged to explain the domino oxa-Michael–Michael reaction occurring between 3-methylbut-2-enal and (*E*)-2-(2-nitrovinyl)-benzene-1,4-diol upon catalysis with chiral diphenylprolinol trimethylsilyl ether, which afforded the corresponding almost enantiopure oxa-Michael–Michael cycloadduct in 76% yield, as shown in Scheme 3.7.[29] This product was further implicated in a Michael–aldol sequence induced by the same catalyst through reaction with crotonaldehyde to give the corresponding hexahydro-6*H*-benzo[*c*]chromene in 74% yield. These two domino reactions have constituted the key steps of the first asymmetric total synthesis of the natural biologically active product (+)-conicol.

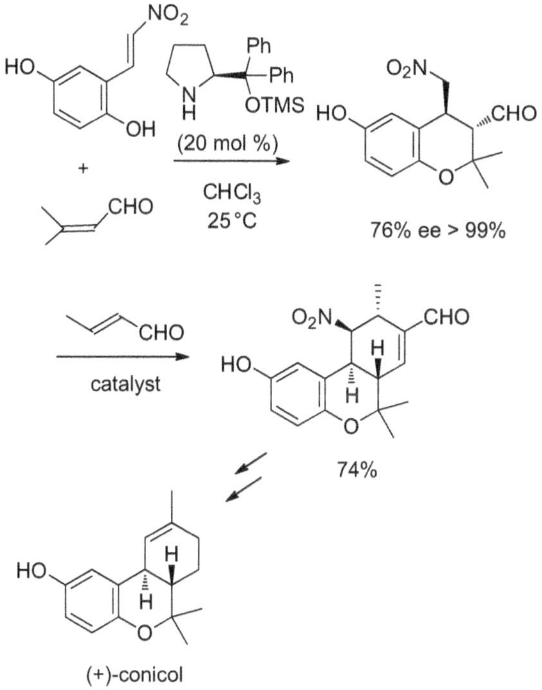

Scheme 3.7 Domino oxa-Michael–Michael reaction of 3-methylbut-2-enal with (*E*)-2-(2-nitrovinyl)-benzene-1,4-diol as key-step in the synthesis of (+)-conicol.

As an extension of the iminium–enamine activation mode, Wang *et al.* have recently developed a novel asymmetric domino oxa-Michael–Michael reaction catalysed by a closely related chiral catalyst, such as diphenylprolinol *tert*-butyldimethylsilyl ether (Scheme 3.8), involving an unprecedented chiral iminium–allenamine cascade.[30] This process was a viable one-pot approach to produce synthetically and biologically significant chiral 4*H*-chromenes in high yields, and with generally excellent enantioselectivities of at least 99% ee. A broad substrate scope has been successfully employed in this methodology, including aromatic and aliphatic alkynals as Michael acceptors, and 2-(*E*)-(2-nitrovinyl)-phenols as Michael donors/acceptors with significant structural variation, as summarised in Scheme 3.8. Indeed, the use of an alkynal substrate, instead of an enal as in the conventional iminium–enamine activation mode, in the initial oxa-Michael addition to the iminium ion formed *in situ* afforded a chiral allenamide intermediate. This allenamide further acted as the nucleophile in a subsequent enantioselective Michael addition to the nitro-alkene moiety to give the final product.

The iminium–enamine activation mode may also be involved in a highly diastereo- and enantioselective domino Michael–Michael-aldol reaction of α,β-unsaturated aldehydes with (*E*)-7-oxooct-5-enal or (*E*)-7-oxonon-5-enal

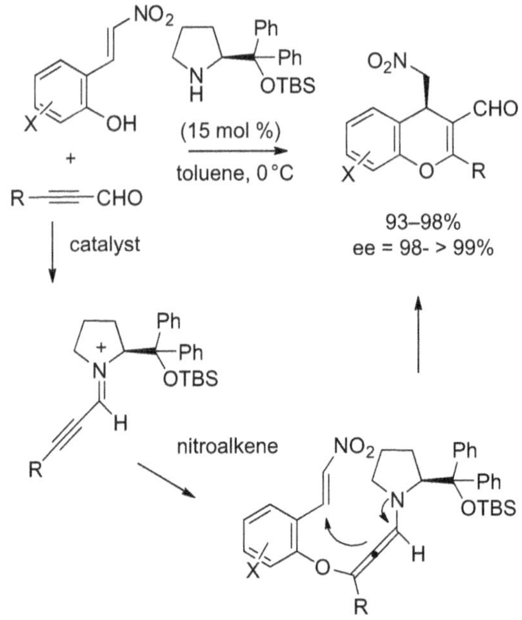

Scheme 3.8 Domino oxa-Michael–Michael reaction of alkynals with 2-(*E*)-(2-nitrovinyl)-phenols through iminium–allenamine activation mode.

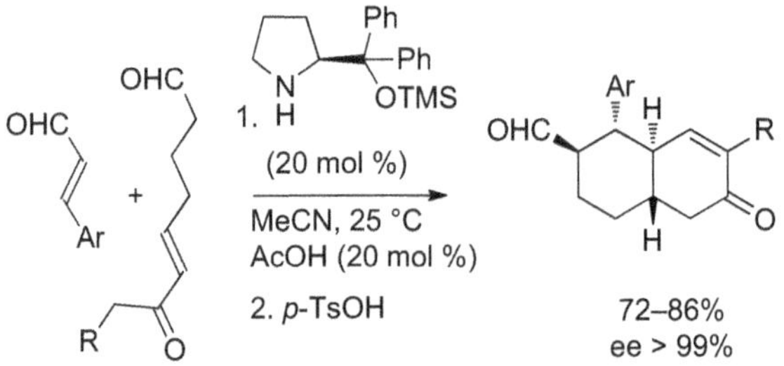

Scheme 3.9 Domino Michael–Michael–aldol reaction of α,β-unsaturated aldehydes with (*E*)-7-oxooct-5-enal, or (*E*)-7-oxonon-5-enal through iminium–enamine activation mode.

developed by Hong *et al.*, in 2009.[31] Upon catalysis with chiral diphenylprolinol trimethylsilyl ether, this process provided expedited access to various highly functionalised and almost enantiopure octahydro-6-oxo-1-phenylnaphthalene-2-carbaldehydes bearing four stereogenic centres in high yields and excellent enantioselectivities of at least 99% ee, as shown in Scheme 3.9. In 2012, these authors applied this methodology to (*E*)-3-methyl-7-oxooct-5-enal as the ketoaldehyde component, reacting under the same conditions with an aromatic enal to give the corresponding domino product, which constituted a key inter-mediate in a first enantioselective total synthesis of (+)-galbulin.[32]

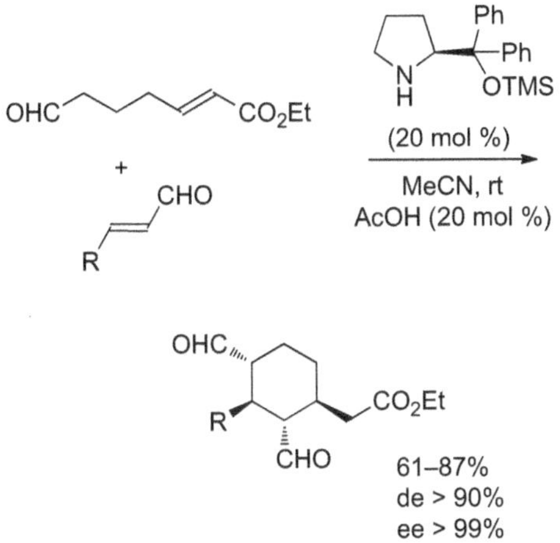

Scheme 3.10 Domino Michael–Michael reaction of α,β-unsaturated aldehydes with ethyl (*E*)-7-oxohept-2-enoate through iminium–enamine activation mode.

In 2011, the same authors reported a nice synthesis of chiral cyclohexa-nedicarbaldehydes on the basis of enantioselective domino Michael–Michael reaction of α,β-unsaturated aldehydes with ethyl (*E*)-7-oxohept-2-enoate, catalysed by the same catalyst.[33] Remarkably, almost diastereo- and enantiopure domino products were achieved in all cases of aromatic as well as α,β-unsaturated aldehydes studied, as shown in Scheme 3.10.

Along with α,β-unsaturated aldehydes, enones have been also widely employed as Michael acceptor components in recent enantioselective organo-catalysed domino Michael–Michael processes. In this context, Dend *et al.* developed the first example of an asymmetric domino reaction catalysed by chiral primary amines, in 2007.[34] In this work, a chiral primary aminocatalyst derived from quinine promoted a Michael–Michael–retro-Michael cascade, occurring between α,α-dicyanoalkenes and α,β-unsaturated ketones such as benzylideneacetone, which afforded the corresponding almost enantiomerically pure polysubstituted 2-cyclohexen-1-one derivatives. This highly efficient domino process evolved through iminium–enamine activation mode, and involved two reagents which acted alternatively and selectively as the Michael donor and acceptor under readily controllable conditions. The reaction scope was quite substantial and excellent stereoselectivity was generally observed along with moderate to good yields, as shown in Scheme 3.11.

Asymmetric organocatalytic domino reactions are not limited to amine catalysis. Indeed, significant contributions have also been made in the field of hydrogen-bonding and Bronsted-acid catalysis.[4d,4h,35] These catalysts activate the substrates by forming a hydrogen bond (LUMO-lowering) and are able to promote several C–C and C–heteroatom bond-forming reactions. The inter-action between the catalyst and the substrate is non-covalent, and the chiral ion pair is the intrinsic activated species. In particular, chiral thiourea-based

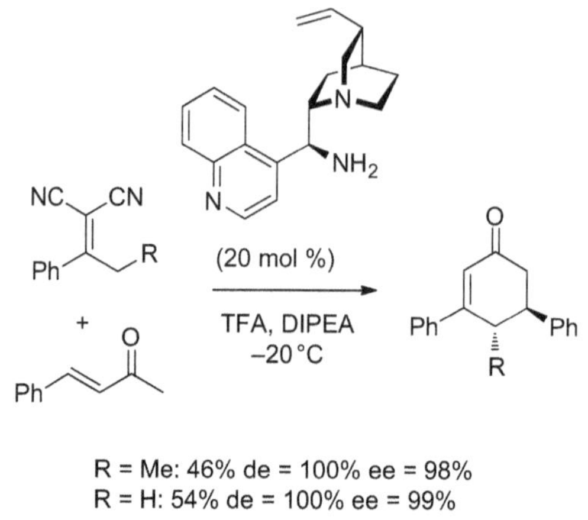

R = Me: 46% de = 100% ee = 98%
R = H: 54% de = 100% ee = 99%

Scheme 3.11 Domino Michael–Michael–retro-Michael reaction of benzylide-neacetone with α,α-dicyanoalkenes through iminium–enamine activation mode.

derivatives and phosphoric acid derivatives are well known for their application as effective Bronsted acid organocatalysts. A recent example of this type of activation was reported by Gong and Wei, who investigated a series of Takemoto type catalysts to promote the domino Michael–Michael reaction of Nazarov reagents with methyleneindolines to provide the corresponding spiro[4-cyclohexanone-1,3'-oxindoline] derivatives.[36] Among a series of bifunctional (thio)urea catalysts studied, the chiral urea depicted in Scheme 3.12 was selected as the optimal catalyst, allowing the formal [4 + 2] cycloadducts to be obtained in excellent yields, diastereo- and enantios-electivities of up to 98% or 96% de and 97% ee, respectively, as shown in Scheme 3.12. The activation through hydrogen-bond formation of the substrates by the catalyst possessing Bronsted acid (BH) and Lewis base (LB) functionalities is depicted in Scheme 3.12.

In the same area, Wang et al. have disclosed a new enantioselective domino Michael–oxa-Michael–tautomerisation reaction induced by a novel indane amine-thiourea catalyst via hydrogen-bonding catalysis.[37] This sequence afforded highly functionalised chiral pyranochromenes in high to excellent enantioselectivities of up to 99% ee combined with good to high yields by reaction of malononitrile with enones such as (E)-3-benzylidenechroman-4-ones. As shown in Scheme 3.13, malononitrile was used both as nucleophile and electrophile. The efficiency of this catalyst allowed the reaction of a range of benzylidenechroman-4-one substrates to occur, such as those possessing neutral functionalities as well as those having electron-withdrawing groups and electron-donating groups in the backbone.

In another context, N-heterocyclic carbenes,[38] prepared in situ from diarylimidazolium salts, are known as highly effective catalysts for the generation of reactive homoenolates from α,β-unsaturated aldehydes.[39]

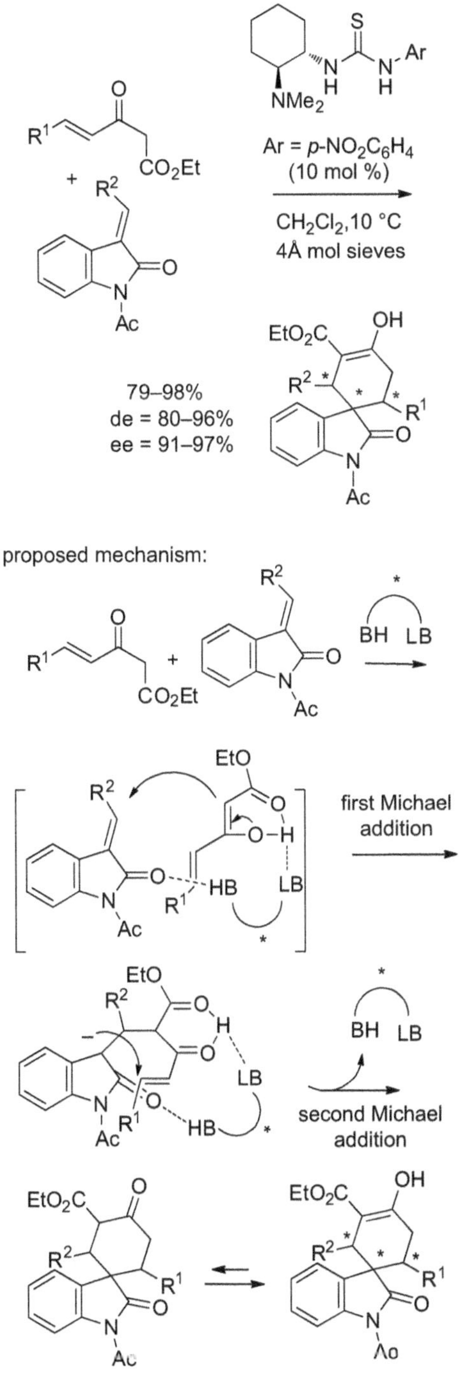

proposed mechanism:

Scheme 3.12 Domino Michael–Michael reaction of Nazarov reagents with methyl-eneindolines through hydrogen-bonding activation mode.

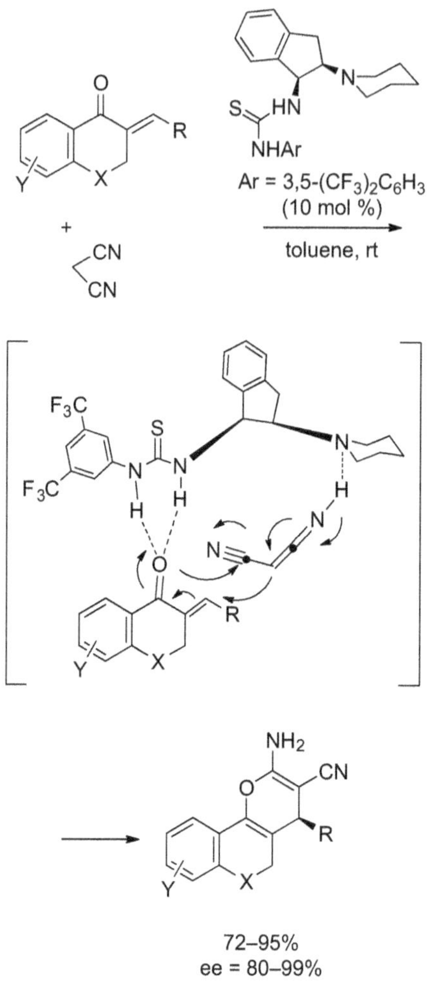

Scheme 3.13   Domino   Michael–oxa-Michael–tautomerisation   reaction   of   (*E*)-
3-benzylidenechroman-4-ones with malononitrile through hydrogen-
bonding activation mode.

A recent application of this type of catalyst was reported by Chi *et al.* with a
new annulation of benzodi(enone)s and α,β-unsaturated aldehydes mediated by
a chiral aminoindanol-derived triazolium catalyst.[40] In the presence of DBU
combined with MgSO₄ as an additive, the reaction afforded the corresponding
tricyclic domino Michael–Michael products with excellent diastereo- and
enantioselectivities of up to 90% de and 99% ee, respectively, as shown in
Scheme 3.14. In order to explain the results, the authors have proposed the
initial formation of a Breslow intermediate as a homoenolate equivalent,
followed by a Michael-type addition to one branch of the di(enone) to give an
enolate intermediate. Protonation and tautomerisation of this intermediate

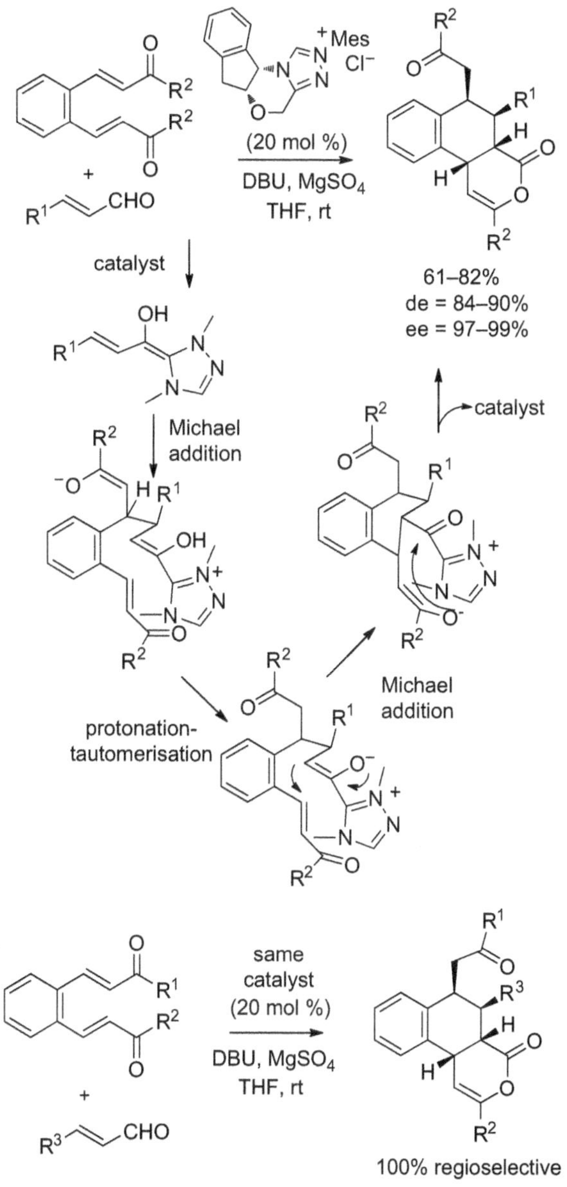

Scheme 3.14 *N*-Heterocyclic carbene-catalysed annulations of α,β-unsaturated aldehydes with benzodi(enone)s.

R^1 = R^3 = *p*-BrC$_6$H$_4$, R^2 = Ph: 73% de = 90% ee = 97%
R^1 = *p*-BrC$_6$H$_4$, R^2 = *p*-Tol, R^3 = Ph: 81% de = 90% ee = 98%
R^1 = 3,4-Cl$_2$C$_6$H$_3$, R^2 = *p*-Tol, R^3 = Ph: 84% de = 90% ee = 99%

gave another enolate intermediate, which exclusively underwent intramolecular Michael addition of the enal α-carbon atom to the other branch of the di(enone) to afford a third enolate intermediate. The last step of the catalytic cycle was the formation of an enol ester to generate the final product with the release of the N-heterocyclic carbene catalyst. The scope of the reaction of α,β-unsaturated aldehydes could be extended to unsymmetrical benzo(dienone)s bearing two different substitutives. In these cases of substrates, the corresponding cascade products were obtained with an exceptionally high regioselectivity of 100% in almost all cases of the substrates used, combined with excellent diastereo- and enantioselectivities of up to 90% de and 99% ee, respectively (Scheme 3.14).

In addition to α,β-unsaturated aldehydes and α,β-unsaturated ketones or related compounds, nitroalkenes have also been employed as Michael acceptor components in a number of enantioselective organocatalytic domino Michael–Michael reactions. In this context, one of the milestones in organocatalysis was the first enantioselective addition of malonates to nitrostyrenes,[41] catalysed by chiral thioureas reported by Takemoto, in 2003.[42] In this work, the authors disclosed the domino Michael–Michael reaction of γ,δ-unsaturated-β-ketoesters with nitroalkenes, catalysed by a chiral bifunctional thiourea and 1,1,3,3-tetramethylguanidine. Interestingly, the conjugate addition of the ketoesters to nitroalkenes took place first and then an intramolecular Michael addition catalysed by base provided the corresponding final cyclohexane derivatives. As shown in Scheme 3.15, the process afforded highly functionalised chiral cyclohexanones in high yields combined with diastereo- and enantioselectivities of up to >99% de, and 92% ee, respectively. Furthermore, the reaction was applied to the total synthesis of naturally occurring and biologically active (–)-epibatidine.

In 2008, Zhong et al. developed powerful domino Michael–Michael reactions of diethyl 5-acetylhex-2-enedionate with nitrostyrenes, catalysed by a chiral cinchona alkaloid.[43] The process consisted in a Michael addition of ketoester to nitrostyrene followed by a subsequent intramolecular cyclisation via a second Michael addition of nitro compound to α,β-unsaturated ester. This reaction was possible due to the low reactivity as Michael acceptors of unsaturated esters in comparison with nitrostyrenes. As shown in Scheme 3.16, the process furnished, through hydrogen-bonding activation mode, the corresponding tetrasubstituted cyclopentanes with high yields and in almost diastereo- and enantiopure form.

In the same year, Wang et al. reported the synthesis of chiral thiochromenes on the basis of an enantioselective domino Michael–Michael reaction catalysed by another chiral cinchona alkaloid, thiourea, at a low catalyst loading of 2 mol%.[44] The process occurred through dynamic kinetic resolution[45] between nitroalkenes and trans-3-(2-mercapto)henyl)-2-propenoic acid ethyl ester derivatives, providing the corresponding chiral thiochromanes in good to excellent yields, generally excellent diastereoselectivity of >94% de, and also excellent enantioselectivities ranging from 92 to 99% ee, as shown in Scheme 3.17.

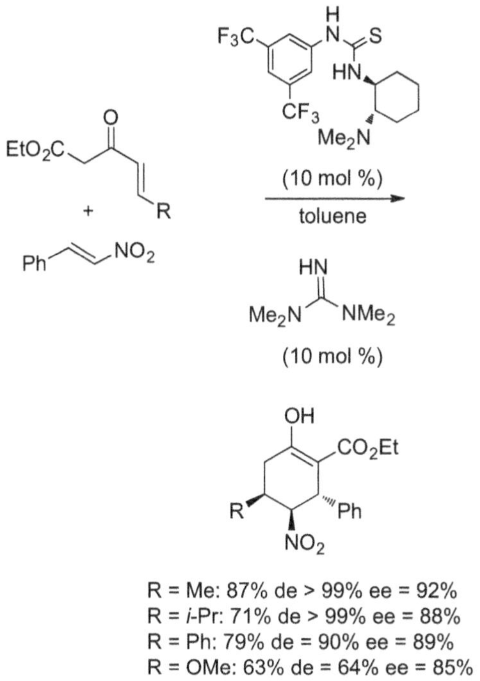

Scheme 3.15 Domino Michael–Michael–reaction of nitroalkenes with γ,δ-unsaturated-β-ketoesters through hydrogen-bonding activation mode.

In 2010, Xiao *et al.* reported the asymmetric domino thia-Michael–Michael reaction of thiols with nitroolefin enoates, providing the corresponding polyfunctionalised chroman derivatives in a highly stereoselective manner.[46] This reaction, catalysed by a bifunctional chiral thiourea employed at a low catalyst loading of 3 mol%, allowed three consecutive stereogenic centres including one quaternary to be generated with high enantioselectivities of up to 96% ee combined with excellent diastereoselectivities of >90% de, as shown in Scheme 3.18. The authors have proposed that the catalyst activated nitroolefin enoates through hydrogen-bonding activation, and its tertiary amino moiety activated the nucleophilic thiols, forming an intermediate which underwent the intermolecular thia-Michael addition, to form another intermediate which was submitted to another intramolecular Michael addition, leading to the final products.

Later, the same authors described a remarkable enantioselective domino aza-Michael–Michael reaction of nitroolefin enonates with benzotriazoles catalysed with the same bifunctional thiourea, which afforded the corresponding densely functionalised chiral chromans containing a quaternary stereogenic centre in high yields, diastereo- and enantioselectivities, as shown in Scheme 3.19.[47]

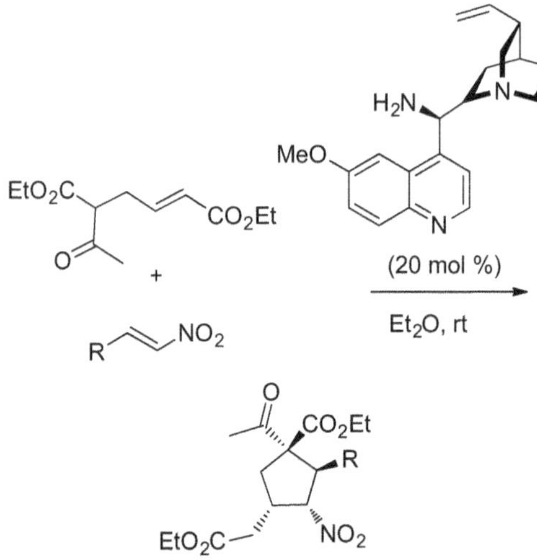

R = Ph: 91% de > 98% ee = 97%
R = *p*-Tol: 89% de = 94% ee = 95%
R = *m*-MeOC$_6$H$_4$: 85% de = 92% ee = 95%
R = *p*-MeOC$_6$H$_4$: 83% de = 92% ee = 94%
R = *o*-MeOC$_6$H$_4$: 81% de = 96% ee = 90%
R = *p*-BrC$_6$H$_4$: 92% de > 98% ee = 95%
R = *p*-ClC$_6$H$_4$: 88% de = 96% ee = 95%
R = 2-Naph: 84% de = 90% ee = 95%
R = 1-Naph: 87% de = 94% ee = 97%
R = 2-Fu: 87% de = 90% ee = 96%
R = 3-Fu: 86% de = 96% ee = 97%
R = *p*-NO$_2$C$_6$H$_4$: 81% de = 92% ee = 95%

Scheme 3.16   Domino   Michael–Michael–reaction   of   nitroalkenes   with γ,δ-unsaturated-β-ketoesters through hydrogen-bonding activation mode.

In addition, these authors developed asymmetric domino aza-Michael–Michael reactions of anilines with nitroolefin enoates upon catalysis with another chiral bifunctional thiourea, which afforded the corresponding chiral 4-aminobenzopyrans bearing three consecutive stereogenic centres including one quaternary.[48] As shown in Scheme 3.20, these products were isolated with high yields, combined with excellent diastereo- and enantioselectivities of up to >90% de and >99% ee, respectively.

In 2011, another asymmetric domino thia-Michael–Michael reaction was described by Wang *et al.*, which involved catalysis by a chiral bifunctional thiourea.[49] This novel hydrogen-bonding-mediated highly stereoselective cascade reaction of *trans*-ethyl-4-mercapto-2-butenoate with (*E*) – β-nitro-styrenes led to the corresponding biologically significant trisubstituted tetrahy-drothiophenes bearing three stereogenic centres with high enantio- and diastereoselectivities of up to 97% ee and >94% de, respectively (Scheme 3.21).

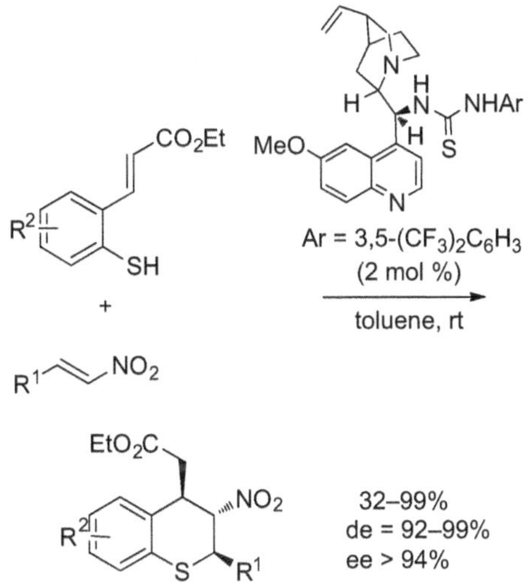

Scheme 3.17 Domino Michael–Michael–reaction of nitroalkenes with *trans*-3-(2-mercapto)phenyl-2-propenoic acid ethyl esters through hydrogen-bonding activation mode.

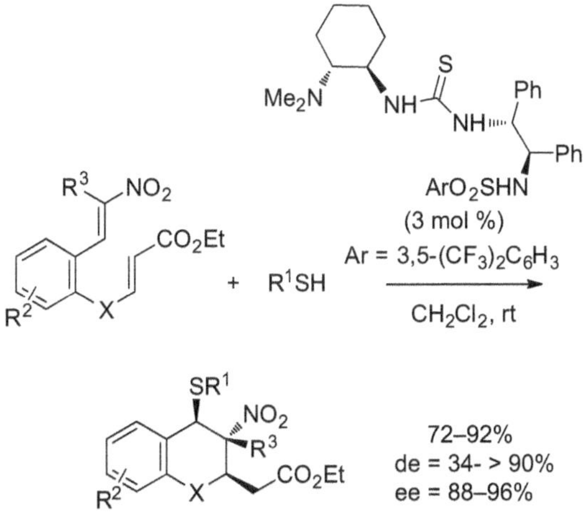

Scheme 3.18 Domino thia-Michael–Michael reaction of nitroolefin enoates with thiols through hydrogen-bonding activation mode.

Aromatic nitroolefins bearing electron-neutral, electron-withdrawing, as well as electron-donating groups, were well tolerated. The authors have demonstrated that a dynamic kinetic resolution (DKR) occurred in the cascade reaction in addition to the direct stereocontrol of the substrates by the bifunctional catalyst through hydrogen bonds.

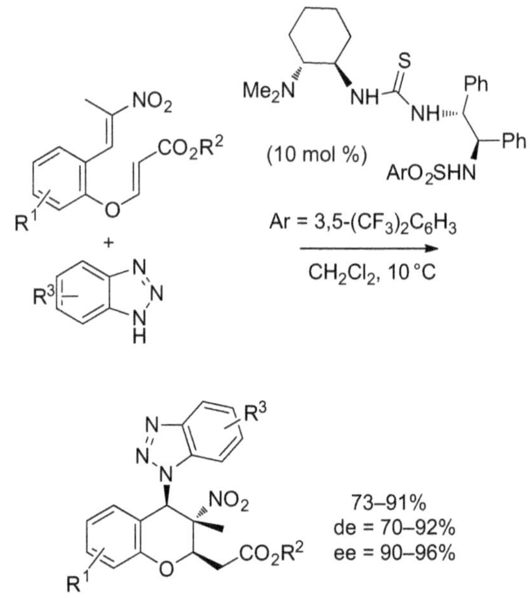

Scheme 3.19 Domino aza-Michael–Michael reaction of nitroolefin enoates with benzotriazoles through hydrogen-bonding activation mode.

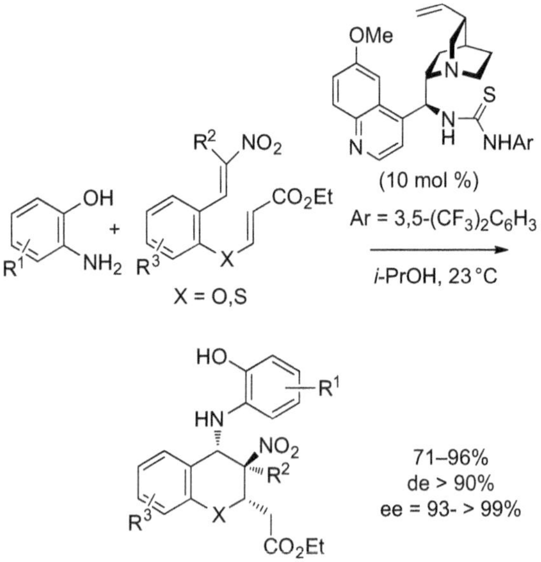

Scheme 3.20 Domino aza-Michael–Michael reaction of nitroolefin enoates with anilines through hydrogen-bonding activation mode.

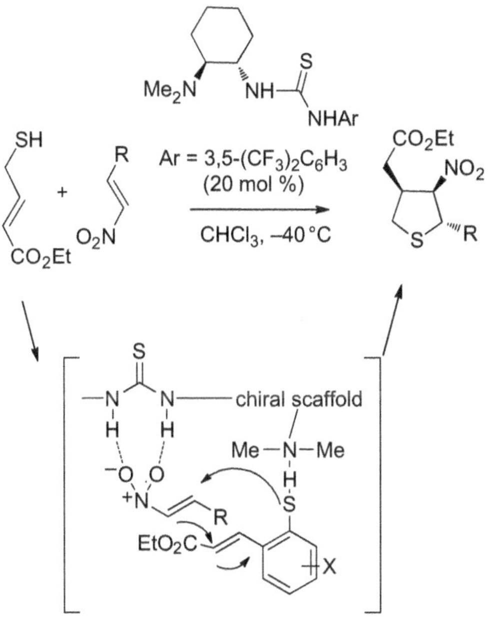

R = Ph: 82% de = 84% ee = 93%
R = *p*-FC$_6$H$_4$: 93% de = 86% ee = 96%
R = 2,5-Cl$_2$C$_6$H$_3$: 50% de = 88% ee = 93%
R = *m*-BrC$_6$H$_4$: 62% de = 88% ee = 97%
R = *m*-MeOC$_6$H$_4$: 60% de = 92% ee = 96%
R = Fu: 75% de > 94% ee = 94%

Scheme 3.21 Domino thia-Michael–Michael reaction of *(E)*-β-nitrostyrenes with *trans*-ethyl-4-mercapto-2-butenoate through cooperative DKR/hydrogen-bonding activation mode.

In 2012, Cobb *et al.* developed a novel enantioselective domino Michael–Michael reaction between nitrohex-4-enoates and nitroolefins, catalysed by a chiral bifunctional thiourea.[50] The process generated the corresponding chiral highly functionalised cyclohexanes containing four contiguous stereocentres, including one quaternary one. As shown in Scheme 3.22, these products were achieved in moderate to good yields and moderate diastereoselectivities, but almost enantiopure, in almost all cases of substrates studied.

On the other hand, Melchiorre *et al.* have used chiral primary amines to activate α,β-unsaturated enones towards a well-defined enamine–iminium activation mode.[51] Thus, these authors have developed asymmetric domino Michael–Michael reactions between enones and nitroalkenes which, upon catalysis by a chiral primary amine in combination with 2-fluorobenzoic acid as an additive, afforded the corresponding formal Diels–Alder adducts bearing three or four stereogenic centres, with very high enantioselectivities

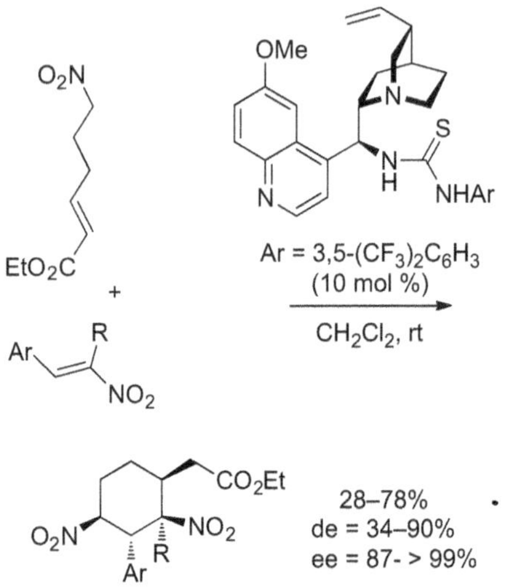

Scheme 3.22 Domino Michael–Michael reaction of nitroalkenes with nitrohex-4-enoates through hydrogen-bonding activation mode.

of up to 99% ee. The authors assumed the initial formation of an enamine generated *in situ* by condensation of the catalyst onto the enone. The nitroalkene acted as a Michael acceptor through intercepting this enamine, which provided the resulting carbon nucleophile. The latter then selectively engaged in an intramolecular, iminium-catalysed conjugate addition to afford the final cyclohexanones. As shown in Scheme 3.23, in all cases of the substrates studied, the preferential formation of the formal *endo trans*-products was observed with high diastereoselectivities of up to >90% de. The extension of this methodology to other Michael acceptors, such as *trans*-α-cyanocinnamate and *N*-benzyl maleimide, led to the corresponding cyclohexanones and bicyclic compounds, respectively, which were obtained in excellent enantioselectivities of up to >99% ee, as shown in Scheme 3.23. A closely related methodology was previously developed by Cordova *et al.*, in 2007.[52] In this study, the authors employed chiral diamine (*S*)-(+)-1-(2-pyrrolidinylmethyl)pyrrolidine as organocatalyst to induce the domino Michael–Michael reaction of α,β-unsaturated cyclic ketones with nitroolefins, which afforded the corresponding bicyclic Diels–Alder products containing four stereogenic centres in good to excellent yields (69–95%), excellent diastereoselectivities (>92% de), combined with good to high enantioselectivities (65–86% ee). The formation of these functionalised bicyclic products was explained by the authors through the *in situ* generation of chiral enamines of α,β-unsaturated cyclic ketones, according to an enamine–iminium activation mode.

On the other hand, Wang *et al.* have recently proposed a novel activation mode to explain the highly enantioselective domino oxa-Michael–Michael

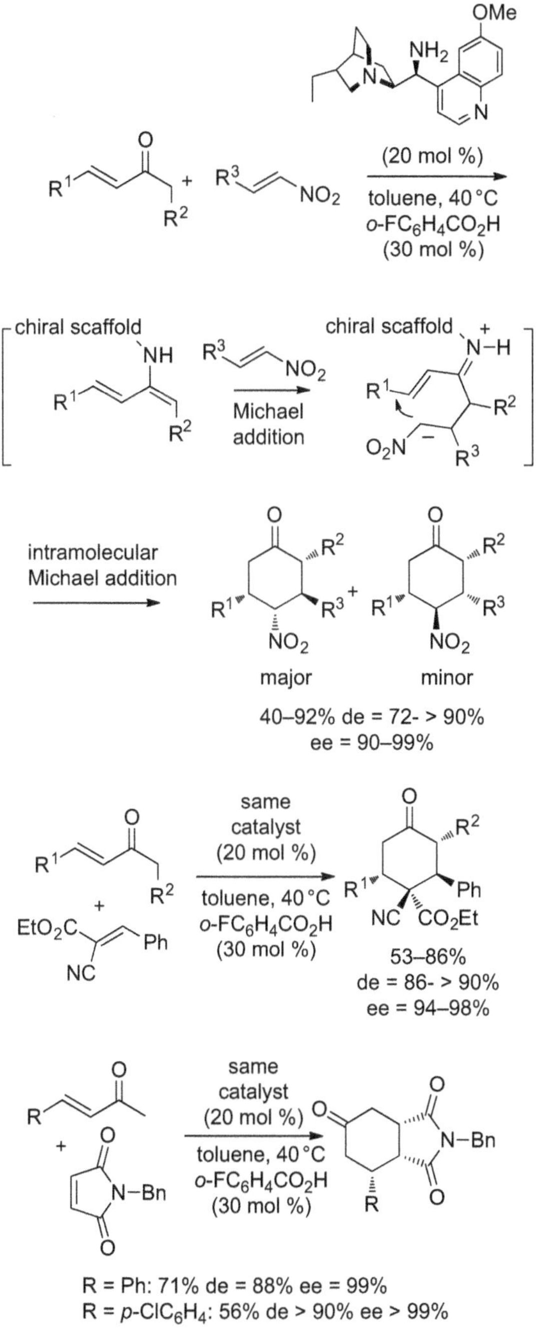

Scheme 3.23 Domino Michael–Michael reactions of enones with various Michael acceptors through enamine–iminium activation mode.

reaction of 2-hydroxy cinnamaldehydes with nitroalkenes, catalysed with chiral diphenylprolinol trimethylsilyl ether.[53] This reaction provided a series of highly functionalised useful chiral chromanes through an unprecedented activation mode, involving an aminal, which was formed very quickly from the reaction of 2-hydroxy cinnamaldehydes with the catalyst in an almost quantitative yield. The authors assumed that this aminal was produced through an intramolecular *O*-addition to an intermediate iminium. This aminal served as a nucleophile for the first Michael addition to the nitroalkene to produce an iminium intermediate, which released the catalyst for the next cycle reaction and generated an intermediate enal. Finally, an intramolecular Michael reaction in this enal gave rise to the final products, which were obtained in moderate to high yields (60–85%) and diastereoselectivities (34–82% de) combined with excellent enantioselectivities of up to 98% ee, as shown in Scheme 3.24.

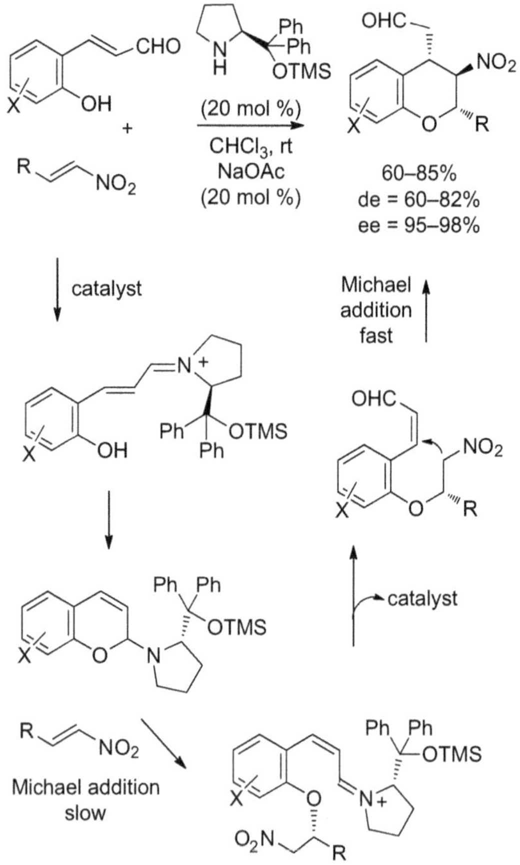

Scheme 3.24 Domino oxa-Michael–Michael reaction of nitroalkenes with 2-hydroxy cinnamaldehydes through iminium–other activation mode.

3.2.1.2 Domino Michael–Aldol Reactions

In recent years, a number of enantioselective organocatalysed domino Michael–aldol reactions have been developed, employing α,β-unsaturated aldehydes or α,β-unsaturated ketones or derivatives as Michael acceptor components. Among the reactions of α,β-unsaturated aldehydes, are those using 2-mercapto-, 2-hydroxy-, or 2-aminobenzaldehydes as Michael donors. As an example, the groups of Wang[54] and Cordova[55] developed, almost at the same time in 2006, an organocatalytic asymmetric synthesis of chiral thio-chromenes, evolving through a domino thia-Michael–aldol reaction. This process was based on the reaction occurring between α,β-unsaturated aldehydes and 2-mercaptobenzaldehydes, upon catalysis with a chiral diaryl-prolinol trimethylsilyl and benzoic acid as an additive. Carrying out the reaction in toluene at room temperature or in CHCl₃ at −15 °C, chiral thiochromenes derived from aromatic and alkyl enals were achieved in good to excellent yields of up to 97% combined with good to excellent enantios-electivities of up to 98% ee. Furthermore, various substituents in the benzene ring of mercaptobenzaldehydes did not significantly reduce the excellent outcome of the reaction, as shown in Scheme 3.25.

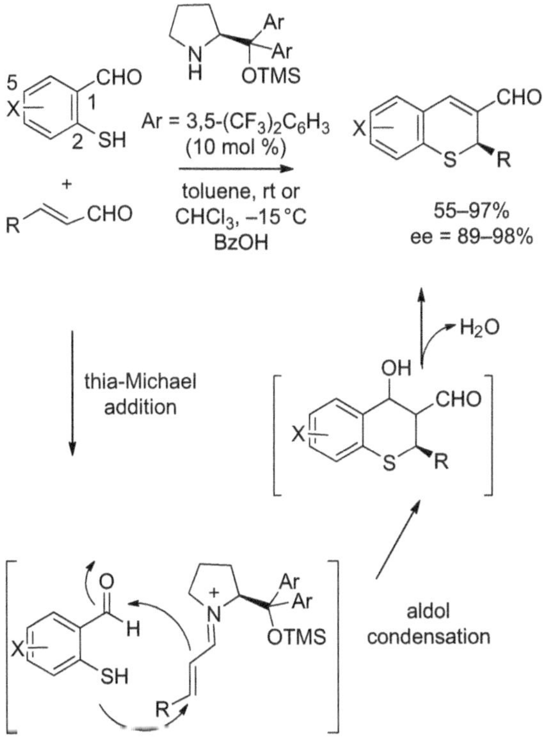

Scheme 3.25 Domino thia-Michael–aldol reaction of α,β-unsaturated aldehydes with 2-mercaptobenzaldehydes through iminium–enamine activation mode.

Soon after, the same groups of Cordova[56] and Wang[57] described related processes that gave access to a range of chiral chromanes. As shown in Scheme 3.26, the oxygenated analogous compounds were synthesised using the same approach, based on a domino oxa-Michael–aldol reaction of enals with 2-hydroxybenzaldehydes. Upon catalysis with 20 mol% of chiral diphenylprolinol trimethylsilyl ether, Cordova *et al.* obtained moderate to high yields (20–92%) combined with excellent enantioselectivities (92–98% ee) by performing the reaction in toluene at room temperature in the presence of 2-nitrobenzoic acid as an additive (Scheme 3.26). On the other hand, Wang *et al.* employed chiral diphenylprolinol triethylsilyl ether at a catalyst loading of 30 mol% in the presence of benzoic acid as an additive in 1,2-dichloroethane at room temperature, which also allowed an excellent outcome of the reaction, because the chiral chromanes were achieved in good to excellent yields (64–98%) combined with good to excellent enantioselectivities (75–99% ee), as summarised in Scheme 3.26. In 2009, Wang *et al.* extended the scope of this domino process to other enals and 2-hydroxybenzaldehydes, demonstrating that the reaction could also be performed in diethylether at room temperature in the presence of 20 mol% of chiral diphenylprolinol trimethylsilyl ether, providing high yields and enantioselectivities of up to 99% ee.[58] In 2011, Xu *et al.* investigated the same reactions by using a recyclable sterically hindered chiral diarylprolinol trimethylsilyl ether catalyst containing two tertiary amine groups.[59] The chiral chromenes were achieved in moderate to excellent yields (40–95%) and enantioselectivities (53–94% ee).

In addition, Cordova *et al.* have reported the application of their domino thia-Michael–aldol reaction to mercaptobenzofenone.[60] Indeed, the reaction of this substrate with α,β-unsaturated aldehydes in the presence of a chiral diarylprolinol trimethylsilyl ether as catalyst with 2-nitrobenzoic acid as an additive in toluene at −25 °C afforded, by avoiding the dehydration step, the corresponding chiral thiochromenes bearing three contiguous stereocentres with good to high yields (63–98%), excellent enantioselectivities (96–99% ee), and good diastereoselectivities (82–88% de). This methodology constituted a simple, catalytic, and highly stereoselective entry to pharmaceutically valuable

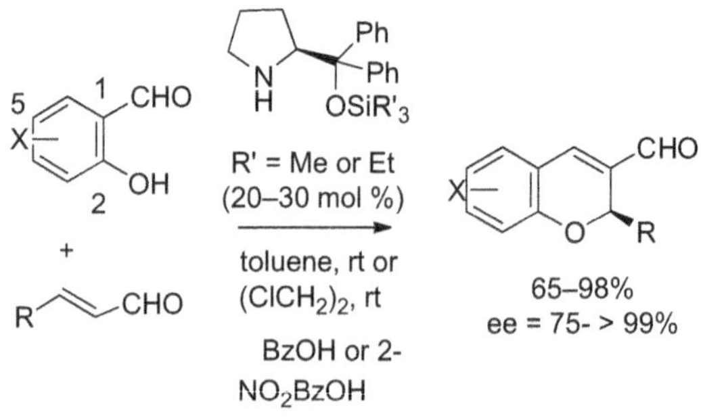

Scheme 3.26   Domino oxa-Michael–aldol reaction of α,β-unsaturated aldehydes with 2-hydroxybenzaldehydes through iminium–enamine activation mode.

benzothiopyran derivatives. The same authors have also adapted this process for the synthesis of chiral 1,2-dihydroquinolidines.[61] The development of conjugate addition of an amine to an electron-deficient α,β-unsaturated aldehyde represented an unprecedented organocatalytic process because, generally, an amine is a much weaker nucleophile than a thiol or an alcohol. In fact, this methodology represented the first asymmetric organocatalytic domino aza-Michael–aldol reaction of primary amines with α,β-unsaturated aldehydes. Indeed, the domino aza-Michael–aldol reaction of 2-aminobenzaldehydes and α,β-unsaturated aldehydes provided the corresponding chiral 1,2-dihydro-quinolidines in moderate to high yields (31–90%) and generally excellent enantioselectivities of up to 99% ee, when employing chiral diphenylprolinol trimethylsilyl ether as catalyst in DMF at −25 °C (Scheme 3.27). Some months later, Wang *et al.* reported the same sequence, employing 2-*N*-protected-aminobenzaldehydes in a basic medium (NaOAc), also rendering good results.[62] In this case, the process was catalysed by 20 mol% of (*S*)-diphenylprolinol triethylsilyl ether in 1,2-dichloroethane, and provided a range of biologically useful chiral 1,2-dihydroquinolines in high yields (83–98%) and enantioselectivities (88–95% ee). In 2008, Hamada *et al.* developed a related domino reaction induced by the same catalyst.[63] In this case, the reaction occurred between 4-methoxycarbonylanthranilaldehyde and 6-*N*-Boc-2-hexenal in acetonitrile at −20 °C in the presence of acetic acid as an additive, to give the corresponding 1,2-dihydroquinolidine in 88% yield and excellent enantio-selectivity of up to 98% ee.

In 2010, Wang *et al.* reported the synthesis of other chiral chromene derivatives on the basis of a highly enantioselective domino oxa-Michael–aldol reaction, evolving through iminium–allenamine activation mode induced by chiral diphenylprolinol *tert*-butyldimethylsilyl ether.[64] As shown in Scheme 3.28, remarkably high yields and enantioselectivities were observed for the products arising from the reaction of a range of alkynals with ethyl 2-(2-hydroxyphenyl)-2-oxoacetate.

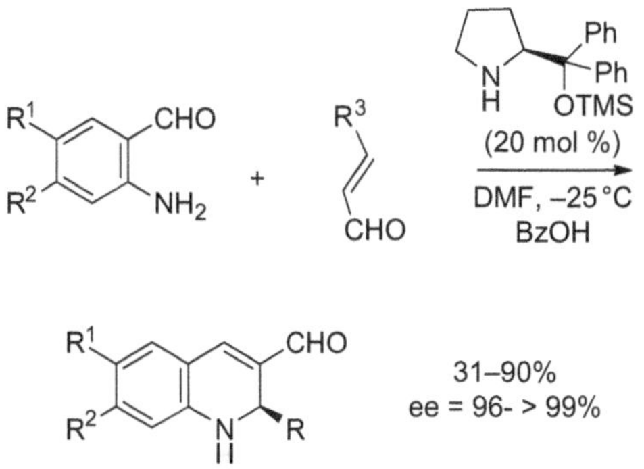

Scheme 3.27 Domino thia-Michael–aldol reaction of α,β-unsaturated aldehydes with 2-aminobenzaldehydes through iminium–enamine activation mode.

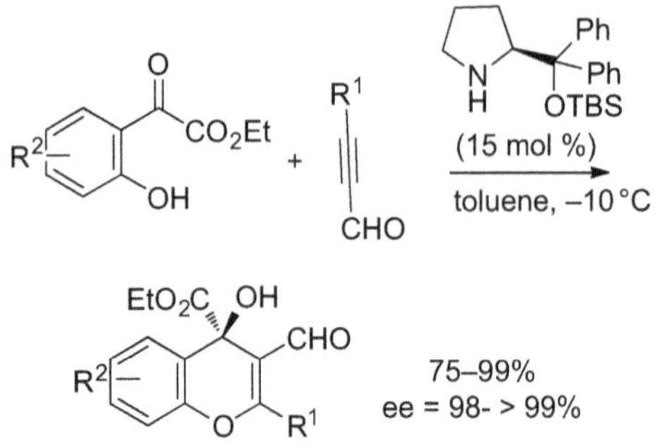

Scheme 3.28 Domino oxa-Michael–aldol reaction of alkynals with ethyl 2-(2-hydroxyphenyl)-2-oxoacetates through iminium–allenamine activation mode.

The concept of hetero-Michael–aldol reaction was also put into practice by Jorgensen *et al.* through the formation of highly functionalised chiral tetrahydrothiophenes, which constitute a family of compounds very useful in biochemistry, pharmaceutical science, and nanoscience.[65] Interestingly, Jorgensen *et al.* demonstrated that an appropriate choice of the additive allowed control of the regioselectivity of the reaction (aldol step). When aliphatic α,β-unsaturated aldehydes and a thiol were put under the effect of 10 mol% of a chiral diarylprolinol trimethylsilyl ether as catalyst and benzoic acid as additive, the corresponding tetrahydrothiophene carbaldehydes were obtained with moderate yields, excellent enantioselectivities (90–96% ee), and a total diastereocontrol, as shown in Scheme 3.29. This outcome involved the normal pathway in this type of domino reactions (thia-Michael addition over the iminium-ion, and a subsequent aldol reaction between the just formed enamine and ketone moiety). On the other hand, when the reaction was carried out in basic conditions (NaHCO$_3$) instead of benzoic acid, the aldol step was thermodynamically controlled by the substrate, without catalyst induction, affording the corresponding (tetrahydrothiophen-2-yl)phenyl methanones as single diastereomers in good yields (43–66%) and good enantioselectivities (64–82% ee), as shown in Scheme 3.29.

Another contribution of Wang *et al.* was focused on the synthesis of chiral cyclopentenes.[66] Based on a domino Michael addition–aldol reaction catalysed by chiral diphenylprolinol triethylsilyl ether followed by dehydration, occurring between aromatic enals and dimethyl 2-oxoethylmalonate, a range of densely functionalised chiral cyclopentenes were achieved in good to high yields of up to 89% and excellent enantioselectivities (91–97% ee), as shown in Scheme 3.30.

In 2008, a highly diastereo- and enantioselective domino Michael–aldol–dehydration reaction between α,β-unsaturated aldehydes and 5-oxoalkanals was successfully developed by Hong *et al.*[67] Evolving through iminium–enamine activation mode by using chiral diphenylprolinol trimethylsilyl ether in the

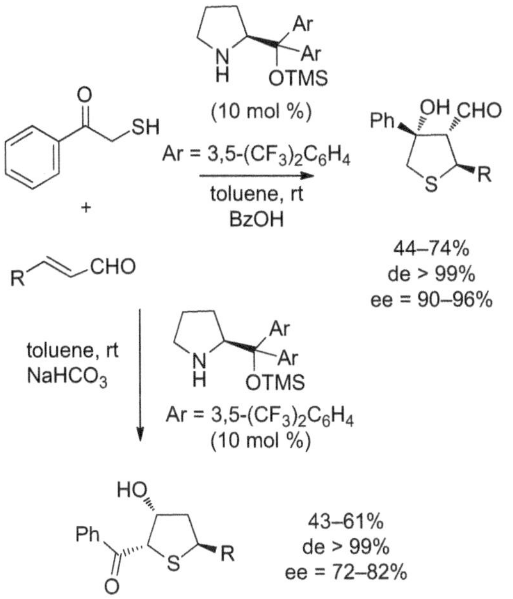

Scheme 3.29 Domino thia-Michael–aldol reactions of α,β-unsaturated aldehydes with a thiol through iminium–enamine activation mode.

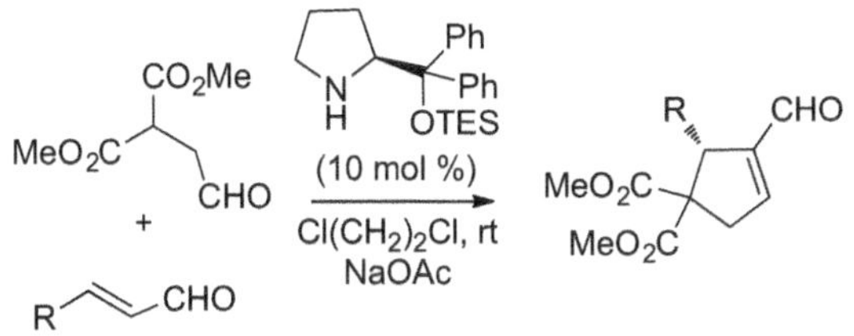

R = Ph: 81% ee = 93%
R = 2-Fu: 89% ee = 91%
R = 2-Fu: 89% ee = 91%
R = *p*-$NO_2C_6H_4$: 88% ee = 92%
R = *p*-FC_6H_4: 73% ee = 93%
R = *p*-CNC_6H_4: 63% ee = 94%
R = *p*-$MeOC_6H_4$: 80% ee = 96%
R = *o*-$MeOC_6H_4$: 78% ee = 97%

Scheme 3.30 Domino Michael–aldol reaction of α,β-unsaturated aldehydes with dimethyl 2-oxoethylmalonate through iminium–enamine activation mode.

presence of acetic acid as an additive, this process afforded the corresponding densely functionalised cyclohexenes in good yields, general trans-diastereoselectivity of >99% de, and excellent enantioselectivities of up to >99% ee, as shown in Scheme 3.31.

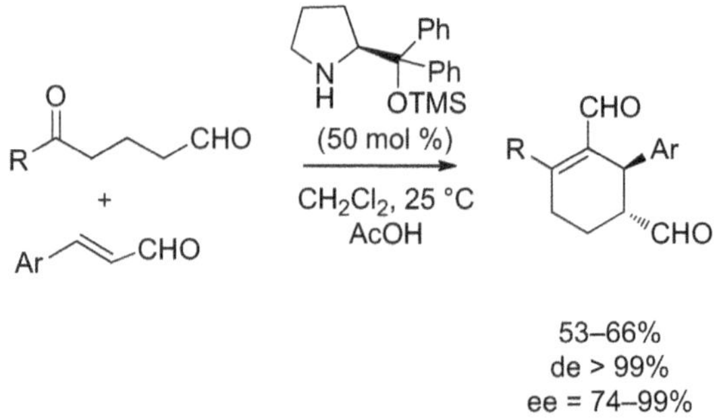

Scheme 3.31 Domino Michael–aldol–dehydration reaction of α,β-unsaturated aldehydes with 5-oxoalkanals through iminium–enamine activation mode.

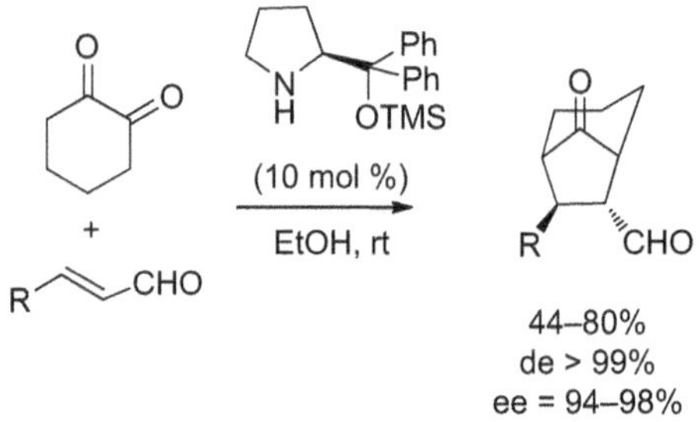

Scheme 3.32 Domino Michael–aldol reaction of α,β-unsaturated aldehydes with 1,2-cyclohexadione through iminium–enamine activation mode.

The iminium–enamine activation mode was also applied by Rueping *et al.* to the domino Michael–aldol reaction of α,β-unsaturated aldehydes with 1,2-cyclohexadione, which provided access upon catalysis with chiral diphenylprolinol trimethylsilyl ether to the corresponding chiral bicyclo[3.2.1]octane-6-carbaldehydes in good yields and with excellent enantioselectivities of up to 98% ee.[68] These bicyclic products were obtained as single diastereomers in all cases of the substrates studied, as shown in Scheme 3.32.

Asymmetric domino oxa-Michael–aldol reactions have also been recently developed by several groups. For example, Vicario *et al.* have developed asymmetric domino oxa-Michael–aldol-hemiacetalisation reactions of α,β-unsaturated aldehydes with dihydroxyacetone dimer, which led to the enantioselective formation of the corresponding hexahydrofuro [3,4-*c*]furans.[69] This process consisted of an initial oxa-Michael reaction, a subsequent intramolecular aldol reaction, and lastly a hemiacetalisation step, and proceeded with the generation of four new stereocentres upon catalysis with chiral diphenylprolinol trimethylsilyl ether. Remarkably, the intramolecular aldol reaction step involved the participation of a ketone as internal electrophile, therefore generating a quaternary stereocentre. As shown in Scheme 3.33,

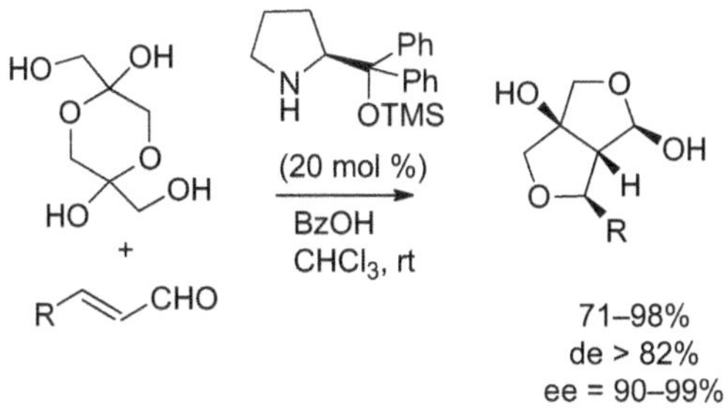

Scheme 3.33 Domino oxa-Michael–aldol reaction of α,β-unsaturated aldehydes with dihydroxyacetone dimer through iminium–enamine activation mode.

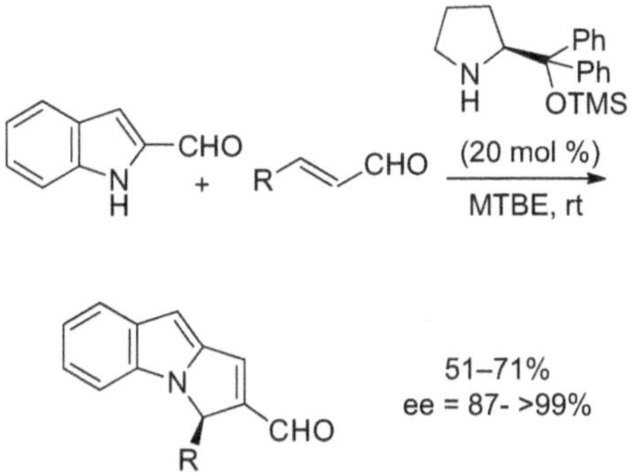

Scheme 3.34 Domino aza-Michael–aldol reaction of α,β-unsaturated aldehydes with 1*H*-indole-2-carbaldehyde through iminium–enamine activation mode.

generally excellent enantioselectivities (90–99% ee) were obtained in combination with high diastereoselectivity of >82% de in almost all cases of the substrates studied.

The iminium–enamine activation mode has also been applied by Enders *et al.* to asymmetric domino aza-Michael–aldol reactions catalysed with the same catalyst.[70] These novel processes opened efficient and enantioselective entry to the tricyclic pyrrolo indole core, which is a characteristic structural unit of many bioactive natural products. As shown in Scheme 3.34, a series of chiral 3-substituted 3*H*-pyrrolo[1,2-*a*]indoles were achieved in moderate to good yields and high to excellent enantioselectivities of up to >99% ee by reaction of 1*H*-indole-2-carbaldehyde and α,β-unsaturated aldehydes. Later, these reactions were reinvestigated by Wang *et al.*, using toluene as the solvent and in the presence of molecular sieves at room temperature. Under these conditions, these authors obtained the same products, albeit in better yields (61–84%) combined with comparable enantioselectivities (81 to >99% ee).[71]

Furthermore, several groups have successfully developed asymmetric domino thia-Michael–aldol reactions on the basis of the iminium–enamine activation mode. For example, Wang *et al.* have used a chiral diarylprolinol trimethylsilyl ether as catalyst to promote the synthesis of chiral highly functionalised tetrahydrothiophenes by reaction of α,β-unsaturated aldehydes with ethyl-mercapto-2-oxopropanoate.[72] Notably, three consecutive stereogenic centres including one chiral quaternary carbon centre were efficiently created in this process. It must be noted that a subsequent reduction of the aldehyde function of the domino process was necessary for convenient chromatography of the products. The corresponding diols were obtained with moderate to good yields associated with excellent general enantioselectivities (91–97% ee) and high diastereoselectivities (78 to >90% de), as shown in Scheme 3.35.

In 2010, Xu *et al.* condensed 2-mercaptoacetaldehyde, generated from 1,4-dithiane-2,5-diol under equilibrium conditions, to α,β-unsaturated aldehydes in the presence of chiral diphenylprolinol trimethylsilyl ether as catalyst associated with *p*-nitrobenzoic acid as an additive, which provided through iminium–enamine activation mode the corresponding chiral domino thia-Michael–aldol products in high yields and excellent enantioselectivities of up to >99% ee, as shown in Scheme 3.36.[73]

In 2010, Cho *et al.* investigated the enantioselective domino Michael–aldol reaction of α,β-unsaturated aldehydes with 2-trihaloacetylpyrroles by using a chiral diarylprolinol trimethylsilyl ether as catalyst.[74] The process employed benzoic acid as an additive and provided the corresponding chiral functionalised pyrrolizines in good yields, generally excellent diastereoselectivity of >90% de, and excellent enantioselectivities (90–98% ee), as shown in Scheme 3.37.

Highly diastereo- and enantioselective domino Michael–aldol reactions of α,β-unsaturated aldehydes with imidazole derivatives evolving through iminium–enamine activation mode were reported by Ye *et al.*, in 2011.[75] As

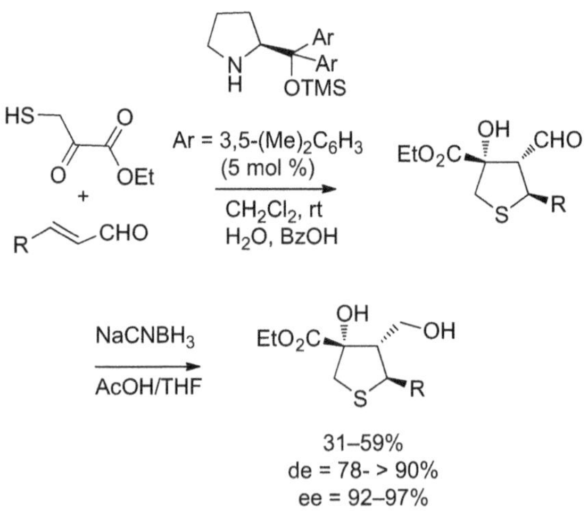

Scheme 3.35 Domino thia-Michael–aldol reaction of α,β-unsaturated aldehydes with ethyl-mercapto-2-oxopropanoate through iminium–enamine activation mode.

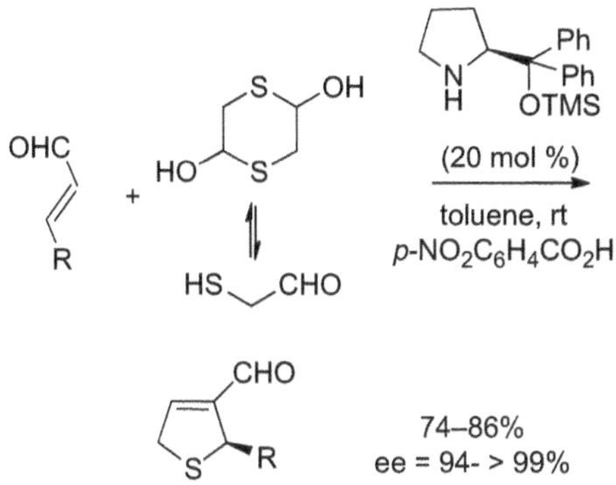

Scheme 3.36 Domino thia-Michael–aldol reaction of α,β-unsaturated aldehydes with 1,4-dithiane-2,5-diol through iminium–enamine activation mode.

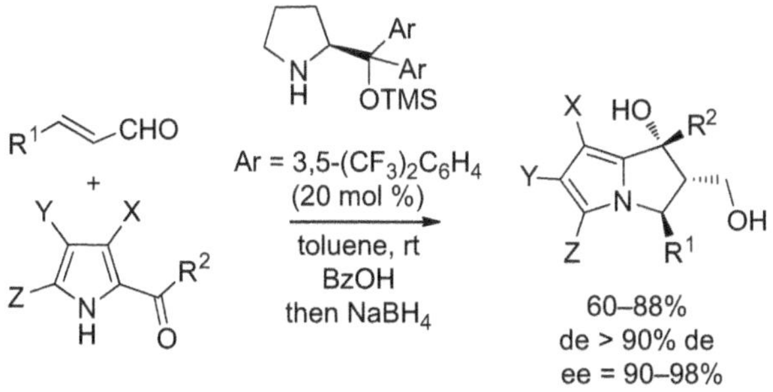

Scheme 3.37 Domino thia-Michael–aldol reaction of α,β-unsaturated aldehydes with 2-trihaloacetylpyrroles through iminium–enamine activation mode.

shown in Scheme 3.38, the corresponding chiral domino bicyclic products were achieved in high yields, good to excellent diastereoselectivities of up to 94% de, and enantioselectivities of >99% ee in all cases of the substrates studied, and by using chiral diphenylprolinol trimethylsilyl ether as catalyst.

On the other hand, a number of organocatalytic enantioselective domino Michael–aldol reactions have involved the condensation of nitroalkanes onto α,β-unsaturated aldehydes. For example, Enders *et al.* have shown that chiral diphenylprolinol trimethylsilyl ether could, highly efficiently, induce the domino nitroalkane-Michael addition–aldol reaction.[76] This process, based on the iminium–enamine activation mode, provided the efficient synthesis of chiral trisubstituted 3,4-dihydronaphthalenes in moderate to good yields (40–75%) and excellent diastereo- and enantioselectivities of up to 98% de and >99% ee, respectively, starting from 2-(nitromethyl)benzaldehyde and various α,β-unsaturated aldehydes, as shown in Scheme 3.39.

Moreover, the condensation of β-nitroketones onto α,β-unsaturated aldehydes in the presence of the same catalyst, combined with acetic acid as an

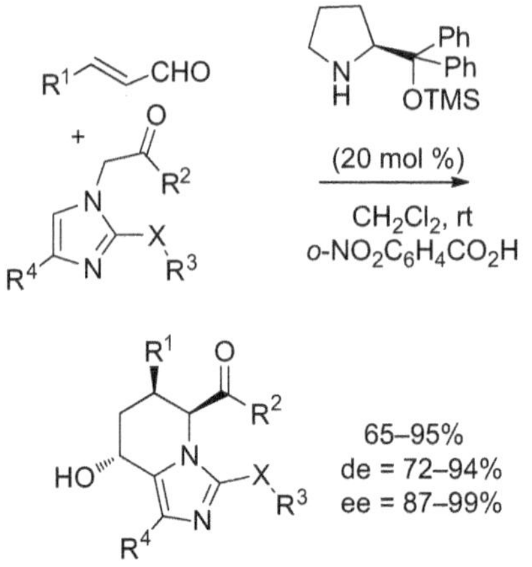

Scheme 3.38 Domino Michael–aldol reaction of α,β-unsaturated aldehydes with imidazoles through iminium–enamine activation mode.

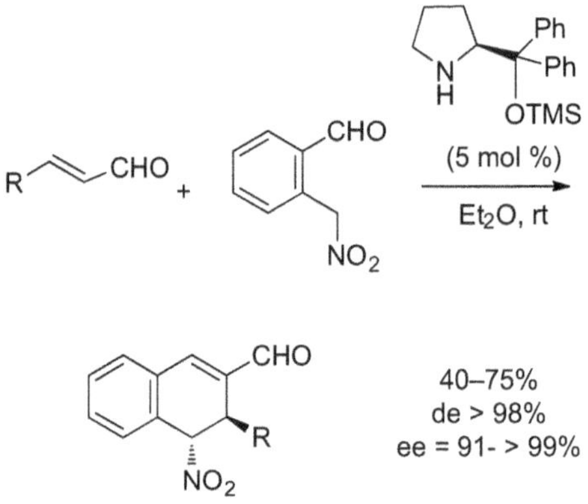

Scheme 3.39 Domino Michael–aldol reaction of α,β-unsaturated aldehydes with 2-(nitromethyl)benzaldehyde through iminium–enamine activation mode.

additive, was shown by Hong *et al.* to afford the corresponding domino Michael–aldol products through iminium–enamine activation mode.[77] As shown in Scheme 3.40, these fully substituted cyclopentene derivatives were obtained in moderate to good yields, albeit as single diastereomers, and with excellent enantioselectivities of up to 99% ee.

In addition to α,β-unsaturated aldehydes, α,β-unsaturated ketones and derivatives have also been widely employed as Michael acceptors in recent enantioselective organocatalysed domino Michael–aldol reactions. As an example, Cordova *et al.* have developed a simple catalytic synthesis of tetrahydrothioxanthenones and tetrahydroxanthenones by applying the iminium–enamine activation mode to a domino Michael–aldol of enones.[78] As

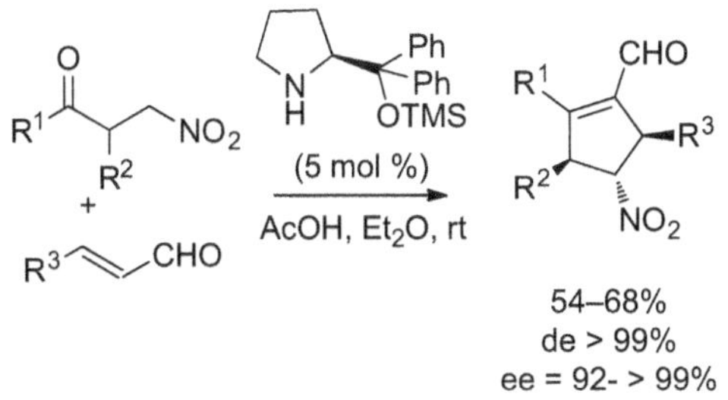

Scheme 3.40 Domino Michael–aldol reaction of α,β-unsaturated aldehydes with β-nitroketones through iminium–enamine activation mode.

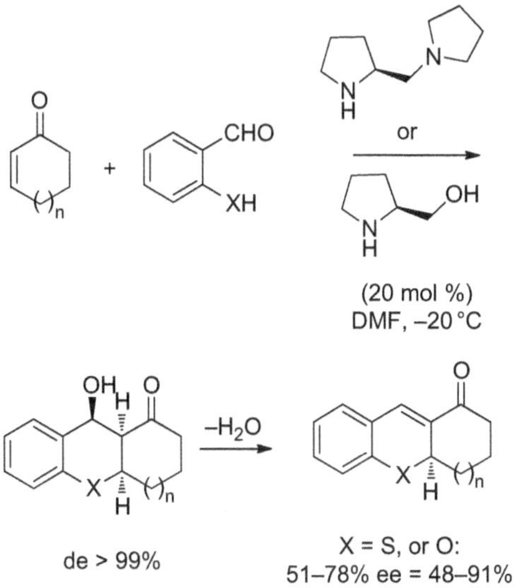

Scheme 3.41 Domino Michael–aldol reaction of cyclic enones with 2-mercapto-benzaldehydes or salicylic aldehyde derivatives.

shown in Scheme 3.41, the organocatalytic domino Michael–aldol reaction of 2-mercaptobenzaldehydes or salicylic aldehyde derivatives with cyclic enones proceeded in a highly chemoselective fashion, furnishing the corresponding tetrahydrothioxanthenones or tetrahydroxanthenones, respectively, in high yields and moderate to good enantioselectivities (48–91% ee). The intermediate chiral alcohols were isolated as single diastereomers when a rapid column chromatography eluent system was employed. The mechanism proposed involved the iminium activation of the cyclic enone by a chiral pyrrolidine catalyst, depicted in Scheme 3.41. Stereoselective nucleophilic conjugate attack on the β-carbon by the thiol or alcohol resulted in a chiral enamine inter-mediate, which performed an intramolecular 6-*exo*-trig aldol addition from the same face as the incoming thiol or alcohol. Hydrolysis of the resulting iminium

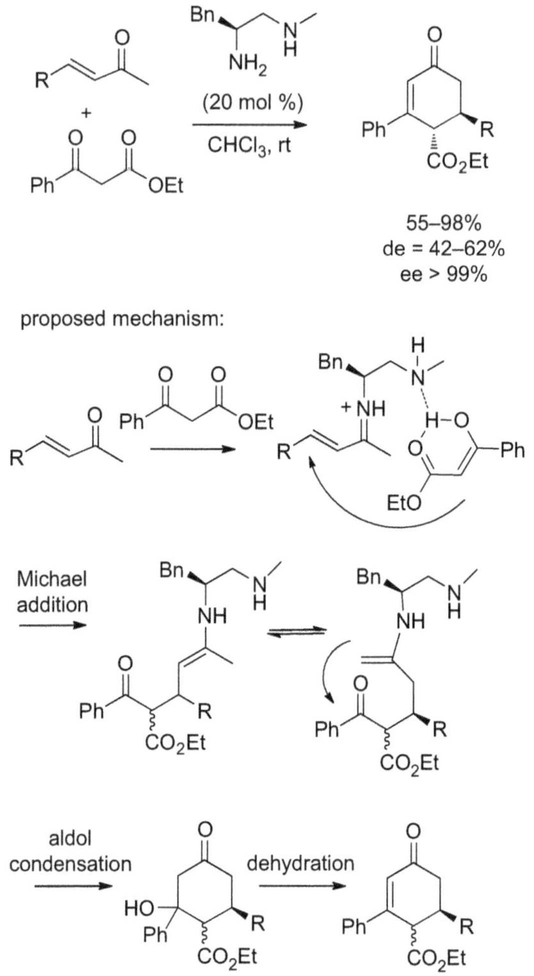

Scheme 3.42 Domino Michael–aldol reaction of α,β-unsaturated ketones with ethyl benzoylacetate through iminium–enamine activation mode.

intermediate gave the intermediate chiral alcohol. Elimination of water afforded tetrahydrothioxanthenones or tetrahydroxanthenones.

A simple chiral primary–secondary diamine catalyst (Scheme 3.42) was successfully employed by Zhao *et al.* to induce the highly enantioselective domino Michael–aldol–dehydration reaction between benzoylacetate and α,β-unsaturated ketones to furnish the corresponding functionalised chiral cyclohexenones.[79] General enantioselectivities of at least 99% ee for both major and minor diastereomers were obtained, combined with high yields and moderate to high diastereoselectivities as shown in Scheme 3.42. The authors have explained the formation of these products through the mechanism based on iminium–enamine activation mode that is depicted in Scheme 3.42.

In the first step, the α,β-unsaturated ketone was activated by the formation of its corresponding iminium ion by the catalyst. The *Re* face of the enone in this intermediate was shielded by the bulky group of the chiral catalyst, allowing the benzoylacetate to attack the *Si* face of the enone. Next, the generated enamine performed an intramolecular aldol addition to form the cyclohexane ring. After dehydration, the final chiral cyclohexenones were achieved.

The enantioselective catalytic construction of chiral fluorinated quaternary carbon centres, which are a class of versatile and important monofluorinated synthons used in organic synthesis, is still a very challenging subject in organic chemistry. In this context, the same authors have used another chiral primary–secondary amine as catalyst (Scheme 3.43) in combination with *p*-nitrobenzoic acid as co-catalyst to catalyse a closely related domino Michael–aldol–dehydration reaction occurring between α-fluoro-β-keto esters and α,β-unsaturated ketones, which provided the corresponding functionalised fluorinated cyclohexenones with excellent diastereo- and enantioselectivities of up to >98% de and >99% ee, combined with moderate to good yields, as shown in Scheme 3.43.[80] A similar mechanism to that shown in Scheme 3.43, based on iminium–enamine activation mode, was also proposed by the authors.

In order to give access to chiral spiro[cyclohex-2-enone-oxindole] derivatives, Wang *et al.* have developed a novel domino Michael–aldol reaction of oxindole derivatives with α,β-unsaturated ketones based on iminium–enamine activation mode, induced by a chiral cinchona-derived primary amine (Scheme 3.44).[81] The best results were observed when the reaction was performed in the presence of an additive such as TFA, providing the domino products in high yields of up to 99% and with excellent levels of diastereoselectivity of >90% de in all cases of the substrates studied, along with excellent enantioselectivities of up to 96% ee, as shown in Scheme 3.44.

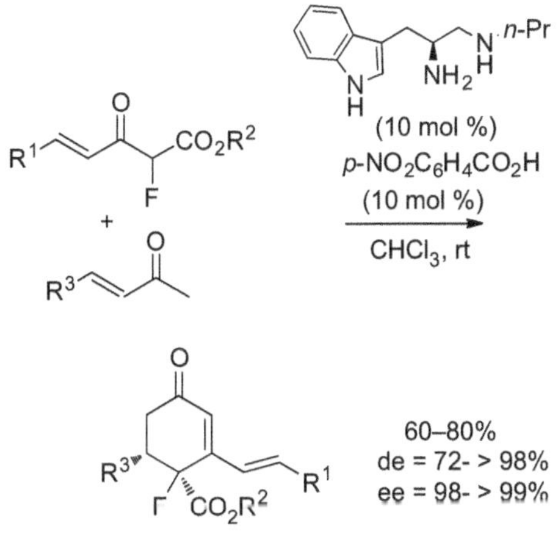

Scheme 3.43 Domino Michael–aldol reaction of α,β-unsaturated ketones with α-fluoro-β-keto esters through iminium–enamine activation mode.

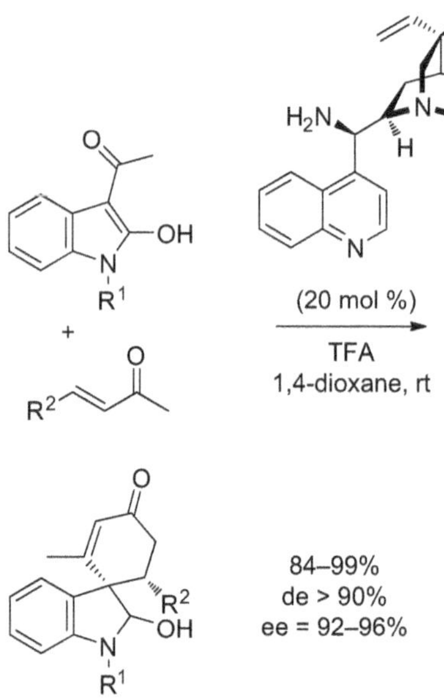

Scheme 3.44 Domino Michael–aldol reaction of α,β-unsaturated ketones with oxindole derivatives through iminium–enamine activation mode.

Hydrogen-bonding mediated catalysis was used by Wang *et al.* in 2007 in order to perform highly enantio- and diastereoselective domino Michael–aldol reactions.[82] They were efficiently catalysed by a cinchona alkaloid thiourea, using as low as 1 mol% of catalyst loading, *via* synergistic non-covalent hydrogen-bonding activation of both the Michael donor and acceptor. The domino reaction of 2-mercaptobenzaldehydes with α,β-unsaturated oxazolidinones afforded the corresponding chiral thiochromanes bearing three stereogenic centres in good to excellent yields and enantioselectivities of up to 99% ee, as shown in Scheme 3.45.

Soon after, the same group reported another organocatalytic, enantioselective domino Michael–aldol reaction, occurring this time between 2-mercaptobenzaldehydes and maleimides, these last being much less explored substrates.[83] They managed to incorporate succinimides into complex benzothiopyrans, generating three stereogenic centres. The domino process was catalysed by a bifunctional chiral amine thiourea through a hydrogen-bonding mediated activation mechanism. The domino products were achieved in good to high yields and enantioselectivities of up to 94% ee and moderate to high diastereoselectivities, as shown in Scheme 3.46.

In 2010, another asymmetric domino thia-Michael–aldol reaction was described by Wang *et al.*, which involved catalysis by a chiral bifunctional thiourea (Scheme 3.47).[84] This novel hydrogen-bonding mediated highly stereoselective cascade reaction of various benzylidenechroman-4-ones with 2-mercaptobenzaldehyde allowed an efficient construction of chiral spiro

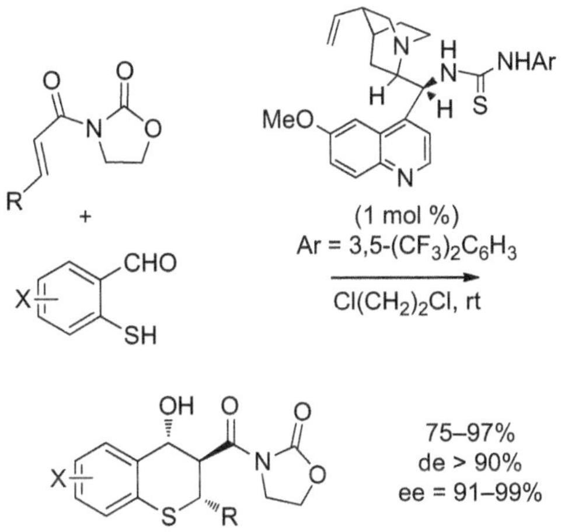

Scheme 3.45 Domino Michael–aldol reaction of α,β-unsaturated oxazolidinones with 2-mercaptobenzaldehydes through hydrogen-bonding activation mode.

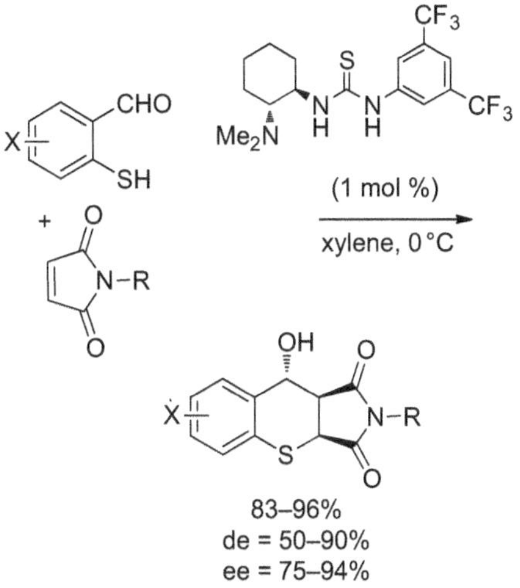

Scheme 3.46 Domino Michael–aldol reaction of maleimides with 2-mercaptobenz-aldehydes through hydrogen-bonding activation mode.

chromanone-thiochromans. As shown in Scheme 3.47, these potentially biologically interesting products were obtained in high yields and generally excellent enantioselectivities of up to 99% ee combined with moderate to excellent diastereoselectivities of up to 96% de.

The hydrogen-bonding activation mode was also recently involved by Barbas *et al.* to explain the formation of chiral complex (1*S*,2*S*,3*R*,4*S*)-bispirooxindoles on the basis of a novel asymmetric domino Michael–aldol

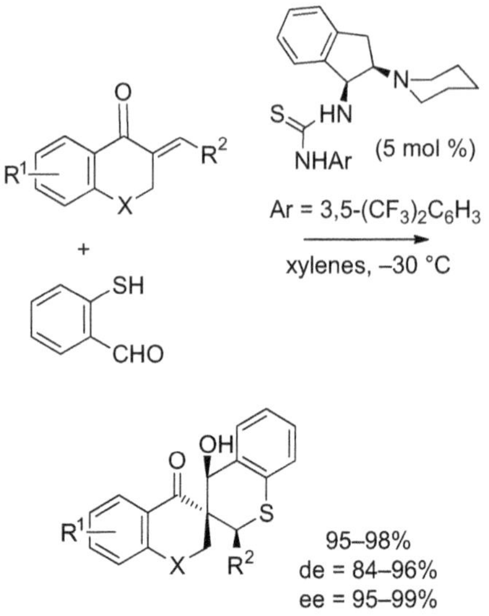

Scheme 3.47 Domino thia-Michael–aldol reaction of benzylidenechroman-4-ones with
2-mercaptobenzaldehyde through hydrogen-bonding activation mode.

reaction occurring between 3-substituted oxindoles and a range of methyleneindolinones.[85] This process was catalysed by a novel multifunctional
organocatalyst (Scheme 3.48) containing tertiary and primary amines and
thiourea moieties to activate substrates simultaneously, which provided
extraordinary levels of stereocontrol over four stereocentres, three of which
were quaternary carbon centres. As shown in Scheme 3.48, this novel methodology provided facile access to a number of potent biologically active multi-
substituted bispirocyclooxindole derivatives, with high yields and excellent
diastereo- and enantioselectivities of up to >98% de and 96% ee, respectively.
Even more interestingly, these authors have proposed the possibility of
achieving the opposite enantiomers of these novel multifunctionalised
products. Indeed, they demonstrated that performing the reaction in the
presence of a reconfigured catalyst, depicted in Scheme 3.48, in which the
S-diamine component was kept and the tertiary amine and the thiourea
configurations were changed when compared with the catalyst first employed,
gave rise to the expected enantiomer (1*R*,2*R*,3*S*,4*R*)-bispirooxindole in
good yield, combined with high diastereoselectivity of 88% de, and high
enantioselectivity of 90% ee, as shown in Scheme 3.48. The possibility of
achieving both enantiomers of these novel products will be highly useful in
the investigation of their biological activity. Furthermore, these studies have
highlighted the growing potential of reaction and catalyst design in
organocatalysis.

   In 2012, a novel enantioselective domino thia-Michael–aldol reaction of 2-
mercaptobenzaldehyde with α,β-unsaturated *N*-acylpyrazoles was reported by
Wang *et al.*[86] This highly efficient protocol, catalysed by a chiral cinchona

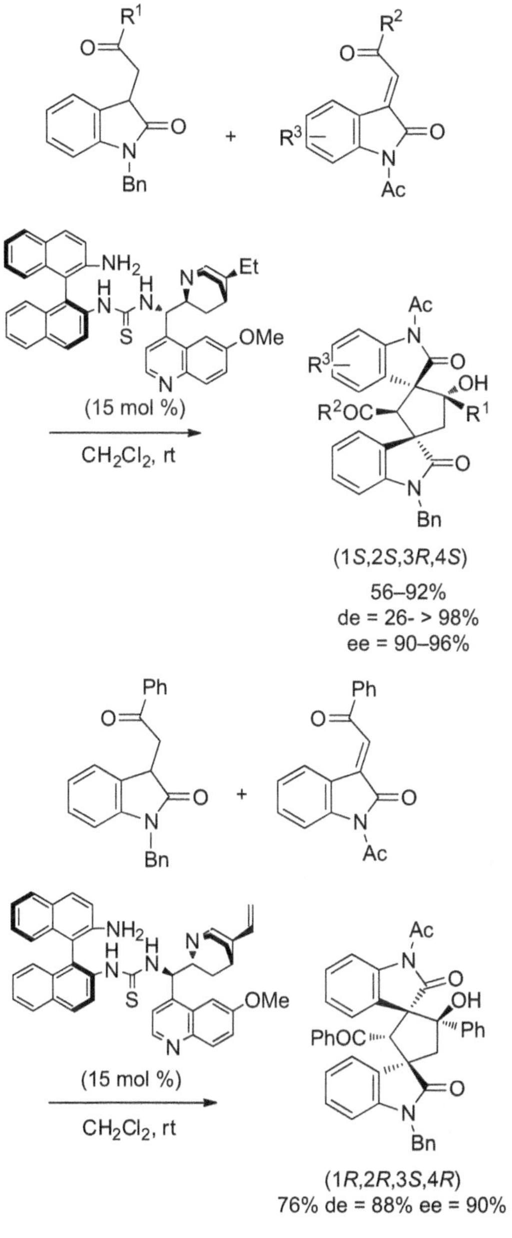

Scheme 3.48 Domino Michael–aldol reaction of methyleneindolinones with 3-substituted oxindoles through hydrogen-bonding activation mode.

alkaloid thiourea used at as low as 1 mol% of catalyst loading, allowed a range of highly substituted and biologically active thiochromanes to be achieved in high yields and excellent diastereo- and enantioselectivities, as shown in Scheme 3.49.

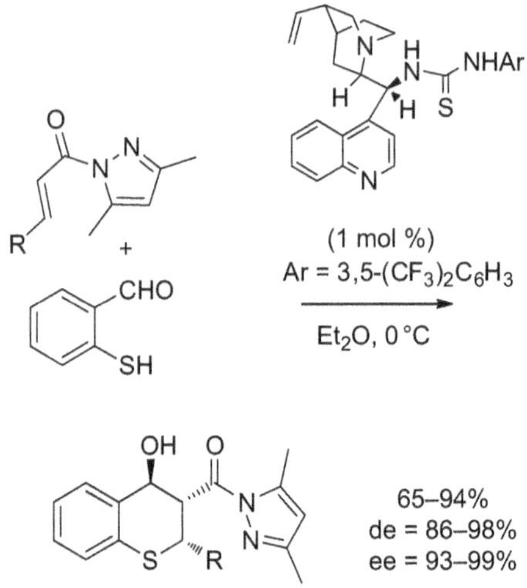

Scheme 3.49 Domino Michael–aldol reaction of α,β-unsaturated *N*-acylpyrazoles with 2-mercaptobenzaldehyde through hydrogen-bonding activation mode.

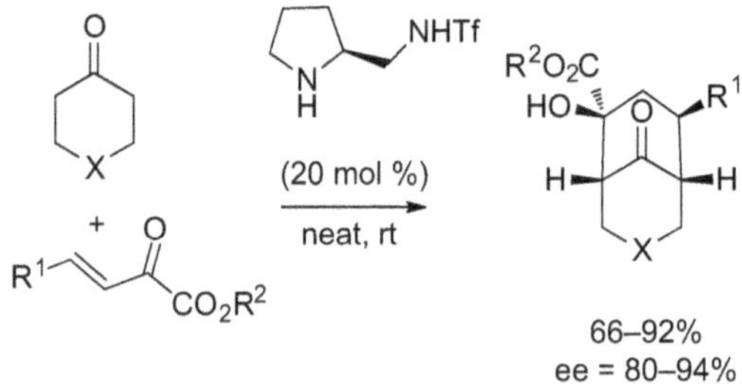

Scheme 3.50 Domino Michael–aldol reaction of enones with cyclic ketones through enamine–iminium activation mode.

In 2007, Tang *et al.* reported the first example of an asymmetric formal [3 + 3] annulation of cyclic ketones with enones.[87] As shown in Scheme 3.50, the enantioselective domino Michael–aldol reaction, which evolved through enamine–iminium activation mode, was catalysed by chiral *N*-(pyrrolidin-2-ylmethyl)trifluoromethanesulfonamide and produced a range of chiral bicyclic [3.3.1] adducts in moderate to good yields and good to high enantio-selectivities of up to 94% ee under mild conditions.

On the other hand, asymmetric domino Michael–aldol reactions have also been developed on the basis of enamine activation. As an example, Cordova *et al.* have reported a highly enantioselective domino Michael–aldol reaction catalysed by chiral diphenylprolinol trimethylsilyl ether, giving access to highly functionalised cyclohexanes bearing four stereogenic centres.[88] These chiral products were generated with high yields and enantioselectivities (83–98% ee),

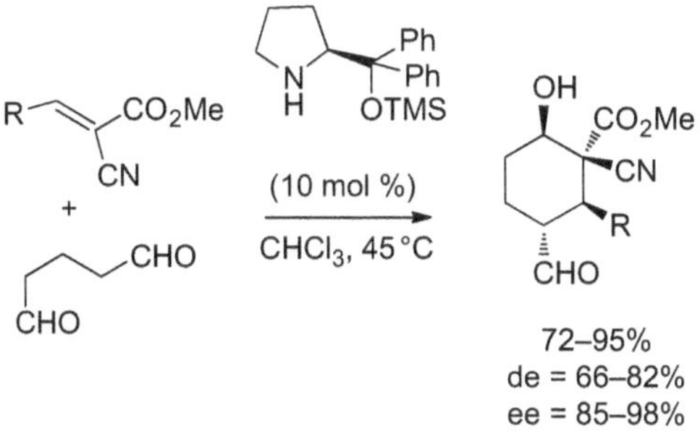

Scheme 3.51 Domino Michael–aldol reaction of alkylidenemalonates with glutar-aldehyde through enamine catalysis.

albeit with moderate to good diastereoselectivities (66–82% de) by reaction of glutaraldehyde with various alkylidenemalonates, as shown in Scheme 3.51.

3.2.1.3 Domino Michael–Intramolecular Heterocyclisation Reactions

A wide number of enantioselective domino Michael–intramolecular hetero-cyclisation reactions have been successfully developed in the last six years, using three types of Michael acceptor component, α,β-unsaturated aldehydes, α,β-unsaturated ketones or derivatives, and nitroolefins. The reactions of these three possible types of reactant will be successively developed in this section. In the context of reactions involving α,β-unsaturated aldehydes as Michael acceptors, Jorgensen *et al.* reported, in 2005, the first asymmetric organo-catalytic epoxidation of α,β-unsaturated aldehydes, employing as an oxygen source simple peroxides, such as H_2O_2.[89] This reaction proceeded *via* an iminium–enamine activation mode, using a chiral diarylprolinol trimethylsilyl ether as catalyst. First, the chiral iminium ion formed was attacked by nucleophilic oxygen at the electrophilic β-carbon, generating the first carbon–oxygen bond, and leading to an enamine intermediate. Afterwards, this nucleophilic enamine attacked the electrophilic peroxygen atom, forming the α,β-epoxy aldehyde and regenerating the catalyst. It was noteworthy that the reaction worked well in a wide range of solvents at room temperature, obtaining the best results when dichloromethane was used with 10 mol% of a chiral diaryl prolinol silylether. The domino reaction tolerated a broad range of β-substituents in the enal moiety such as variously substituted aromatic rings, alkylic substituents, and functionalised carbons, such as esters or protected alcohols. The epoxy aldehydes were achieved in high yields (75–98%) and enantioselectivities of up to 99% ee. Soon after, Cordova *et al.* performed a similar reaction using chiral diphenyl prolinol trimethylsilyl ether as organocatalyst, which provided excellent results in terms of conversion (59–82% yields), and enantioselectivities (91–97% ee).[90] In 2007, the same authors reported a one-pot combination of amine and heterocyclic carbene catalysis which enabled the synthesis of β-hydroxy and β-aminoesters from α,β-unsaturated aldehydes with high enantioselectivities.[91] Indeed, iminium activation of enals followed by enantioselective Michael addition of nucleophiles such as H_2O_2 or CbzNHOAc (where Cbz is benzoyloxycarbonyl), followed by cyclisation, furnished the corresponding epoxy, or aziridine aldehydes. Next, the base-generated hetero-cyclic carbene catalyst catalysed the C–O, or C–N bond-cleavage ring-opening, followed by concomitant oxidation of aldehyde and subsequent esterification. The final β-hydroxy and β-aminoesters were achieved in good yields (59–82%) and high enantioselectivities (91–97% ees). In 2008, Wang and List published a nice epoxidation of α,β-unsaturated aldehydes through asymmetric counterion-directed catalysis.[92] This asymmetric induction mode involved an achiral secondary amine which formed a cationic achiral iminium ion with the enal. The interaction of this cation with an anionic chiral phosphoric acid (the chiral counterion) created a chiral environment. Then, *tert*-butyl hydroperoxide performed an asymmetric epoxidation through an iminium–enamine mechanism, as previously described. Aromatic enals were epoxidated with excellent diastereo- and enantioselectivities of up to >98% de and 96% ee, respectively. However, the reaction of aliphatic enals provided the corresponding epoxides with high diastereoselectivities of up to 88% de, albeit moderate enantioselectivities (≤70% ee). In 2007, Cordova *et al.* developed the first asymmetric organo-catalytic synthesis of aziridines between aliphatic enals and acylated hydroxyc-arbamates, which was catalysed by chiral diphenylprolinol trimethylsilyl ether in chloroform.[93] As occurred with epoxidation, the choice of the nitrogen source was crucial for the success of the domino reaction. It was demonstrated that acylated hydroxycarbamates were the best substrates, providing the corresponding chiral 2-formylaziridines in moderate to good yields (54–78%) and diastereoselectivities (60–82% de), combined with high enantioselectivities (84–99% ee). More recently, Hamada *et al.* reported an interesting variation on enantioselective aziridination of α,β-unsaturated aldehydes, based on the use of *N*-arenesulfonylcarbamates as the nitrogen source and using chiral diphenylprolinol triethylsilyl ether as catalyst in the presence of 3 equivalents of a base such as NaOAc or Na_2CO_3 (Scheme 3.52).[94] This new protocol improved the yields (51–99%) and the

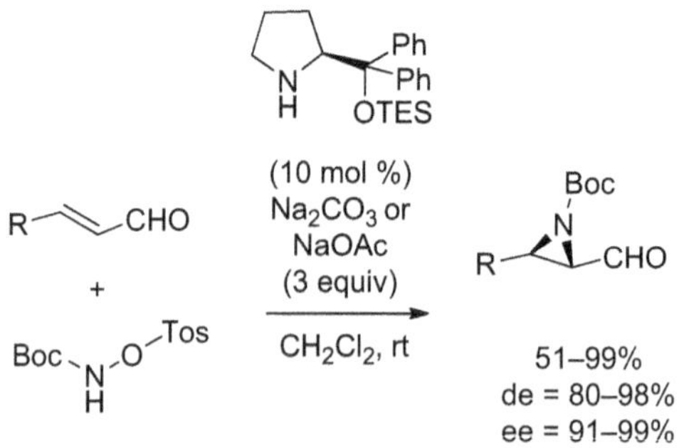

Scheme 3.52 Aziridination of α,β-unsaturated aldehydes through iminium–enamine activation mode.

diastereoselectivity (80–98% de), maintaining the excellent enantioselectivity (91–99% ee), in comparison with the previous methodology reported by Cordova. Moreover, it was also noteworthy that this methodology expanded the aldehyde scope, allowing the aziridination of aromatic enals. On the other hand, Melchiorre *et al.* developed, in 2008, the aziridination of α,β-unsaturated ketones.[95] This domino process was catalysed by a chiral primary amine derived from cinchona alkaloids as a salt with D-*N*-Boc-phenylglycine employed at 20 mol% of catalyst loading in chloroform at 23 °C, and occurred efficiently with both linear and cyclic substrates, providing the corresponding chiral aziridines in high yields (82–96%) combined with general excellent diastereoselectivity of >90% de, and good to excellent enantioselectivities (73–99% ee).

In recent years, several groups have demonstrated that asymmetric domino Michael–hemiacetal (or hemiaminal) formation reactions constituted powerful methodologies to afford a number of various chiral heterocycles. As an early example, Cordova *et al.* developed a very elegant synthesis of 5-hydroxyisoxazolidines based on the addition of *N*-protected hydroxyamines to α,β-unsaturated aldehydes, in 2007.[96] The authors disclosed that, in a first step, the amine attacked the β-position of the iminium ion, this reaction being in equilibrium. This equilibrium was pushed to the final products by hemiacetal formation between the hydroxyl moiety at the nitrogen atom and the aldehyde. The reaction worked well with any unsaturated aldehyde (aromatic and aliphatic), achieving the final compounds in high yields and enantioselectivities. Moreover, the importance of this reaction was clearly shown by the synthesis of chiral β-aminoacids from α,β-unsaturated aldehydes in only two steps, as shown in Scheme 3.53.

In the same year, Rios and Cordova developed a synthesis of chiral highly functionalised pyrrolidines, following a similar approach.[97] As shown in Scheme 3.54, 2-amidomalonates reacted with α,β-unsaturated aldehydes owing to a malonate addition, followed by hemiaminal formation between the corresponding amide and aldehyde. This domino process furnished the corresponding chiral pyrrolidines from aromatic α,β-unsaturated aldehydes in good

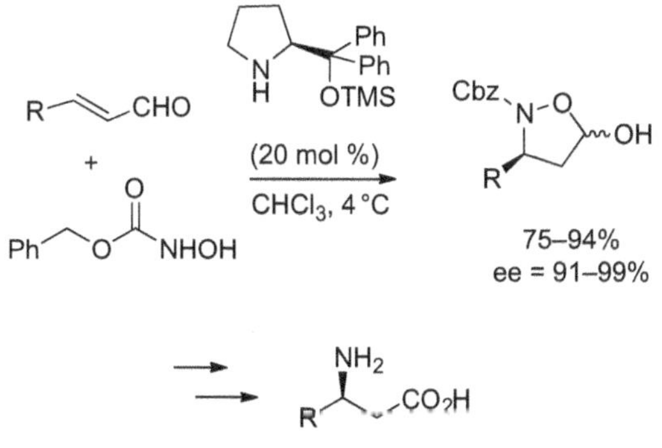

Scheme 3.53 Synthesis of 5-hydroxyisoxazolidine derivatives through iminium activation mode.

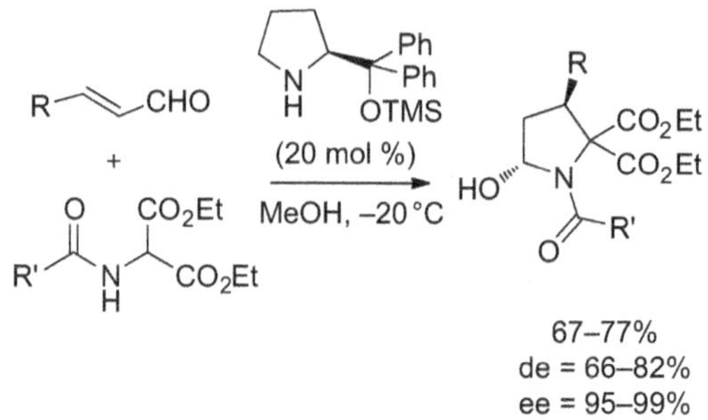

67–77%
de = 66–82%
ee = 95–99%

Scheme 3.54   Synthesis of pyrrolidines through iminium activation mode.

48–95%
de = 44–90%
ee = 82–97%

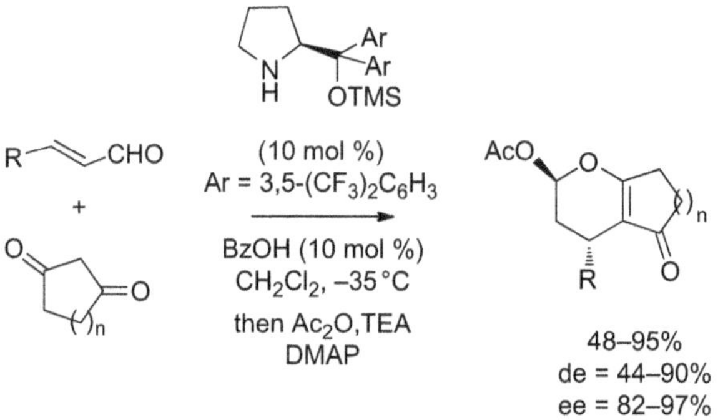

Scheme 3.55   Synthesis of 3,4-dihydropyrans through iminium activation mode.

yields, moderate to good diastereoselectivities ($\leq 82\%$), combined with excellent enantioselectivities of up to 99% ee (Scheme 3.54).

In 2008, Jorgensen[98] and Rueping[99] independently reported the enantio-selective addition of 1,3-cycloalkanediones to α,β-unsaturated aldehydes catalysed by a chiral diarylprolinol trimethylsilyl ether, which furnished the corresponding 3,4-dihydropyrans through a domino Michael–hemiketalisation reaction. These domino products were achieved in good to high yields and diastereoselectivities of up to >90% de, combined with high enantios-electivities of up to 97% ee, as shown in Scheme 3.55.

The same year, Rueping et al. developed the enantioselective synthesis of 1,4-pyranonaphthoquinones on the basis of a domino Michael–hemiacetali-sation reaction of 2-hydroxy-1,4-naphthoquinone with α,β-unsaturated aliphatic as well as aromatic aldehydes induced by the same catalyst.[100] These biologically interesting quinines were achieved in moderate to high yields and excellent enantioselectivities of up to 99% ee, as shown in Scheme 3.56. Furthermore, these products could be easily converted into the corresponding valuable 1,4-naphthoquinones without loss of enantiopurity.

In 2009, Franzen and Fisher reported an easy entry to the synthesis of chiral quinolizidine derivatives bearing three stereogenic centres by reaction of

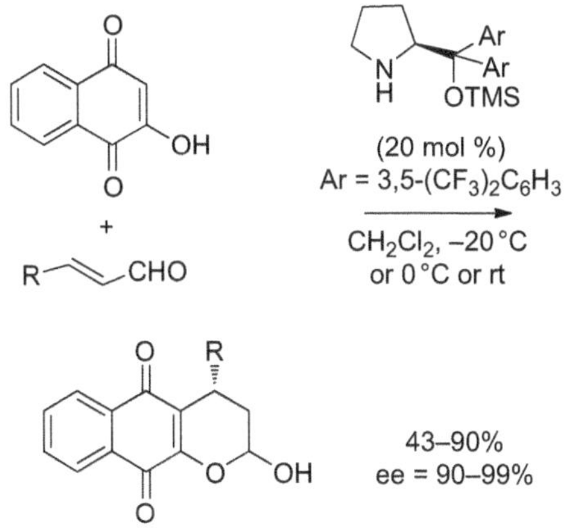

Scheme 3.56 Synthesis of 1,4-pyranonaphthoquinones through iminium activation mode.

activated indol substituted amide and α,β-unsaturated aldehydes.[101] The authors disclosed that, in a first step, the α,β-unsaturated aldehydes were activated by chiral diphenylprolinol trimethylsilyl ether to form the corresponding iminium intermediates which added, according to a Michael process, the activated amide to form corresponding intermediates. Subsequently, these intermediates cyclised spontaneously under the reaction conditions to give hemiaminals. A further epimerisation of their stereochemically labile stereocentre at C3 established the more thermodynamically stable *trans* configuration. These domino products were further treated with added catalytic amounts of HCl, resulting in the formation of the corresponding acyliminium ions which cyclised to provide the final indolo[2,3a]quinolizidines with high yields and diastereoselectivities of up to 80% de, combined with excellent enantioselectivities (87–95% ee), as shown in Scheme 3.57. The scope of this methodology was extended to another activated amide depicted in Scheme 3.57 in which the 3-indolyl moiety was replaced with an electron-rich phenyl group (3,4-dimethoxyphenyl). The reaction of this amide with α,β-unsaturated aldehydes furnished in the same conditions the corresponding benzo[a]quinolizidines in excellent enantioselectivities (89–98% ee) combined with moderate to good diastereoselectivities (24–70% de), as shown in Scheme 3.57.

In the same context, Rios *et al.* have developed the Michael addition of more simple nucleophiles, such as amidomalonates, to a range of α,β-unsaturated aldehydes by using the same catalyst, followed by formation of the corresponding hemiaminals in high yields and excellent enantioselectivities of up to 99% ee, as shown in Scheme 3.58.[102] In all cases of the substrates studied, these piperidines were produced as a mixture of two diastereomers with 3:1 to 5:1 ratio. This diastereoselectivity was demonstrated by the authors to correspond to the equatorial or axial position of the hemiaminal hydroxyl of the *trans* piperidine. The thermodynamically more stable *trans* configuration was established through epimerisation of the stereochemically labile stereocentre at

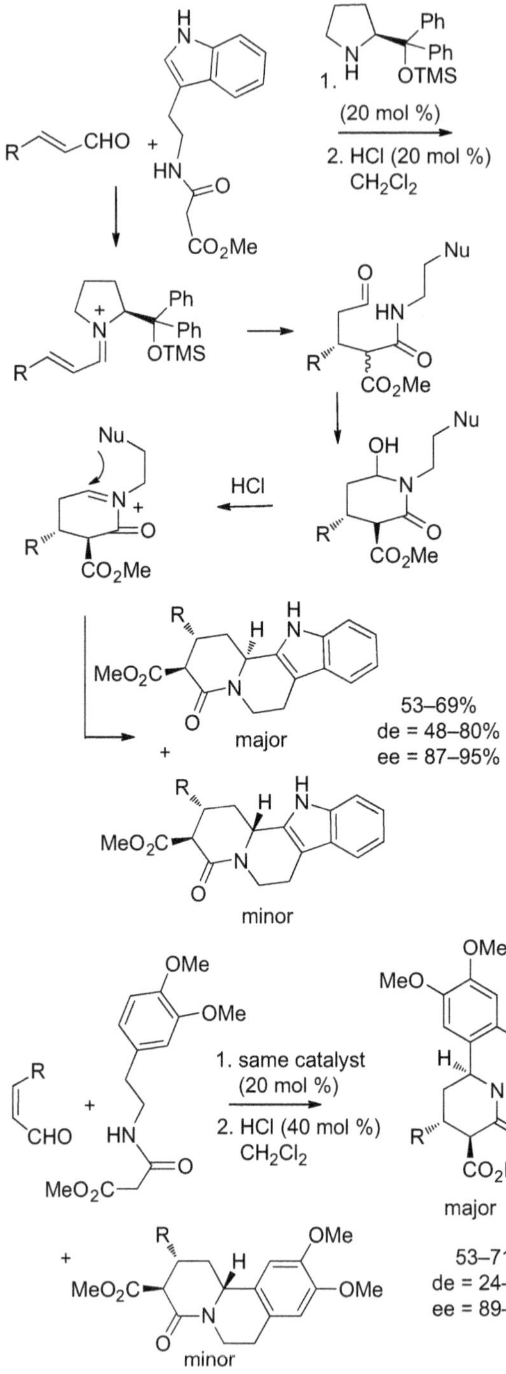

Scheme 3.57 Synthesis of quinolizidine derivatives through iminium activation mode.

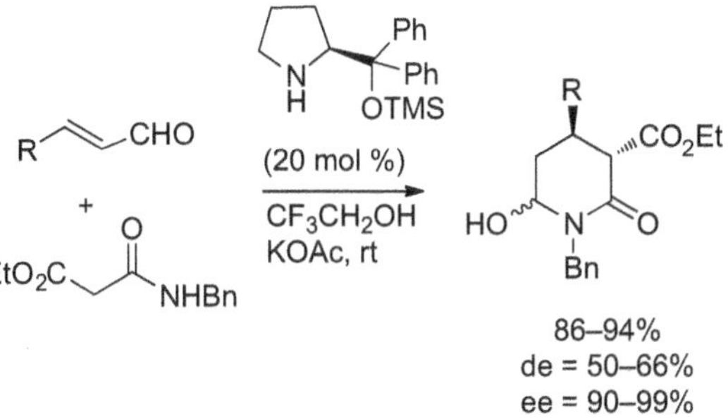

Scheme 3.58 Synthesis of piperidines through iminium activation mode.

C3 of the hemiaminal. One of the chiral piperidines formed was applied as an intermediate in the synthesis of (−)-paroxetine, an important antidepressive drug.

In 2010, Chen *et al.* reported a facile method for the asymmetric synthesis of pyrimidinone derivatives *via* a domino aza-Michael–hemiaminal formation reaction of α,β-unsaturated aldehydes with *N*,*N*′-dibenzyloxyurea as dinitrogen source.[103] The domino reaction, induced by the same catalyst in the presence of acetic acid, afforded the corresponding hemiaminals, which were subsequently submitted to dehydroxylation by treatment with Et₃SiH/BF₃ which was directly added to the reaction, thus providing the corresponding pyrimidinones in both excellent yields and enantioselectivities of up to 97% ee, as shown in Scheme 3.59.

In 2011, chiral highly functionalised quinolizidines were achieved by Zhao *et al.* on the basis of an enantioselective organocatalysed cascade reaction including a domino Michael–intramolecular cyclisation of β-ketoamides and α,β-unsaturated aldehydes.[104] The domino products were not isolated but were subsequently submitted to a Pictet–Spengler cyclisation by treatment with TFA to give the corresponding chiral indolo[2,3-α]quinolizidines and benzo[α]quinolizidines in moderate to good yields and good to excellent enantioselectivities, as shown in Scheme 3.60. The authors have previously reported the synthesis of related chiral quinolizidines through a two-step sequence, including an enantioselective domino Michael–intramolecular hemiaminalisation–dehydration reaction of β-ketoesters and α,β-unsaturated aldehydes catalysed by a chiral diarylprolinol trimethylsilyl ether as a first step followed, after addition of tryptamine and benzoic acid, by a Pictet–Spengler cyclisation as a second step, which provided the corresponding chiral indolo-quinolizidines as single diastereomers in good to high yields (51–95%) and high to excellent enantioselectivities (67–96% ee).[105]

A total synthesis of the natural product diazonamide A was accomplished by MacMillan *et al.*, in 2011.[106] This work featured a highly stereoselective synthesis of the C-10 quaternary centre and the central furanoindoline core enabled by an iminium-catalysed Michael–intramolecular cyclisation reaction, depicted in Scheme 3.61. The process, which involved propynal as Michael

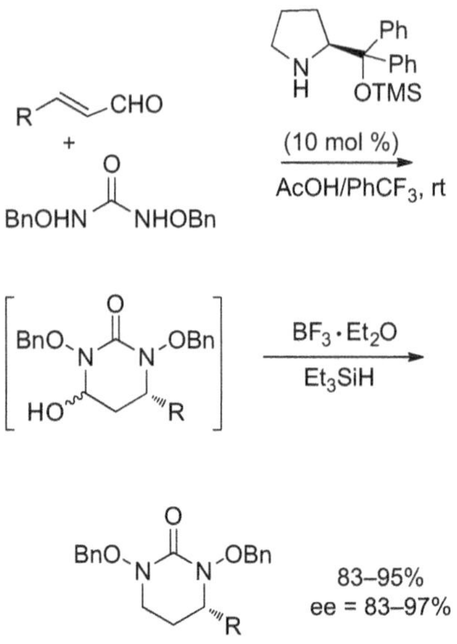

R⌒CHO

+

BnOHN—C(=O)—NHOBn

Ph
Ph
OTMS
(pyrrolidine catalyst)
(10 mol %)
AcOH/PhCF₃, rt

[BnO-N—C(=O)—N-OBn, HO···, ···R]

BF₃·Et₂O
Et₃SiH

BnO-N—C(=O)—N-OBn
···R

83–95%
ee = 83–97%

Scheme 3.59 Synthesis of pyrimidinones through iminium activation mode.

Ar⌒CHO

+

(indole ethyl) HN—C(=O)CH₂C(=O)CH₃

Ar
Ar
OTMS
(10 mol %)
Ar = 3,5-(CF₃)₂C₆H₃

CH₂Cl₂, rt
p-NO₂C₆H₄CO₂H
(10 mol %)

[(indole) HN, N⁺=, O, O, ···Ar]

TFA
Pictet-Spengler

(quinolizidine fused indole) OH

56–90%
de > 99%
ee = 77–97%

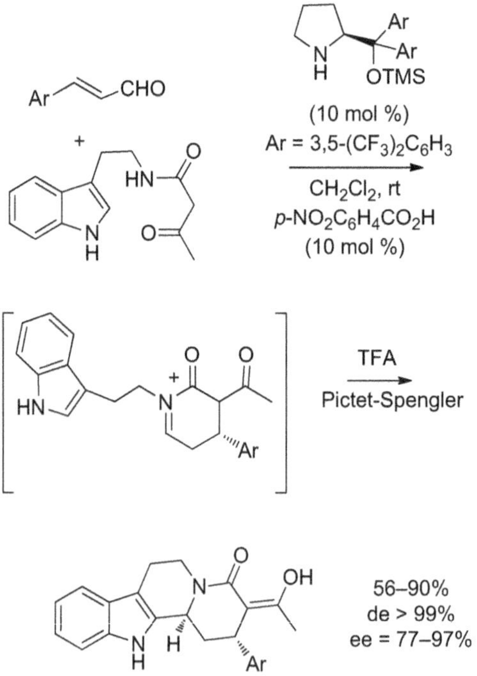

Scheme 3.60 Synthesis of quinolizidines through iminium activation mode.

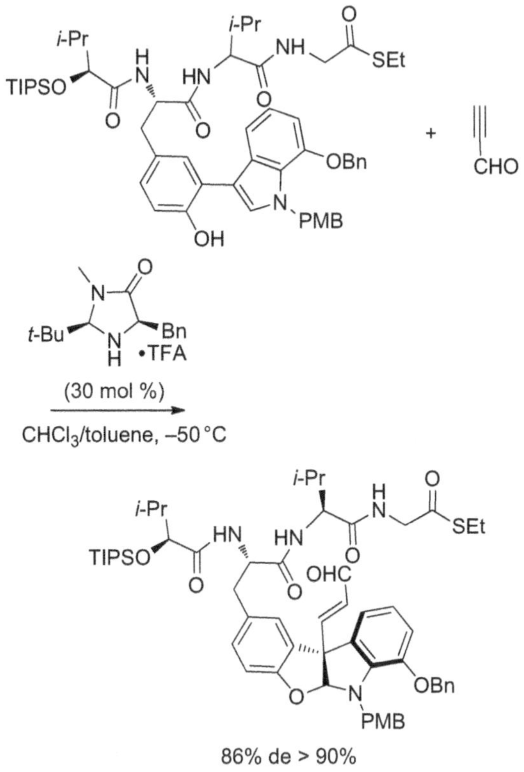

Scheme 3.61 Synthesis of a furanoindoline through iminium activation mode.

acceptor, was catalysed by a chiral imidazolidinone salt, affording the corresponding chiral tetracyclic densely functionalised domino product in 86% yield and diastereoselectivity of >90% de.

On the other hand, a novel general methodology based on iminium activation mode was recently developed by Kanger *et al.* to access highly enantiomerically enriched 1,4-dihydropyridines, which are a well-known class of biologically active heterocycles as well as analogues of nicotinamide adenine dinucleotide (NADH) coenzymes.[107] These products were generated by reaction of aliphatic and aromatic α,β-unsaturated aldehydes with enaminones (or β-enaminoesters) employed as nucleophiles. Instead of being Michael acceptors, the amine-activated double bond of enaminones (or β-enaminoester) enabled a conjugate addition to α,β-unsaturated aldehydes *via* iminium intermediates, followed by proton transfer and spontaneous iminium–enamine transformation, as depicted in Scheme 3.62. The following cascade of reactions involved the hydrolysis of the iminium intermediates, leading to an intramolecular cyclisation, which gave rise to six-membered heterocycles. After a proton transfer had taken place, these intermediates dehydrated to give the energetically more favoured final 1,4-dihydropyridines. This cascade of

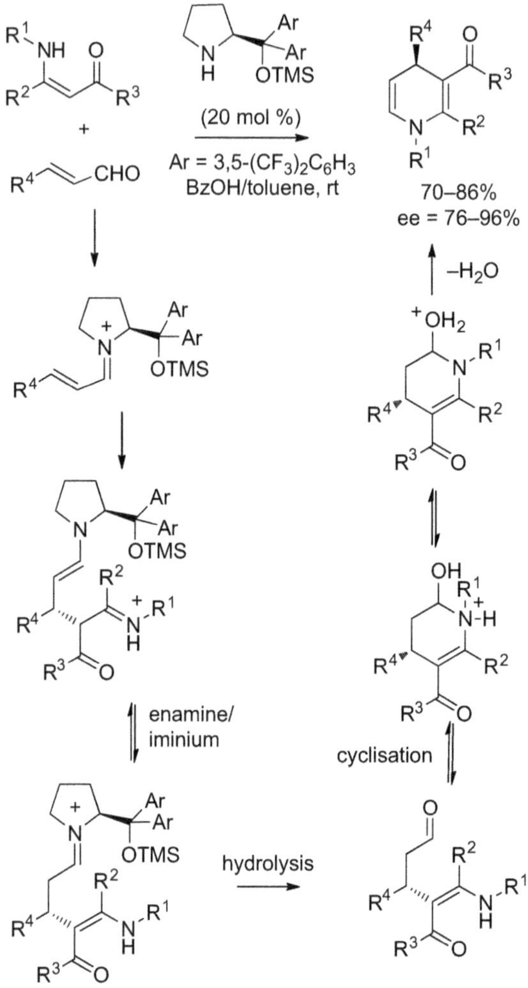

Scheme 3.62 Synthesis of 1,4-dihydropyridines through iminium activation mode.

reactions was induced by bulky 3,5-ditrifluoromethylphenylprolinol trimethylsilyl ether combined with benzoic acid as an additive, and provided good to excellent enantioselectivities of up to 96% ee and high yields of up to 86%. The process had a wide scope, affording a range of 1,4-dihydropyridines with various substituents at four different positions.

In 2010, Zlotin *et al.* demonstrated that a novel chiral ionic liquid bearing an α,α-diphenylprolinol unit (Scheme 3.63) was able to act as an efficient organo-catalyst in asymmetric domino reactions between α,β-unsaturated aldehydes and N-protected hydroxylamines, involving an aza-Michael addition followed by an intramolecular hemiacetalisation reaction.[108] Indeed, the corresponding 5-hydroxy-3-arylisoxazolidines were achieved in excellent yields of up to 94% and with moderate to high enantioselectivities (64 to >99% ee), as shown in

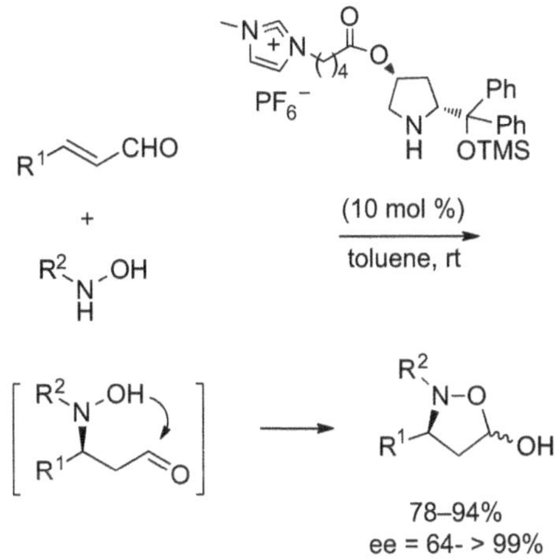

Scheme 3.63 Synthesis of 5-hydroxy-3-arylisoxazolidines catalysed by ionic liquid.

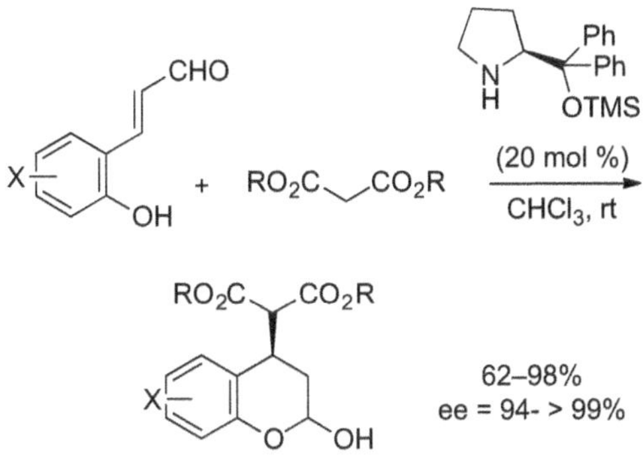

Scheme 3.64 Synthesis of 4-substituted chroman-2-ols through iminium activation mode.

Scheme 3.63. The ionic liquid supported catalyst could be easily recycled and reused at least four times without significant loss of chemical yield or enantioselectivity.

In 2011, Kim *et al.* developed the enantioselective domino Michael–hemiacetalisation reaction of 2-hydroxybenzalacetones with malonates, catalysed by chiral diphenylprolinol trimethylsilyl ether.[109] Evolving through iminium activation mode, this powerful process afforded a range of almost enantiopure 4-substituted chroman-2-ols in good to excellent yields, as shown in Scheme 3.64.

A novel approach to chiral pyrazolidines has been very recently reported by Vicario *et al.*, employing α,β-unsaturated aldehydes and *N,N'*-disubstituted hydrazines as components.[110] It consisted in an enantioselective domino aza-Michael–hemiaminalisation reaction catalysed by a chiral diarylprolinol

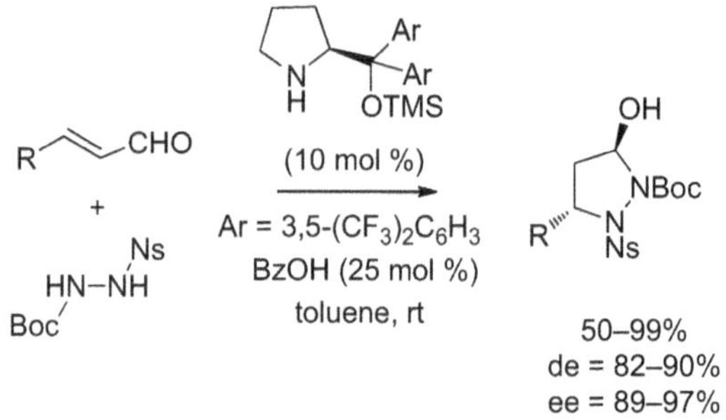

Scheme 3.65   Synthesis of pyrazolidines through iminium activation mode.

trimethylsilyl ether, providing through iminium activation the corresponding pyrazolidin-3-ols in moderate to excellent yields, high enantioselectivities of up to 97% ee, and good to high diastereoselectivities, as shown in Scheme 3.65. These chiral products could be further converted into the corresponding pyrazolines through sequential deprotection/dehydration or oxidised to the corresponding pyrazolidin-3-ones.

Dienamine activation mode consists in the formation *in situ* of dienamines from α,β-unsaturated aldehydes or ketones in the presence of a chiral amine organocatalyst.[111] This activation mode has been recently applied by Enders *et al.* to the asymmetric synthesis of functionalised dihydrocoumarins and chromans on the basis of a domino reaction of 2-hydroxy-nitrostyrenes with α,β-unsaturated aldehydes.[112] Indeed, this domino Michael–intramolecular hemiacetalisation reaction evolved through the *in situ* formation of the dienamine of the α,β-unsaturated aldehyde in the presence of chiral diphenylprolinol trimethylsilyl ether, employed as organocatalyst. This dienamine underwent a Michael addition onto 2-hydroxy-nitrostyrene to give the corresponding nitromethane intermediate, which was subsequently submitted to an intramolecular hemiacetalisation to give the domino product. This product was further oxidised by treatment with pyridinium chlorochromate (PCC) to provide the corresponding dihydrocoumarin in good yield and diastereoselectivity combined with excellent enantioselectivity, as shown in Scheme 3.66.

A number of recent works have also involved enones or derivatives as Michael acceptors in asymmetric domino reactions involving Michael and intramolecular heterocyclisation reactions. As an early example, Mukaiyama *et al.* described chiral quaternary ammonium phenoxides derived from cinchona alkaloids and proved them to be useful novel asymmetric organocatalysts.[113] As shown in Scheme 3.67, a chiral cinchonidine-derived catalyst bearing both a sterically hindered N1-9-anthracenylmethyl group and a strongly electron-withdrawing 9-O-3,5-bis(trifluoromethyl)benzyl group was found to be highly effective for the Michael addition of ketene silyl acetals derived from phenyl carboxylates and α,β-unsaturated enones followed by lactonisation. The corresponding optically active 3,4-dihydropyran-2-one

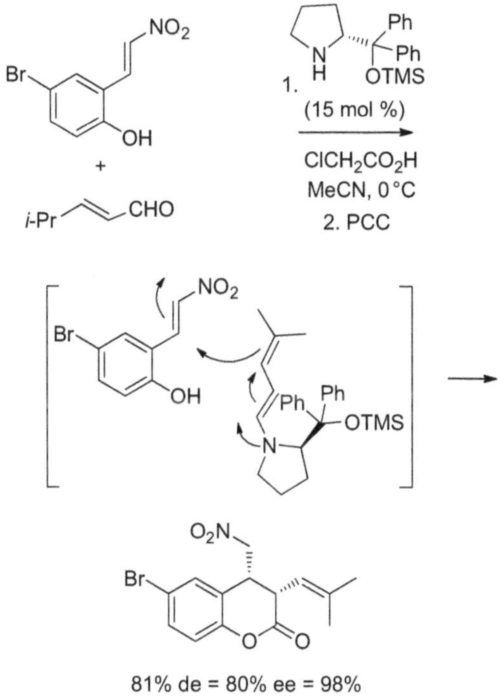

81% de = 80% ee = 98%

Scheme 3.66 Synthesis of a dihydrocoumarin through dienamine activation mode.

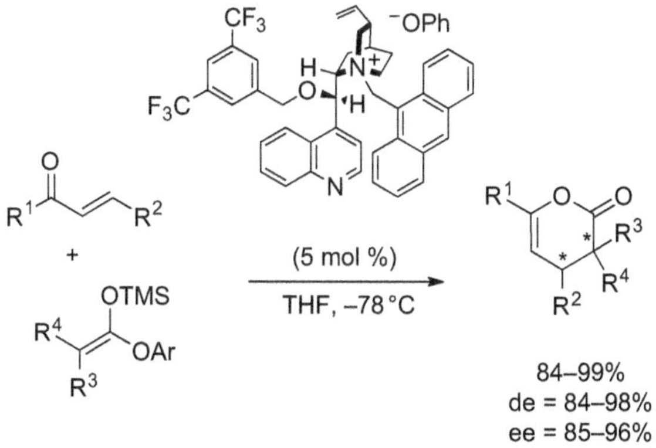

84–99%
de = 84–98%
ee = 85–96%

Scheme 3.67 Synthesis of 3,4-dihydropyran-2-one derivatives through domino Michael–lactonisation reaction.

derivatives were achieved in high yields and excellent control of enantio- and diastereoselectivity.

In 2009, Xie *et al.* developed an enantioselective synthesis of highly functionalised chiral 2-amino-2-chromene derivatives using a domino reaction occurring between α,β-unsaturated enones, such as 2-hydroxybenzalacetones, and malononitrile.[114] This novel process was catalysed by a cinchona alkaloid-derived primary amine, such as 9-amino-9-deoxyepiquinine, in combination with (*R*)-1,1'-binaphth-2,2'-diyl hydrogen phosphate ((*R*)-BDHP). As shown in Scheme 3.68, excellent enantioselectivities of up to 96% ee were obtained in combination with high yields of up to 84% for a range of β-substituted 2-hydroxybenzalacetones. The scope of the reaction could be extended to other 2-hydroxychalcones, providing the corresponding 2-amino-2-chromene derivatives in good yields and enantioselectivities (75–95% ee), as shown in Scheme 3.68. Once again, the iminium activation mode was proposed by the authors to explain the results. As depicted in Scheme 3.68, a chiral primary amine was an effective catalyst for the formation of the iminium from the enone, and an internal acidic proton of OH activated the imine moiety of this iminium intermediate, stabilising the transition state by hydrogen bonding. Subsequently, the higher reactivity of this iminium was used to facilitate the Michael reaction between the enone and malononitrile to produce the corresponding enamine intermediate, and the final products were obtained from the following Knoevenagel condensation ($R^2 = Me$) and the nucleophilic addition of the phenolic OH group on the cyano moiety and proton transfer. On the other hand, when the iminium intermediate bore a bulkier imine moiety ($R^2 =$ Et, Ar), ketones were obtained instead of alkylidene malononitriles, arising from Knoevenagel condensation under the same reaction conditions.

In the same year, Zhao *et al.* reported the first enantioselective domino Michael–hemiketalisation reaction of β,γ-unsaturated α-keto esters with α-substituted cyano ketones.[115] This process employed a novel tyrosine-derived chiral bifunctional thiourea–tertiary amine catalyst in a low catalyst loading of 2 mol%, and provided the corresponding chiral densely functionalised dihydropyrans in high yields and excellent enantioselectivities of up to 96% ee, as shown in Scheme 3.69. These authors have also studied the enantioselective domino Michael–hemiketalisation–dehydration reaction of β,γ-unsaturated α-keto esters with malononitrile which provided, when catalysed by 5 mol% of another chiral bifunctional thiourea bearing a tertiary amine in toluene at − 30 °C, the corresponding highly functionalised 4*H*-pyrans in good yields (50–68%) and enantioselectivities (72–88% ee).[116]

On the other hand, Calter and Wang have reported the synthesis of novel cinchona alkaloid-derived pyrimidines which were shown to be effective chiral catalysts for the domino Michael–hemiketalisation reaction of cyclic diketones with β,γ-unsaturated α-ketoesters.[117] For example, the employment of the chiral cinchona alkaloid-derived pyrimidine depicted in Scheme 3.70 as catalyst allowed the corresponding bicyclic lactols to be achieved as an equilibrating mixture of anomers in good to high yields and excellent enantioselectivities of up to 99% ee (Scheme 3.70). Some of these lactols could be subsequently converted upon treatment with $SOCl_2$ into the corresponding dehydrated dihydropyrans, albeit in low yields.

Another asymmetric domino Michael–hemiacetalisation reaction was reported by Ma *et al.*, in 2010.[118] In this case, it involved the reaction of cyclic

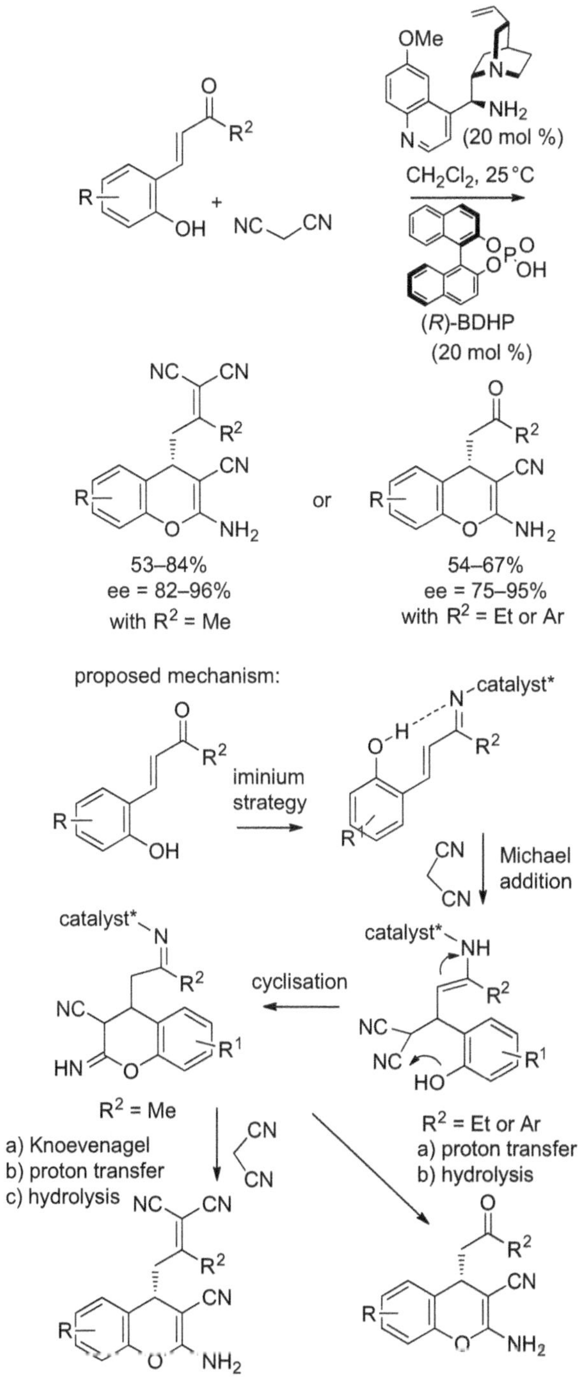

Scheme 3.68 Synthesis of 2-amino-2-chromenes through iminium activation mode.

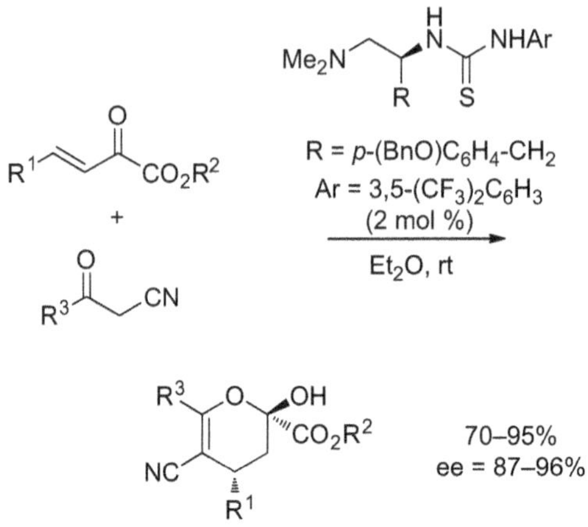

Scheme 3.69 Synthesis of dihydropyrans through domino Michael–hemiketalisation reaction.

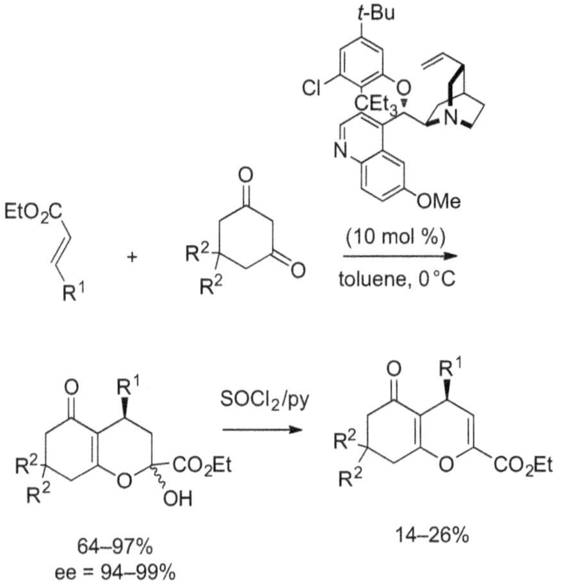

Scheme 3.70 Synthesis of dihydropyrans through domino Michael–hemiketalisation reaction followed by dehydration.

β-oxo aldehydes with β,γ-unsaturated α-keto esters, resulting in the formation of the corresponding spiro-dihydropyrans as mixtures of two anomers in good yields of up to 99% upon catalysis with dihydroquinidine (DHQD)$_2$PYR. The obtained mixtures of anomers were subsequently oxidised by treatment with

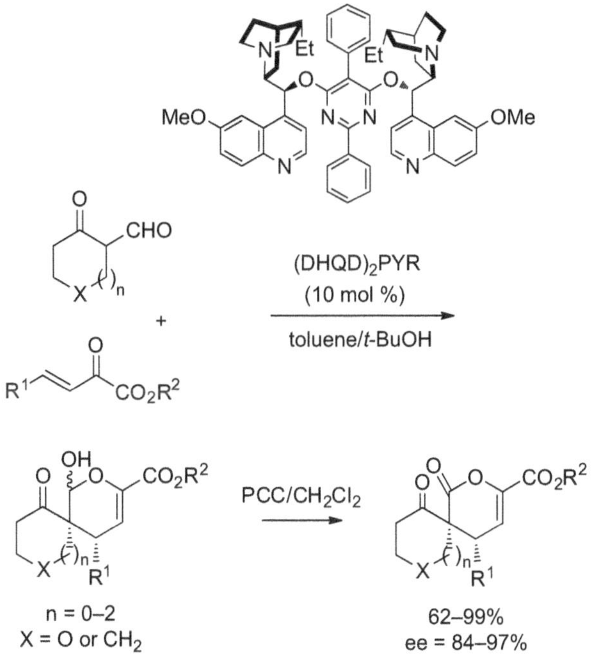

Scheme 3.71 Synthesis of spiro-dihydropyrans through domino Michael–hemiace-talisation reaction.

PCC to give the corresponding α-spirolactones as single diastereomers in high enantioselectivities of up to 97% ee, as shown in Scheme 3.71.

In 2010, Brière *et al.* described the first organocatalytic enantioselective synthesis of biologically important pyrazolines based on a domino aza-Michael–cyclocondensation reaction of chalcones with *N*-Boc hydrazines, catalysed by a chiral cinchona alkaloid.[119] The reaction was performed under phase transfer catalysis, providing the domino products in good to high yields and enantioselectivities of up to 92% ee, as shown in Scheme 3.72.

Given that heterocycles containing a trifluoromethyl group are represen-tatives of a major structure type in agricultural and medicinal chemistry, Shibata *et al.* have developed a novel enantioselective synthesis of trifluoromethyl-substituted 2-isoxazolines on the basis of a domino Michael–cyclisation–dehydration reaction of hydroxylamine with a range of (*E*)-trifluoromethylated enone derivatives.[120] This process, which employed *N*-3,5-bis(trifluoro-methylbenzyl)quinidinium bromide as a chiral phase-transfer catalyst combined with CsOH as a base, provided a series of trifluoromethyl-substituted 2-isoxazolines in high yields and enantioselectivities of up to 94% ee, as shown in Scheme 3.73. The authors have proposed that the oxygen anion of hydroxylamine added in Michael fashion to the enones to furnish the corresponding hydroxyamine intermediates, which was followed by

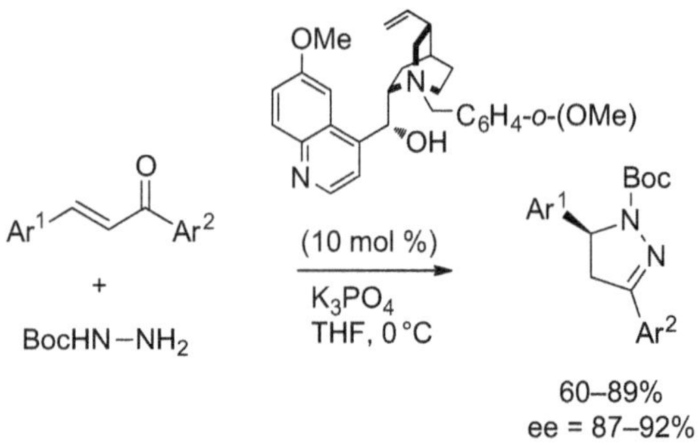

Scheme 3.72 Synthesis of pyrazolines through domino aza-Michael–cyclo-condensation reaction.

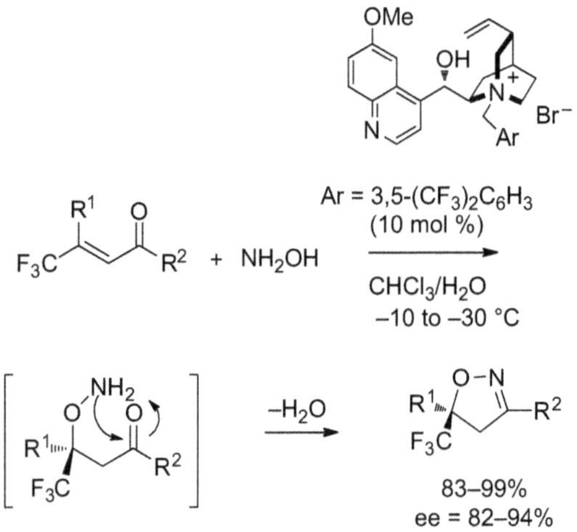

Scheme 3.73 Synthesis of trifluoromethyl-substituted 2-isoxazolines through domino Michael–cyclisation-dehydration reaction.

an intramolecular ring-closure cascade from the nitrogen atom onto the ketone moiety. Subsequently, elimination of water generated the final isoxazolines.

In 2011, Wang *et al.* reported the use of a chiral simple bifunctional indane amine–thiourea catalyst to induce an enantioselective domino Michael–hemiketalisation–retro-Henry sequence of β,γ-unsaturated ketoesters with α-nitroketones.[121] This process provided a new route to chiral 5-nitro-pent-2-enoates, which were obtained in both excellent yields and enantio-selectivities of up to 97% ee, as summarised in Scheme 3.74. The catalyst was supposed by the authors to activate the β,γ-unsaturated ketoester using the amine group and thiourea group through enolisation and hydrogen bond formation, as depicted in Scheme 3.74.

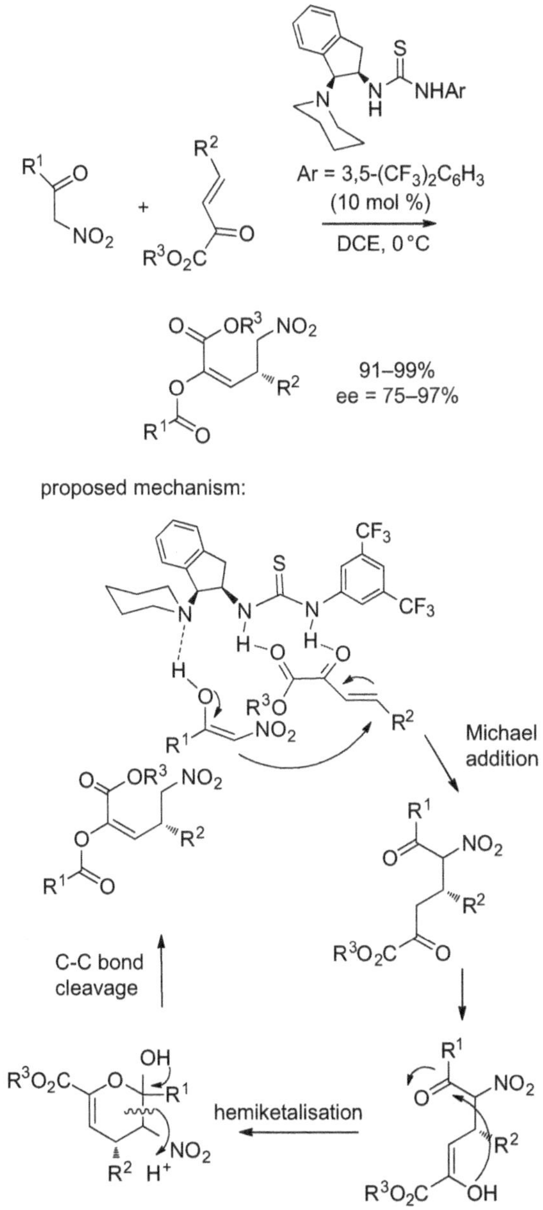

Scheme 3.74 Synthesis of 5-nitro-pent-2-enoates through domino Michael–hemi-ketalisation-dehydration reaction.

In addition, another chiral piperidine-based thiourea–tertiary amine catalyst (Scheme 3.75) was applied by Yan *et al.* to induce the conjugate addition of malononitriles to conformationally restricted dienones, which was followed by an intramolecular cyclisation, providing the corresponding chiral bicyclic pyran

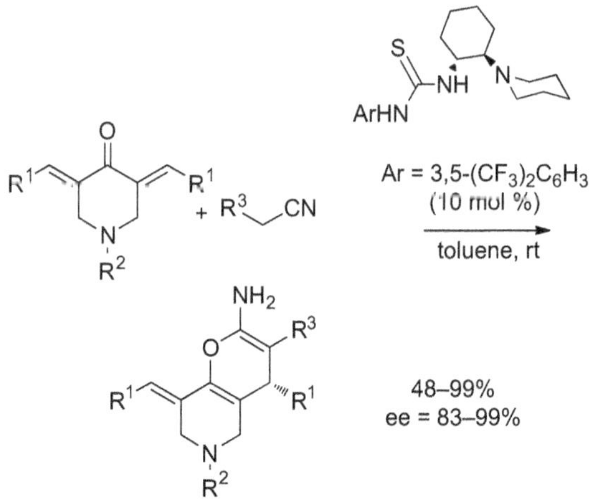

Scheme 3.75 Synthesis of bicyclic pyrans through domino Michael–intramolecular cyclisation reaction.

derivatives in excellent yields and enantioselectivities of up to 99% ee, as shown in Scheme 3.75.[122] It must be noted that these products constituted potential inhibitors against *Mycobacterium tuberculosis*. Furthermore, the results obtained with conformationally restricted dienones were significantly different from those obtained for the reaction of conformationally flexible dienones, which generated upon catalysis with 9-amino-9-deoxyepiquinine the corresponding cyclohexanones in moderate to high yields of up to 99%, and excellent enantioselectivities of up to 99% ee, as previously reported by the same authors.[123]

A nice asymmetric synthesis of trifluoromethyl substituted dihydropyrans was reported by Yan *et al.*, in 2011.[124] It was based on an enantioselective domino Michael–hemiketalisation reaction of ethyl 4,4,4-trifluoroacetoacetate and other trifluoromethyl substituted nucleophiles with β,γ-unsaturated α-ketoesters, catalysed by a chiral quinine-derived thiourea. By using this methodology, a number of chiral trifluoromethyl substituted dihydropyrans bearing three contiguous stereocentres could be produced at remarkable levels of yield, diastereo-, and enantioselectivity, as shown in Scheme 3.76.

On the other hand, a highly enantioselective synthesis of chiral multi-substituted polyfunctionalised 2,3-dihydropyrroles was described by Wang *et al.*, in 2011.[125] This unique approach was catalysed by a novel tertiary amine–thiourea based on rosin and evolved through enantioselective domino Michael–cyclisation reaction of 2-aminomalonates with α,α-dicyanoolefins. These densely functionalised heterocyclic compounds were achieved in high yields and enantioselectivities of up to 97% ee, as shown in Scheme 3.77.

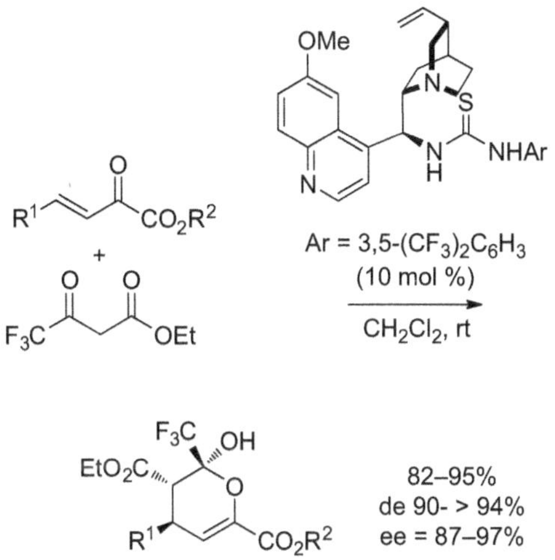

Scheme 3.76 Synthesis of trifluoromethyl substituted dihydropyrans through domino Michael–hemiketalisation reaction.

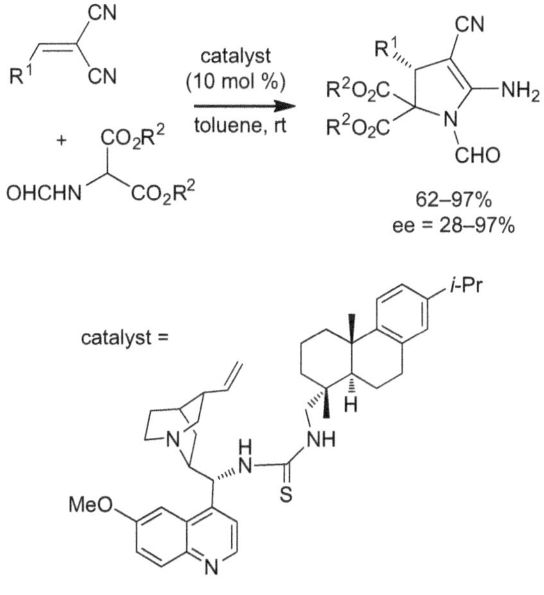

Scheme 3.77 Synthesis of functionalised 2,3-dihydropyrroles through hydrogen bonding activation.

Asymmetric domino Michael–hemiketalisation reactions have also been developed starting from nitroalkenes as Michael acceptors. As an example, Chandrasekhar *et al.* have reported an enantioselective synthesis of chiral cycloalkane fused tetrahydropyrans on the basis of the reaction of

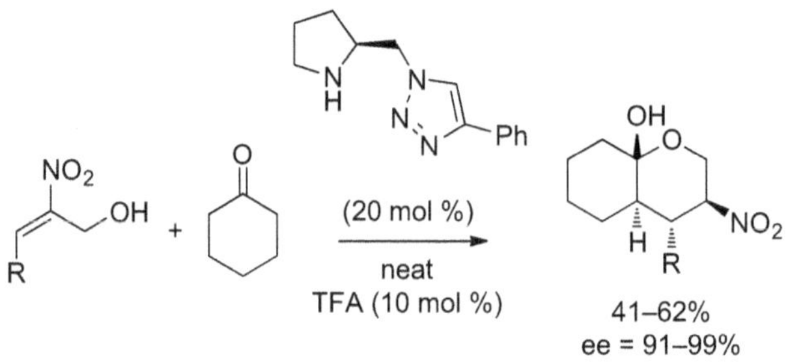

Scheme 3.78 Synthesis of cycloalkane fused tetrahydropyrans through enamine activation mode.

cyclohexanone with various nitroalkenes.[126] This process was induced through enamine-activation mode by chiral pyrrolidine-triazole, affording the corresponding domino Michael–hemiketalisation products in moderate to good yields and high to excellent enantioselectivities of up to 99% ee, as shown in Scheme 3.78.

In the same area, Gong et al.[127] and Hong et al.[128] have independently reported asymmetric domino Michael–hemiacetalisation reactions of 2-hydroxynitrostyrenes with aliphatic aldehydes, catalysed by simple chiral diphenylprolinol trimethylsilyl ether combined with benzoic acid in 95% EtOH or water only as solvent. These reactions afforded, after a subsequent oxidation with PCC, the corresponding 3,4-disubstituted dihydrocoumarins as a mixture of *cis* and *trans* diastereomers. In all cases of the substrates studied, the *cis* diastereomers were obtained as major products with moderate to good diastereoselectivities of up to 76% de combined with high yields and high to excellent enantioselectivities of up to >99% ee, as shown in Scheme 3.79. Interestingly, the reaction of 2-hydroxynitrostyrenes with glutaraldehyde provided the corresponding tetrahydro-6*H*-benzo-[c]chromen-6-ones with excellent enantioselectivities of >99% ee combined with moderate to good *cis*-diastereoselectivities of up to 76% de (Scheme 3.79). Similar reactions have also been investigated by Ramachary et al. using the same catalyst, which afforded the corresponding enantiomeric 3,4-disubstituted dihydrocoumarins in good to high yields (65 to >95%) and excellent *cis*-diastereo- and enantioselectivities of up to >99% de and >99% ee, respectively. In this case, the reactions were performed in dichloromethane as solvent in the presence of 20 mol% of catalyst, combined with benzoic acid as co-catalyst.[129]

Furthermore, Enders et al. developed, in the same year, an enantioselective domino Michael–hemiketalisation reaction of 2-hydroxynitrostyrenes with acyclic β-keto esters to afford the corresponding functionalised chiral 4*H*-chromenes, after subsequent dehydration by treatment of the domino products with PTSA (where PTSA is *p*-toluenesulfonic acid).[130] As shown in

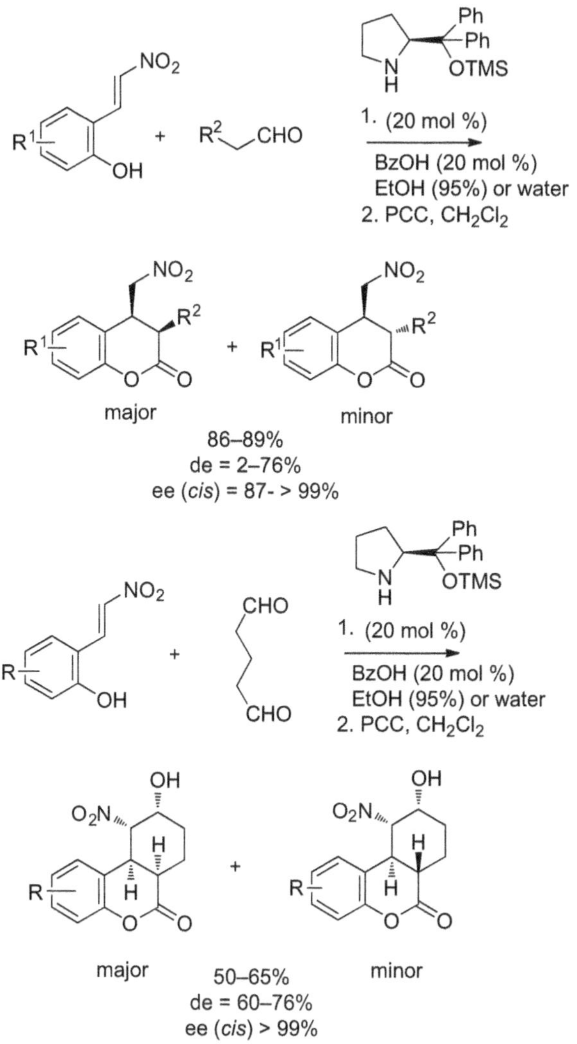

Scheme 3.79 Synthesis of 3,4-disubstituted dihydrocoumarins and tetrahydro-6*H*-benzo[*c*]chromen-6-ones through enamine activation mode.

Scheme 3.80, this remarkable process was catalysed by a chiral thiourea derived from norpseudoephedrine, providing the oxidised domino products in high yields and moderate to excellent enantioselectivities.

In 2012, these authors investigated the related domino Michael–hemiketalisation reaction of 2-hydroxynitrostyrenes with cyclic β-keto esters as well as cyclic 1,3-diketones.[131] A chiral quinine thiourea was demonstrated to be the most effective catalyst to afford the corresponding tricyclic polyfunctionalised

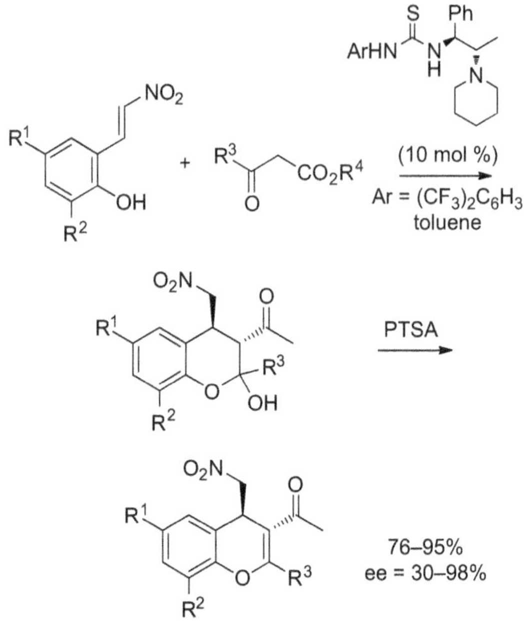

Scheme 3.80 Synthesis of 4*H*-chromenes through domino Michael–hemiketalisation reaction followed by dehydration.

chromans, after subsequent dehydration. As shown in Scheme 3.81, a range of tricyclic complex products were achieved in good to excellent yields and high to excellent enantioselectivities of up to 99% ee.

In the same area, chiral 1′,3-spiro-2′-oxocyclohexan-3,4-dihydrocoumarins have been achieved by Hong *et al.* on the basis of a remarkable enantioselective domino Michael–hemiacetalisation reaction of 2-hydroxynitrostyrenes with 2-oxocyclohexanecarbaldehyde, catalysed by a chiral bifunctional thiourea bearing a tertiary amine.[132] As shown in Scheme 3.82, the corresponding domino products were further oxidised by treatment with PCC to give the corresponding lactones, which were obtained in good to high yields and excellent diastereo- and enantioselectivities of up to >90% de, and >99% ee, respectively.

Ma *et al.* have recently reported an organocatalytic approach to chiral polysubstituted piperidines through enantioselective Michael addition of various aldehydes onto trisubstituted nitroalkenes, followed by intramolecular hemiaminalisation catalysed by chiral diphenylprolinol trimethylsilyl ether associated with benzoic acid as an additive.[133] This domino reaction performed in water allowed four contiguous stereocentres to be formed in excellent enantioselectivities of up to >99% ee, combined with high yields and diastereoselectivities of up to >90% de, as shown in Scheme 3.83. The authors have

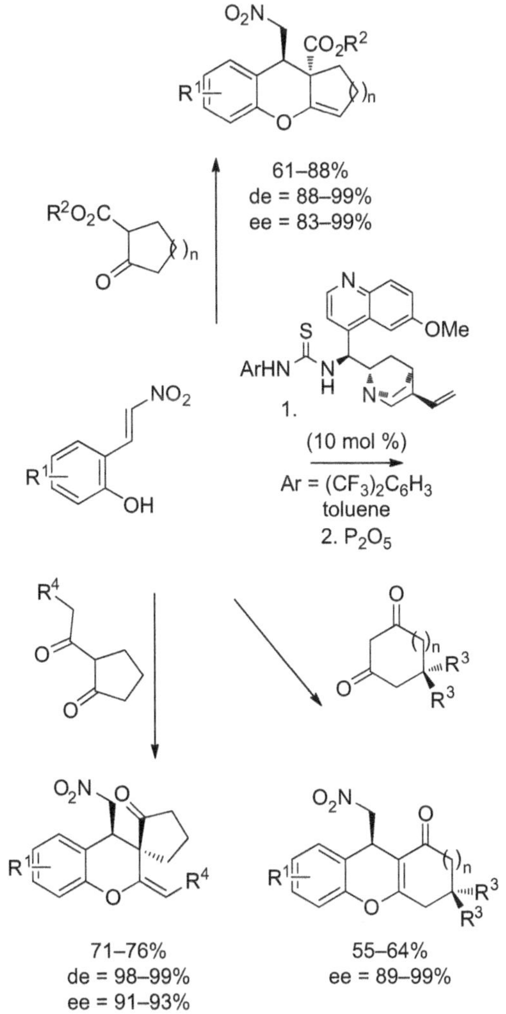

Scheme 3.81 Synthesis of tricyclic chromans through domino Michael–hemiketali-sation reaction followed by dehydration.

demonstrated that the intramolecular hydrogen bond between the nitro and the NHX groups of the alkene seemed to play an important role in activating the nitro group.

In 2011, Enders *et al.* developed a highly efficient enantioselective domino Michael–hemiaminalisation reaction of aldehydes and (*E*)-2-(2-nitrovinyl)-1*H*-indoles, which provided the corresponding almost diastereo- and enantiopure 1*H*-pyrrolo[1,2*a*]indol-3(2*H*)-ols in moderate to good yields

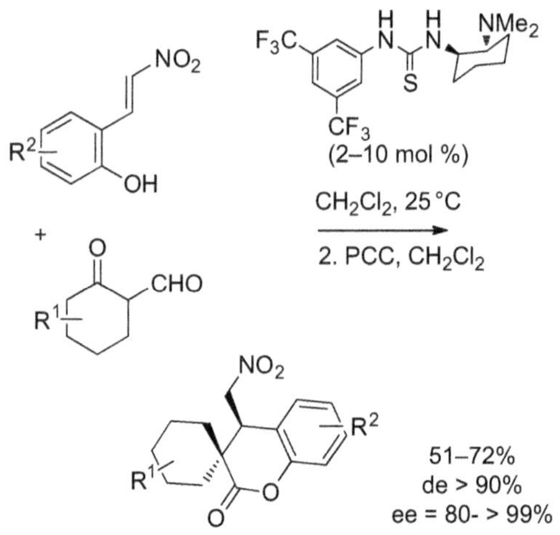

Scheme 3.82 Synthesis of 1′,3-spiro-2′-oxocyclohexan-3,4-dihydrocoumarins through domino Michael–hemiacetalisation reaction followed by oxidation.

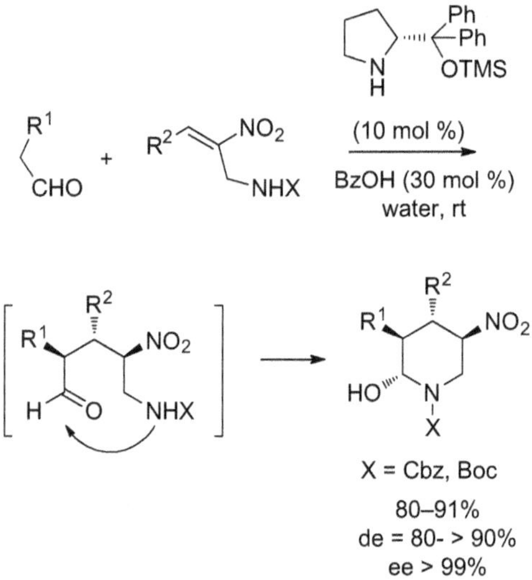

Scheme 3.83 Synthesis of polysubstituted piperidines through enamine activation mode.

(Scheme 3.84).[134] The domino products obtained through catalysis by chiral diphenylprolinol trimethylsilyl ether in the presence of acetic acid as an additive were subsequently oxidised by treatment with PCC to give the corresponding 1*H*-pyrrolo[1,2*a*]indol-3(2*H*)-ones.

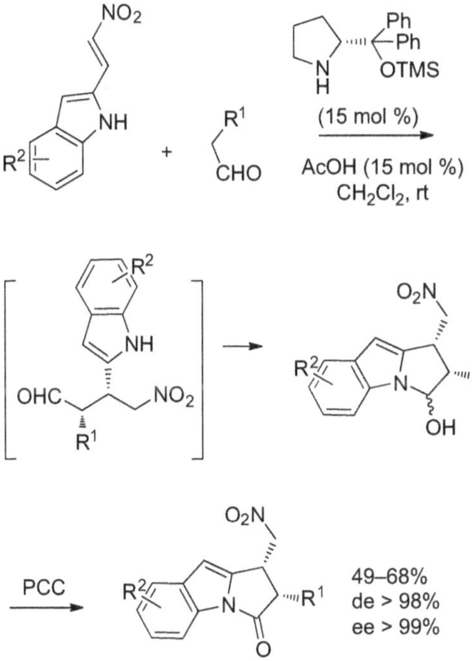

Scheme 3.84 Synthesis of 1*H*-pyrrolo[1,2*a*]indol-3(2*H*)-ones through domino Michael–hemiaminalisation reaction.

3.2.1.4 Domino Michael–Intramolecular Alkylation Reactions

The groups of Cordova and Wang have developed, at nearly the same time, asymmetric organocatalytic domino reactions involving a Michael addition followed by an intramolecular alkylation for the synthesis of chiral functionalised cyclopropanes.[135] Indeed, highly diastereo- and enantioselective cyclopropanation of α,β-unsaturated aldehydes and 2-bromomalonates was achieved on the basis of Michael addition followed by α-alkylation of the enamine intermediate to furnish the final cyclopropanes. The domino process, evolving through iminium–enamine activation mode, was catalysed by chiral diphenylprolinol trimethylsilyl ether, which provided the best results with aromatic enals, leading to the corresponding cyclopropanated aldehydes in total *trans*-diastereoselectivity, good to high yields (50–87%), and excellent enantioselectivities (93–99% ee), as shown in Scheme 3.85. On the other hand, when aliphatic α,β-unsaturated aldehydes were used, the *trans/cis* ratio diminished up to 90:10–94:6, while maintaining the high enantiocontrol.

In 2009, Rios *et al.* reported a nice variation of this domino process, employing 2-bromoketoesters instead of 2-bromomalonates. This change allowed the synthesis of chiral cyclopropanes containing a quaternary stereocentre with high diastereocontrol of up to >92% de, high yields, and excellent

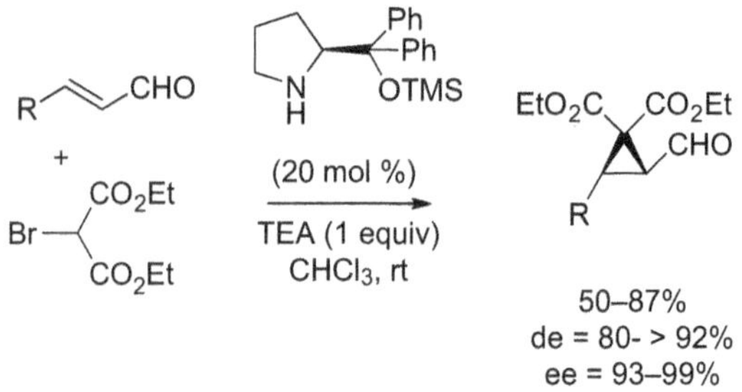

Scheme 3.85 Synthesis of cyclopropanes through iminium–enamine activation mode.

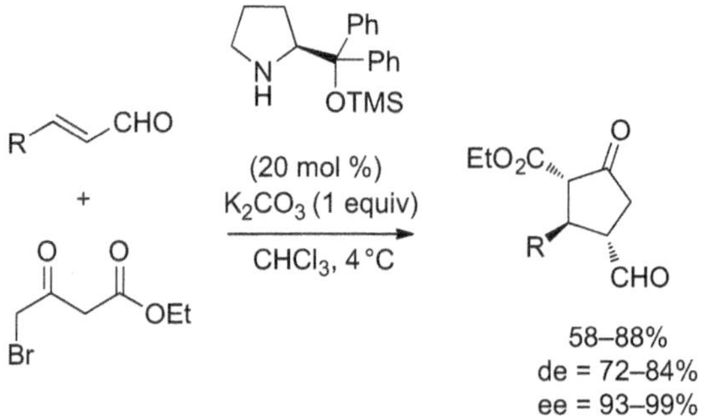

Scheme 3.86 Synthesis of cyclopentanones through iminium–enamine activation mode.

enantioselectivities of up to 99% ee.[136] In 2008, Cordova *et al.* reported a novel nitrocyclopropanation of α,β-unsaturated aldehydes employing bromonitromethane.[137] The domino process was catalysed by 20 mol% of chiral diphenylprolinol trimethylsilyl ether in the presence of 1 equivalent of a base, such as TEA, and afforded the corresponding chiral 1-nitro-2-formylcyclopropane derivatives in moderate to good yields (29–63%), zero to moderate diastereoselectivities of up to 50% de, and high enantioselectivities of up to 99% ee. Employing the same domino Michael–α-alkylation reaction, these authors developed an enantioselective synthesis of cyclopentanones.[138] Indeed, replacing 2-bromomalonate by 4-bromoacetoacetate, under the effect of 20 mol% of the same catalyst combined with one equivalent of potassium carbonate, the cyclopentanones containing three novel stereocentres were achieved in good to high yields, good diastereoselectivities and excellent enantioselectivities, as shown in Scheme 3.86. In 2010, de Figueiredo *et al.* revisited the previously reported enantioselective cyclopropanation of α-substituted α,β-unsaturated aldehydes with bromomalonate through a domino Michael–alkylation reaction catalysed by simple chiral diphenylprolinol trimethylsilyl ether.[139] This method allowed the efficient formation of chiral cyclopropanes bearing a quaternary stereogenic centre at the α-position of the aldehydes by using iminium–enamine activation mode, and gave a nice extension to this reaction with very good yields of up to 81% and excellent enantioselectivities of up to 97% ee. It is interesting to note the ambivalent character of the bromomalonate in this process, which successively played the role of both nucleophile and electrophile.

Finally, Kim *et al.* have reported an enantioselective domino intramolecular 1,5-hydride transfer–ring closure reaction catalysed by a chiral diarylprolinol triethylsilyl ether associated with an equimolecular quantity of (–)-camphorsulfonic acid (CSA).[140] Indeed, the intramolecular reaction of a series of cinnamaldehyde derivatives provided the corresponding ring-fused tetrahydroquinolines in moderate to excellent enantioselectivities of up to 99% ee, combined with generally moderate yields and diastereoselectivities. A remarkable result was obtained for a tetrahydroisoquinoline. As shown in Scheme 3.87, the catalyst activated the substrate through the formation of an iminium, which was then submitted to an intramolecular hydride transfer leading to an enamine intermediate. This intermediate subsequently cyclised to form the final tetrahydroquinoline.

In addition to enals, enones or derivatives have also been successfully employed in asymmetric cyclopropanations through domino Michael–intramolecular alkylation reactions. As an early example, Ley *et al.* reported the enantioselective organocatalytic intermolecular cyclopropanation reaction of 2-cyclohexen-1-one with bromonitromethane, catalysed by chiral 5-(pyrrolidin-2-yl)-1*H*-tetrazole employed at 15 mol% of catalyst loading.[141] The corresponding chiral cyclopropane was achieved in 80% yield and good enantioselective control (77% ee). In 2008, the scope of this reaction was expanded using a variety of cyclic and acyclic enones.[142] In 2009, Wang *et al.* reported the enantioselective synthesis of functionalised nitrocyclopropanes through domino Michael–alkylation reaction of bromonitromethane with various cyclic α,β-unsaturated ketones.[143] This process, induced by chiral amino-9-deoxyepiquinine combined with *p*-methyl mandelic acid as co-catalyst in the presence of *N*-methyl morpholine (NMM) as base, allowed complete diastereoselectivity and high levels of enantioselectivities of up to >99% ee to be reached through iminium–enamine activation mode (Scheme 3.88).

In 2011, Lattanzi *et al.* developed the first enantioselective cyclopropanation to spirocyclopropanes from the domino Michael–alkylation reaction of 2-arylidene-1,3-indanediones and dimethyl bromomalonate, using chiral diphenylprolinol trimethylsilyl ether as catalyst and potassium carbonate as an additive.[144] The chiral spirocyclopropanes were achieved in high yields (73–96%) and moderate to good enantioselectivities (70–85%). In the same year, Bartoli *et al.* reported the enantioselective nitrocyclopropanation of oxindoles with bromonitromethane, which was induced through hydrogen-bonding activation mode by chiral 9-*epi*-9-thiourea-9-deoxydihydroquinidine (Scheme 3.89) in the presence of NaHCO$_3$.[145] As shown in Scheme 3.89, the corresponding spiro nitrocyclopropyl oxindoles were achieved in good to high

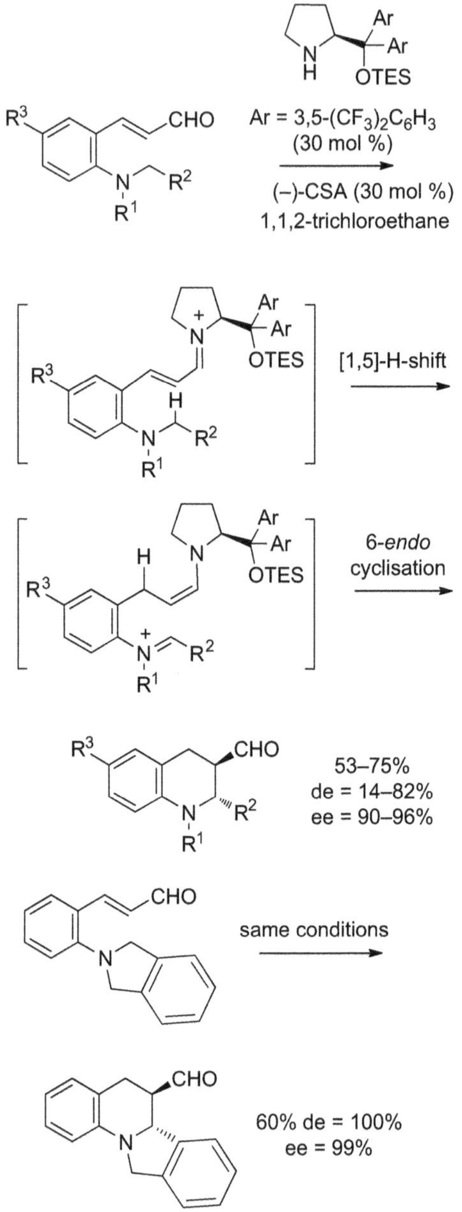

Scheme 3.87 Synthesis of ring-fused tetrahydroquinolines through domino intramolecular 1,5-hydride transfer–ring closure reaction.

yields, moderate to high diastereoselectivities, and excellent enantioselectivities
of up to 98% ee.

An intramolecular version of enantioselective domino Michael–intramolecular alkylation reaction was developed by Gaunt *et al.* as a rare example

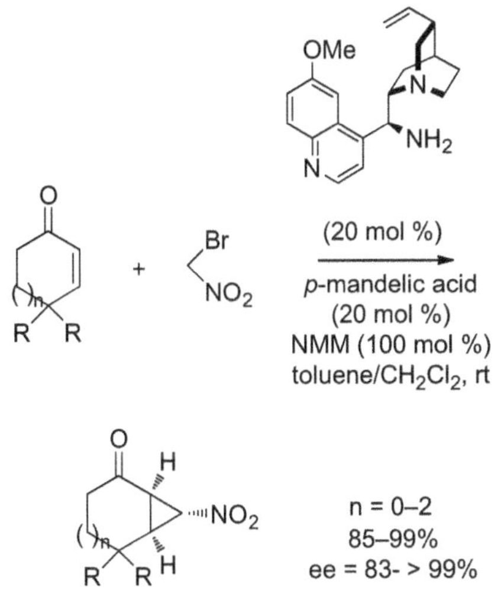

Scheme 3.88 Synthesis of nitrocyclopropanes through iminium–enamine activation mode.

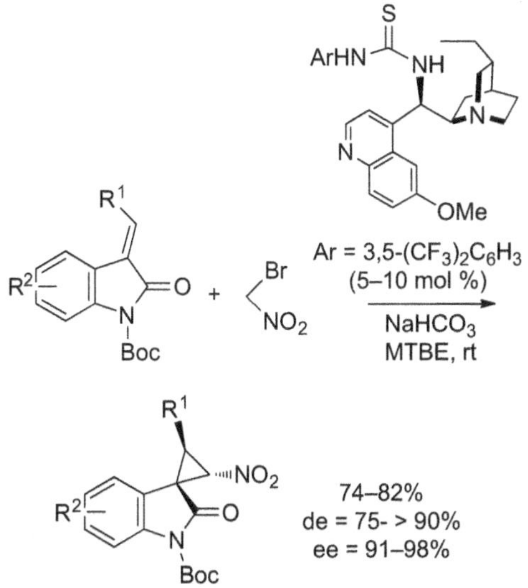

Scheme 3.89 Synthesis of spiro nitrocyclopropyl oxindoles through hydrogen-bonding activation mode.

with an enantioselective intramolecular cyclopropanation using chiral modified cinchona alkaloid organocatalysts.[146] This domino Michael–intramolecular alkylation reaction afforded chiral functionalised [4.1.0]-bicycloheptanes in high yields and excellent enantioselectivities, as shown in Scheme 3.90. The

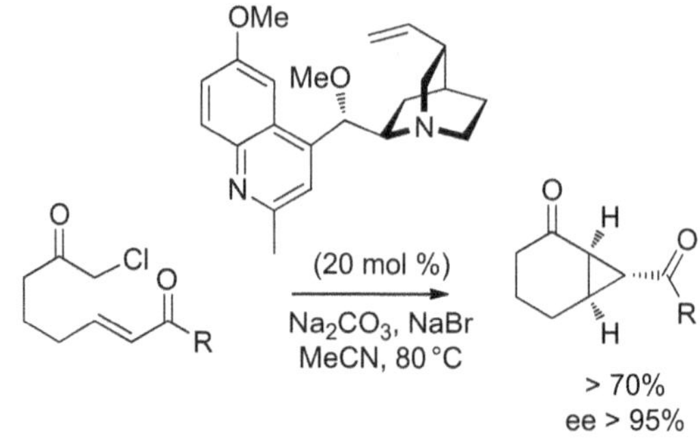

Scheme 3.90 Synthesis of [4.1.0]-bicycloheptanes.

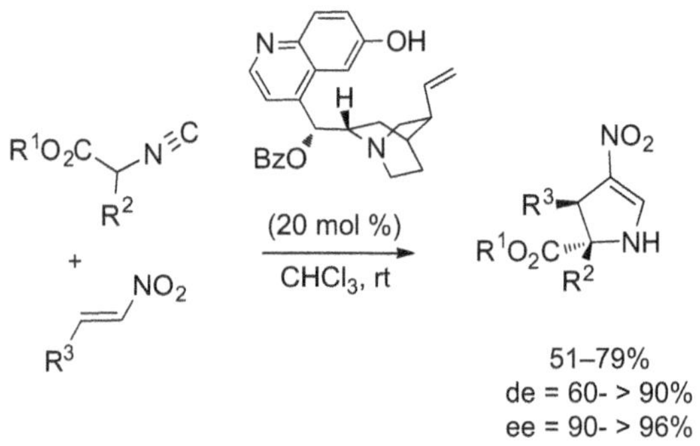

Scheme 3.91 Synthesis of dihydropyrroles.

novel catalysts contained an alkyl substituent at the C-2' position, thus preventing the quinoline nitrogen atom from interfering in the reaction.

In addition to α,β-unsaturated aldehydes, α,β-unsaturated ketones and derivatives, nitroalkenes have also been used as Michael acceptor components in enantioselective domino Michael–intramolecular alkylation reactions. For example, Yan *et al.* developed, in 2009, a novel methodology for the cyclopropanation of nitroalkenes, based on the addition of 2-bromomalonates to nitroalkenes catalysed by chiral cinchona alkaloids.[147] The domino process proceeded with good yields (47–78%), and almost complete diastereo- and enantioselectivities in all cases of the substrates studied. In another context, Gong *et al.* have developed an enantioselective formal [3 + 2] cycloaddition reaction of isocyanoesters to nitroolefins catalysed by chiral cinchona alkaloids.[148] In this approach, isocyanoesters underwent a Michael addition to nitroalkenes, and a subsequent intramolecular alkylation afforded the dihydropyrroles after protonation. As shown in Scheme 3.91, the chiral highly functionalised domino products were achieved in good yields, good to excellent diastereoselectivities (60 to >90% de) and generally excellent enantioselectivity of up to 96% ee.

Another asymmetric domino Michael–alkylation methodology has been independently developed by Xie *et al.*[149] and Rueping *et al.*[150] for the enantioselective synthesis of a range of chiral 2,3-dihydrofurans from reaction of α-bromonitroalkenes with 1,3-dicarbonyl compounds. In both cases, the catalysts were chiral bifunctional thioureas associated with a base, such as DIPEA or tetramethylethylenediamine (TMEDA). For example, the reaction of bicyclic 1,3-dicarbonyl compounds, such as 4-hydroxyl(thio)coumarins, led, in the presence of the chiral thiourea depicted in Scheme 3.92, to the corresponding tricyclic 2,3-dihydrofurans in high yields of up to 98%, excellent diastereoselectivities of up to >99% de, and high enantio-selectivities (82–92% ee). The scope of this process was extended to the synthesis of chiral bicyclic 2,3-dihydrofurans by reaction of α-bromon-itroalkenes with cyclohexane-1,3-diones, which provided comparable diaste-reoselectivities (>99% de) combined with generally excellent yields (96–99%), and high enantioselectivities of up to 89% ee, as shown in Scheme 3.92. The scope of the domino reaction was further extended to acyclic 1,3-dicarbonyl compounds, which afforded the corresponding chiral monocyclic 2,3-dihydrofurans in comparable diastereoselectivities (>99% de), high yields (86–89%), and good enantioselectivities (76–87% ee), as shown in Scheme 3.92.

In 2011, Lu *et al.* described an enantioselective synthesis of 3(2*H*)-furanones based on a domino Michael–alkylation reaction of ethyl 4-bromoacetoacetate and nitroolefins, induced by a novel chiral tertiary amine thiourea catalyst derived from L-threonine.[151] As shown in Scheme 3.93, the biologically important 3(2*H*)-furanones were achieved in good to high yields and generally high enantioselectivities of up to 96% ee.

3.2.1.5   Domino Michael–aza-Henry Reactions

The Henry reaction has been also associated with the Michael reaction in successful asymmetric domino sequences.[152] For example, Jorgensen *et al.* reported, in 2007, an asymmetric organocatalytic synthesis of pentasubstituted cyclohexanes based on a Michael reaction of nitroalkanes with α,β-unsaturated aldehydes, followed by an intramolecular Henry reaction.[153] This domino process, evolving through iminium activation, was catalysed by a chiral diarylprolinol trimethylsilyl ether in the presence of DABCO and led to the formation of pentasubstituted cyclohexanols, with remarkable control over five contiguous stereocentres. As shown in Scheme 3.94, the domino products were achieved in moderate to good yields and diastereoselectivities of up to 72% de, combined with high enantioselectivities of up to 94% ee.

Later, Ruano *et al.* described a sequence that constituted of an enantios-elective organocatalysed Michael addition of 1,3-diones to α,β-unsaturated aldehydes as a first step, which was followed, after the addition of nitromethane and TBAF, by a domino intermolecular Henry–intramolecular Henry reaction, constituting the second step of the sequence.[154] As shown in Scheme 3.95, this sequential Michael–domino intermolecular Henry–intramolecular Henry

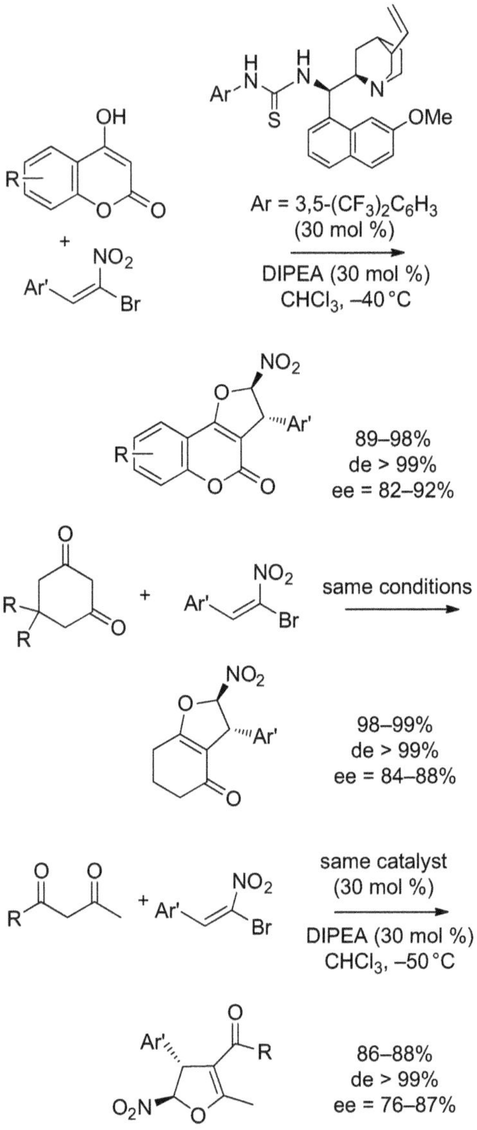

Scheme 3.92 Synthesis of monocyclic, bicyclic, and tricyclic 2,3-dihydrofurans.

reaction induced by the same catalyst produced the corresponding cyclohexanes bearing five stereogenic centres with high enantio- and diastereoselectivities for a wide range of substrates.

In 2010, Lu and Liu developed asymmetric domino aza-Henry reactions by
employing a quinidine-derived tertiary amine–thiourea catalyst.[155] The authors
have shown that installing an electron-withdrawing sulfone group on the amino
group of 2-aminobenzaldehydes allowed for their activation, because the

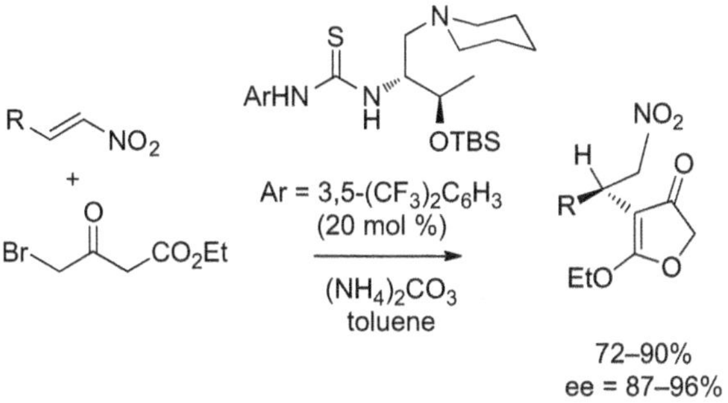

Scheme 3.93   Synthesis of 3(2*H*)-furanones.

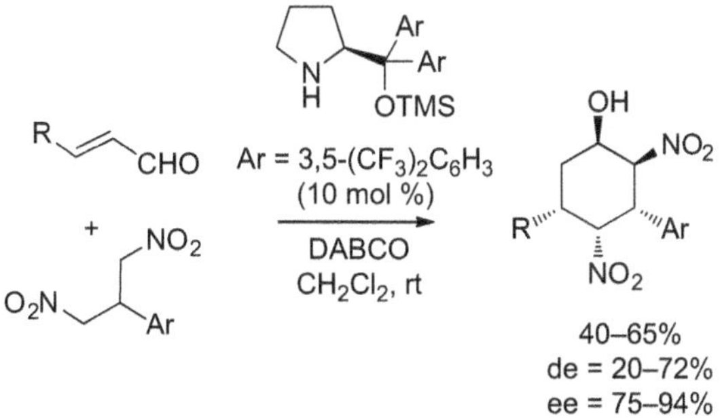

Scheme 3.94   Synthesis of pentasubstituted cyclohexanols.

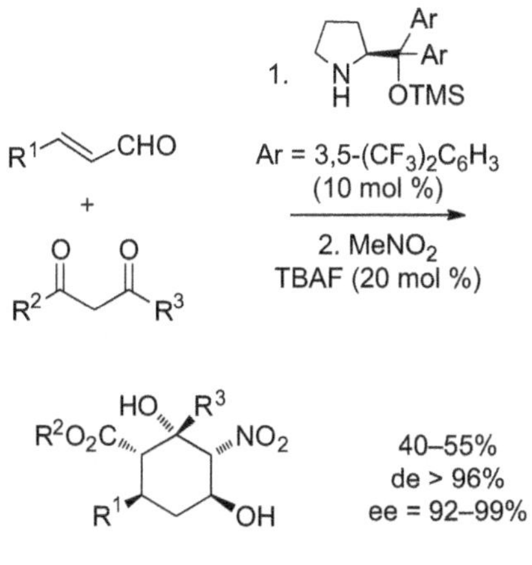

Scheme 3.95   Synthesis of pentasubstituted cyclohexanes.

corresponding sulfonated 3-nitro-1,2-dihydroquinolines were generally obtained in better yields (75–92%), and enantioselectivities (70–90% ee). More recently, Xu *et al.* developed access to 2,3,4-trisubstituted tetrahydroquinolines on the basis of asymmetric domino Michael–aza-Henry reactions involving

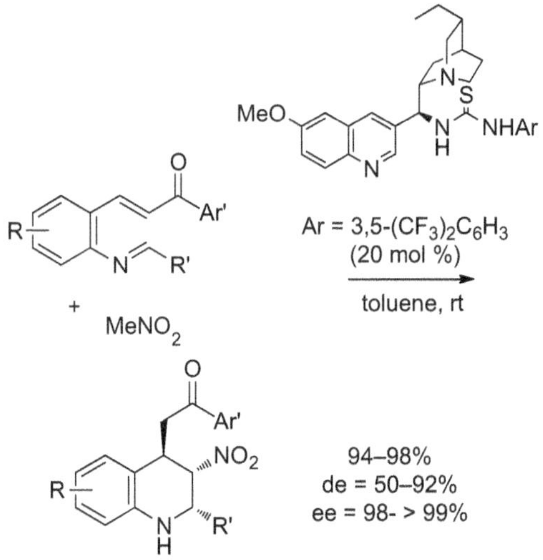

Scheme 3.96   Synthesis of 2,3,4-trisubstituted tetrahydroquinolines.

nitromethane and chalcones.[156] When these reactions were catalysed with a chiral bifunctional thiourea, they furnished the potentially biologically active polysubstituted tetrahydroquinolines in generally both excellent yields (94–98%) and enantioselectivities (98 to >99% ee), combined with moderate to high diastereoselectivities of up to 92% de, as shown in Scheme 3.96.

On the other hand, a number of organocatalytic enantioselective domino Michael–aza-Henry reactions have employed nitroolefins as Michael acceptor components. For example, Takemoto *et al.* reported, in 2008, the use of a chiral thiourea at 10 mol% of catalyst loading in toluene to induce the domino Michael–aza-Henry reaction of α-amino malonate imines with nitroolefins to give the corresponding highly functionalised pyrrolidines with moderate to high yields and good to excellent diastereoselectivities, combined with moderate to high enantioselectivities, as shown in Scheme 3.97.[157]

In 2008, Xu *et al.* developed a novel organocatalytic domino oxa-Michael–Henry reaction of 2-hydroxybenzaldehydes with nitroalkenes, providing the corresponding 3-nitro-2*H*-chromenes in high yields and enantioselectivities of up to 91% ee.[158] The process was efficiently catalysed by a chiral pyrrolidine derivative in the presence of salicylic acid, as shown in Scheme 3.98. In the first step, the alcohol added to the nitroalkene according to a Michael addition, and a subsequent Henry reaction followed by dehydration furnished the final domino products. In 2007, Zhao *et al.* had developed a similar reaction for the synthesis of chiral thiochromanes through domino oxa-Michael–Henry reaction of 2-mercaptobenzaldehydes with nitrostyrenes, catalysed by a simple chiral cupreine.[159] The corresponding chiral 2-aryl-3-nitrothiochroman-4-ols were achieved in high yields (84–97%) with enantioselectivities of up to 86% ee and diastereoselective ratios of up to 78:22.

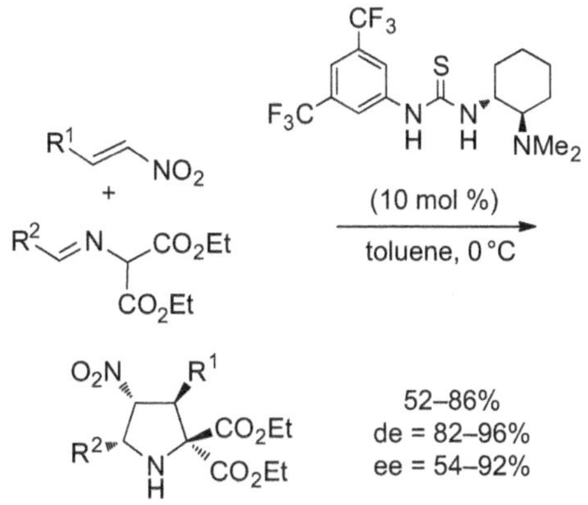

Scheme 3.97 Synthesis of pyrrolidines.

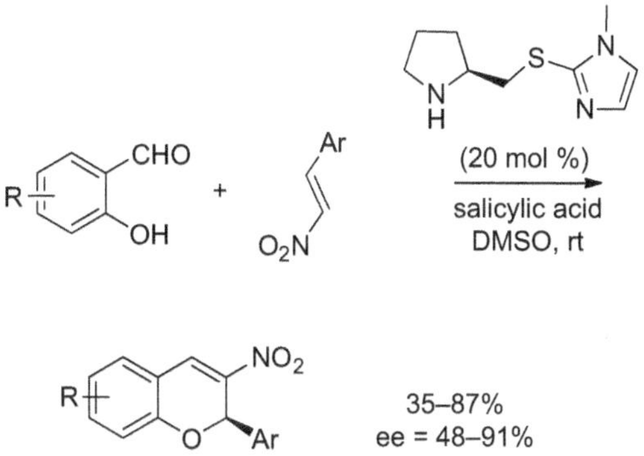

Scheme 3.98 Synthesis of 3-nitro-2*H*-chromenes.

In adition, several groups have developed asymmetric domino Michael–aza-Henry reactions catalysed by chiral thioureas. For example, Xu *et al.* have reported the reaction of 2-aminobenzaldehydes with various nitroolefins upon catalysis with a chiral primary amine thiourea catalyst to provide a variety of 3-nitro-1,2-dihydroquinolines in moderate yields (37–70%) and moderate to high enantioselectivities (52–90% ee).[160] The authors assumed synergistic activation of the reactants through stereoselective covalent activation and hydrogen-bonding interactions. This mechanism was confirmed by electrospray ionisation (ESI) mass spectrometric detection of the reaction intermediates. In another context, a highly diastereo- and enantioselective synthesis of multisubstituted and densely functionalised cyclopentanes bearing four contiguous stereogenic centres was described by Zhong *et al.*, in 2008.[161] It was based on the enantioselective domino Michael–Henry reaction of ethyl 2-acetyl-4-oxo-4-phenylbutanoate and nitroolefins induced by a chiral cinchona alkaloid-derived diamine. As shown in Scheme 3.99, remarkable

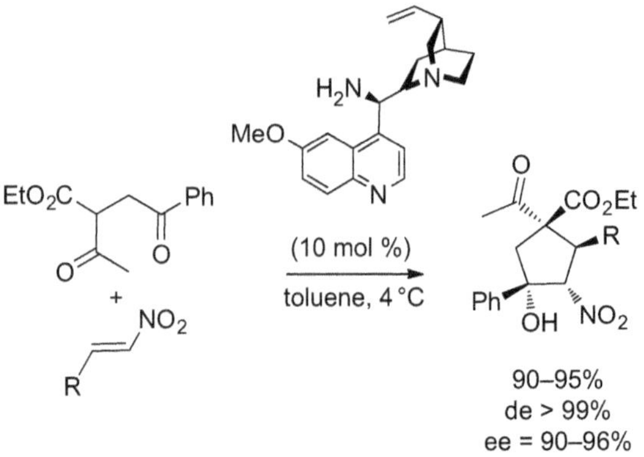

Scheme 3.99 Synthesis of cyclopentanes.

results were obtained because the yields in domino products ranged from 90 to 95%, the enantioselectivities were up to 96% ee, and the diastereoselectivity was >99% de in all cases of the substrates studied. The scope of this methodology was applied to the synthesis of chiral functionalised cyclohexanes from the reaction of ethyl 2-acetyl-5-oxohexanoate and nitroolefins.[162] This enantioselective domino Michael–Henry reaction provided excellent enantioselectivities (97–99% ee) combined with high diastereoselectivities (86–98% de) and high yields (88–94%).

Later, the groups of Zhao[163] and Rueping[164] independently developed a novel asymmetric domino Michael–Henry reaction induced by chiral bifunctional cinchona thioureas, which provided access to a range of chiral synthetically unique and medicinally important bicyclo[3.2.1]octan-8-ones. The reaction occurred between cyclohexane-1,2-diones and nitroalkenes. The best results collected in Scheme 3.100 were obtained by using a chiral quinine-derived thiourea as catalyst. These complex polyfunctionalised products were obtained as mixtures of two diastereomers bearing four stereogenic centres with good diastereoselectivities of up to 90% de, combined with high yields, and enantioselectvities of up to 99% ee. It is interesting to note that methylcyclohexane-1,2-dione was also a good substrate for this reaction. Indeed, given that this compound is not symmetric, it could enolize either at the less substituted 6-position (to give a kinetic enolate) or the more substituted 3-position (to give a thermodynamic enolate). However, only the thermodynamic enolate participated in the reaction, to give a single regioisomer with a good diastereoselectivity of 76% de, as shown in Scheme 3.100. This major diastereomer was isolated in 97% ee and 67% yield. Nevertheless, this process was limited to cyclohexane-1,2-diones, because the attempted reactions of both butane-2,3-dione and 3-methylcyclopentane-1,2-dione failed under the same conditions.

In the same area, Zhong *et al.* have developed highly enantio- and diastereoselective domino Michael–Henry reactions of nitroalkenes with

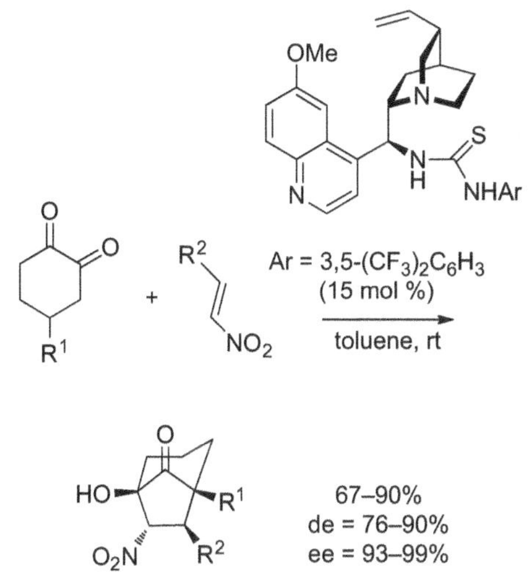

Scheme 3.100 Synthesis of bicyclo[3.2.1]octan-8-ones.

1,3-dicarbonyl compounds, such as 2-carboxymethyl-1,4-cyclohexanedione, which provided a range of other chiral bicyclo[3.2.1]octan-8-ones bearing four stereogenic centres, including two quaternary stereocentres.[165] These products were produced in high yields, and excellent enantioselectivities of up to 96% ee, with a remarkable general diastereoselectivity of >98% de in all cases of the substrates studied, as shown in Scheme 3.101. The process was induced by a chiral cinchona alkaloid thiourea catalyst, which was employed at a low catalyst loading of 5 mol%. Theoretical DFT calculations on the transition states have been carried out to reveal the origins of the stereoselectivities. They have shown that the induction of chirality was achieved by using a novel type of activation mode in which the thiourea group and an acidic proton in the phenyl ring of the catalyst activated the 1,3-dicarbonyl substrate together, and at the same time a tertiary amine activated the nitro group, which promoted the domino reaction smoothly with excellent stereoselectivity (Scheme 3.101).

In 2011, a new type of pyrrolidine-based chiral organocatalyst was developed by Headley *et al.* and found to be very effective for inducing the domino Michael–Henry reaction of nitroolefins with pentane-1,5-dial.[166] Performed in aqueous media and in the presence of benzoic acid as an additive, the domino process provided the corresponding tetrasubstituted and densely functionalised cyclohexanes in high yields and excellent enantioselectivities of >98% ee in all cases of the substrates studied, albeit moderate diastereoselectivities of 50% de (Scheme 3.102). Furthermore, the catalyst could be recycled up to four times resulting in good yields, and up to seven times with good enantioselectivity.

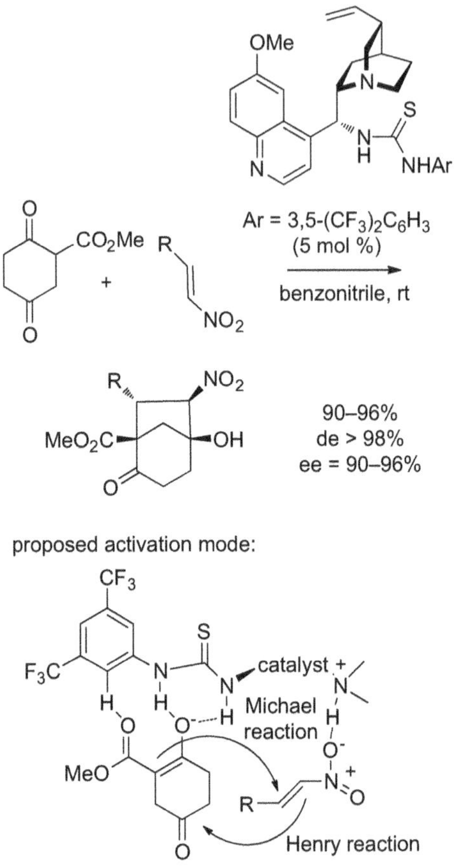

Scheme 3.101 Synthesis of tetrasubstituted bicyclo[3.2.1]octan-8-ones.

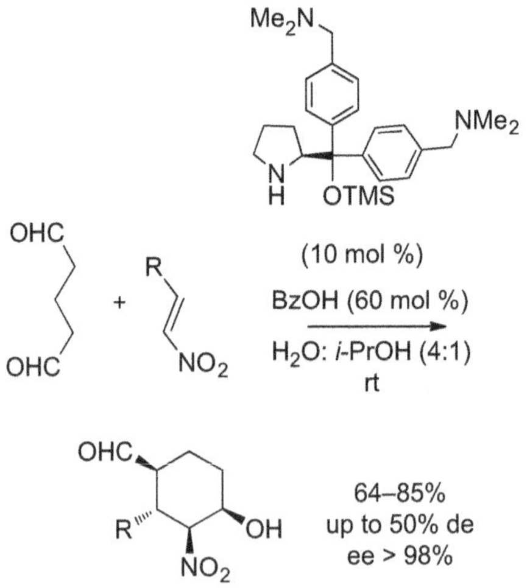

Scheme 3.102 Synthesis of tetrasubstituted cyclohexanes.

3.2.1.6 Domino Michael–Knoevenagel Reactions

An early example of asymmetric domino Michael–Knoevenagel reaction was reported by Jorgensen *et al.* in 2006, providing the synthesis of chiral cyclohexenones from the reaction of *tert*-butyl 3-oxobutyrate and α,β-unsaturated aldehydes, catalysed by chiral diphenylprolinol silyl ether derivative.[167] Soon after, the same authors developed a similar domino reaction starting from 4-chloro-ketoesters.[168] This process furnished highly functionalised epoxycyclohexanone derivatives with good yields, excellent diastereoselectivity of >99% de in all cases of substrates studied, combined with high enantioselectivities (85–97% ee). In this case, the domino Michael–aldol products were directly submitted to an intramolecular S_N2 reaction by treatment with a base, such as K_2CO_3, to give the final epoxycyclohexanone derivatives. In 2009, Hayashi *et al.* developed a highly enantioselective formal [3 + 3] cycloaddition reaction of α,β-unsaturated aldehydes and dimethyl 3-oxopentanedioate catalysed by chiral diphenylprolinol *t*-butyldimethylsilyl ether in combination with benzoic acid as an additive *via* an asymmetric domino Michael–Knoevenagel reaction.[169] This reaction proceeded with good yields and constituted a clean process to afford chiral substituted cyclohexenone derivatives after a subsequent reduction performed with sodium borohydride. As shown in Scheme 3.103, excellent enantioselectivities (95–99% ee) were obtained for a range of aromatic and heteroaromatic aldehydes. In 2011, Pericas *et al.* reported the suitability of polymer-supported diarylprolinol silylethers for these reactions.[170] Indeed, comparable results were achieved with these catalysts immobilized onto polystyrenes both under batch and continuous

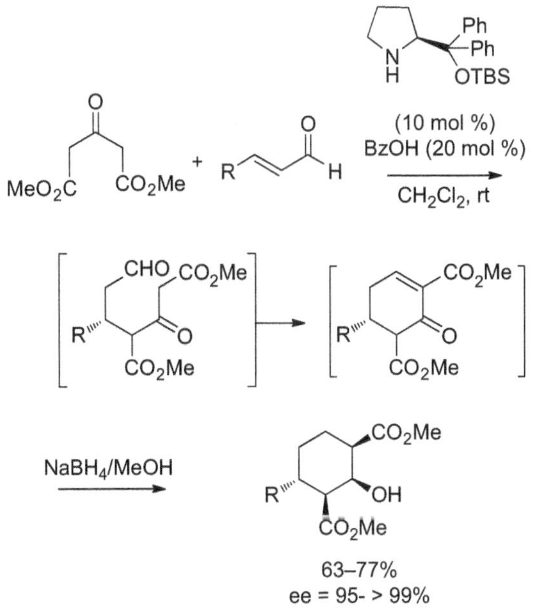

Scheme 3.103   Synthesis of tetrasubstituted cyclohexanols.

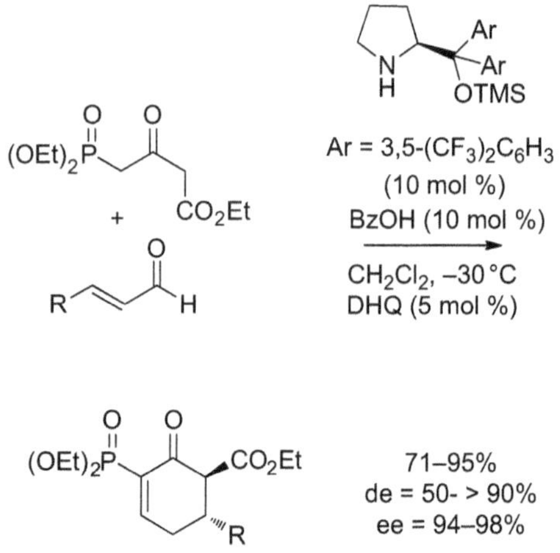

Scheme 3.104 Synthesis of 6-substituted 3-diethoxyphosphoryl-2-oxocyclohex-3-enecarboxylates.

flow conditions. Excellent enantioselectivities of up to >99% ee were obtained for aromatic aldehydes and, moreover, this methodology was extended to aliphatic aldehydes, which provided the corresponding chiral substituted cyclohexanols with enantioselectivities of up to 87% ee (with R = *n*-Bu).

Independent of the work published by Hayashi *et al.*, Jorgensen *et al.* have reported the synthesis of chiral 6-aryl 3-diethoxyphosphoryl-2-oxocyclohex-3-enecarboxylates on the basis of asymmetric domino Michael–Knoevenagel reactions of ethyl 4-diethoxyphosphoryl-3-oxobutanoate with aromatic α,β-unsaturated aldehydes.[171] These reactions were catalysed by a chiral diarylprolinol trimethylsilyl ether in combination with benzoic acid as co-catalyst. Excellent diastereo- and enantioselectivities of up to >90% de and up to 98% ee, respectively, were achieved for the corresponding products in association with high yields in the case of aromatic aldehydes. Surprisingly, aliphatic aldehydes turned out to be unreactive under these conditions. However, the use of an additive such as hydroquinine (DHQ) allowed the corresponding 6-alkylated 3-diethoxyphosphoryl-2-oxocyclohex-3-enecarboxylates to be afforded in excellent enantioselectivities (94–96% ee), albeit in moderate diastereoselectivities (50–60% de), as shown in Scheme 3.104.

3.2.1.7 Domino Michael–(aza)-Morita–Baylis–Hillman Reactions

Since the early example of proline-catalysed asymmetric domino Michael–aldol reaction reported by Bui and Barbas in 2000,[18] a number of asymmetric organocatalysed domino Michael–Morita–Baylis–Hillman reactions have been successfully developed by several groups. In 2006, Hong *et al.* reported an interesting example of iminium–enamine sequence catalysed by unmodified L-proline.[172] This process involved 2 equivalents of crotonaldehyde, which reacted through a domino Michael–aldol reaction to afford the corresponding chiral cyclohexenecarbaldehydes as a 53:47 mixture of two diastereomers obtained in 69% yield (Scheme 3.105). While the diastereoselectivity of the reaction was low (6% de), the two epimers were obtained in high optical purity of 80% and 95% ee, respectively. In order to explain the obtention of the products, the authors have proposed a Michael–Morita–Baylis–Hillman sequence. First of all, proline activated one molecule of crotonaldehyde by iminium formation, and the other molecule of crotonaldehyde, forming a dienamine. Then, this dienamine promoted a conjugated-type addition over the iminium activated crotonaldehyde, forming a novel iminium intermediate. Afterwards, this intermediate underwent an intramolecular

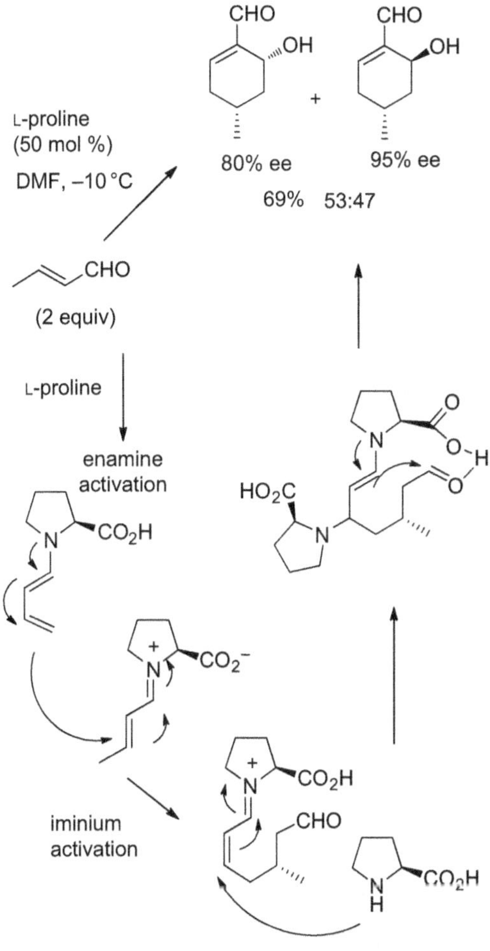

Scheme 3.105 Synthesis of cyclohexenecarbaldehydes.

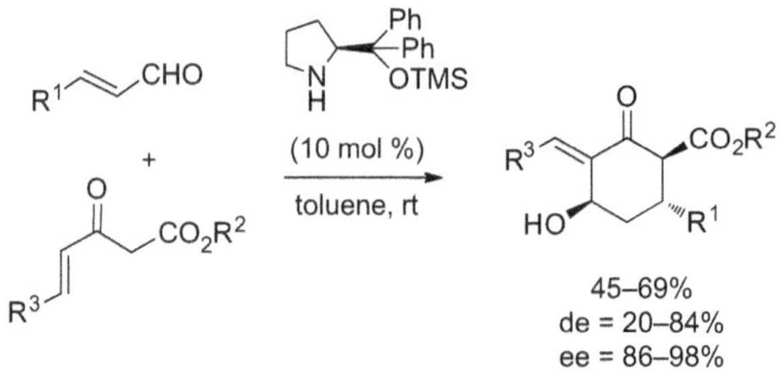

Scheme 3.106 Synthesis of cyclohexanones.

Morita–Baylis–Hillman-like reaction promoted by free proline, providing the final six-membered enal ring. It must be noted, however, that the aldehyde scope of this transformation was rather limited. Unlike crotonaldehyde, all other enals tested under the same reaction conditions gave rise to the corresponding chiral cyclohexadienes *via* formal [4 + 2] reactions.

In 2008, Jorgensen *et al.* reported the enantioselective domino Michael–Morita–Baylis–Hillman reaction of Nazarov reagents with α,β-unsaturated aldehydes, catalysed with chiral diphenylprolinol trimethylsilyl ether in the presence of benzoic acid as an additive.[173] As shown in Scheme 3.106, the corresponding chiral highly substituted cyclohexanones were produced in good yields, low to good diastereoselectivities, and generally excellent enantioselectivities of up to 98% ee.

Given that 4-aminochromanes are a class of structure integrated in hundreds of natural and bioactive compounds, the above authors later proposed novel access to this type of product based on an asymmetric domino Michael–aza-Baylis–Hillman reaction occurring between alkynals and salicyl *N*-tosylimine.[174] This highly enantioselective reaction was catalysed by a chiral diarylprolinol trimethylsilyl ether, which was supposed to activate the alkynals, forming iminium intermediates, that underwent an oxa-Michael addition with the salicyl *N*-tosylimine. The resulting alenamine intermediate reacted with the imine in an intramolecular fashion leading to the final 4-amino-4*H*-chromenes and regenerating the catalyst. It must be noted that this reaction took place in less than 2 h with high yields and generally excellent enantioselectivities (94–98% ee), as shown in Scheme 3.107. Importantly, the catalytic loading could be reduced to 5 mol% with a slight increase in reaction times.

3.2.1.8   Domino Michael–Mannich Reactions

In 2007, Vicario *et al.* reported an enantioselective [3 + 2] cycloaddition reaction of azomethine ylides with α,β-unsaturated aldehydes, which can also be described as an enantioselective domino Michael–Mannich reaction of *in situ* generated azomethine ylides from imines with α,β-unsaturated aldehydes, catalysed by chiral diphenylprolinol in the presence of water in

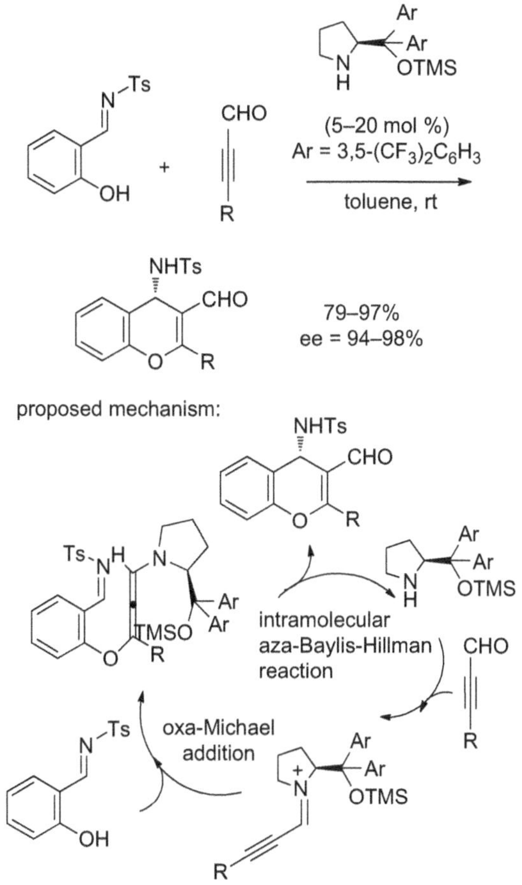

Scheme 3.107 Synthesis of 4-amino-4*H*-chromenes.

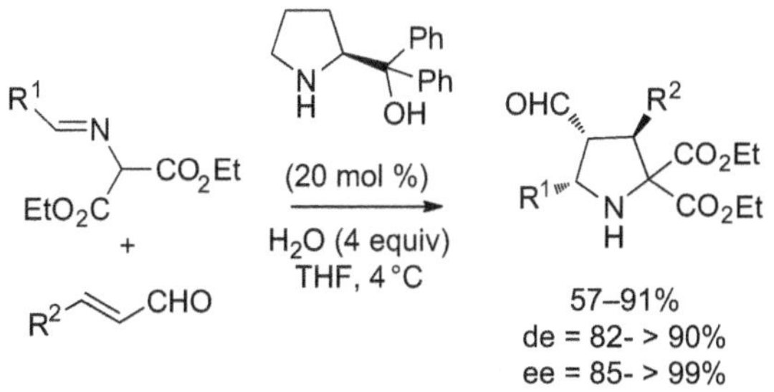

Scheme 3.108 Synthesis of pyrrolidines.

THF.[175] This remarkable process, evolving through iminium–enamine activation mode, afforded the corresponding highly functionalised polysubstituted
almost diastereo- and enantiopure pyrrolidines in good to high yields in almost
all cases of the substrates studied (Scheme 3.108). Linear and branched

aliphatic aldehydes as well as different imine substrates were tolerated in the domino reaction.

Later, Xu *et al.* reported a novel type of bifunctional catalyst generated by the self-assembly of a chiral pyrrolidine, depicted in Scheme 3.109, and readily available primary amino acid D-*t*-leucine.[176] The protonation of the aromatic nitrogen atom of the chiral pyrrolidine by this amino acid spontaneously led to ion-pair assemblies possessing dual activation centres which, similar to catalysis

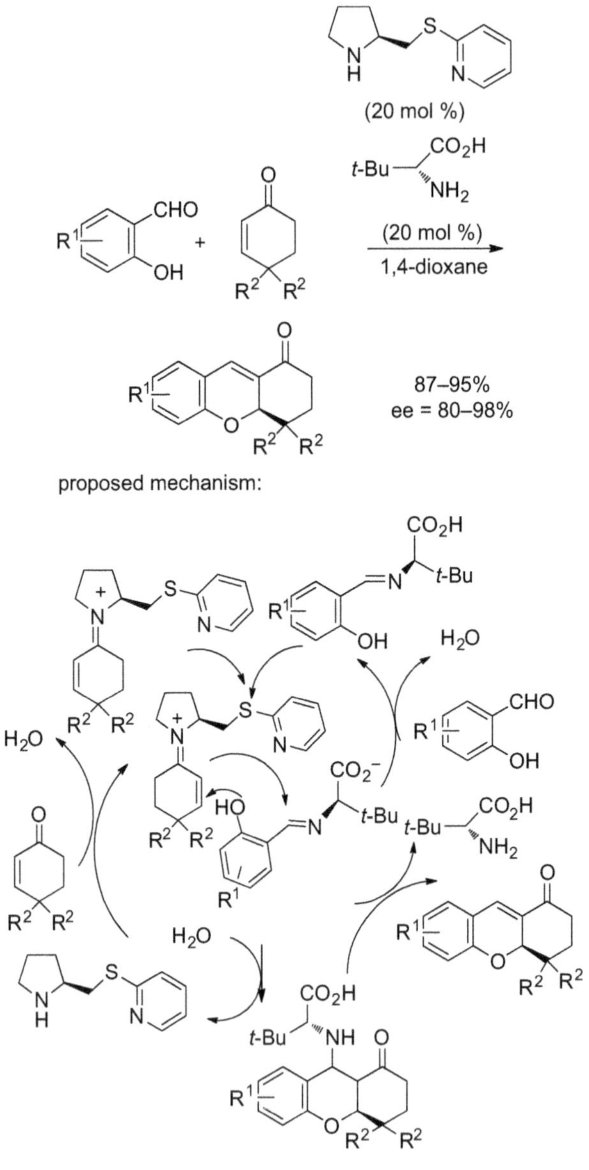

Scheme 3.109 Multicatalytic synthesis of tetrahydroxanthenones.

by enzymes, enabled the catalysis of the electrophilic and nucleophilic substrates simultaneously. This system was applied to induce novel asymmetric domino oxa-Michael–Mannich reactions of salicylic aldehydes with cyclohexenones to afford the corresponding tetrahydroxanthenones. The authors have proposed the mechanism depicted in Scheme 3.109, in which the simultaneous activation of cyclohexenones and salicyladehydes by pyrrolidine catalyst and the amino acid, respectively, generated the transient ion pairs through iminiums and imines. The resulting assemblies ensured that the enantioselective domino oxa-Michael addition and intramolecular Mannich reaction proceeded to afford, upon hydrolysis, the corresponding final products and released the chiral pyrrolidine catalyst. Finally, elimination of the resulting Mannich bases led to the desired tetrahydroxanthenones, concurrently regenerating the amino acid. As summarised in Scheme 3.109, the products were obtained in high yields and enantioselectivities of up to 98% ee.

A synthesis of highly substituted chiral tetrahydroimidazopyrimidine derivatives containing three stereogenic centres was described by Hu *et al.*, in 2011.[177] These biologically important products were achieved for the first time through organocatalytic enantioselective domino aza-Michael–Mannich reaction of α,β-unsaturated aldehydes and *N*-arylidene-1*H*-imidazol-2-amines. The reaction was efficiently catalysed by chiral diphenylprolinol trimethylsilyl ether in the presence of benzoic acid as an additive in a mixture of dichloromethane and methanol as solvent. It provided the domino products in good yields and general excellent diastereo- and enantioselectivities of up to >90% de and >99% ee, respectively, as shown in Scheme 3.110.

In the same year, You *et al.* reported an intramolecular enantioselective domino Michael–Mannich reaction of indolyl methyl enones catalysed by a chiral quinine-derived primary amine, affording a series of highly enantioenriched tetracyclic compounds in moderate to high yields, low to good diastereoselectivities, and low to excellent enantioselectivities (Scheme 3.111).[178] This novel polycyclisation, evolving through iminium–enamine activation mode, has been applied to the total synthesis of an analogue of the natural product (+)-kreysiginine.

In addition, Barbas *et al.* have recently reported the enantioselective synthesis of carbazole spirooxindoles based on a domino Michael–Mannich-type

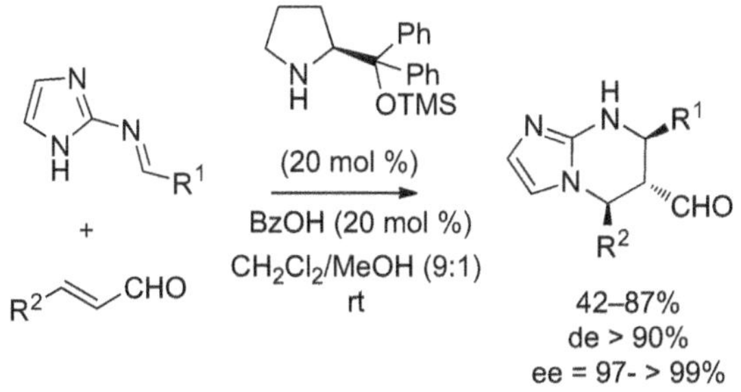

Scheme 3.110 Synthesis of tetrahydroimidazopyrimidines.

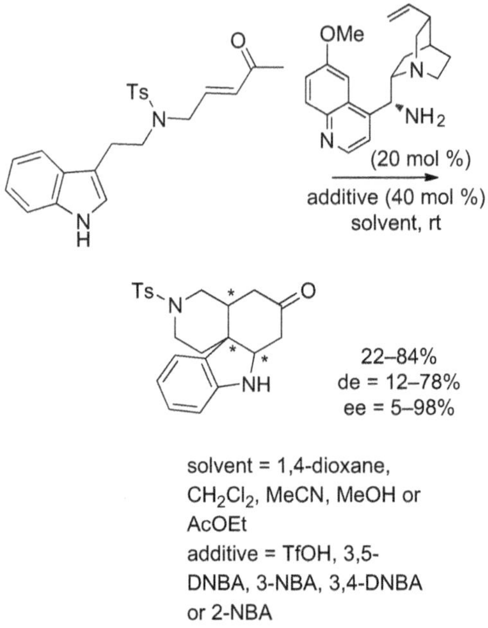

Scheme 3.111   Synthesis of tetracyclic products.

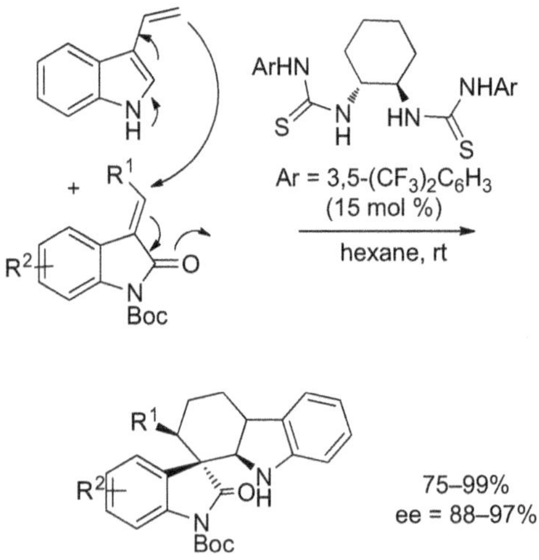

Scheme 3.112   Synthesis of carbazole spirooxindoles.

reaction occurring between two indole derivatives, upon catalysis with a chiral bifunctional thiourea, depicted in Scheme 3.112.[179] While it was unclear whether this reaction proceeded in a concerted Diels–Alder cycloaddition, or in a stepwise fashion, a possible pathway involved the indolenine intermediate generated from the vinylogous enamine to be intercepted in a Mannich-type reaction to provide the final spirocyclic products in high yields and enantioselectivities.

3.2.1.9 Other Domino Reactions initiated by the Michael Reaction

In 2008, Cordova *et al.* reported an enantioselective domino aminosulfenylation reaction of α,β-unsaturated aldehydes catalysed by chiral diphenylprolinol trimethylsilyl ether.[180] The innovation of this method was the efficient incorporation of all components of the reacting substrate in an intermolecular catalytic domino process. For the success of this reaction, the addition of a catalytic amount of succinimide as initiator was necessary. This promoted the first reversible conjugated addition over the enal. Subsequently, the attack of the formed enamine on the electrophilic counterpart pushed the equilibrium towards the formation of the domino product, releasing free succinimide which restarted the catalytic cycle. Although the diastereocontrol was fairly low for aromatic enals, they were easily separated by simple flash chromatography, and both *syn* and *anti* adducts were obtained in excellent enantioselectivities (93–99% ee), as shown in Scheme 3.113.

In 2009, Chen *et al.* reported the synthesis of a newly designed bulky chiral secondary amine derived from proline (Scheme 3.114), which was applied as catalyst for the asymmetric domino Michael–Wittig reaction of various α,β-unsaturated aldehydes with (3-carboxy-2-oxopropylidene)triphenylphosphorane.[181] The process, performed in the presence of LiClO$_4$ and DABCO, afforded the corresponding formal [3 + 3] cycloadducts in good to high yields (56–85%) and moderate to excellent diastereoselectivities (50–96% de), combined with high enantioselectivities (86–99% ee), as shown in Scheme 3.114. The best

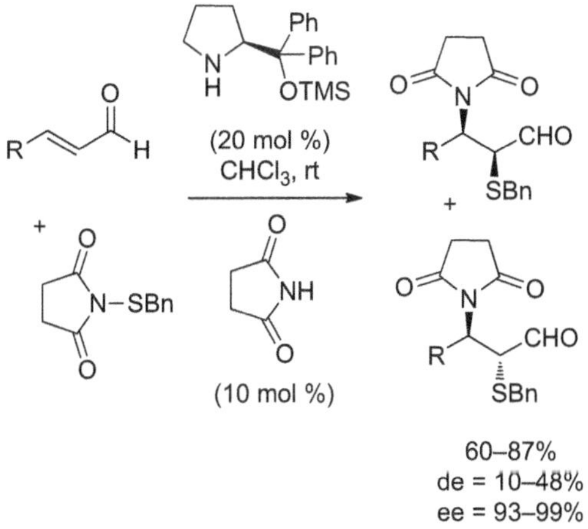

60–87%
de = 10–48%
ee = 93–99%

Scheme 3.113 Domino aminosulfenylation reaction of α,β-unsaturated aldehydes.

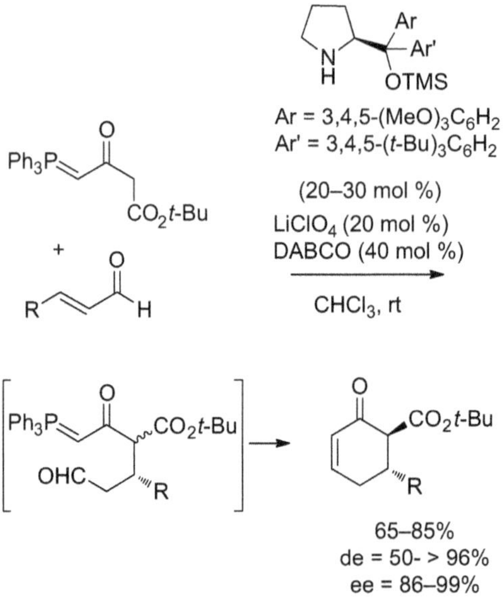

Scheme 3.114 Synthesis of cyclohexenones through domino Michael–Wittig reaction.

results were generally obtained for aromatic α,β-unsaturated aldehydes. Indeed, α,β-unsaturated aldehydes with alkyl substitutions could be successfully used, while higher catalytic loadings and longer reaction times were required (for example 30 mol% instead of 20 mol% with aromatic substitutions).

In another context, Zhao and Gogoi have developed the first enantioselective synthesis of biologically active 6-amino-5-cyanodihydropyrano[2,3-*c*]pyrazoles through a cinchona alkaloid-catalysed domino Michael–Thorpe–Ziegler type reaction between 2-pyrazolin-5-ones and benzylidenemalononitriles.[182] The authors have investigated a series of cinchona alkaloid catalysts, such as quinine, cupreine, 9-*epi*-cupreine, quinidine, and cupreidine, to induce this process. They have found that the enantioselectivity of the reaction was highly dependent on the structure of the catalysts and also that of the substrates, in addition to the reaction conditions. The best results were obtained using cupreine, which provided high yields combined with mediocre to excellent enantioselectivities (26–96% ee), as shown in Scheme 3.115.

A novel asymmetric domino reaction of cyclic ketones with (*E*)-2-nitroallylic acetates catalysed by a chiral pyrrolidine-thiourea (Scheme 3.116) combined with an equal amount of 4-methoxybenzoic acid as co-catalyst was developed by Li *et al.*[183] The process afforded the corresponding bicyclic cyclohexanones bearing four stereogenic centres with enantioselectivities of up to 98% ee and moderate to high yields (27–78%). The cooperative effects of both the enamine catalyst and the urea catalyst were demonstrated to be essential for the high reactivity, diastereoselectivity, and enantioselectivity of the reaction by both experimental data and theoretical calculations. The authors have proposed the mechanism depicted in Scheme 3.116, in which an enamino intermediate

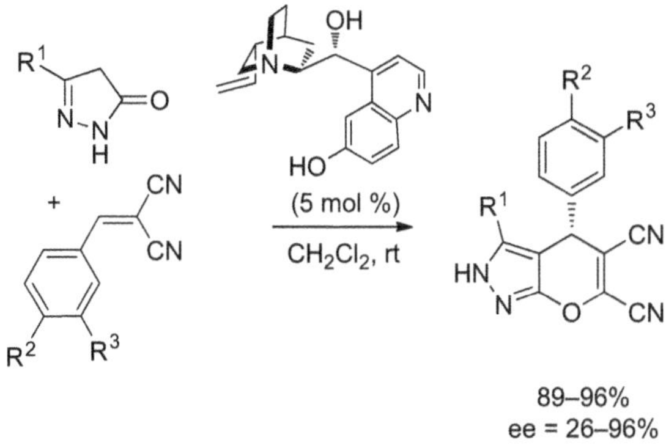

Scheme 3.115 Synthesis of 6-amino-5-cyanodihydropyrano[2,3-*c*]pyrazoles through domino Michael–Thorpe–Ziegler reaction.

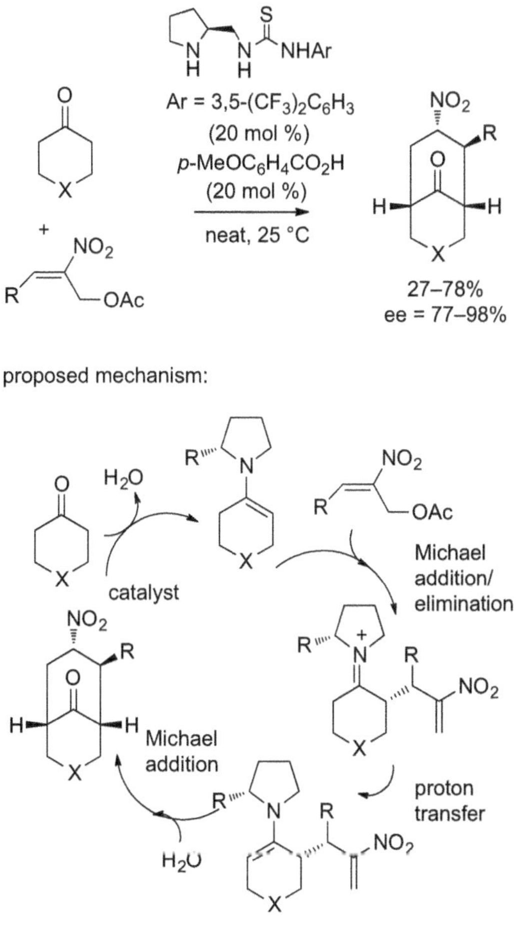

proposed mechanism:

Scheme 3.116 Synthesis of bicyclic cyclohexanones through domino Michael–elimination–Michael reaction.

derived from the cyclic ketone and the catalyst added to the (*E*)-2-nitroallylic acetate to give the corresponding Michael product, which was further submitted to dehydratation to produce a nitroalkene intermediate. The latter was then isomerised to the corresponding enamine intermediate, which underwent an intramolecular Michael addition, affording the final products and regenerating the catalyst to finish the catalytic cycle.

In 2011, Wang *et al.* demonstrated that cyclohexan-1,2-dione could be used as a dual C–O nucleophile towards α,β-unsaturated pyruvates acting as dual electrophiles.[184] As shown in Scheme 3.117, they developed an enantioselective domino Michael–enolisation–cyclisation reaction catalysed by a chiral thiourea, providing chiral dihydro-2*H*-pyranes in high yields and excellent enantioselectivities.

Although a number of examples of asymmetric domino reactions have been developed, most rely on a single catalyst to perform generally two sequential operations. More recently, multiple catalyst systems for this type of reaction have been proposed.[185] Although these reactions showcase the potential power in this field, relatively few asymmetric versions exist.[186] In this context, Rovis and Lathrop have developed a multicatalytic asymmetric domino reaction for the preparation of chiral α-hydroxycyclopentanones containing three contiguous stereocentres.[187] This process was based on the reaction of 1,3-diketones with α,β-unsaturated aldehydes, which generated through Michael addition induced by a chiral diarylprolinol trimethylsilyl ether, followed by hydrolysis, the corresponding aldehydes bearing a tethered ketone (Scheme 3.118). These intermediates then underwent an intramolecular crossed benzoin reaction in the presence of the second carbene catalyst depicted in Scheme 3.118 to afford the final cyclopentanones *via* a formal [3 + 2] process. In spite of moderate diastereoselectivities, these highly functionalised products were obtained in high enantioselectivities of up to 97% ee, as shown in Scheme 3.118.

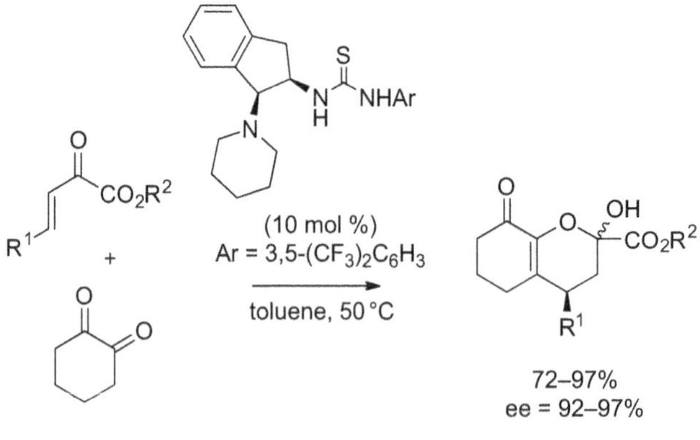

(10 mol %)
Ar = 3,5-(CF$_3$)$_2$C$_6$H$_3$

toluene, 50 °C

72–97%
ee = 92–97%

Scheme 3.117   Synthesis of dihydro-2*H*-pyranes through domino Michael–enolisation–cyclisation reaction.

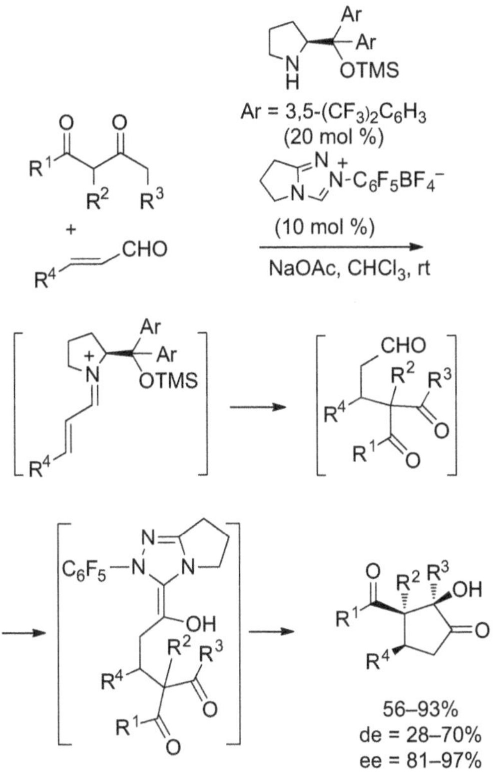

Scheme 3.118 Synthesis of cyclopentanones through multicatalytic domino Michael–crossed benzoin reaction.

These authors have developed another multicatalytic asymmetric domino reaction occurring between salicylaldehydes and electron-deficient alkynes, such as dimethyl acetylenedicarboxylate (DMAD).[188] In this case, the reaction proceeded *via* amine-catalysed Michael addition followed by an intramolecular Stetter reaction promoted by a *N*-heterocyclic carbene. As shown in Scheme 3.119, a tertiary amine, such as quinuclidine or DABCO, activated DMAD *via* the corresponding ammonium intermediate towards nucleophilic attack to give an intermediate aldehyde. Subsequent chiral carbene-promoted Stetter reaction set a quaternary stereocentre and yielded the corresponding final functionalised benzofuranones in good yields and good to excellent enantioselectivities of up to 94% ee. In addition, a *N*-heterocyclic carbene catalyst was very recently combined by the same authors with chiral diphenylprolinol trimethylsilyl ether to promote the diastereo- and enantioselective domino Michael–benzoin cyclisation reaction between aliphatic aldehydes and activated enones, providing the corresponding complex cyclopentanones in moderate to excellent yields (35–99%), moderate to high diastereoselectivities (56–94% de), and moderate to excellent enantioselectivities (51–98% ee).[189]

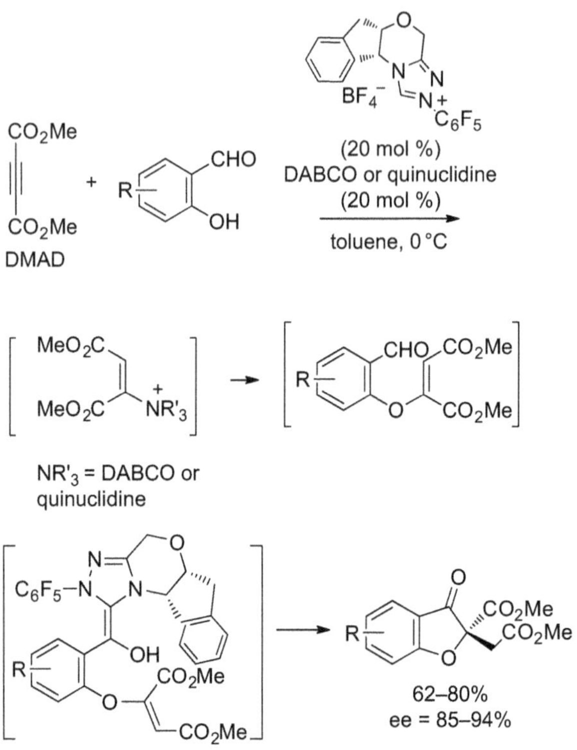

Scheme 3.119   Synthesis of benzofuranones through multicatalytic domino Michael–
Stetter reaction.

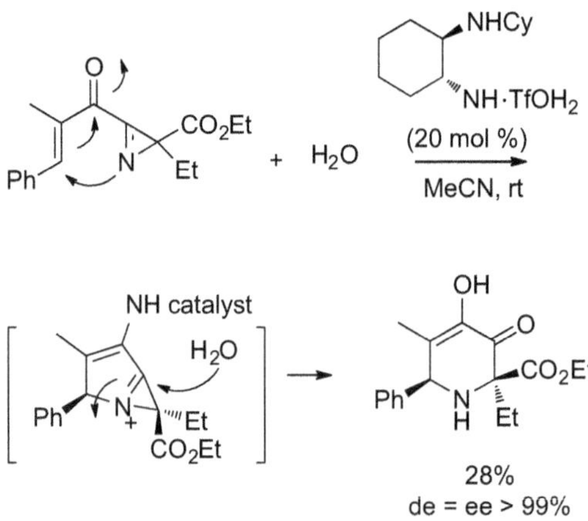

Scheme 3.120   Synthesis of piperidone through domino aza-Michael–water addition
reaction.

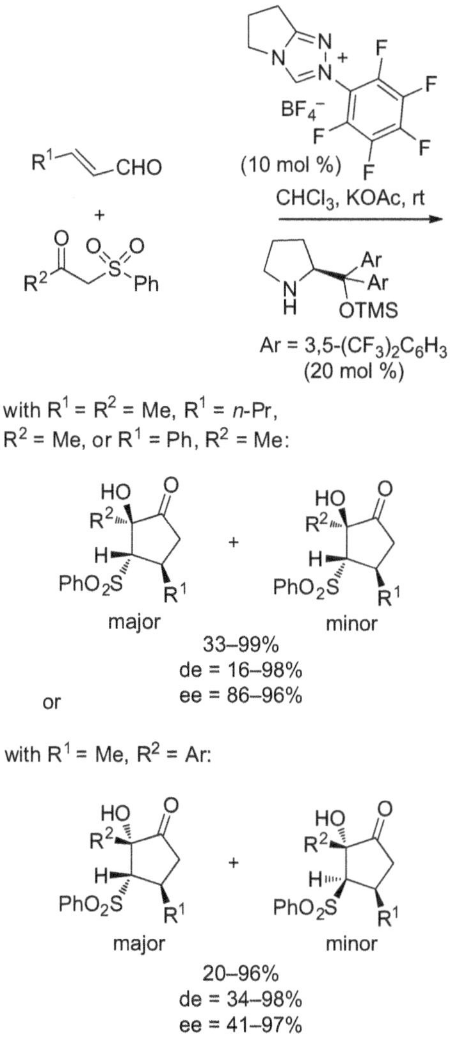

Scheme 3.121 Synthesis of cyclopentanones through multicatalytic domino Michael–cross benzoin reaction.

An enantioselective formal aza-Nazarov cyclisation reaction was described by Tius *et al.*, in 2010.[190] As shown in Scheme 3.120, it involved the domino aza-Michael–water condensation reaction of an azirine catalysed by a chiral 1,2-diaminocyclohexane derivative in the presence of water, which led to the corresponding functionalised piperidone in low yield (28%), albeit with excellent enantioselectivity of 99% ee. Indeed, the first step of the sequence dealt with intramolecular aza-Michael addition to give an iminium inter-mediate, which was then trapped by a molecule of water.

In 2011, Enders *et al.* reported the synthesis of chiral polyfunctionalised cyclopentanones with three contiguous stereogenic centres on the basis of an enantioselective domino Michael–cross-benzoin reaction of β-oxo sulfones with α,β-unsaturated aldehydes.[191] This process was efficiently catalysed by a dual chiral secondary amine/N-heterocyclic carbene catalytic system, which provided the corresponding cyclopentanones in good yields, modest to excellent diastereoselectivities, and very good enantiomeric excesses, as shown in Scheme 3.121. These products have the advantages of bearing a synthetically useful β-(phenylsulfonyl) group, an α-hydroxy function, and three contiguous stereocentres.

3.2.2 Domino Reactions Initiated by Other Reactions

3.2.2.1 Domino Reactions initiated by the indirect Mannich Reaction

The classic direct Mannich reaction, discovered in 1912,[192] is an aminoalkylation of carbonylic compounds involving ammonia (or a primary or secondary amine derivative), a non-enolisable aldehyde (usually formaldehyde) or a ketone, and an enolisable carbonyl compound, leading to β-aminocarbonyl derivatives.[193] The indirect version of the Mannich reaction is the corresponding two-component reaction of a preformed imine and an enolisable carbonyl compound. An early example of asymmetric domino reactions initiated by the Mannich reaction was successfully developed by Ohsawa *et al.*, in 2003.[194] In 2006, the same authors reported an enantioselective domino Mannich–Michael reaction catalysed by L-proline, which occurred between a 3,4-dihydro-β-carboline and 3-ethyl-3-buten-2-one.[195] Remarkably, the corresponding *ent*-dihydrocorynantheol was formed in 85% yield as a single stereomer, as shown in Scheme 3.122.

In 2008, Xu *et al.* described a proline-catalysed enantioselective construction of 2,3-disubstituted tetrahydropyridines on the basis of a domino

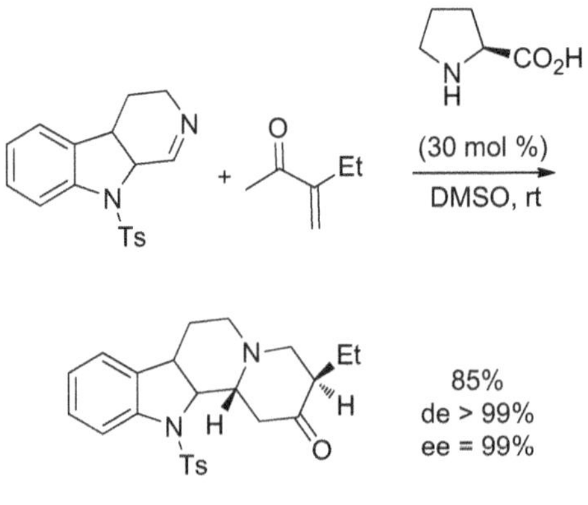

Scheme 3.122 Synthesis of *ent*-dihydrocorynantheol through domino Mannich–Michael reaction.

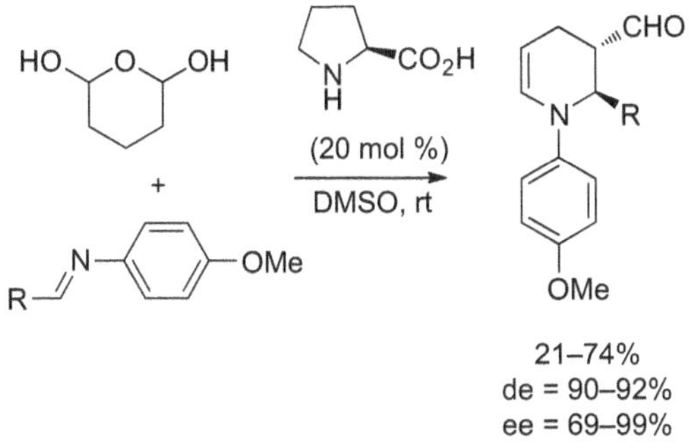

21–74%
de = 90–92%
ee = 69–99%

Scheme 3.123 Synthesis of tetrahydropyridines through domino Mannich-type–intramolecular hemiaminalisation–dehydration reaction.

Mannich-type–intramolecular hemiaminalisation–dehydration reaction.[196] This process occurred between preformed *N-p*-methoxyphenyl aldimines and inexpensive aqueous tetrahydro-2*H*-pyran-2,6-diol in DMSO at room temperature. It afforded the corresponding chiral tetrahydropyridines as almost single diastereomers (>92% de) in moderate to good yields and good to excellent enantioselectivities of up to >99% ee, as shown in Scheme 3.123.

A more recent example was reported by Carter *et al.*, allowing a new route to azabicyclo[2.2.2]octanes, which have generated considerable attention owing to their presence in numerous natural alkaloid products. Indeed, this work dealt with a synthesis of these chiral products on the basis of an asymmetric domino Mannich–aza-Michael reaction of cyclohexenone and a series of aromatic imines, promoted by a *p*-dodecylphenylsulfonamide-modified L-proline catalyst.[197] The process performed in the absence of solvent strongly favoured the *exo* products, which were achieved in high levels of enantioselectivities of up to 99% ee combined with moderate to good yields, as shown in Scheme 3.124. The use of other ketones in place of cyclohexenone, such as cyclopentenone or acyclic enones, led to complex mixtures of products. Furthermore, attempts to extend the scope of the reaction to aliphatic imines did not result in the formation of the corresponding azabicyclo[2.2.2]octanes.

In 2010, Enders *et al.* developed another asymmetric domino Mannich–aza-Michael reaction, involving γ-malonate-substituted α,β-unsaturated ester and *N*-Boc-protected arylaldimines as the substrates.[198] As shown in Scheme 3.125, the reaction catalysed by a chiral bifunctional thiourea afforded the corresponding 2,5-*cis*-configured polysubstituted pyrrolidines in good to excellent yields (75–99%) and enantioselectivities of up to 94% ee combined with an excellent diastereoselectivity of >95% de in all cases of the substrates studied.

Another recent example of asymmetric domino reaction initiated by the Mannich reaction was reported by Rueping and Lin, in 2010.[199] This work constituted the first catalytic asymmetric synthesis of 4-aminobenzopyrans and was based on a domino Mannich–ketalisation reaction of electron-rich alkenes with *o*-hydroxy benzaldimines induced by a chiral *N*-triflylphosphoramide derived from (*S*)-BINOL. For example, the reaction of 2,3-dihydro-2*H*-furan

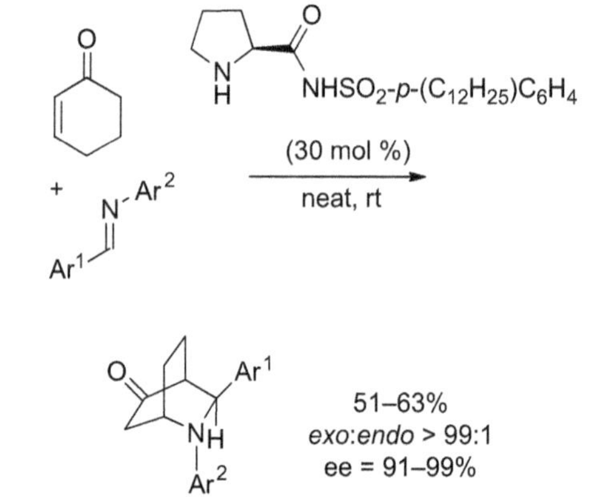

Scheme 3.124 Synthesis of azabicyclo[2.2.2]octanes through domino Mannich–aza-Michael reaction.

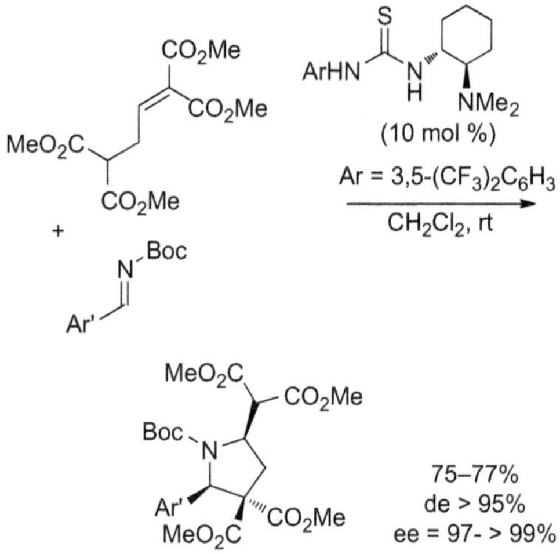

Scheme 3.125 Synthesis of pyrrolidines through domino Mannich–aza-Michael reaction.

with o-hydroxy benzaldimines provided the corresponding furanobenzopyrans in good yields of up to 89% and high enantioselectivities of up to 96% ee, combined with moderate to high diastereoselectivities, as shown in Scheme 3.126. The scope of this methodology was extended to 3,4-dihydro-2H-pyran, which gave rise to the corresponding pyranobenzopyrans with comparable results (Scheme 3.126).

In 2011, Wang *et al.* developed an expeditious assembly of a 2-amino-4H-chromene skeleton on the basis of an enantioselective domino elimination–Mannich–intramolecular ring cyclisation–tautomerisation reaction catalysed by a chiral indane-amine-thiourea.[200] A base, such as Li_2CO_3, was

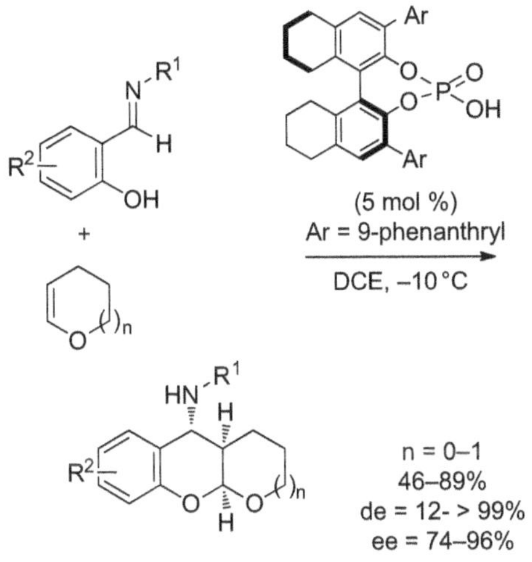

Scheme 3.126 Synthesis of 4-aminobenzopyrans through domino Mannich–ketalisation reaction.

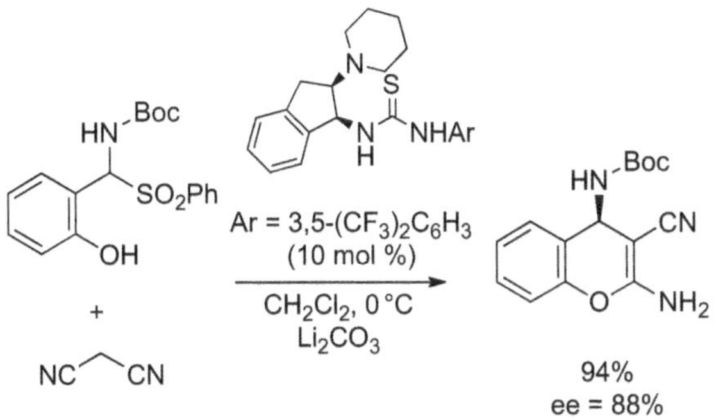

Scheme 3.127 Synthesis of a 2-amino-4*H*-chromene through domino elimination–Mannich–intramolecular cyclisation–tautomerisation reaction.

employed to generate the Mannich substrate from *tert*-butyl (2-hydroxyphenyl)-(phenylsulfonyl) methylcarbamate through elimination. Then, the Mannich reaction of the *in situ* generated imine with malononitrile occurred to give the intermediate Mannich product, which was subsequently submitted to an intramolecular oxa-nucleophilic addition of a nitrile group. The intermediate imine formed finally tautomerised to provide the corresponding 2-amino-4*H*-chromene in high yield and enantioselectivity, as shown in Scheme 3.127. Even if the domino reaction did not really begin with a Mannich reaction, it was decided to include it in this section dealing with domino reactions initiated by the indirect Mannich reaction.

In 2011, Feng *et al.* developed a highly enantioselective synthesis of 3,4-diaminochroman-2-ones through the domino Mannich-type reaction of

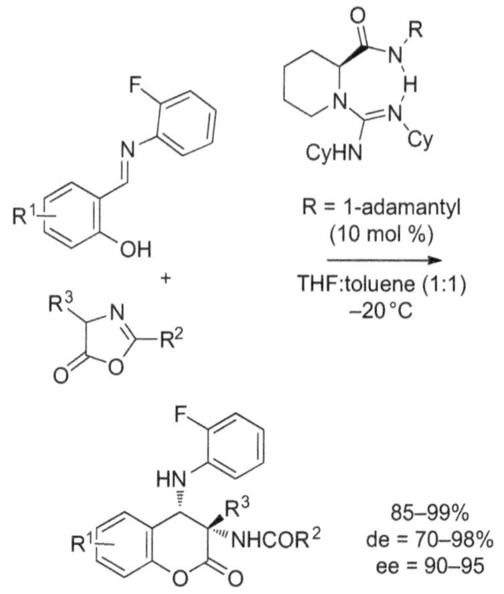

Scheme 3.128   Synthesis   of   *cis*-3,4-diaminochroman-2-ones   through   domino
Mannich-type reaction.

o-hydroxy aromatic aldimines and azlactones.[201] As shown in Scheme 3.128,
the process was efficiently catalysed by an (*S*)-pipecolic acid-derived guanidine,
leading to a range of *cis*-3,4-diaminochroman-2-ones in high yields, and
excellent diastereo- and enantioselectivities. Furthermore, the authors have
demonstrated that changing the catalyst to a chiral bisguanidium salt,
3·HBAr$^F_4$, allowed the corresponding *trans*-domino products to be achieved
with comparable excellent results.

3.2.2.2   Domino Reactions Initiated by the (aza)-Morita–Baylis–Hillman Reaction

The aza-Morita–Baylis–Hillman reaction is known to be a useful and atom-
economical C–C bond-forming reaction of electron-deficient alkenes with
imines, usually catalysed by Lewis bases.[202] It formally involves a sequence of
reactions including a Michael addition, Mannich reaction, proton-transfer, and
retro-Michael reaction (β-elimination). Although there are many reports in the
field of the enantioselective aza-Morita–Baylis–Hillman reaction, rare examples
of asymmetric domino reactions initiated by this reaction have been reported.
In 2010, Sasai *et al.* developed the first organocatalysed asymmetric domino
aza-Morita–Baylis–Hillman–aza-Michael reaction of α,β-unsaturated carbonyl
compounds with *N*-tosylimines, allowing easy access to chiral *cis-1,3*-disub-
stituted isoindolines as single diastereomers.[203] The process was induced by an
H$_8$-BINOL-derived catalyst and provided these products in high yields and
enantioselectivities of up to 93% ee, as shown in Scheme 3.129.

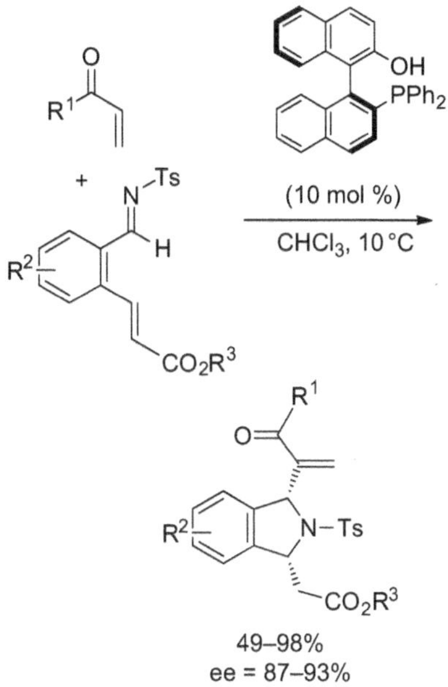

Scheme 3.129 Synthesis of isoindolines through domino aza-Morita–Baylis–Hillman–aza-Michael reaction.

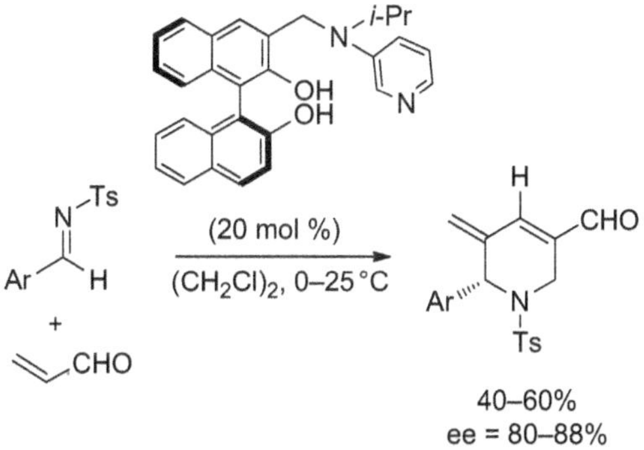

Scheme 3.130 Synthesis of tetrahydropyridines through domino aza-Morita–Baylis–Hillman–aza-Michael–aldol–dehydration reaction.

Later, Sasai *et al.* developed another example of asymmetric domino reaction initiated by the aza-Morita–Baylis–Hillman reaction followed successively by an aza-Michael reaction, an aldol reaction, and a dehydration.[204] This novel process, induced by a chiral acid–base organocatalyst, involved acrolein and various *N*-tosylarylimines as the substrates, and led to the corresponding highly functionalised tetrahydropyridines in moderate yields and good enantio-selectivities of up to 88% ee (Scheme 3.130). Attempts to extend the scope of the reaction to other activated alkenes, such as methylvinylketone, failed because only the corresponding normal aza-Morita–Baylis–Hillman product was formed and no desired domino product was obtained.

3.2.2.3   Domino Reactions Initiated by the Friedel–Crafts Reaction

The asymmetric Friedel–Crafts reaction is one of the most powerful methods to synthesize optically active aromatic compounds, and chiral Bronsted acids have recently emerged as efficient catalysts.[205] In 2009, You *et al.* reported the synthesis of chiral fluorene derivatives by an asymmetric domino double Friedel–Crafts reaction catalysed with a chiral phosphoric acid. This reaction involved indoles and 3′,5′-dimethoxybiphenyl-2-carbaldehyde as the substrates, and afforded the corresponding fluorenes in moderate to high yields and enantioselectivities of up to 94% and 96% ee, respectively. As shown in Scheme 3.131, the authors have proposed that the first Friedel–Crafts reaction occurred between the two substrates upon catalysis with a chiral phosphoric acid to afford the corresponding secondary alcohol. This intermediate was unstable under the reaction conditions and led to the formation of a close counterion, in which the chiral phosphate anion created a chiral environment to control the enantioselectivity of the second Friedel–Crafts reaction, which provided the final fluorenes.

   Another asymmetric domino reaction initiated by a Friedel–Crafts reaction promoted by chiral diphenylprolinol trimethylsilylether was developed by Wang *et al.*, affording chiral chromanes.[206] These products were generated in good yields, albeit with moderate diastereo- and enantioselectivities, by reaction of α,β-unsaturated aldehydes with 1-naphthol, according to a domino Friedel–Crafts–cyclisation reaction depicted in Scheme 3.132. The authors have proposed that aldehydes were activated by the catalyst through the formation of an iminium intermediate, which then added 1-naphthol to give a novel iminium intermediate. Subsequent hydrolysis and half acetalisation provided the final chromane.

   In addition, a novel asymmetric domino Friedel–Crafts–Michael–aromatisation reaction, recently reported by Xiao *et al.*, provided a range of chiral highly functionalised tetrahydrocarbazoles upon catalysis with a chiral diarylprolinol trimethylsilyl ether.[207] These products were obtained by reaction of α,β-unsaturated aldehydes with 2-vinylindoles, evolving through iminium–enamine activation mode, as shown in Scheme 3.133. First, aldehydes were activated by the amine catalyst through an iminium intermediate, which enantioselectively intercepted the nucleophilic indole through a Friedel–Crafts type alkylation. By doing so, a new active intermediate, containing both iminium and enamine components, was produced to accomplish the intramolecular conjugate addition to form a novel iminium–enamine intermediate. Subsequently, a rapid [1,3]-H migration regenerated the aromaticity of the indole ring, and the hydrolysis could release the catalyst and afford the final

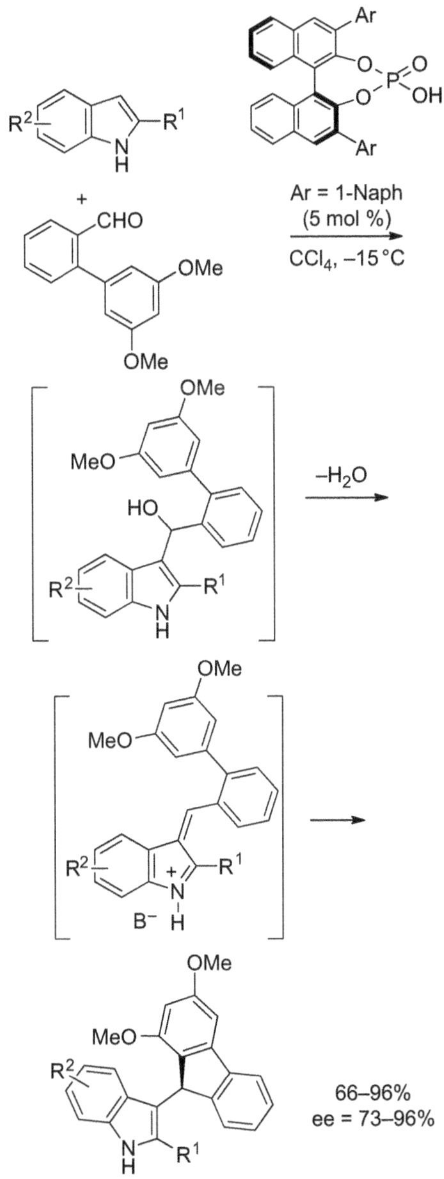

Scheme 3.131 Synthesis of fluorene derivatives through domino double Friedel–Crafts reaction.

tetrahydrocarbazoles in good to high yields, and excellent diastereo- and enantioselectivities of up to >98% de, and >99% ee, respectively. Furthermore, these authors have previously developed a route to this type of product through another asymmetric domino reaction, initiated by the

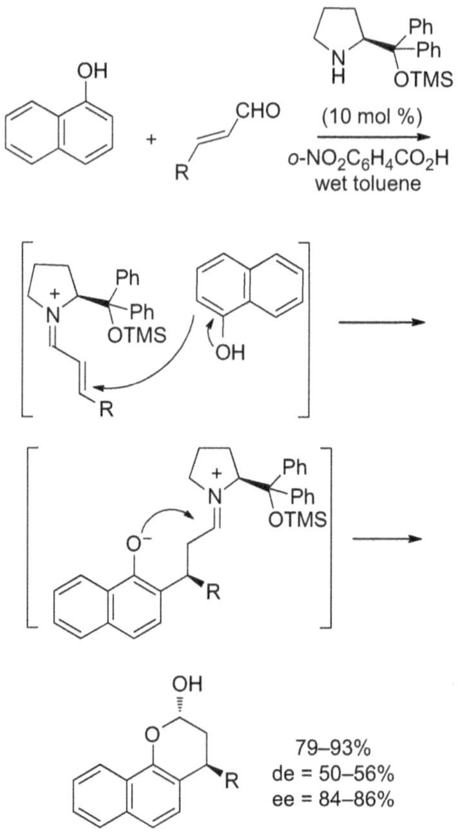

Scheme 3.132   Synthesis of chromanes through Friedel–Crafts–cyclisation reaction.

Friedel–Crafts reaction occurring between nitroolefins and 2-propenylindoles. In this case, the reaction was induced by a chiral bis-sulfonamide catalyst which, through hydrogen-bonding activation, afforded the corresponding highly substituted tetrahydrocarbazoles in moderate to good yields (42–86%) combined with good to high diastereo- and enantioselectivities (66–98% de, 82–98% ee).[208]

Enantioselective domino reactions exploiting the electrophilic properties of indole intermediates have been developed by several groups.[209] For example, Gouverneur *et al.* demonstrated an enantioselective approach to chiral tricyclic indoles by using fluorine electrophiles to quaternarise indole C-3 and trapping the C-2 intermediate iminium species by a C-3-tethered nucleophile.[210] As shown in Scheme 3.134, the domino fluorination–intramolecular cyclisation reaction was catalysed by (DHQD)$_2$PHAL (1,4-phthalazinediyl) in the presence of K$_2$CO$_3$, providing the corresponding tricyclic indoles in moderate to excellent yields, and excellent diastereo- and enantioselectivities.

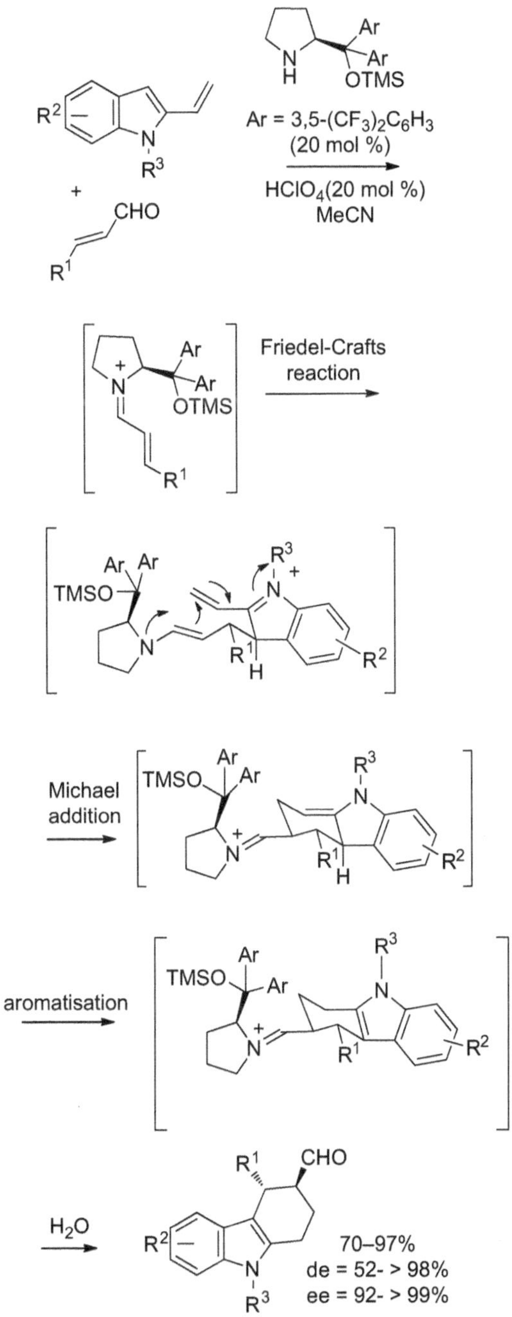

Scheme 3.133 Synthesis of tetrahydrocarbazoles through domino Friedel–Crafts–Michael–aromatisation reaction.

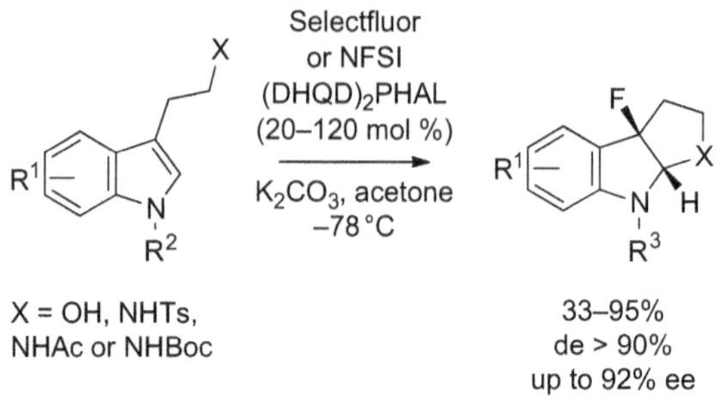

X = OH, NHTs,   33–95%
NHAc or NHBoc   de > 90%
                                                up to 92% ee

Scheme 3.134 Synthesis of tricyclic indoles through domino Friedel–Crafts type–intramolecular cyclisation reaction.

3.2.2.4 Miscellaneous Domino Reactions

In 2007, Landais *et al.* reported an enantioselective domino aldolisation–lactonisation reaction of α-oxocarboxylic acids and various aldehydes, which provided, when catalysed by a chiral benzimidazole proline-derived catalyst, the corresponding chiral isotetronic derivatives in moderate to high yields, and good to high enantioselectivities, as shown in Scheme 3.135.

In the same year, Yamamoto *et al.* reported an enantioselective domino *N*-nitroso aldol–Michael reaction of a dienamine with aromatic nitroso compounds catalysed by a chiral BINOL derivative.[211] The process afforded the corresponding chiral 3-aza-2-oxa-bicycloketones with excellent regioselectivity and very good enantioselectivity, as shown in Scheme 3.136. In addition, Ramachary and Sakthidevi later successfully developed the asymmetric domino aldol–hemiacetalisation reaction of simple ketones with a range of 2-hydroxybenzaldehydes catalysed by *trans*-4-OH-L-proline, which was synthesised for the first time. The reaction provided the corresponding chiral lactols in moderate yields and enantioselectivities (26–90% ee).[212]

Various enantioselective domino reactions have been initiated by an aldolisation reaction. As a recent example, a domino aldolisation–cyclisation reaction was developed by Zhao and Guang, involving α-isothiocyanato imides and activated carbonyl compounds such as isatin derivatives as components.[213] The reaction was efficiently catalysed by a chiral amine thiourea derived from cinchona alkaloids, providing the corresponding spirobicyclic thiocarbamates in high yields and enantioselectivities of up to 98% ee, combined with moderate to good diastereoselectivities of up to 72% de, as shown in Scheme 3.137.

An enantioselective vinylogous aldol–oxa-Michael reaction constituted the key step of total syntheses of natural products diversonol and lachnone C13706, which were reported by Bräse *et al.*, in 2011.[214] This domino process, occurring between salicylaldehyde and prenal, was induced by a chiral diarylprolinol trimethylsilyl ether and gave the corresponding enantiopure tricyclic lactol in 63% yield, as shown in Scheme 3.138.

In 2008, Jorgensen *et al.* developed a novel organocatalytic domino vinylic substitution–cyclisation reaction for the synthesis of optically pure

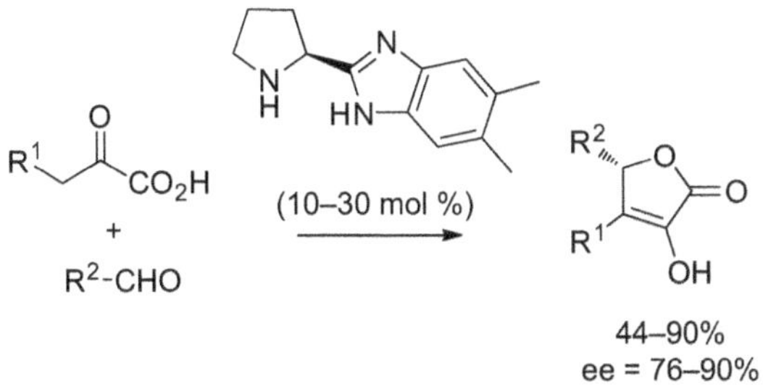

Scheme 3.135 Synthesis of isotetronic derivatives through domino aldolisation–lactonisation reaction.

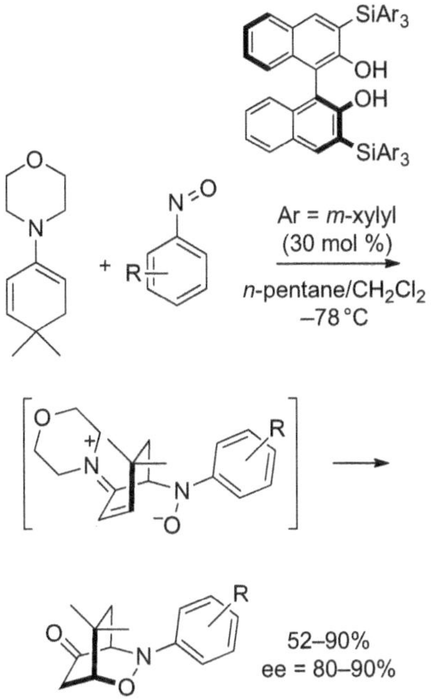

Scheme 3.136 Synthesis of 3-aza-2-oxa-bicycloketones through domino *N*-nitroso aldol–Michael reaction.

halo-substituted pyrrolidin-2-ones, which constitute flexible starting materials for the preparation of chiral γ-lactams.[215] As shown in Scheme 3.139, the reaction occurred between α,β-dihalogenated acrylate esters as the electrophiles and 1,2-nucleophiles in the presence of a chiral dihydrocinchonine-derived phase-transfer catalyst, to provide the corresponding almost enantiopure chiral C-5 quaternary 3-halo-3-pyrrolin-2-ones in good yields. The domino process evolved through the stereospecific substitution of the chlorine atom with retention of configuration by the 1,2-nucleophile, followed by immediate ring closure, resulting in the selective formation of the corresponding halo-substituted pyrrolidin-2-ones.

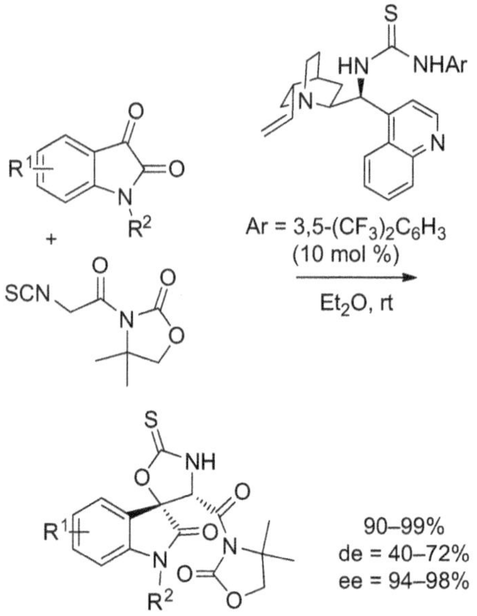

Scheme 3.137 Synthesis of spirobicyclic thiocarbamates through domino aldolisation–cyclisation reaction.

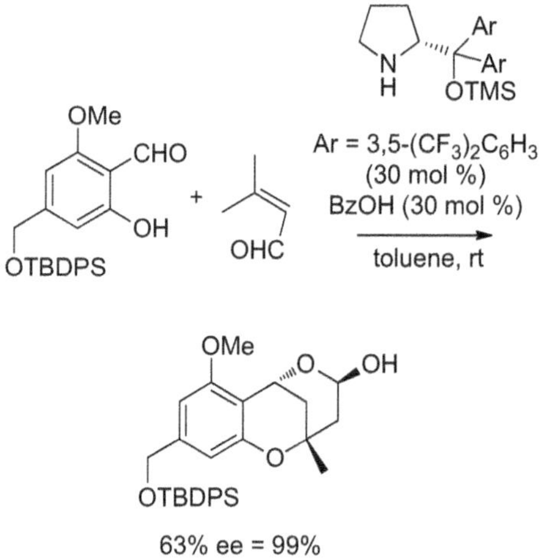

63% ee = 99%

Scheme 3.138 Synthesis of a tricyclic lactol through domino aldolisation–oxa-Michael reaction.

In 2008, Zhong *et al.* developed a novel, practical and highly enantio- and diastereoselective domino α-aminoxylation–aza-Michael reaction for the synthesis of chiral functionalised tetrahydro-1,2-oxazines by using simple L-proline as organocatalyst in the presence of one equivalent of triethylammonium bicarbonate (TEAB) under mild conditions.[216] The reaction occurred

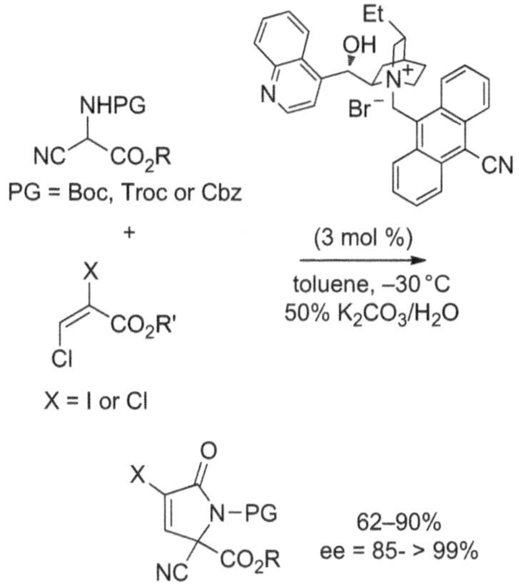

Scheme 3.139 Synthesis of halo-substituted pyrrolidin-2-ones through domino vinylic substitution–cyclisation reaction.

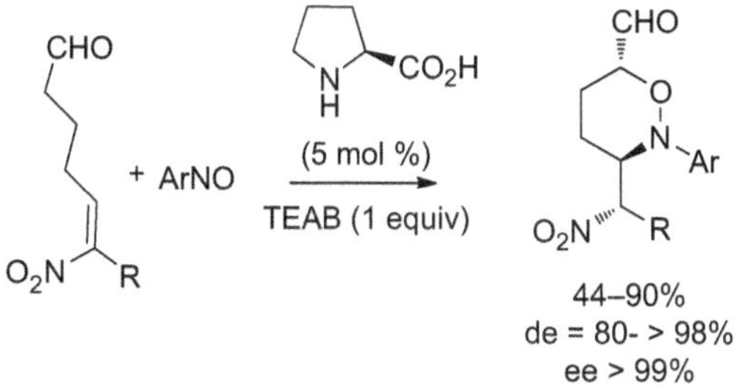

Scheme 3.140 Synthesis of tetrahydro-1,2-oxazines through domino α-aminoxylation–aza-Michael reaction.

between aryl nitroso compounds and bifunctional enemalonate aldehydes. In the first step, *O*-alkylation took place in the α-position of the aldehyde; then an intramolecular aza-Michael reaction took place closing the ring and affording the corresponding chiral tetrahydro-1,2-oxazines with high yields, and excellent diastereo- and enantioselectivities, as shown in Scheme 3.140. In this process, the bifunctional enemalonate aldehydes acted successively as nucleophile, through the enamine intermediate, and as electrophile, through the enemalonate moiety.

In the same year, Chen *et al.* reported the first organocatalytic formal inverse-electron-demand aza-Diels–Alder reaction of *N*-tosyl-1-aza-1,3-butadienes and aldehydes.[217] Actually, this remarkable reaction evolved through the formation of an enamine from the aldehyde component and chiral diphenylprolinol trimethylsilyl ether employed as catalyst. Subsequently, this enamine underwent an aza-Diels–Alder cycloaddition to give the corresponding chiral piperidine derivatives as single diastereomers in good to excellent yields and generally excellent enantioselectivities, as shown in Scheme 3.141. Later, the same authors developed a related enantioselective domino inverse-electron-demand aza-Diels–Alder–hemiacetalisation reaction of N-tosyl-1-aza-1,3-butadienes with aliphatic dialdehydes, such as glutaraldehyde, induced by the same catalyst.[218] This powerful process allowed the corresponding almost diastereo- and enantiopure δ-lactone[2,3-*b*]piperidine derivatives to be achieved in good yields, as shown in Scheme 3.141.

The first organocatalytic enantio- and diastereoselective 1,3-dipolar cycloaddition[219] of azomethine ylides with nitroalkenes was reported by Zhang *et al.*, in 2008.[220] This domino azomethine ylide formation–1,3-dipolar cycloaddition process was efficiently catalysed by thiourea alkaloid derivatives. Indeed, benzophenone imines reacted with various nitroalkenes, providing the corresponding pyrrolidines in good to high yields (49–77%) and diastereoselectivities (70 to >98% de) combined with moderate enantioselectivities (≤65% ee). It must be noted that the relative configuration of the substituents in the pyrrolidine ring was *trans*. A number of asymmetric domino reactions initiated by pericyclic reactions have been reported.[221] A recent example was developed by MacMillan *et al.* in the context of a total synthesis of (+)-minfiensine.[222] As shown in Scheme 3.142, the key step of this synthesis was

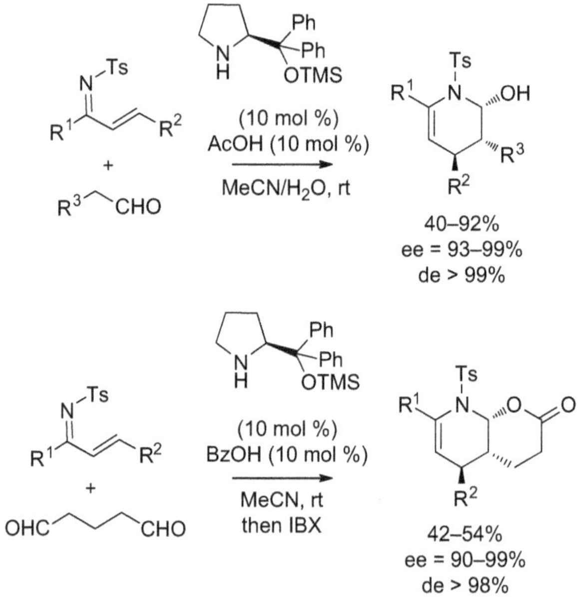

Scheme 3.141   Synthesis of piperidines through domino enamine formation–inverse-electron-demand aza-Diels–Alder reaction, and synthesis of δ-lactone[2,3-*b*]piperidines through domino enamine formation–inverse-electron-demand aza-Diels–Alder–hemiacetalisation reaction.

a novel organocatalytic domino Diels–Alder–amine cyclisation reaction, which allowed rapid and enantioselective access to the tetracyclic carbazole framework of this natural product. The reaction involved propynal and 2-vinylindole as the substrates and a chiral secondary amine (Scheme 3.142) as the catalyst associated with tribromoacetic acid (TBA). The authors assumed that this catalyst activated propynal through the formation of the corresponding iminium ion with the acetylenic group being partitioned away from the bulky *t*-butyl substituent of the catalyst framework (transition state depicted in Scheme 3.142). In this conformation, the aryl ring shielded the top face of the reactive alkyne, facilitating an *endo*-selective Diels–Alder cycloaddition with 2-vinylindole in a regioselective manner to produce a tricyclic diene intermediate. Subsequent protonation of the enamine moiety gave rise to an iminium ion, facilitating a 5-*exo* amine heterocyclisation to deliver the corresponding tetracyclic pyrroloindoline, which was then converted by reduction with sodium borohydride into the final product in 87% yield combined with 96% ee. As a key design element, the dienyl substructure of indole incorporated a 1-methyl sulfide substituent, accelerating the Diels–Alder cycloaddition while providing a latent handle for radical formation, as required in the final ring-forming step. The tetracyclic product was further converted into the expected (+)-minfiensine in five steps. It is important to note that catalyst loadings as low as 5 mol% were sufficient to effect the cascade while maintaining high levels of efficiency (80% yield combined with 94% ee instead of 87% yield combined with 96% ee by using 15 mol% of catalyst loading).

In another context, Dixon *et al.* have reported that the reaction of enol lactones and tryptamines performed in the presence of a chiral phosphoric acid led to the formation of intermediate ketoamides.[223] The latter were subsequently submitted to a dehydrative condensation which led to the corresponding *N*-acyliminium ions, which finally cyclised to provide the corresponding chiral tetracyclic products in moderate to excellent yields (63–99%) and enantioselectivities (72–99% ee), as shown in Scheme 3.143.

In 2009, Bode and Kaeobamrung reported that a chiral triazolium-derived *N*-heterocyclic carbene catalyst promoted the asymmetric annulation of α,β-unsaturated aldehydes and an α-hydroxy enone to afford the corresponding cyclopentane-fused lactones with excellent enantioselectivity of 99% ee, high yields, and moderate diastereoselectivities (≤40% de), as shown in Scheme 3.144.[224] Surprisingly, the authors found that using an imidazolium-derived *N*-heterocyclic carbene catalyst to induce the reaction of α,β-unsaturated aldehydes with the same α-hydroxy enone achieved structurally different cyclopentane-fused lactones with the same enantioselectivity of 99% ee, and moderate yields and diastereoselectivities, as shown in Scheme 3.144. These results represented a dramatic demonstration of the stereodivergency of chiral imidazolium *vs.* triazolium-derived *N*-heterocyclic carbene catalysts.

An organocatalytic enantioselective synthesis of highly biologically active dihydroquinazolinones was described by Rueping *et al.* on the basis of an asymmetric domino condensation–amine addition reaction involving the

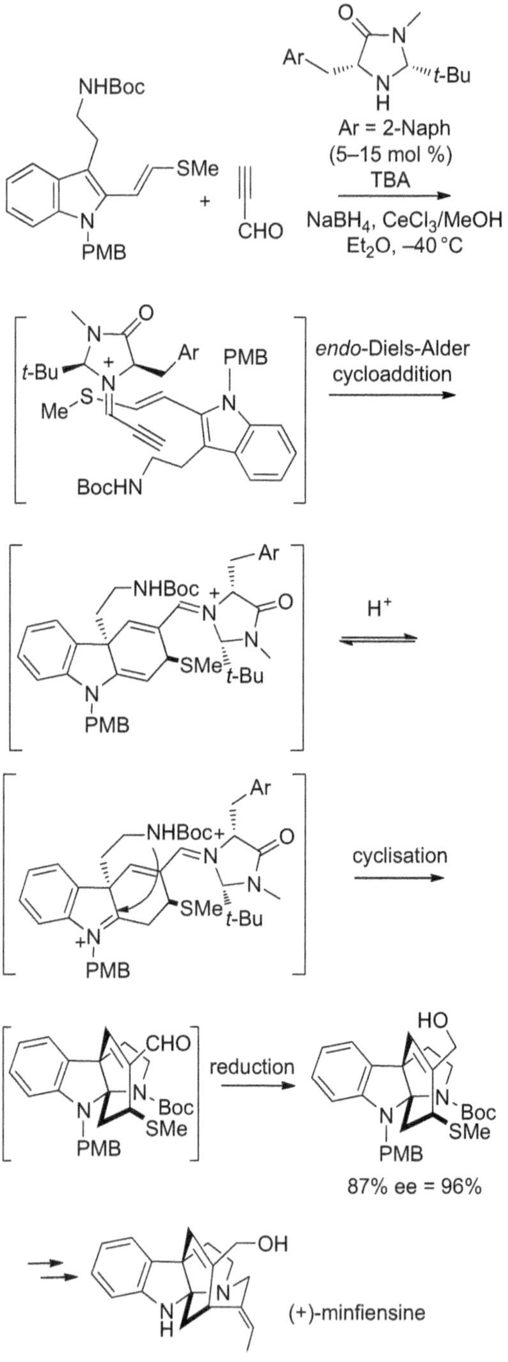

Scheme 3.142 Synthesis of (+)-minfiensine through domino Diels–Alder–amine cyclisation reaction.

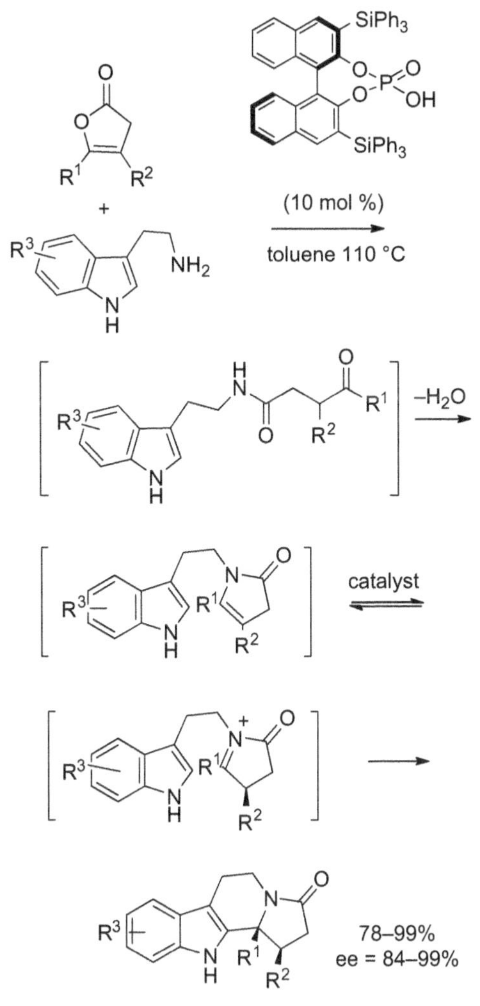

Scheme 3.143 Synthesis of tetracyclic products through domino amidation–
N-acyliminium cyclisation reaction.

simplest and readily available starting materials, such as 2-aminobenzamide
and various benzaldehydes.[225] This extremely simple and practical process was
catalysed by a chiral phosphoric acid derived from BINOL, providing the
required products in good to high yields and enantioselectivities of up to 92%
ee, as shown in Scheme 3.145.

On the other hand, a range of chiral 4,5-disubstituted isoxazoline-*N*-oxide
products have been achieved by Jorgensen *et al.* through two novel interesting
asymmetric domino Henry reaction cyclisation reactions of *in situ* generated
α-brominated aldehydes with ethyl nitroacetate, and of *in situ* generated
epoxides with *t*-butyl nitroacetate, respectively (Scheme 3.146).[226] Remarkable
enantioselectivities of up to 99% ee were obtained combined with moderate

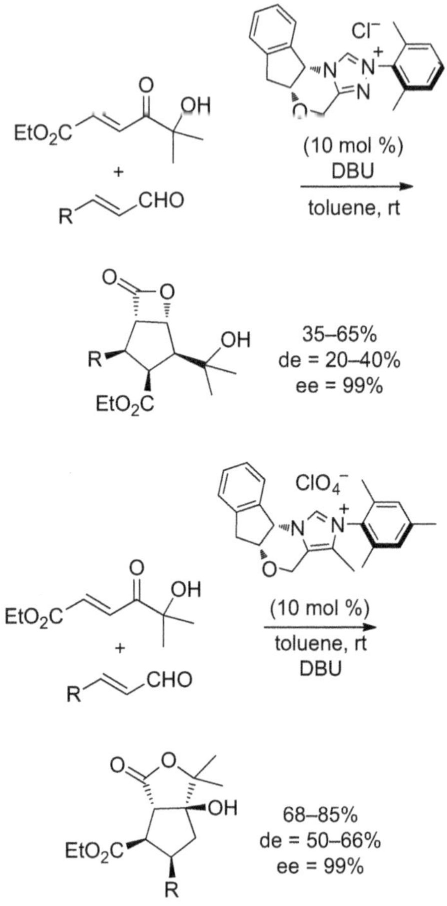

Scheme 3.144 Synthesis of cyclopentane-fused lactones through domino annulation
reactions.

yields in both cases of domino reactions, which involved a chiral diarylprolinol
trimethylsilyl ether as catalyst. Diastereoselectivity of >90% de was obtained
for the domino reaction of α-brominated aldehydes, whereas moderate
diastereoselectivities (≤56% de) were obtained for the domino reaction of
epoxides. It must be noted that the domino reaction of epoxides was performed
under phase-transfer conditions with CsOH as a base combined with a chiral
ammonium salt, because the use of weak bases, such as the imidazole employed
for the domino reaction of α-brominated aldehydes, provided poor or no
selectivity in the Henry reaction.

In 2009, Jacobsen and Klausen described an enantioselective domino imine
formation–Pictet–Spengler reaction of aliphatic as well as aromatic aldehydes
with tryptamines.[227] This process was catalysed by a chiral thiourea combined
with benzoic acid as an additive, and provided the corresponding chiral

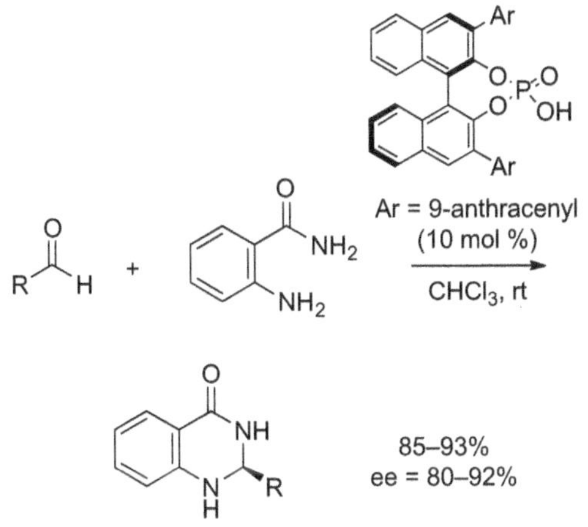

Scheme 3.145 Synthesis of 2,3-dihydroquinazolinones through domino condensation–amine addition reaction.

tetrahydro-β-carbolines which constituted ubiquitous structural motifs in
biologically active natural products. As shown in Scheme 3.147, these products
were achieved in good to high yields and high enantioselectivities of up to
99% ee.

Even if this reaction involved the use of a metal oxidant such as Cu(OTf)$_2$
and, consequently, is somewhat out of the range of this chapter, which does not
include dual catalysis, it is important to remember the first and remarkable
enantioselective catalytic cyclisation strategy for accessing steroidal and
terpenoidal frameworks using organocatalysis, which was reported by
MacMillan and Rendler in 2010.[228] This SOMO-activated polyene cyclisation
of unsaturated aldehydes provided, in the presence of a chiral imidazolidinone
as co-catalyst, the corresponding chiral polycycles in high yields and enantioselectivities of up to 92% ee. An example is depicted in Scheme 3.148. The
complex chiral formed products arose from radical bi- and tricyclisation. The
organocatalyst was supposed to activate the radical aldehyde through the
formation of the corresponding α-imino radical intermediate upon oxidation
with Cu(OTf)$_2$. The authors have extended the scope of this powerful
procedure to extended ring systems, such as tetra-, penta-, and hexacycles,
bearing up to 11 contiguous stereocentres, of which five were all-carbon
quaternary centres.

While iminium catalysis, involving the transformation of iminium ions to
enamines, has been extensively studied in the context of a number of organocatalytic processes, the reverse of this process, involving the conversion of
enamines to iminium species, has been achieved only very recently by Wang
et al.[229] In this study, the authors developed an unprecedented organocatalytic
method for enantioselective β-functionalization of simple aldehydes, which
involved the oxidative conversion of enamines derived from aldehydes to
iminium species, which then underwent enantioselective Michael addition to a

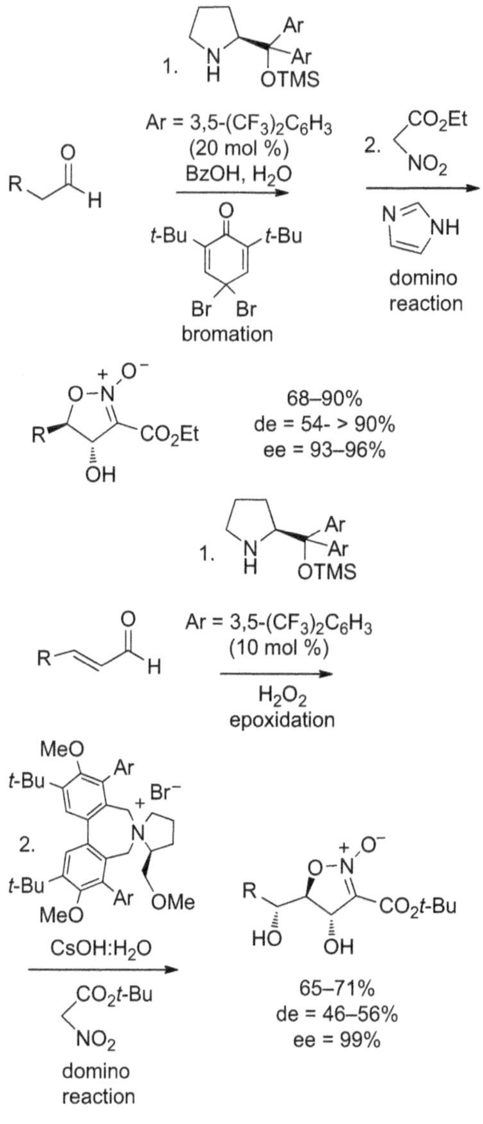

Scheme 3.146 Synthesis of 4,5-disubstituted isoxazoline-*N*-oxides through domino Henry reaction–cyclisation reactions.

nucleophile. This process employed chiral diphenylprolinol trimethylsilyl ether as catalyst combined with the non-toxic *o*-iodoxybenzoic acid (**IBX**) as oxidant. This methodology has been employed to develop a novel enantioselective domino oxidation–Michael–cyclisation reaction occurring between aliphatic aldehydes and 4-hydroxycoumarin as carbon nucleophile, which afforded the corresponding biologically interesting coumarin-containing chiral pyrans in good yields and useful levels of enantioselectivity of up to 91% ee, as shown in

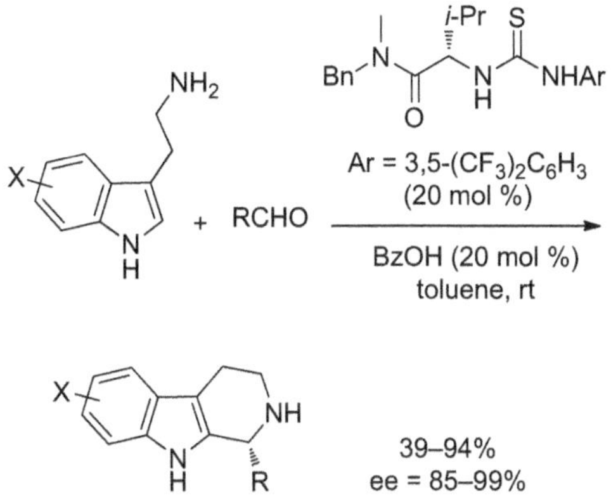

Scheme 3.147 Synthesis of tetrahydro-β-carbolines through domino imine formation–Pictet–Spengler reaction.

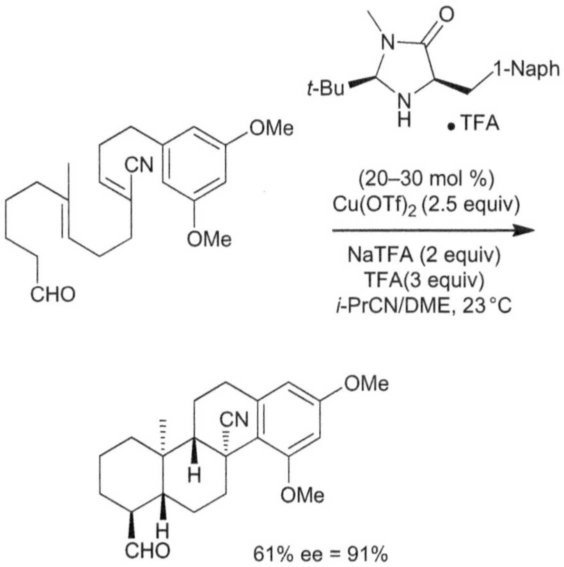

Scheme 3.148 Polyene cyclisation *via* organo-SOMO catalysis.

Scheme 3.149. In addition to carbon nucleophiles, *O*- and *N*-centered nucleophilic reactants (Scheme 3.149) participated by reaction with aldehydes, such as 3-phenylpropionaldehyde, in enantioselective domino oxidation–Michael–aldol–dehydration processes, leading to the corresponding chiral bicyclic products in moderate yields and good enantioselectivities, as shown in Scheme 3.149.

In the same area, Rueping *et al.* developed, in 2012, enantioselective domino oxidation–Michael–intramolecular alkylation reaction of allylic alcohols with

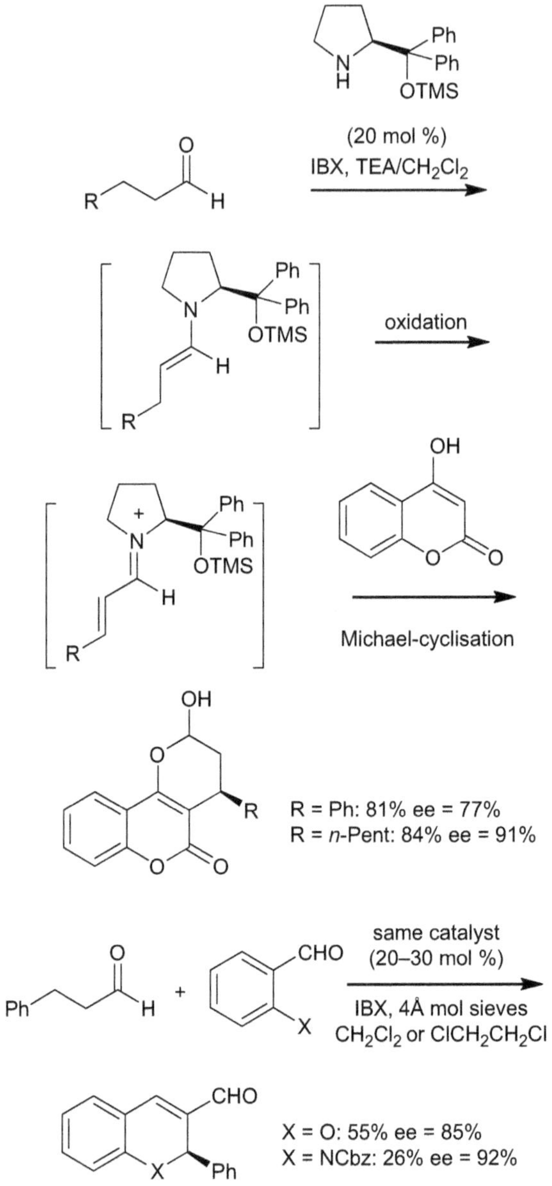

Scheme 3.149 Domino oxidation–Michael–cyclisation and domino oxidation–Michael–aldol–dehydration reactions.

diethyl bromomalonate, providing the corresponding cyclopropanes in high yields and enantioselectivities of up to 98% ee.[230] As shown in Scheme 3.150, the reaction was performed in the presence of MnO_2 as the oxidant, and chiral diphenylprolinol trimethylsilyl ether as organocatalyst.

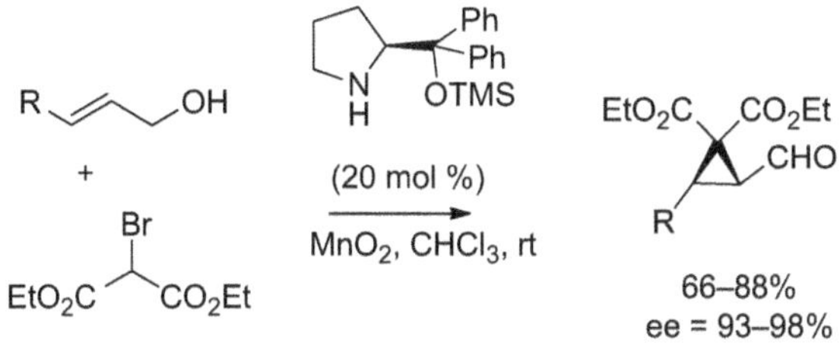

Scheme 3.150 Synthesis of cyclopropanes through domino oxidation–Michael–intramolecular alkylation reaction.

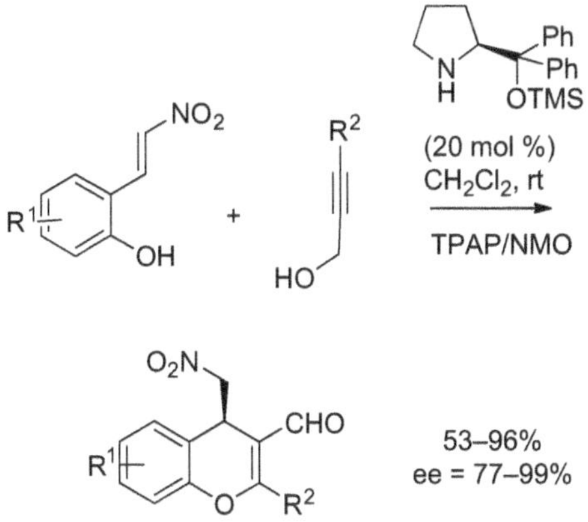

Scheme 3.151 Synthesis of bicyclic chromenes through domino oxidation–oxa-Michael–Michael reaction.

Moreover, an enantioselective domino oxidation–oxa-Michael–Michael reaction could be employed by the same authors in order to synthesise bi- and tricyclic 4*H*-chromenes.[231] The domino process occurred between propargylic alcohols and (*E*)-2-(2-nitrovinyl)-phenols in the presence of the same catalyst, and tetrapropylammonium peruthenate (TPAP)/*N*-methylmorpholine-*N*-oxide (NMO) as the catalytic oxidative system. Excellent results were reached, particularly in the case of bicyclic chromenes, as depicted in Scheme 3.151.

The same authors reported, in 2011, an enantioselective domino Nazarov cyclisation–bromination reaction of dienones with 2,4,4,6-tetrabromo-cyclohexa-2,5-dienone (TBCHD) as a bromine source.[232] The process, catalysed by a chiral BINOL-derived *N*-triflylphosphoramide, yielded the corresponding chiral highly functionalised *trans*-4,5-substituted 5-bromocyclopentenone derivatives in good yields and excellent enantioselectivities, as shown in Scheme 3.152.

In 2006, Akiyama *et al.* employed the pyridinium salt of a chiral BINOL-derived phosphoric acid as organocatalyst to induce a formal aza-Diels–Alder reaction of aldimines with Brassard's diene, providing the corresponding

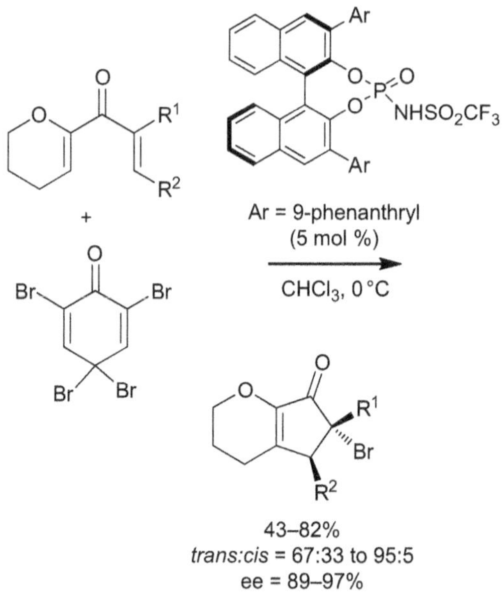

Scheme 3.152   Synthesis of 5-bromocyclopentenones through domino Nazarov cyclisation–bromination reaction.

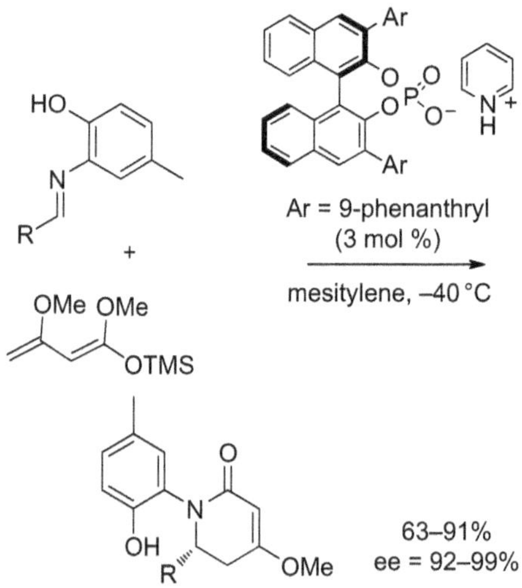

Scheme 3.153   Synthesis of α,β-unsaturated δ-lactams through formal aza-Diels–Alder reaction of aldimines with Brassard's diene.

α,β-unsaturated δ-lactams in high yields and excellent enantioselectivities, as shown in Scheme 3.153.[233]

In addition, the same authors have reported an enantioselective inverse electron-demand aza-Diels–Alder reaction of aldimines with enol ethers,

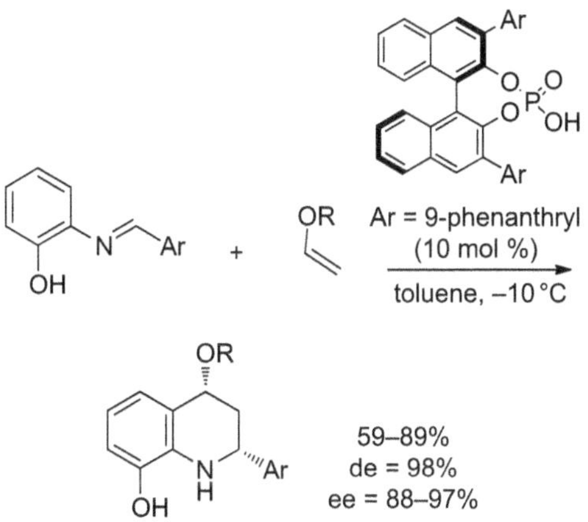

Scheme 3.154 Synthesis of tetrahydroquinolines through formal inverse electron-demand aza-Diels–Alder reaction of aldimines with enol ethers.

employing a BINOL-derived chiral phosphoric acid as catalyst.[234] The domino process gave access to chiral disubstituted tetrahydroquinolines as single *cis*-diastereomers in high yields and enantioselectivities, as shown in Scheme 3.154.

In 2006, Bode *et al.* performed the first *N*-heterocyclic carbene-catalysed formal aza-Diels–Alder reaction.[235] This domino process was performed in the presence of a novel chiral triazolium salt based on the *cis*-1,2-aminoindanol platform, which served as an efficient precatalyst for the *N*-heterocyclic carbene-catalysed redox generation of enolate dienophiles from enals, which were exceptionally reactive. These species underwent LUMO$_{diene}$-controlled Diels–Alder reactions with *N*-sulfonyl-α,β-unsaturated imines in good yields and with exceptional diastereo- and enantioselectivities, providing chiral *cis*-3,4-disubstituted dihydropyridinone products as almost single stereomers, as shown in Scheme 3.155. Additionally, the reaction proceeded at room temperature without stoichiometric reagents, in contrast to uncatalysed variants, and constituted a rare example of a highly enantioselective intermolecular cross-coupling reaction catalysed by an *N*-heterocyclic carbene organocatalyst.

Later, the same authors reported the synthesis of chiral *cis*-1,3,4-trisubstituted cyclopentenes on the basis of an enantioselective *N*-heterocyclic carbene-catalysed domino annulation of α,β-unsaturated aldehydes and 4-oxoenoates.[236] The domino products were achieved in low to high yields, moderate to high diastereoselectivities combined with general excellent enantioselectivities, as shown in Scheme 3.156. The authors have proposed that the reaction evolved through an intermolecular crossed-benzoin reaction, followed by tautomerisation and intramolecular aldol, and, finally, acyl addition and decarboxylation.

In 2007, Scheidt and Chan developed a highly stereoselective formal [3 + 3] cycloaddition between α,β-unsaturated aldehydes and azomethine imines,

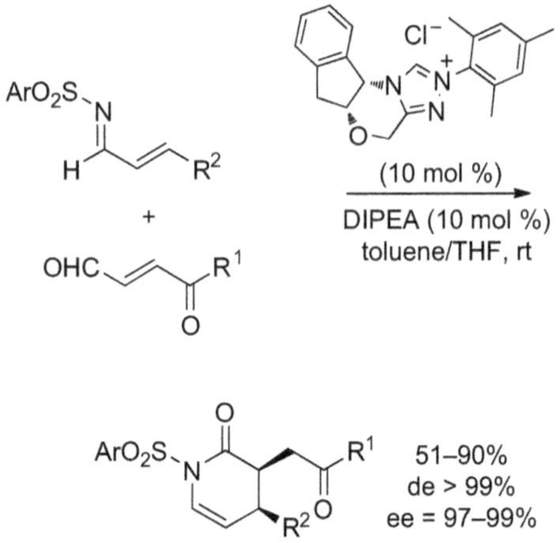

Scheme 3.155 Synthesis of tetrahydroquinolines through *N*-heterocyclic carbene-catalysed formal aza-Diels–Alder reaction.

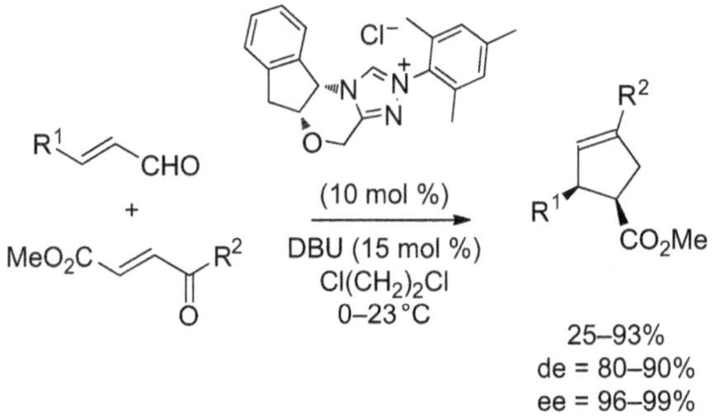

Scheme 3.156 Synthesis of cyclopentenes through *N*-heterocyclic carbene-catalysed domino *cis*-cyclopentannulation.

catalysed by an *N*-heterocyclic carbene derived from *N*-mesityl benzimidatzolium salt.[237] The domino process provided the corresponding substituted pyridazinones in good to excellent yields, and high relative diastereoselectivity, as shown in Scheme 3.157. Again, the *N*-heterocyclic carbene was effective in generating the homoenolate species from α,β-unsaturated aldehyde, which further underwent an addition–acylation sequence with an azomethine imine to afford new bicyclic heterocycles. An intramolecular asymmetric synthesis of α,α-disubstituted cyclopentenes was achieved by the same authors in good yields and good to high enantioselectivities of up to 94% ee, on the basis of *N*-heterocyclic carbene-catalysed desymmetrisation of 1,3-diketones, evolving through enantioselective domino aldol–intramolecular acylation–decarboxylation reaction.[238]

In 2011, Wang *et al.* reported an efficient enantioselective synthesis of medicinally privileged chiral 2-amino-4*H*-chromene-3-carbonitriles, on the

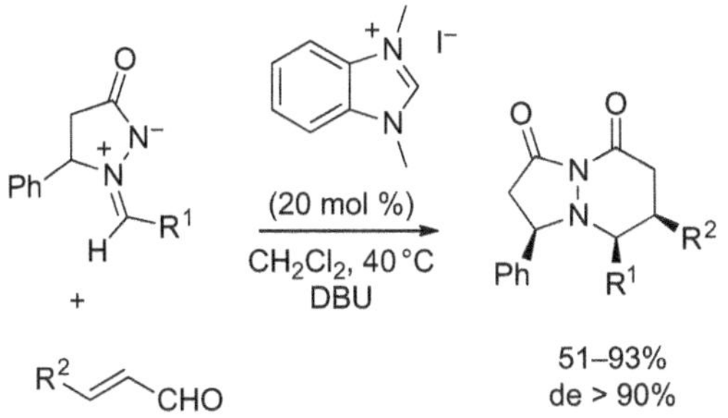

Scheme 3.157 Synthesis of pyridazinones through *N*-heterocyclic carbene-catalysed formal [3 + 3] cycloaddition reaction.

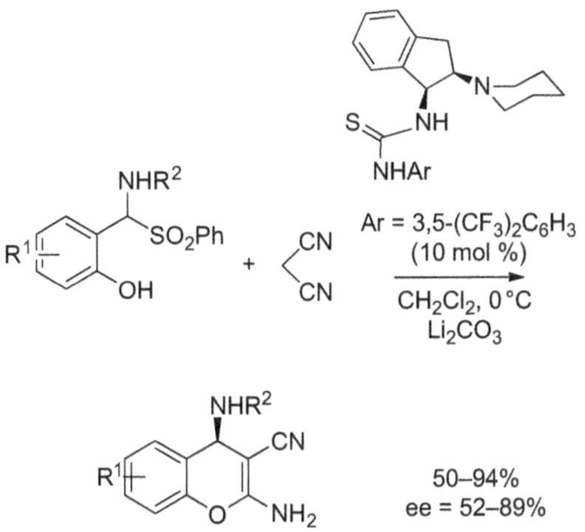

Scheme 3.158 Synthesis of 2-amino-4*H*-chromene-3-carbonitriles.

basis of a domino reaction occurring between *N*-Boc-α-amido sulfones and malononitrile.[239] The process was catalysed by a chiral indane amine-thiourea in the presence of Li_2CO_3 as a base, providing the corresponding 2-amino-4*H*-chromene-3-carbonitriles, in high yields and good to high enantio-selectivities, as shown in Scheme 3.158.

Excellent results in terms of yields, diastereo-, and enantioselectivities were reported by Hayashi *et al.* for a domino desulfonylative *N*-tosyl imine formation–Mannich reaction between an α-amidosulfone and aliphatic aldehydes.[240] As shown in Scheme 3.159, the reaction employed a chiral diarylprolinol trimethylsilyl ether as catalyst combined with $NaHCO_3$ as base, and produced the corresponding chiral domino products in high yields and remarkable levels of stereoselectivities of >90% de and 99% ee for almost all cases of the substrates studied. A further treatment of these domino products by $NaBH_4$ generated the corresponding enantiopure aziridines.

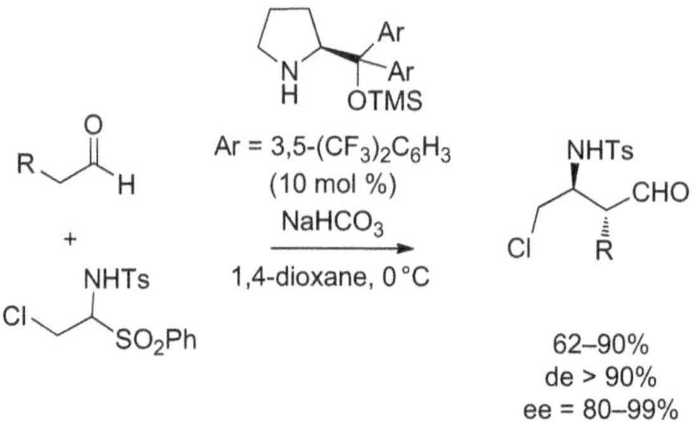

Ar = 3,5-(CF$_3$)$_2$C$_6$H$_3$
(10 mol %)
NaHCO$_3$

1,4-dioxane, 0 °C

62–90%
de > 90%
ee = 80–99%

Scheme 3.159   Synthesis of functionalised amidosulfones through domino desulfonylative imine-formation–Mannich reaction.

3.3   Multicomponent Reactions

One of the challenges in organocatalysis is to implement various reaction
strategies in a multicomponent reaction, which is a domino reaction involving
at least three substrates, to achieve multi-bond formation in one operation.
This strategy is atom economical and avoids the necessity of protecting groups
and isolation of intermediates. Its goal is the resemblance of nature in its highly
selective sequential transformations. Even though the history of multicomponent reactions dates back to the second half of the 19th century with the
reactions of Strecker,[241] Hantzsch,[242] and Biginelli,[243] it was only in the last
decades with the work of Ugi *et al.* that the concept of the multicomponent
reaction has emerged as a powerful tool in synthetic chemistry.[244] The first
organocatalytic enantioselective domino multicomponent reaction was
developed by Barbas *et al.* in 2001, involving an enantioselective
Knoevenagel–Michael reaction between benzaldehyde, acetone, and
diethylmalonate that was catalysed by L-proline.[245]

3.3.1   Multicomponent Reactions Initiated by the Michael Reaction

3.3.1.1   *Michael Reactions of α,β-unsaturated Aldehydes*

Combination of enamine–iminium ion activations in asymmetric organocatalytic domino and multicomponent reactions has been developed to achieve
the enantioselective consecutive formation of two or more bonds in a
stereoselective fashion. The first multicomponent reaction initiated by a
Michael addition combining enamine and iminium ion activations was reported
by Jorgensen *et al.*, in 2005.[246] This domino thia-Michael–amination reaction
proceeded through conjugate addition of nucleophilic thiols to the iminium ion
intermediate of a chiral diarylprolinol trimethylsilyl ether and an
α,β-unsaturated aldehyde, followed by amination of the resulting enamine

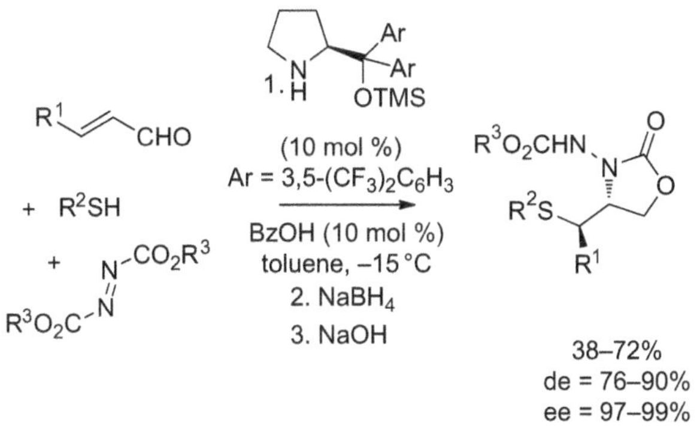

Scheme 3.160   Three-component domino thia-Michael–amination reaction followed by reduction and cyclisation.

intermediate with azodicarboxylates. The domino products were subsequently reduced by treatment with BH$_4$Na, and then base-induced cyclised into final corresponding oxazolidinones. Good results in terms of yields (38–72%), diastereoselectivities (76–90% de), and enantioselectivities (97–99% ee) were achieved in the presence of an additive, such as benzoic acid, for the formation of the corresponding polyfunctionalised oxazolidinones bearing two adjacent stereogenic centres, as shown in Scheme 3.160.

In 2006, Enders *et al.* developed an enantioselective organocatalytic triple cascade reaction for the construction of tetrasubstituted cyclohexene carbaldehydes, as shown in Scheme 161.[247] In this work, the authors paved the way for the sequential creation of three bonds by using a highly enantioselective combination of enamine–iminium–enamine catalysis in a triple cascade domino reaction. This three-component domino process involved a linear aldehyde, a nitroalkene, and an α,β-unsaturated aldehyde as components and chiral diphenylprolinol trimethylsilyl ether as organocatalyst. It proceeded through a Michael–Michael–aldol condensation sequence, affording four stereogenic centres generated in three consecutive carbon–carbon bond formations with high diastereoselectivities and complete enantioselectivities. Indeed, from 16 possible stereomers, only two epimers located in the α-position of the nitro group were formed, in a ratio ranging from 2:1 to 99:1, with the minor isomer easily separated by chromatography. A range of substrates were tolerated, allowing diverse chiral polyfunctional cyclohexenes to be achieved. In the first step, the catalyst activated the linear aldehyde through enamine formation to give the first Michael addition to the nitroolefin. Then, the catalyst was liberated by hydrolysis, being able to form an iminium ion with the enal to catalyse the second Michael addition of the nitroalkane. During this addition, a novel enamine intermediate was formed, which cyclised through an intramolecular aldol condensation to afford the final cyclohexenes with moderate to good yields and complete enantioselectivity.

Later, the methodology of this three-component reaction was extended by these authors, by combination with a highly stereoselective intramolecular

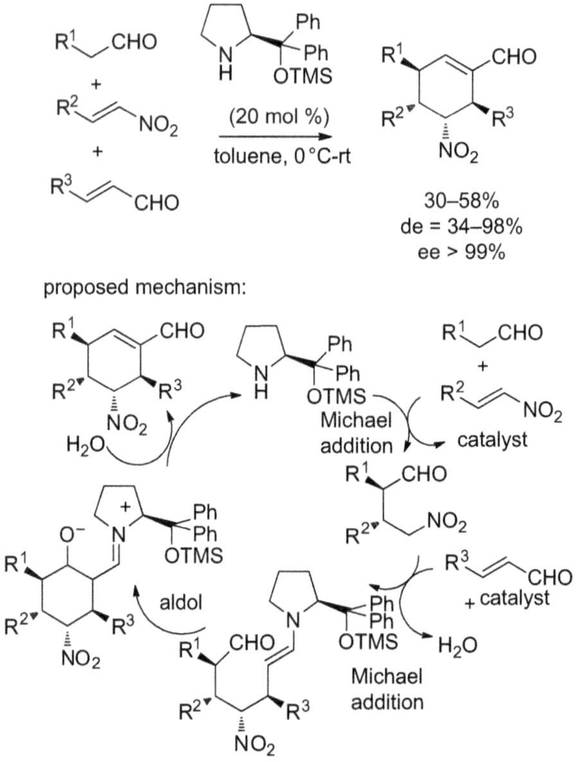

Scheme 3.161   Three-component domino Michael–Michael–aldol reaction.

Diels–Alder reaction.[248] As shown in Scheme 3.162, the domino Michael–Michael–aldol products were submitted to a subsequent intramolecular Diels–Alder cycloaddition through addition of Me₂AlCl to afford the corresponding almost enantiopure chiral tricyclic carbaldehydes in moderate to good yields and diastereoselectivities. In this remarkable domino reaction, five C–C bonds were formed with creation of up to eight new stereogenic centres.

A number of other enantioselective domino Michael–Michael–aldol reactions have been developed by several groups. In 2007, Jorgensen *et al.* disclosed a new approach for an enantioselective three-component reaction evolving through iminium–iminium–enamine catalysis.[249] This process was catalysed by a chiral diarylprolinol trimethylsilyl ether, and involved two α,β-unsaturated aldehydes and an activated methylene compound, such as malononitrile or a cyanoacetate. The reaction began with the Michael addition of the activated methylene compound onto the first α,β-unsaturated aldehyde through iminium activation, which produced the corresponding enamine, which was further hydrolysed into the corresponding aldehyde with regenerating catalyst. This novel intermediate was subsequently added to the second α,β-unsaturated aldehyde activated through iminium catalysis to give the

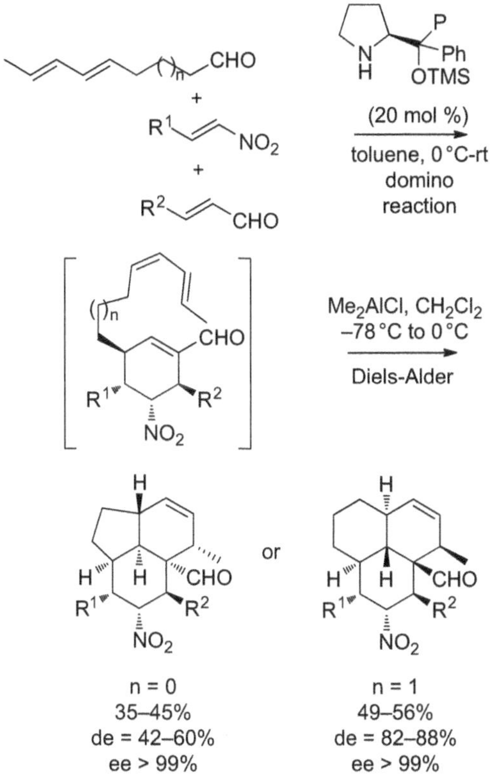

Scheme 3.162 Three-component domino Michael–Michael–aldol reaction followed by intramolecular Diels–Alder cycloaddition.

corresponding enamine. This enamine was then intramolecularly condensed to the aldehyde function to afford, finally, the corresponding cyclohexen-1-ene-carbaldehyde. A range of chiral domino functionalised products could be generated in good yields, good to excellent diastereoselectivities of up to >97% de, and generally excellent enantioselectivities of >97% ee through this novel domino Michael–Michael–aldol reaction, as shown in Scheme 3.163.

A highly efficient enamine–iminium–enamine sequence was described by Melchiorre *et al.*, in 2008.[250] In this work, a three-component domino Michael–Michael–aldol reaction occurred between an aldehyde, an α,β-unsaturated aldehyde, and a *trans*-α-cyanocinnamate, which provided through a triple cascade catalysed by chiral diphenylprolinol trimethylsilyl ether the corresponding tetrasubstituted cyclohexene carbaldehydes in moderate to good yields, moderate to high diastereoselectivities of up to 96% de, and generally excellent enantioselectivities of up to 99% ee, as shown in Scheme 3.164.

The application of polymer-supported pyrrolidine catalysis in directing the asymmetric domino Michael–Michael–aldol three-component reaction of cinnamaldehyde, *trans*-β-nitrostyrene, and propionaldehyde was reported by

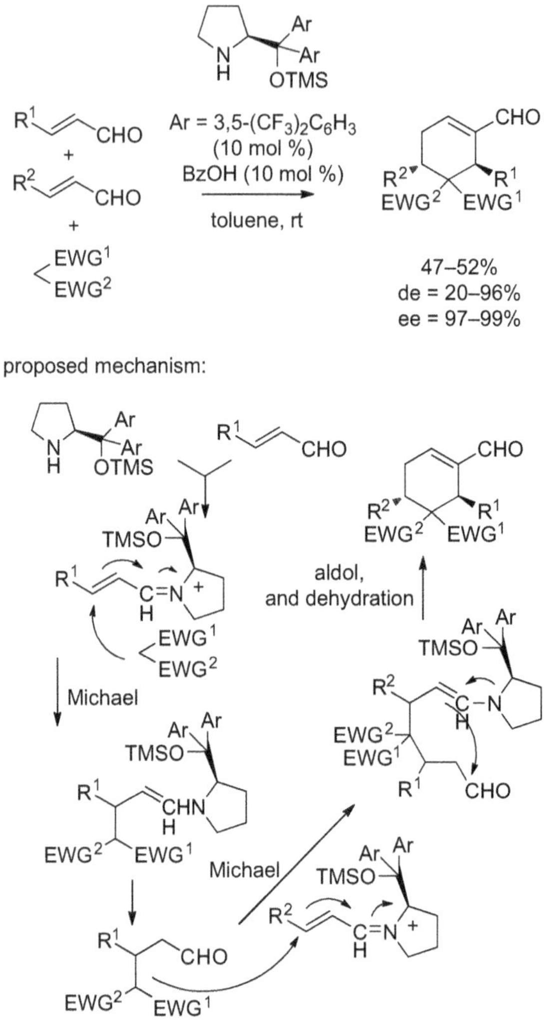

Scheme 3.163 Three-component domino Michael–Michael–aldol–dehydration reaction.

Schore *et al.*, in 2008.[251] The advantages of catalyst immobilisation included the ease of both product separation and purification and reusability of the chiral catalyst. While Enders *et al.* achieved 40% yield and enantioselectivity of >99% ee by using chiral diphenylprolinol trimethylsilyl ether,[247] the solid phase variant proceeded with significantly decreased yield (15%) and selectivity (89% ee), in the case of the non-functionalised polymer-supported catalyst depicted in Scheme 3.165. Superior results in comparison with Enders' work were, however, reached by employing a more sterically constrained and functionalised polymer-supported catalyst, as depicted in Scheme 3.165. Indeed, in this case, 45% yield, combined with enantioselectivity of 99% ee, was achieved.

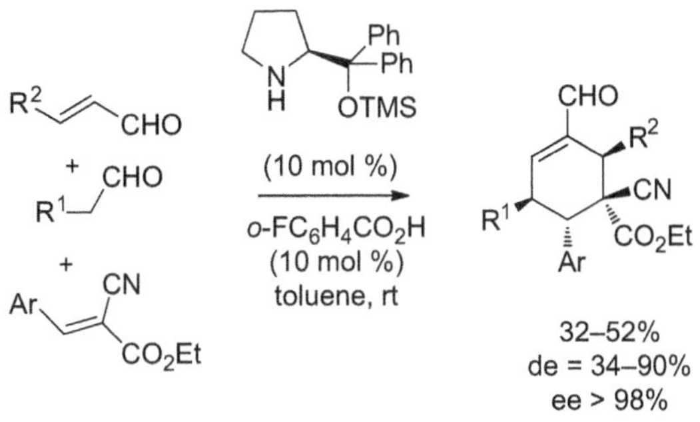

Scheme 3.164   Synthesis of cyclohexene carbaldehydes through three-component domino Michael–Michael–aldol reaction.

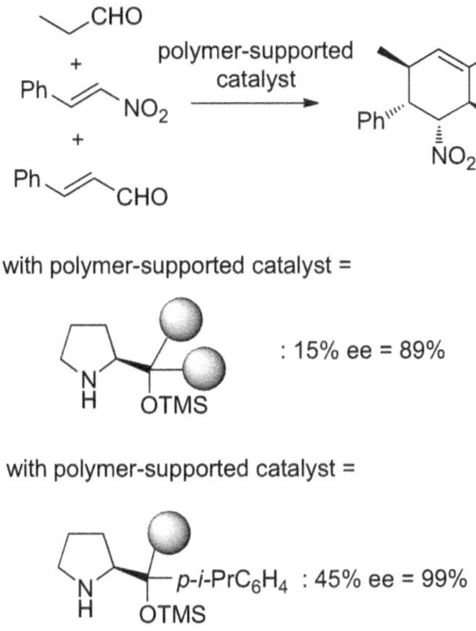

Scheme 3.165   Three-component domino Michael–Michael–aldol reaction induced by polymer-supported catalysts.

A remarkable organocatalytic asymmetric three-component domino reaction of nitromethane and two equivalents of α,β-unsaturated aldehydes was reported by Enders *et al.*, in 2009.[252] Upon catalysis with chiral diphenyl-prolinol trimethylsilyl ether, this novel triple cascade reaction based on two subsequent Michael additions followed by an intramolecular aldol conden-sation provided the corresponding almost diastereo- and enantiopure 5-nitrocyclohexene carbaldehydes in moderate to good yields with diastereo- and enantioselectivities of >98% de and >99% ee, respectively, in all cases of the substrates used, as shown in Scheme 3.166. The mechanism proposed by the authors is depicted in this Scheme. A first equivalent of α,β-unsaturated

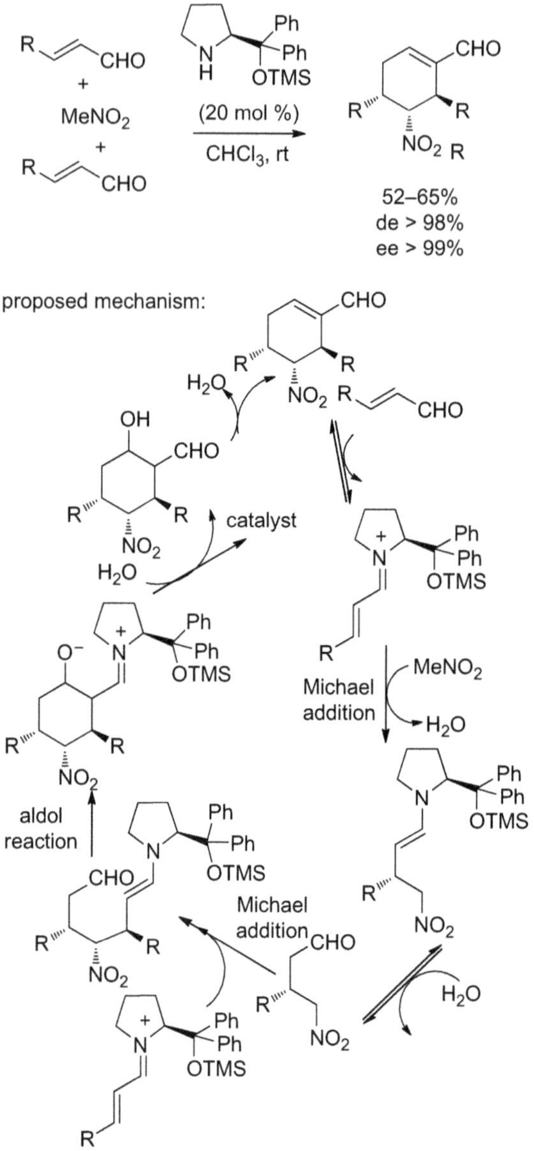

Scheme 3.166 Synthesis of 5-nitrocyclohexene carbaldehydes through three-component domino Michael–Michael–aldol reaction.

aldehyde was activated as its corresponding iminium ion by chiral diphenyl-prolinol trimethylsilyl ether and then underwent the first nitromethane Michael addition to afford a nitroalkane enamine intermediate. Subsequent hydrolysis generated a nitroaldehyde intermediate and the catalyst, which could promote a second Michael addition with a second equivalent of iminium, furnishing an enamine intermediate. This enamine reacted *via* an intramolecular aldol cyclisation to give a novel iminium intermediate. After hydrolysis, the catalyst was regenerated, and the intermediate alcohol could dehydrate to afford the final products.

Another highly enantioselective three-component reaction also based on a domino Michael–Michael–aldol sequence induced by the same catalyst associated with benzoic acid as co-catalyst was reported by Rios *et al.*, in 2010.[253] In this work, 2 equivalents of α,β-unsaturated aldehydes reacted with oxindoles to afford the corresponding almost enantiopure spirooxindoles in good yields and excellent diastereoselectivities of >92% de in almost all cases of the substrates studied, as shown in Scheme 3.167. Remarkably, the scope of this reaction was extended to a range of other heterocycles, such as benzofuranones, pyrazolones, or azlactones, rendering the corresponding spiro compounds in comparable yields and stereoselectivities.

In 2011, Melchiorre *et al.* reported the synthesis of a range of other almost enantiopure spirocyclic benzofuranones in moderate to good yields (52–70%) and high diastereoselectivity of >90% de.[254] These products were generated from a three-component domino Michael–Michael–aldol–dehydration sequence involving aldehydes, α,β-unsaturated aldehydes, and benzofuranone-based compounds, upon catalysis with chiral diphenylprolinol trimethylsilyl ether in combination with *ortho*-fluorobenzoic acid. In this case, the sequence started with the Michael addition of the aldehydes onto the Michael acceptors, evolving through enamine activation to give intermediate aldehydes. Then, a Michael–aldol sequence occurred between these aldehydes and α,β-unsaturated aldehydes through iminium–enamine activation, which led to aldehyde intermediates; these dehydrated subsequently to afford the desired spirocyclic benzofuranones, as shown in Scheme 3.168. These authors previously developed the corresponding three-component domino Michael–Michael–aldol–dehydration sequence from aldehydes, α,β-unsaturated aldehydes, and

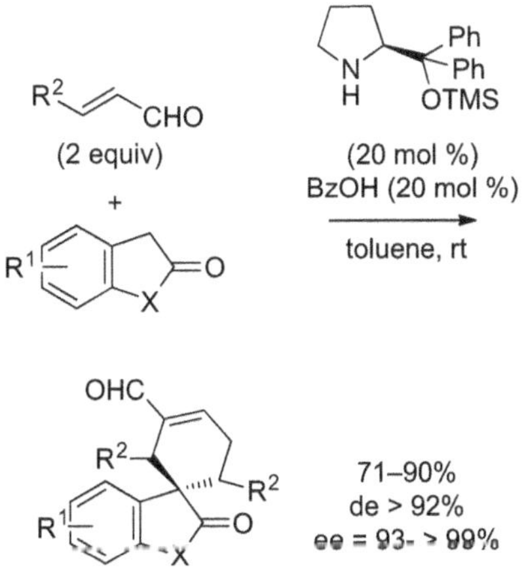

Scheme 3.167 Synthesis of spiro compounds through three-component domino Michael–Michael–aldol reaction.

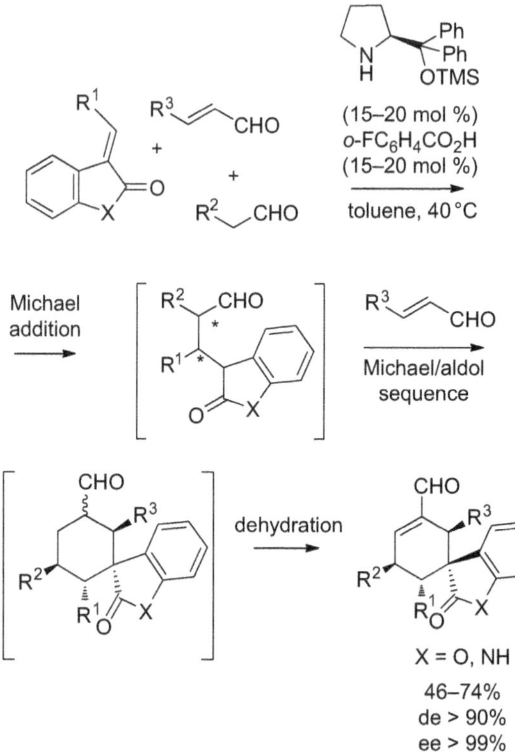

Scheme 3.168 Synthesis of spiro compounds through three-component domino Michael–Michael–aldol–dehydration reaction.

oxindole-based compounds in the presence of comparable conditions.[255] This process provided the corresponding spirocyclic oxindoles in good yields, high diastereoselectivity of >90% de, and excellent enantioselectivity of >99% ee, as shown in Scheme 3.168. Comparable results have also been obtained by Chen *et al.* by using the same catalyst in combination with benzoic acid in acetonitrile as solvent.[256]

Furthermore, remarkable enantioselective domino reactions involving three successive Michael additions followed by an aldol reaction have been recently reported by several groups. As an example, Hong *et al.* have employed chiral diphenylprolinol trimethylsilyl ether to induce a highly diastereo- and enantioselective quadruple-cascade oxa-Michael–Michael–Michael–aldol reaction, constructing four new bonds and five stereocentres in one step, which provided expedited access to highly functionalised and almost enantiopure tetrahydro-6*H*-benzo[*c*]chromene.[257] This process was based on the reaction of 2-(*E*)-(2-nitrovinyl)-phenol with cinnamaldehyde performed in the presence of an additive, such as acetic acid, and occurring through iminium–enamine mode activation. As shown in Scheme 3.169, both the diastereo- and enantioselectivities were excellent, up to >94% de and >99% ee, respectively. The

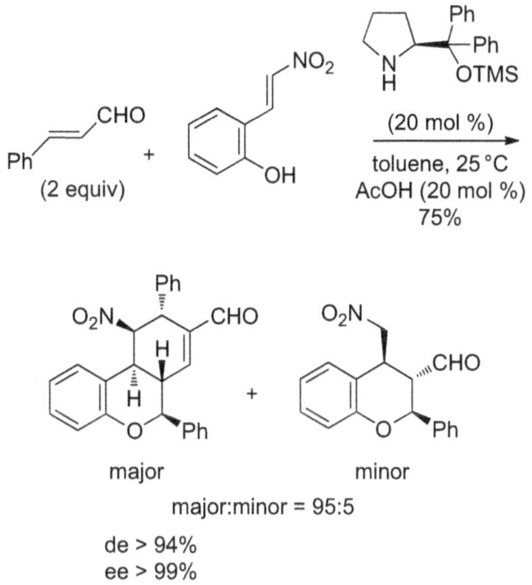

Scheme 3.169 Synthesis of tetrahydro-6*H*-benzo[*c*]chromenes through three-component domino oxa-Michael–Michael–Michael–aldol reaction.

corresponding oxa-Michael–Michael–Michael product was formed as minor product. Although nitroalkenes are also implicated as substrates in this reaction, this work is situated in the section dealing with three-component reactions of α,β-unsaturated aldehydes because the first Michael addition occurred on the latter and not on nitroalkenes.

In 2009, a novel asymmetric four-component oxa-Michael–Michael–Michael–aldol condensation reaction was described by Hong *et al.*, which was based on an iminium–enamine–iminium–enamine sequence, depicted in Scheme 3.170.[258] This process, catalysed by a combination of amine (*S*)-**3** used at a low catalyst loading of 5 mol% and benzoic acid as an additive, afforded a range of almost enantiopure cyclohexene carbaldehydes in moderate to good yields and high diastereoselectivity of >90% de in all cases of tha substrates studied. The first intermolecular oxa-Michael addition of the alcohol to an iminium ion, derived from activation of 1 equivalent of acrolein by chiral diphenylprolinol trimethylsilyl ether, led to the formation of the corresponding enamine intermediate, which underwent a second intermolecular Michael addition onto nitroalkene, providing a nitromethane intermediate. Then, a third intermolecular Michael addition of this nitromethane to a second equivalent of iminium ion produced an enamine intermediate, which was unstable and easily reacted through an intramolecular aldol condensation to give the final trisubstituted cyclohexene carbaldehydes and regenerate the catalyst. Although nitroalkenes are also implicated as substrates in this reaction, this work is situated in this section dealing with three-component

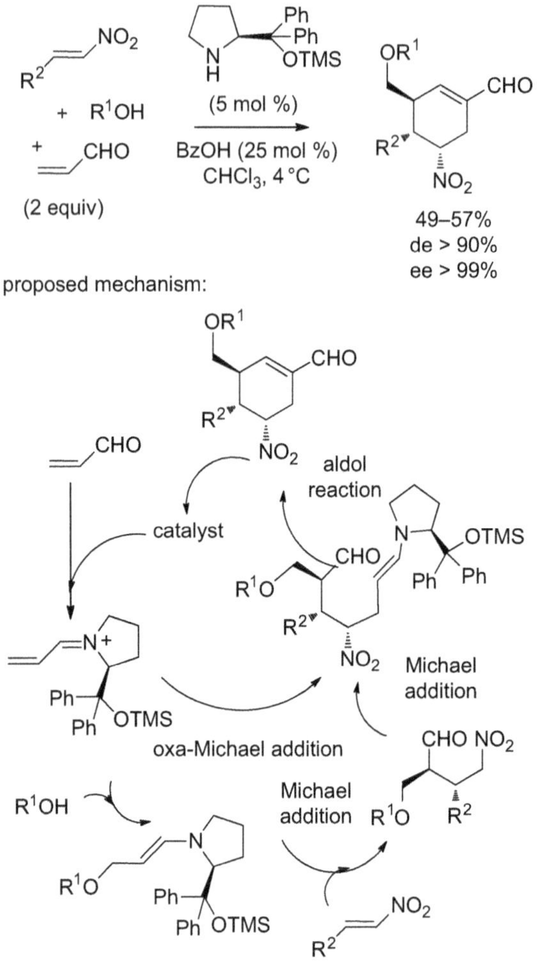

Scheme 3.170 Synthesis of cyclohexene carbaldehydes through four-component domino oxa-Michael–Michael–Michael–aldol reaction.

reactions of α,β-unsaturated aldehydes because the first Michael addition occurred on the latter and not on nitroalkenes.

In 2010, the efficient assembly of hydroindane derivatives incorporating a spirooxindole motif was achieved by Chen *et al.* through another enantio-selective three-component domino reaction of two different molecules of α,β-unsaturated aldehydes with (*E*)-4-(1-methyl-2-oxoindolin-3-ylidene)-3-oxobutanoates, catalysed by the same catalyst combined with benzoic acid as co-catalyst.[259] According to a Michael–Michael–Michael–aldol sequence, the process provided the corresponding almost enantio- and diastereopure complex fused tetracyclic products bearing six contiguous stereogenic centres in moderate to excellent yields (30–97%), as shown in Scheme 3.171. The authors have proposed a quadruple iminium–enamine–iminium–enamine catalysis to

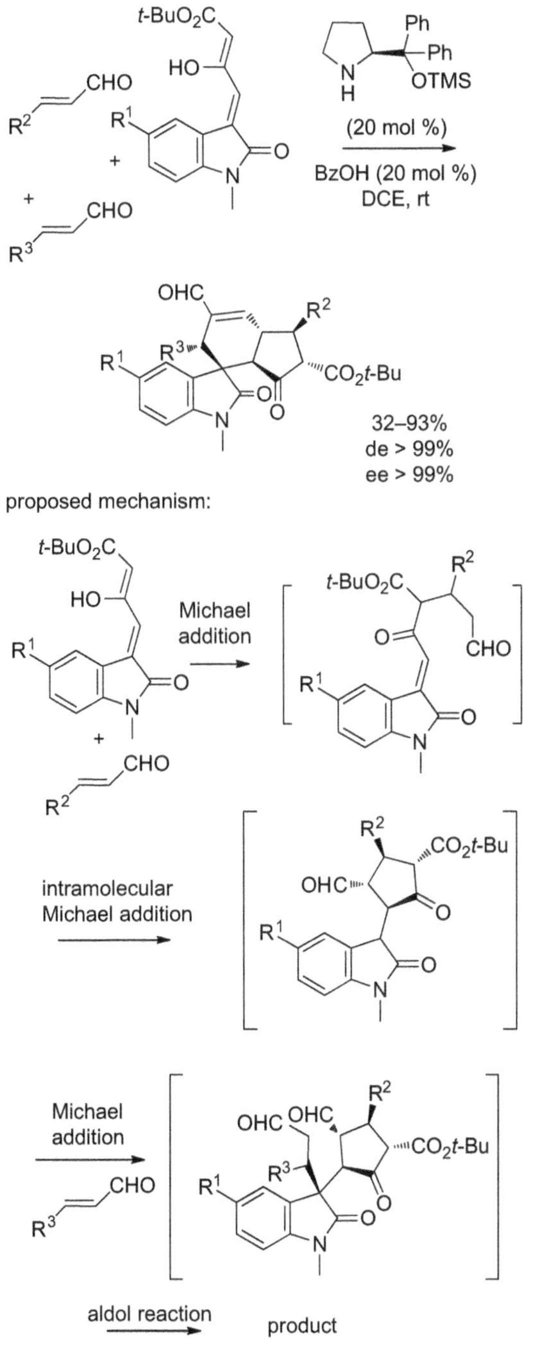

proposed mechanism:

Scheme 3.171 Synthesis of fused tetracyclic products through three-component domino Michael–Michael–Michael–aldol reaction.

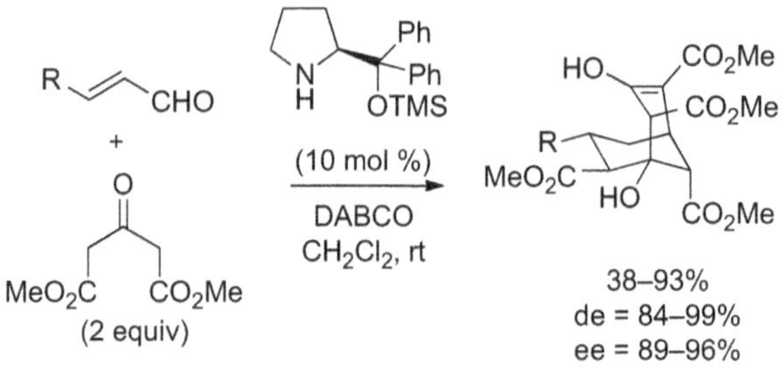

Scheme 3.172 Synthesis of bicyclic cyclohexanones through three-component domino Michael–aldol–dehydration–Michael reaction.

explain the results. As depicted in Scheme 3.171, after the first domino Michael–Michael reaction of butanoate and one equivalent of α,β-unsaturated aldehyde was complete, from which an aldehyde intermediate was generated, the second domino Michael–aldol reaction occurred between this intermediate and the second equivalent of α,β-unsaturated aldehyde, leading to the final product.

In addition, highly efficient simpler enantioselective domino Michael–aldol reactions have been developed. For example, Jorgensen et al. reported the synthesis of chiral bicyclic cyclohexanones on the basis of a three-component reaction of α,β-unsaturated aldehydes with two equivalents of a tricarbonylic compound, catalysed by chiral diphenylprolinol trimethylsilyl ether, in 2008.[260] The domino process provided high levels of diastereo- and enantioselectivities, as shown in Scheme 3.172. It remarkably allowed the formation of 1 out of 64 stereomers by creating six stereocentres. The process seems to proceede through a domino Michael–aldol–dehydration–Michael reaction.

Later, another example was reported by Rios et al., dealing with an interesting asymmetric synthesis of spiropyrazolones induced by the same catalyst.[261] In this case, the domino Michael–aldol reaction occurred between an enolisable aldehyde, an α,β-unsaturated aldehyde, and an unsaturated pyrazolone to afford the corresponding chiral spiropyrazolones bearing four stereocentres in moderate to good yields and excellent diastereo- and enantio-selectivities, as shown in Scheme 3.173.

In another context, Cao et al. have developed a highly enantioselective three-component domino Michael–alkylation–Wittig reaction of α,β-unsaturated aldehydes with two equivalents of arsonium ylides, catalysed by a chiral recyclable dendritic catalyst derived from diphenylprolinol trimethylsilyl ether.[262] The use of this chiral 2-trimethylsilanyloxy-methyl-pyrrolidine-based dendritic catalyst allowed the corresponding almost diastereo- and enantiopure (E)-cyclopropanes in good yields, as shown in Scheme 3.174. It was demonstrated that the catalyst was recoverable and reusable without any loss in activity.

In the same year, Cordova et al. reported the highly enantioselective synthesis of cycloheptane derivatives.[263] This remarkable four-component

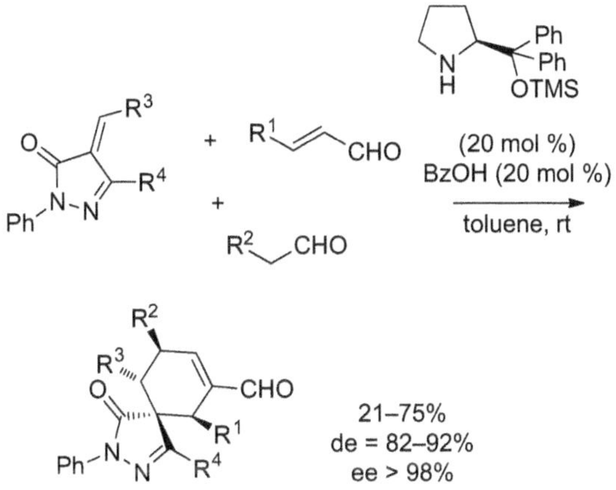

Scheme 3.173 Synthesis of spiropyrazolones through three-component domino Michael–aldol reaction.

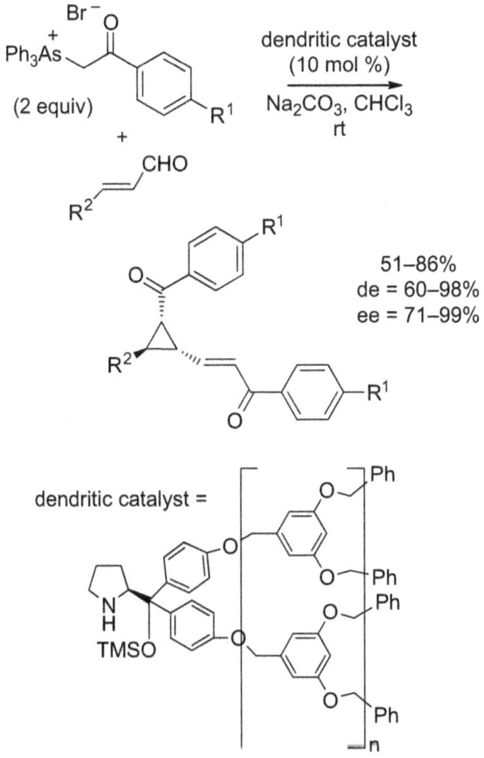

Scheme 3.174 Synthesis of (*E*)-cyclopropanes through three-component domino Michael–alkylation–Wittig reaction.

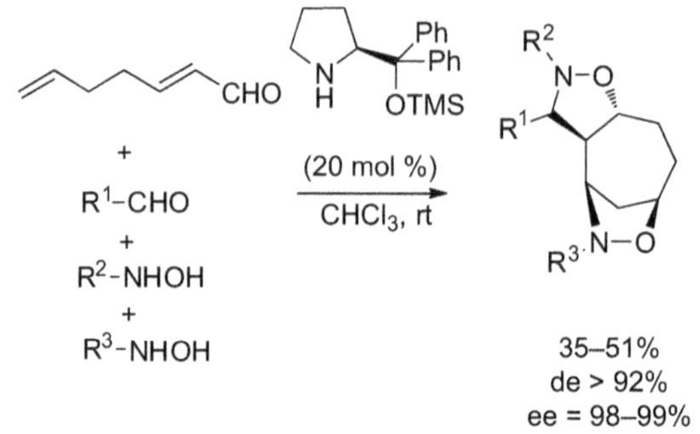

Scheme 3.175 Synthesis of isoxazolidines through four-component aza-Michael-initiated domino reaction.

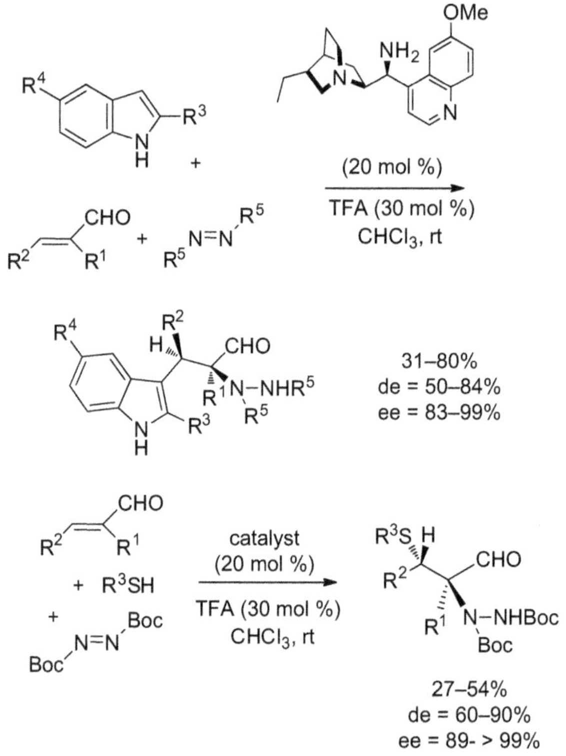

Scheme 3.176 Three-component domino Michael–amination reactions.

process furnished tricyclic products with excellent yields and enantio-selectivities and, moreover, in a diastereopure form, as shown in Scheme 3.175. It was reported as an unprecedented example of a regiospecific and highly chemoselective one-pot organocatalytic cascade synthesis of bis-oxazolidines with a functionalised seven-membered carbocycle core. The reaction was efficiently catalysed by chiral diphenylprolinol trimethylsilyl ether and provided a direct entry to nearly diastereo- and enantiomerically pure cycloheptanes, in which the formation of five new bonds and five stereocentres was controlled.

On the other hand, Melchiorre *et al.* have developed two asymmetric three-component domino Michael–amination reactions catalysed by a chiral cinchona-derived primary amine and evolving through iminium–enamine activation mode.[264] The first sequence involved α,β-disubstituted aldehydes, indoles, and azodicarboxylates, which provided the corresponding chiral highly functionalised indole products in good yields, moderate diastereoselectivities (50–78% de) and excellent enantioselectivities of up to 99% ee, as shown in Scheme 3.176. The second three-component domino thia-Michael–amination reaction occurred between α,β-disubstituted enals, azodicarboxylates, and thiols, affording the corresponding products in moderate yields and moderate to high diastereoselectivities (60–90% de), combined with good to excellent enantioselectivities (72 to >99% ee), as shown in Scheme 3.176.

3.3.1.2 Michael Reactions of other α,β-Unsaturated Carbonyl Compounds

In 2008, Rueping and Antonchick described a highly enantioselective three-component reaction of an enamine, a vinyl ketone, and Hantzsch ester catalysed by a chiral BINOL-derived phosphoric acid.[265] This powerful process provided efficient access to chiral tetrahydropyridines and azadecalinones from simple starting materials, involving five successive types of reaction, starting with Michael addition of the enamine to the enone, intramolecular aza-cyclisation to give an intermediate six-membered hemiaminal, dehydration, isomerisation/protonation, and finally transfer hydrogenation, all catalysed by the same organocatalyst (Scheme 3.177).

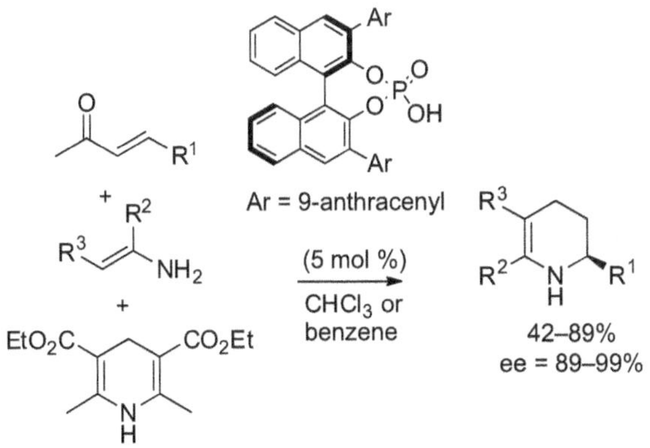

Scheme 3.177 Synthesis of tetrahydropyridines through three-component domino Michael–hemiaminalisation–dehydration–isomerisation–transfer hydrogenation reaction.

Later, Carter and Yang developed the synthesis of chiral highly functionalised bicyclo[2.2.2]-octanes containing up to four contiguous stereocentres starting from cyclic enones such as cyclohexenone.[266] Indeed, the three-component reaction among cyclohexenone, benzylamine, and aldehydes provided, upon catalysis with a proline-derived chiral sulfonamide, the corresponding highly functionalised bicyclo[2.2.2]-octanes in moderate yields, high diastereoselectivity of >90% de in all cases of the substrates studied, and high enantioselectivities of up to 92% ee, as shown in Scheme 3.178. These products arose from a domino Michael–Mannich reaction. For the first time the enamine, generated *in situ* from benzyl amine and aldehyde, was added onto the iminium ion arising from the activation of cyclohexenone by the catalyst to give the corresponding enamine intermediate. After interconversion of this enamine, an intramolecular Mannich reaction cyclisation provided a bicyclic intermediate, which after hydrolysis regenerated the catalyst and furnished the final product, as shown in Scheme 3.178.

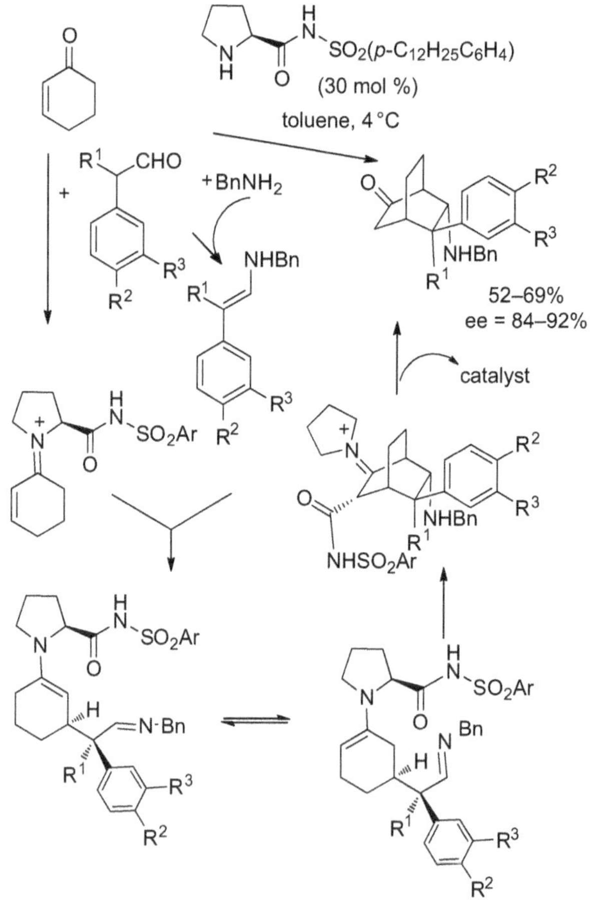

Scheme 3.178 Synthesis of bicyclo[2.2.2]-octanes through three-component domino Michael–Mannich reaction.

These authors have extended the scope of this methodology to various acyclic enones, which afforded by reaction with benzyl amine and aldehydes the corresponding densely functionalised cyclohexenones bearing two contiguous stereogenic centres, arising from the corresponding three-component domino Michael–Mannich reaction.[267] In this case, the reaction was performed in dichloroethane (DCE) as the solvent in the presence of molecular sieves, providing moderate to good yields and diastereoselectivities of up to >90% de combined with enantioselectivities of up to 95% ee, as shown in Scheme 3.179.

In addition, Sugiura *et al.* have developed a diastereo- and enantioselective domino reductive aldol reaction of enones and aldehydes with trichlorosilane

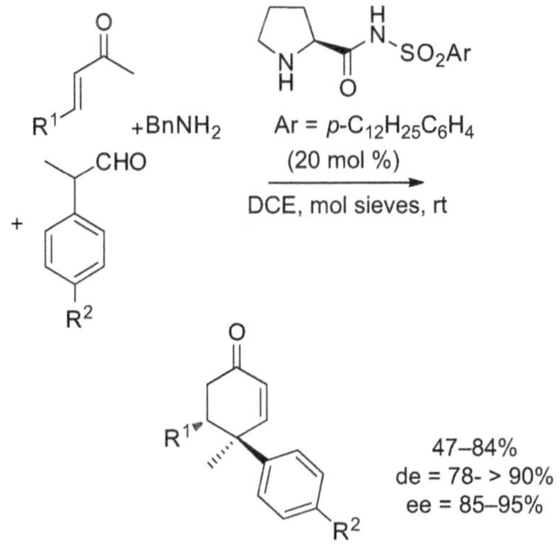

Scheme 3.179 Synthesis of cyclohexenones through three-component domino Michael–Mannich reaction.

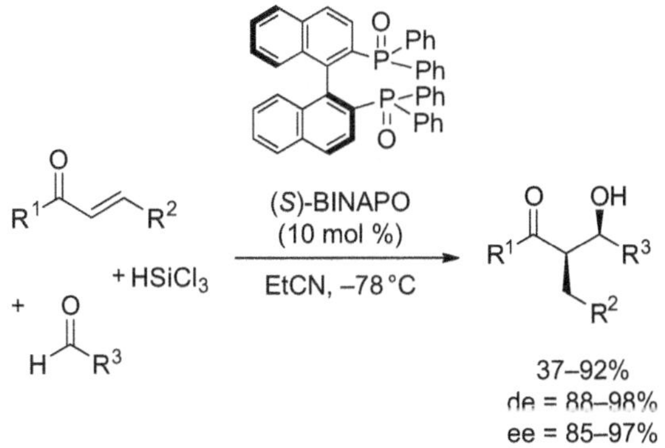

Scheme 3.180 Domino reductive Michael–aldol reaction of enones, aldehydes, and trichlorosilane.

upon catalysis by (*S*)-BINAPO (2-diphenylphosphino-2′-diphenylphosphinyl-1,1′-binaphthalene).[268] This methodology afforded the corresponding chiral β-hydroxy ketones with good to high *syn*-diastereo- and enantioselectivities of up to 98% de and 97% ee, respectively. As shown in Scheme 3.180, the authors have proposed that the catalyst induced the conjugate reduction with trichlorosilane through a six-membered transition state with the enone in the s-*cis*-conformation to give the (*Z*)-trichlorosilyl enolate exclusively. The high observed *syn* selectivity arose from the subsequent aldol process, which proceeded through a chair-like cyclic transition state.

3.3.1.3   Michael Reactions of Nitroolefins

One of the milestones in organocatalysis was the first enantioselective addition of malonates to nitrostyrenes, catalysed by chiral thioureas, reported by Takemoto *et al.* in 2003, which was the starting point for multiple reactions with nitrostyrenes.[42,269] In 2008, the groups of Chen[270] and Rios[271] reported, almost at the same time, the first asymmetric three-component formal 1,3-dipolar cycloaddition of aldehydes, α-aminomalonates, and nitroalkenes catalysed by chiral thioureas. The domino reaction began with the formation of an imine from the aldehyde and the α-aminomalonate. This imine then reacted with the nitroalkene through a Michael addition and subsequently an aza-Henry reaction occurred, affording the corresponding highly substituted pyrrolidines as single diastereomers, in good to high yields and moderate to high enantioselectivities, as shown in Scheme 3.181.

An enantioselective three-component domino Michael–Michael–aldol reaction was employed as a key step by Enders *et al.* for the stereoselective synthesis of polysubstituted mono- and bicyclic cyclohexene-carbaldehydes, in 2008.[272] This domino process involved aldehydes, α,β-unsaturated aldehydes, and nitroalkenes, and was induced by chiral diphenylprolinol trimethylsilyl

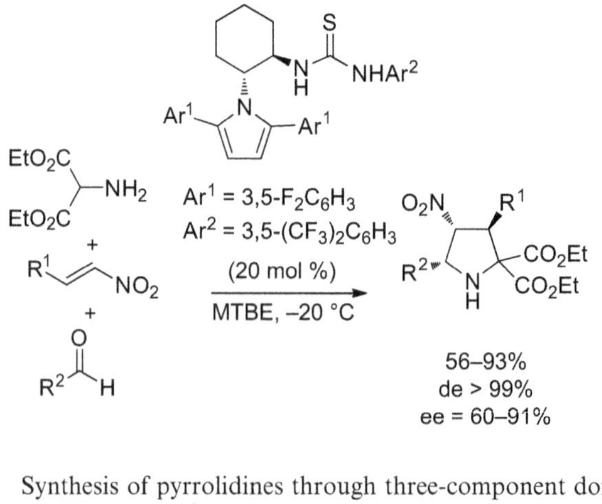

Scheme 3.181   Synthesis of pyrrolidines through three-component domino Michael–aza-Henry reaction.

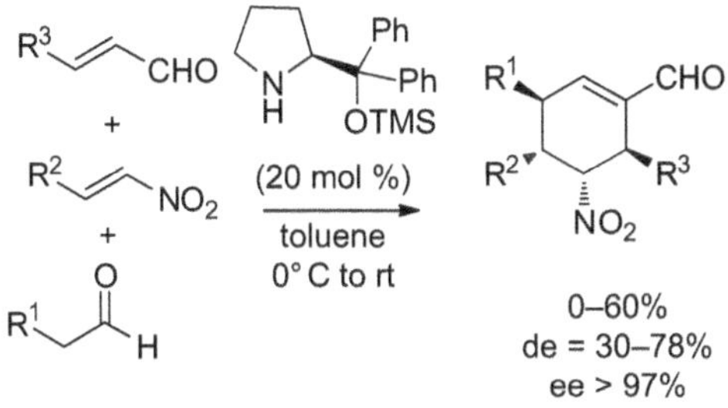

Scheme 3.182 Synthesis of cyclohexene-carbaldehydes through three-component domino Michael–Michael–aldol reaction.

ether. The process, evolving through enamine–iminium–enamine activation mode, began with the formation of the corresponding enamine of the aldehyde. This enamine underwent a Michael addition to the nitroalkene to give an intermediate nitro-aldehyde. The latter was then condensed onto the α,β-unsaturated aldehyde, providing an intermediate which finally cyclised through aldol condensation to give the domino products in zero to good yields, low to good diastereoselectivities, and generally excellent enantioselectivities of >97% ee, as shown in Scheme 3.182.

When nitrostyrenes and dimethylmalonate were reacted with α,β-unsaturated aldehydes in the presence of a combination of chiral diphenylprolinol triethylsilyl ether and a chiral cinchona alkaloid catalyst through a three-component reaction, Xu et al. showed that they afforded, according to a Michael–Michael–aldol domino reaction, the corresponding chiral functionalised cyclohexanes in moderate to good yields and good to excellent enantioselectivities of up to >99% ee, with albeit moderate diastereoselectivities (≤54% de).[273] The first Michael addition occurred between dimethylmalonate and nitroalkenes, affording the corresponding nitroalkanes, which were subsequently added onto α,β-unsaturated aldehydes to give the corresponding aldehydes. These aldehyde intermediates then cyclised into the final cyclohexanes through an intramolecular aldol condensation. Importantly, the authors demonstrated that it was possible to achieve different stereoisomers by changing the combination of catalysts used in the cascade reaction. Some of the best results are collected in Scheme 3.183. Although α,β-unsaturated aldehydes are implicated as substrates in this reaction, this work is situated in this section dealing with three-component reactions of nitroalkenes because the first Michael addition occurred on the latter and not on α,β-unsaturated aldehydes.

In 2010, another enantioselective three-component domino Michael–Michael–aldol reaction was reported by Enders et al., upon catalysis with chiral diphenylprolinol trimethylsilyl ether.[274] In this case, the domino reaction started with a first Michael addition of aldehydes bearing a thioester onto nitroalkenes to give the corresponding nitroalkanes, which then added according to a second Michael addition onto α,β-unsaturated aldehydes to

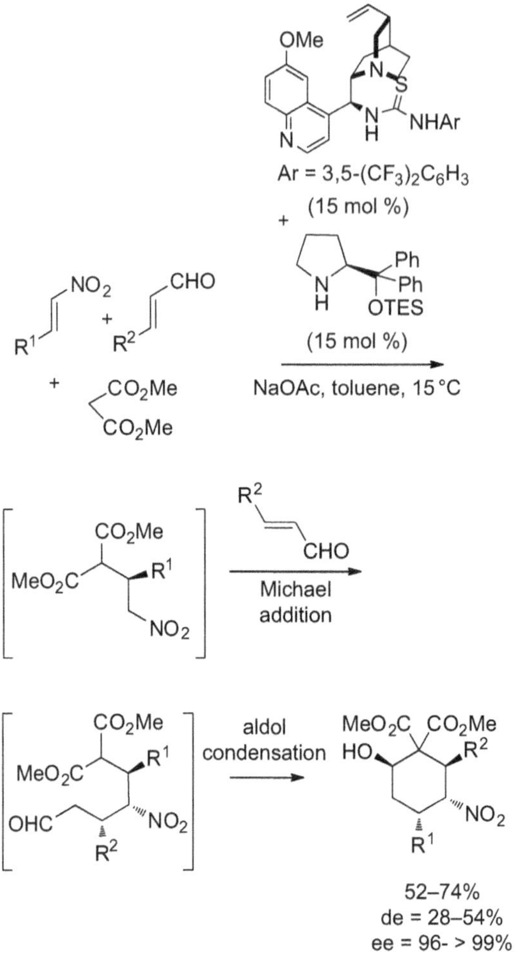

Scheme 3.183 Synthesis of cyclohexanes through three-component domino Michael–Michael–aldol reaction.

afford dialdehydes. These subsequently cyclised through aldol condensation followed by dehydration to give the domino products. The latter were not isolated, but directly submitted to an intramolecular thia-Michael addition which was initiated by the addition of K_2CO_3 finally to provide the corresponding chiral functionalised thiadecalins. As shown in Scheme 3.184, this two-step triple cascade followed by a thia-Michael reaction provided these products in moderate to good yields (32–60%) and excellent diastereo- and enantioselectivities of >94% de and >99% ee, respectively, in almost all cases of the substrates studied. Although α,β-unsaturated aldehydes are also implicated as substrates in this reaction, this work is situated in the section dealing with three-component reactions of nitroalkenes because the first Michael addition occurred on the latter and not on α,β-unsaturated aldehydes.

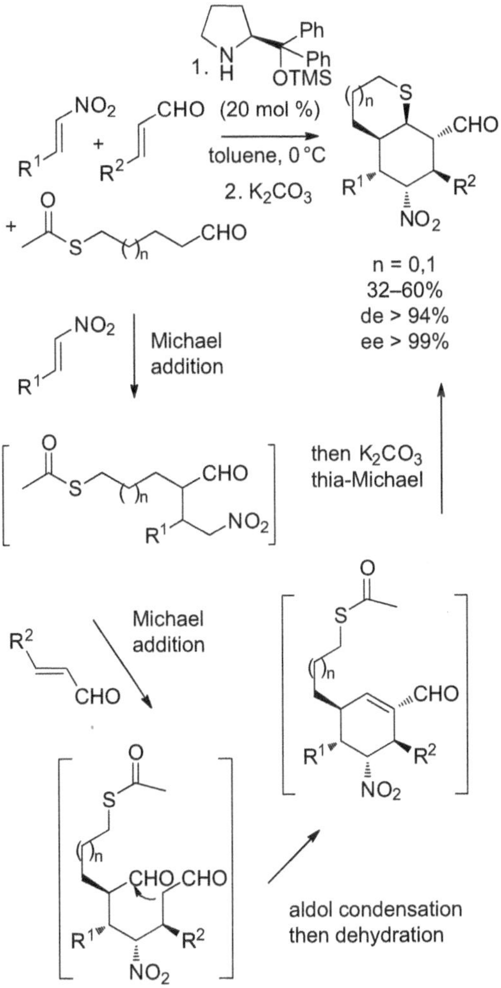

Scheme 3.184 Synthesis of thiadecalins through three-component domino Michael–Michael–aldol reaction followed by thia-Michael reaction.

Polymer-supported organocatalysts are valuable materials for conducting catalytic asymmetric transformations under simple and environmentally benign conditions.[275] A novel polymer-supported diphenylprolinol trimethylsilyl ether was synthesised by Hansen *et al.*, and further applied as catalyst to induce a diastereo- and enantioselective three-component domino Michael–Michael–aldol reaction between cinnamaldehyde, *n*-propanal, and a nitroalkene, affording the corresponding tetrasubstituted cyclohexene carbaldehyde in 32% yield, as shown in Scheme 3.185.[276] Once again, although an α,β-unsaturated aldehyde is also implicated as substrate in this reaction, this work is situated in the section dealing with three-component reactions of nitroalkenes because the

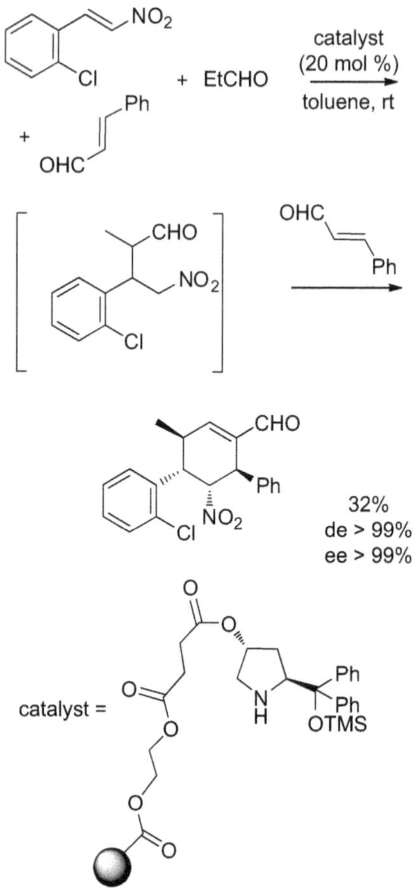

Scheme 3.185   Three-component domino Michael–Michael–aldol reaction induced by a polymer-supported catalyst.

first Michael addition occurred on the latter and not on the α,β-unsaturated aldehyde.

In another context, an unprecedented enantioselective three-component domino Michael–Michael–Wittig reaction was recently reported by Hong *et al.*, providing a direct protocol for the construction of chiral trisubstituted cyclohexenecarboxylates bearing three contiguous stereocentres.[277] These products, all with *cis* stereochemistry, were generated from the reaction of nitroalkenes, α,β-unsaturated aldehydes, and a stabilised Wittig reagent, upon catalysis with chiral diphenylprolinol trimethylsilyl ether in the presence of acetic acid. Excellent enantioselectivities of up to 99% ee were achieved in combination with moderate to good yields and low to moderate diastereoselectivities (0–34% de), as shown in Scheme 3.186. The sequence started with the Michael addition of the Wittig reagent onto nitroalkenes to give intermediate nitroalkanes. These intermediates then added to α,β-unsaturated aldehydes to

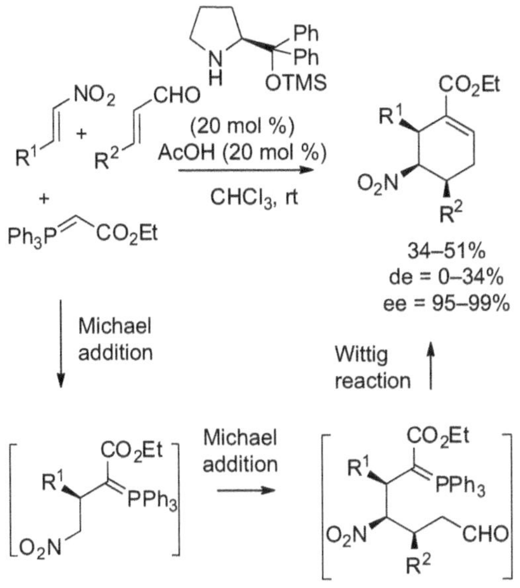

Scheme 3.186 Synthesis of cyclohexenes through three-component domino Michael–Michael–Wittig reaction.

afford aldehydes, which reacted through an intramolecular Wittig reaction to afford the final products. As an extension of this methodology, an enantioselective three-component domino Michael–Michael–Horner–Wadsworth–Emmons reaction was developed by Hayashi *et al.*, in 2010.[278] Employing the same catalyst, this process was applied to the asymmetric total synthesis of the important influenza drug oseltamivir, as well as the drug candidate ABT-341.[279]

In 2010, Rodriguez *et al.* reported an enantioselective synthesis of a poly-functionalised hexasubstituted cyclohexane through a three-component domino Michael–Michael–Henry-type reaction catalysed by a chiral bifunctional amine thiourea.[280] In this process, 2 equivalents of a nitroalkene reacted with a α-ketoanilide to give the product, bearing six contiguous stereogenic centres in 61% yield and excellent diastereo- and enantioselectivity of >90% de and 98% ee, respectively, as shown in Scheme 3.187.

A microwave-assisted four-component enantioselective remarkable domino Michael–Henry–Michael–aldol reaction was developed by Enders *et al.*, in 2010.[281] Employing chiral diphenylprolinol trimethylsilyl ether as catalyst, three equivalents of acetaldehyde reacted with nitroalkenes according to a quadruple cascade to afford the corresponding trisubstituted cyclohexene carbaldehydes in moderate to good yields (25–45%) and good to high enantioselectivities (89 to >99% ee), combined with a general diastereoselectivity of >96% de, as shown in Scheme 3.188. The authors have proposed the mechanism depicted in Scheme 3.188, in which the cascade started with the enamine activation of acetaldehyde by the catalyst and Michael addition to the

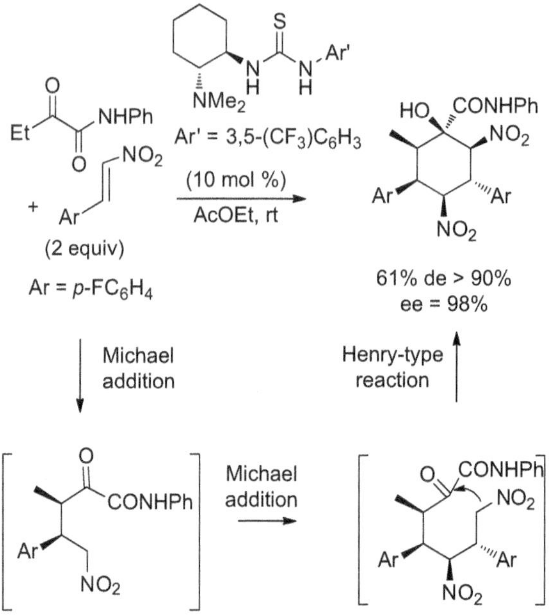

Scheme 3.187 Synthesis of cyclohexanes through three-component domino Michael–
Michael–Henry-type reaction.

nitroalkene. In the next step, the resulting nitroalkane performed a Henry
reaction with a second equivalent of acetaldehyde followed by a condensation,
which led to the newly formed nitroalkene. Enamine activation of a third
equivalent of acetaldehyde and Michael addition of this enamine to the
nitroalkene afforded a dialdehyde. After an enamine-activated intramolecular
aldol condensation reaction followed by hydrolysis, the final products were
achieved and regenerated catalyst. This mechanism was supported by ESI-MS
(mass spectroscopy) measurements.

On the other hand, chiral highly functionalised piperidines were synthesised
by Hayashi *et al.* on the basis of a highly efficient four-component two-step
enantioselective reaction catalysed by the same catalyst.[282] In the first part of
this sequence, a Michael addition of aldehydes onto nitroalkenes occurred,
providing the corresponding nitroalkanes. Subsequently, the latter reacted with
imines through an aza-Henry reaction to give aldehyde intermediates, which
were spontaneously submitted to hemiaminalisation to afford the domino
Michael–aza-Henry–hemiaminalisation products. These domino products were
not isolated, and directly engaged in a subsequent Lewis acid-mediated allylation or cyanation to provide the corresponding 2-allyl and 2-cyano
piperidines, respectively. This last allylation or cyanation step was initiated by a
further addition of $TiCl_4$ and allyltrimethylsilane, or trimethylsilyl cyanide,
respectively. The densely functionalised products bearing five contiguous
stereocentres were isolated as single diastereomers in good to high yields
and excellent enantioselectivities (93–99% ee), as shown in Scheme 3.189.

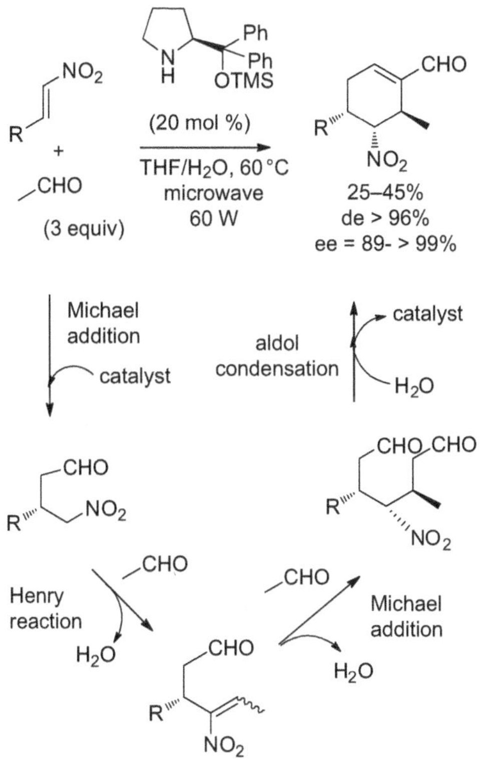

Scheme 3.188   Synthesis of cyclohexenes through four-component domino Michael–Henry–Michael–aldol reaction.

Furthermore, it must be noted that this remarkable process employed a low catalyst loading of only 5 mol%.

It must be noted that Xu *et al.* have previously reported the enantioselective synthesis of fully substituted chiral *N*-tosylpiperidines arisen from the domino Michael–aza-Henry–hemiaminalisation reaction of aldehydes, tosylimines, and nitroalkenes on the basis of a multiple-organocatalyst-promoted three-component reaction.[283] This process was induced by the combination of chiral diphenylprolinol triethylsilyl ether with a chiral thiourea, and afforded mixtures of the corresponding α- and β-diastereomeric hemiaminals, which were obtained in moderate to good yields (47–71%) and enantioselectivity of >99% ee in all cases of the substrates studied (Scheme 3.190).

In 2011, an approach to chiral polyfunctionalised octahydroquinolines incorporating three contiguous stereocentres was reported by Yadav *et al.*[284] These products were generated as single diastereomers in high yields and excellent enantioselectivities through enantioselective three-component domino Michael–aza-Henry–hemiaminalisation–dehydration reaction of 1,3-cyclohexanedione, nitroalkenes, and *N*-tosyl aldimines upon catalysis with chiral diphenylprolinol trimethylsilyl ether (Scheme 3.191).

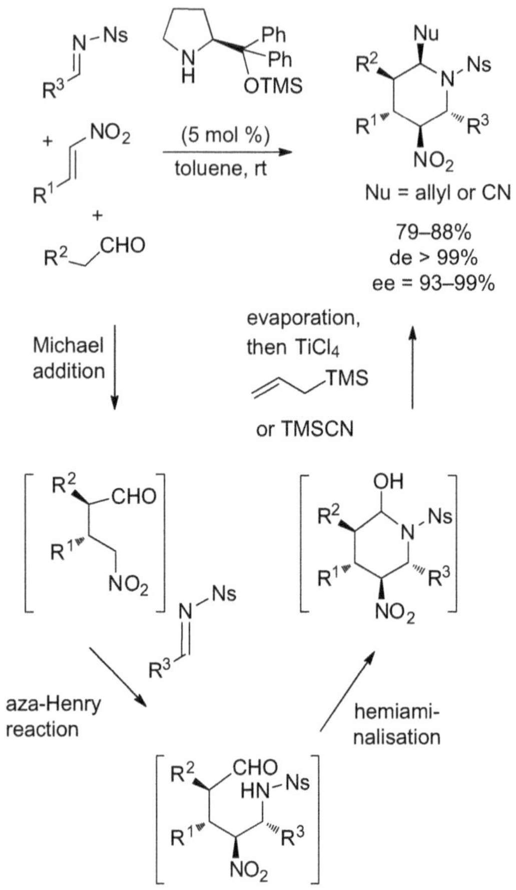

Scheme 3.189 Synthesis of piperidines through three-component domino Michael–aza-Henry–hemiaminalisation reaction followed by TiCl$_4$-mediated allylation/cyanation.

In the same year, Hayashi *et al.* reported an asymmetric synthesis of substituted tetrahydropyranols through a sequence of reactions beginning with a Michael addition of an aldehyde to a nitroalkene, followed by a domino Henry–acetalisation reaction of the formed Michael adduct with ethyl glyoxylate to give the corresponding chiral tetrahydropyranols.[285] When catalysed by chiral diphenylprolinol trimethylsilyl ether, the process provided the domino products in excellent enantioselectivities of up to 99% ee, as α/β diastereomeric mixtures in good to high yields, as shown in Scheme 3.192. The scope of this methodology was extended to the use of aldehydes instead of ethyl glyoxylate, which provided, by reaction with the previously formed Michael adducts through domino Henry–acetalisation–isomerisation reaction, the corresponding tetrahydropyranols in even better yields (50–91%), and enantio-selectivities (95–99% ee), albeit comparable diastereoselectivities (0–86% de).

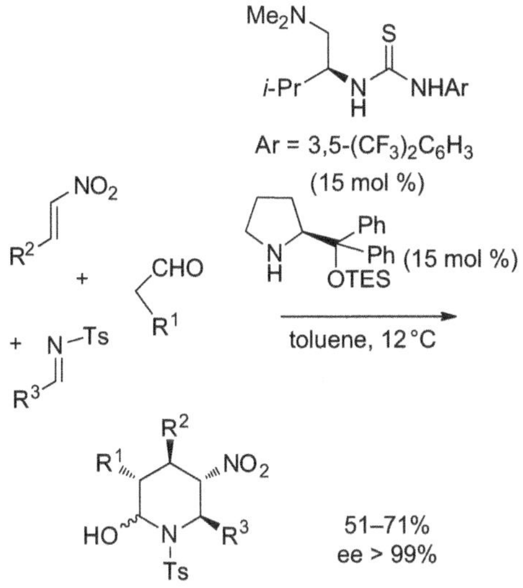

Scheme 3.190 Synthesis of piperidines through three-component domino Michael–aza-Henry–hemiaminalisation reaction.

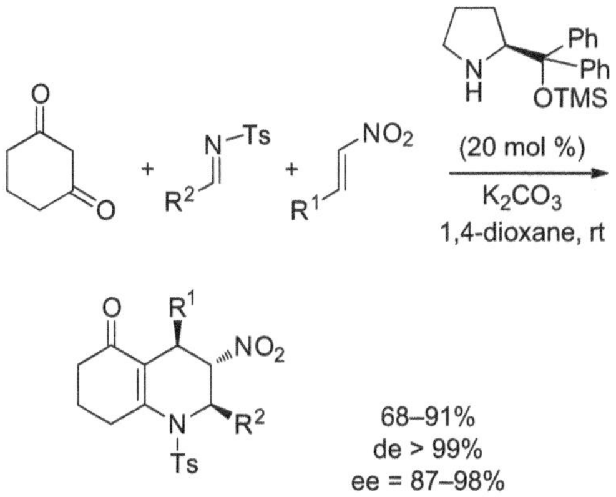

Scheme 3.191 Synthesis of octahydroquinolines through three-component domino Michael–aza-Henry reaction.

A chiral tertiary amine-modified diarylprolinol trimethylsilyl ether was successfully employed by Xu *et al.* to induce an enantioselective four-component domino Michael–Michael–Horner–Wadsworth–Emmons–thia-Michael reaction among ethyl β-nitroacrylate, aldehydes, toluenethiol, and ethyl 2-(diethoxyphosphoryl)-acrylate.[59] Remarkably, the process furnished the corresponding almost stereomerically pure, highly functionalised, poly-substituted cyclohexanes in good yields, as shown in Scheme 3.193.

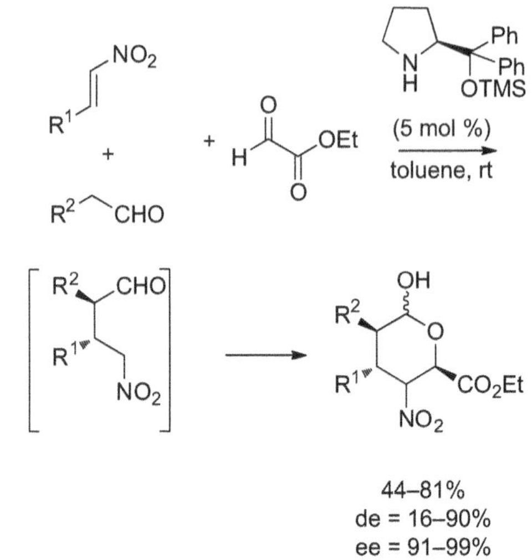

Scheme 3.192   Three-component sequential Michael–domino Henry–acetalisation reaction.

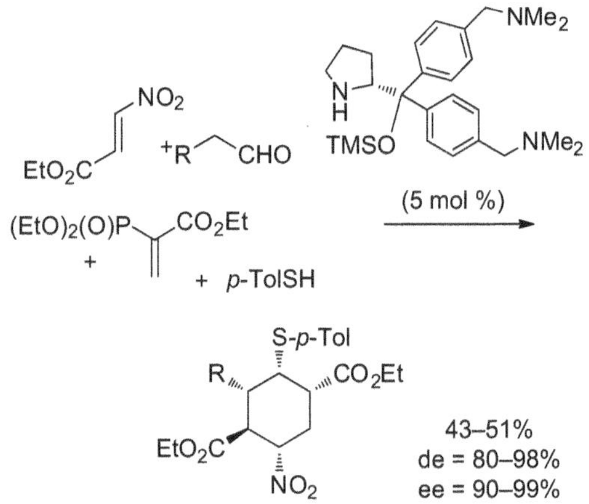

Scheme 3.193   Synthesis of cyclohexanes through four-component domino Michael–Michael–Horner–Wadsworth–Emmons–thia-Michael reaction.

In addition, an enantioselective four-component domino Michael–nitro-Mannich–lactamisation reaction was recently developed by Dixon *et al.*[286] It occurred among a nitroalkene, an activated lactam, formaldehyde, and ally-lamine or a cyclic imine, to give the corresponding chiral bicyclic spirolactam or tetracyclic spirolactam, respectively. As shown in Scheme 3.194, these products were achieved in good yields and high enantioselectivities by using a cinchona-derived chiral bifunctional thiourea as organocatalyst.

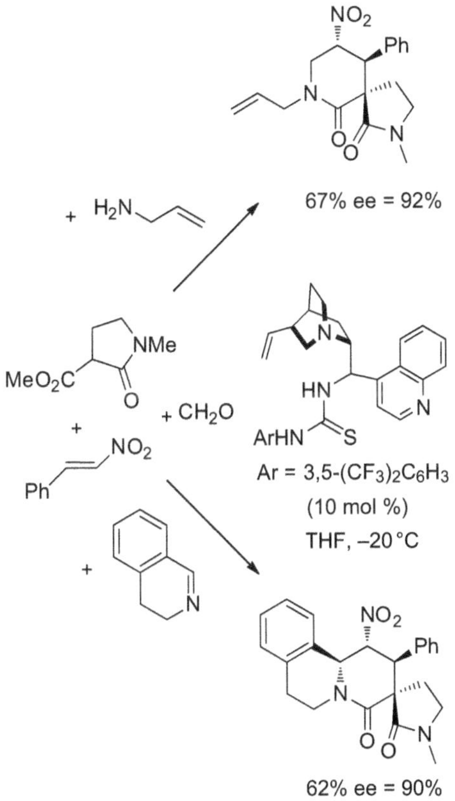

Scheme 3.194 Four-component domino Michael–nitro-Mannich–lactamisation reactions.

3.3.2 Multicomponent Reactions Initiated by the Knoevenagel Reaction

The Knoevenagel condensation was described in 1894 by Emil Knoevenagel.[287] It consists in the condensation of aldehydes or ketones with active methylene compounds in the presence of a base.[288] The first organocatalytic enantioselective domino multicomponent reaction, which was developed by Barbas *et al.* in 2001, was initiated by a Knoevenagel reaction.[245] It involved an enantioselective domino Knoevenagel–Michael reaction among benzaldehyde, acetone, and diethylmalonate, catalysed by L-proline. Ever since, the Knoevenagel reaction has been used to initiate various other types of enantioselective organocatalytic domino multicomponent reaction. In particular, asymmetric domino Knoevenagel hetero-Diels Alder reactions have been extensively studied by Tietze *et al.*[289] and, consequently, this reaction is often referred to as the Tietze multicomponent reaction. Actually, the first organocatalytic diastereospecific and enantioselective direct asymmetric domino

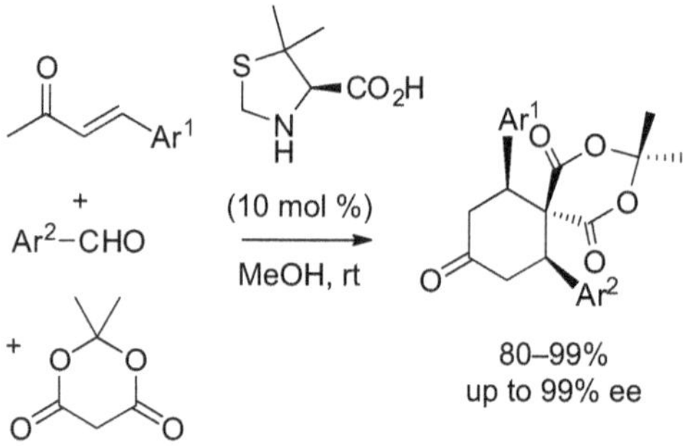

Scheme 3.195   Three-component domino Knoevenagel–Diels–Alder reaction.

Knoevenagel–Diels–Alder reaction was described by Barbas *et al.*, in 2003.[290] This three-component process produced highly substituted spiro [5,5]undecane-1,5,9-triones from 4-substituted-3-buten-ones, aldehydes, and 2,2-dimethyl-1,3-dioxane (Meldrum's acid). A chiral amino acid catalysed the Knoevenagel condensation of the aldehyde to Meldrum's acid to give an intermediate alkylidene derivative of Meldrum's acid, which further underwent a concerted [4 + 2] cycloaddition with the 2-amino-1,3-butadiene generated *in situ* from the enone and amino acid. The corresponding enantiopure spiro-cyclic ketones were obtained in remarkably high yields, as shown in Scheme 3.195. These products constitute attractive starting materials for the synthesis of natural and biologically active products. Later, these authors extended the scope of this reaction by performing hetero-domino Knoevenagel–Diels–Alder-epimerisation reactions which provided access to highly substituted prochiral spiro [cyclohexane-1,2'-indan]-1',3',4-triones, employing 1,3-indanedione instead of Meldrum's acid.[291] These products were achieved in excellent yields (90–98%) and as single stereomers. In 2011, Feng *et al.* reinvestigated these reactions by using chiral 9-amino-9-deoxy-*epi*-quinine as organocatalyst in the presence of 4-bromophenol as an additive in chloroform at room temperature.[292] A range of pharmacological multi-substituted spiro[5,5]undecane-1,5,9-triones were achieved in good yields (53–81%), excellent general diastereoselectivity of >98% de, combined with high to excellent enantioselectivities (80–97% ee). A four-component version of this domino three-component Knoevenagel–Diels–Alder reaction providing identical spirotrione products *via* a domino Wittig-Knoevenagel–Diels–Alder reaction was also previously reported by Barbas and Ramachary.[293] In this case the required α,β-unsaturated aldehyde was generated *in situ*, *via* a Wittig reaction of a phosphorane and an aromatic aldehyde. Altough the versatility of the Knoevenagel–Diels–Alder reaction was improved by this four-component methodology, lower enantioselectivities were observed (42–70% ee).

    In 2008, Ramachary *et al.* reported the asymmetric synthesis of densely functionalised cyclohexanes through dienamine activation mode.[294] This three-component domino Knoevenagel–Diels–Alder reaction occurred among benzaldehyde, methyl cyanoacetate, and an α,β-unsaturated enone in the

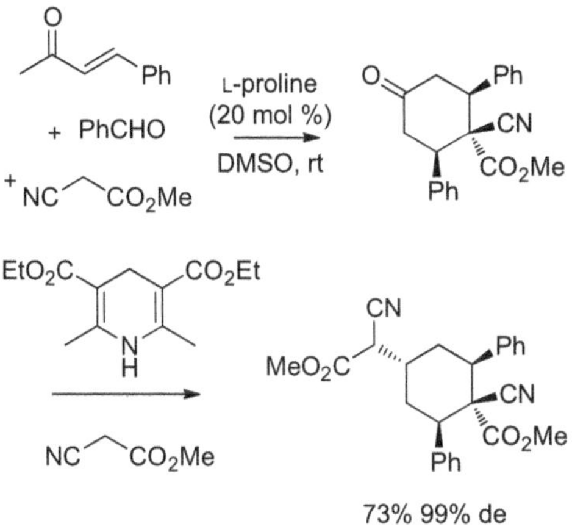

Scheme 3.196 Three-component domino Knoevenagel–Diels–Alder reaction.

presence of L-proline as organocatalyst. The process consisted in the formation *in situ* of the dienamine of the starting α,β-unsaturated enone in the presence of L-proline. This dienamine subsequently underwent a Diels–Alder cycloaddition with the Knoevenagel product generated *in situ* from the starting aldehyde and methyl cyanoacetate. The corresponding highly substituted domino product was not isolated, but directly submitted to reaction with Hantzsch ester and a second equivalent of methyl cyanoacetate finally to afford the corresponding cyclohexane bearing four stereogenic centres in good yield, and almost complete diastereo- and enantioselectivity, as shown in Scheme 3.196.

A more recent example was reported by Yuan *et al.*, who developed the first enantioselective organocatalytic three-component domino Knoevenagel–Michael–cyclisation reaction, giving access to a range of chiral spiro[4*H*-pyran-3,3′-oxindoles].[295] This novel highly efficient sequence was catalysed by cupreine, and involved malononitrile, a 1,3-diketone, and isatylidene malononitrile derivatives as the components. As shown in Scheme 3.197, the spiro heterocyclic products were generally obtained in both high yields and enantioselectivities of up to 96% ee. The authors have proposed the mechanism depicted in Scheme 3.197, in which the sequence started with the Knoevenagel condensation of malononitrile onto isatylidene malononitrile derivatives to give the corresponding alkylidene malononitrile intermediates. Then, Michael addition of the 1,3-diketone to these intermediates led to dicyano intermediates, in equilibrium with enol intermediates, which finally cyclised to afford the products. More recently, Macaev *et al.* have investigated the reaction of isatine ($R^1 = R^2 = H$) with malononitrile and acetylacetone catalysed by brevicolline (10 mol% in CH_2Cl_2 at 0 °C)[296] The corresponding spiro[4*H*-pyran-3,3′-oxindole] was achieved in 62% yield and enantioselectivity of 94% ee.

Another asymmetric four-component reaction initiated by a Knoevenagel condensation was recently reported by Xu *et al.*[297] This process was catalysed

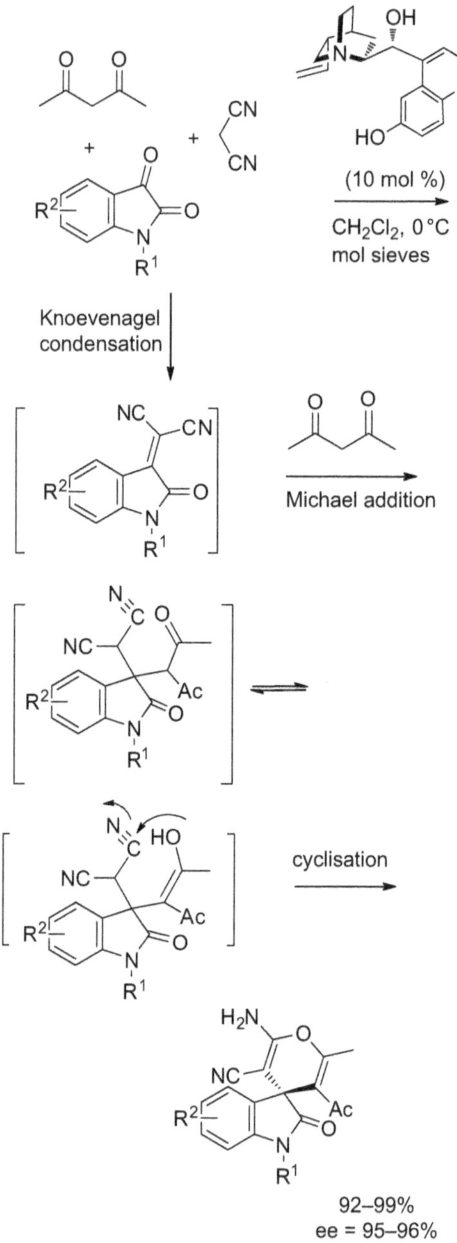

Scheme 3.197 Three-component domino Knoevenagel–Michael–cyclisation reaction.

by simple L-proline and involved formaldehyde, 2 equivalents of a β-keto ester, and *p*-substituted aniline as the substrates. In the first step, the Knoevenagel condensation of the β-keto ester onto formaldehyde formed the corresponding methylene intermediate, which then transformed into the corresponding

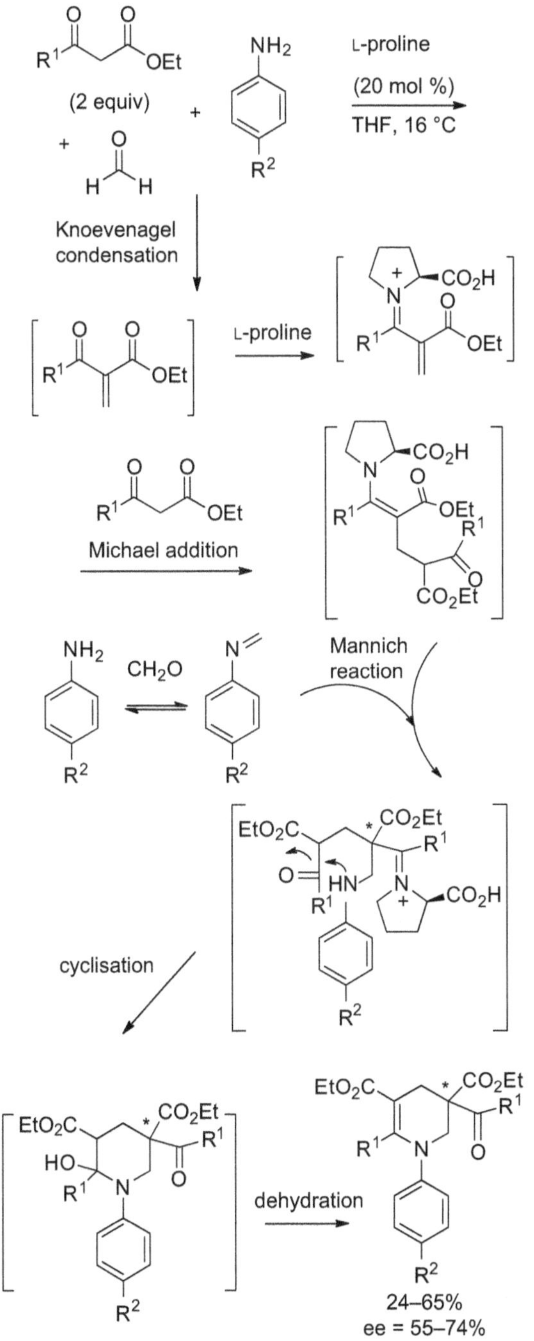

Scheme 3.198 Four-component domino Knoevenagel–Michael–Mannich–cyclisation–dehydration reaction.

iminium ion by activation with L-proline. A second equivalent of β-keto ester attacked this iminium ion to form an enamine intermediate through a Michael addition. This enamine then underwent a Mannich reaction with an imine, which was generated *in situ* through condensation of the starting aniline with formaldehyde, which gave a novel iminium intermediate. The latter underwent an intramolecular cyclisation to provide, after hydrolysis, an intermediate, which produced after dehydration the final piperidine along with regenerated L-proline. As shown in Scheme 3.198, moderate yields and enantioselectivities (≤74% ee) were obtained.

In 2012, a chiral cinchona alkaloid-derived primary amine was associated by Wang *et al.*[298] with an (*R*)-BINOL-derived phosphoric acid to induce a three-component domino Knoevenagel–Michael reaction among isatins, malononitrile, and acetone, providing the corresponding domino products in generally excellent yields and enantioselectivities, as shown in Scheme 3.199.

In addition, chiral thioureas derived from quinine have been identified by Yan *et al.* as the most efficient catalysts to induce a three-component reaction of isatins, malononitrile, and α-aryl-isocyanoacetates.[299] As shown in Scheme 3.200, this process has allowed a number of chiral 3,3′-dihydro-pyrrylspirooxindoles to be achieved in high yields and excellent enantio-selectivities, albeit with moderate diastereoselectivities (≤72% de). These tricyclic products arose from a domino Knoevenagel–Michael–cyclisation reaction.

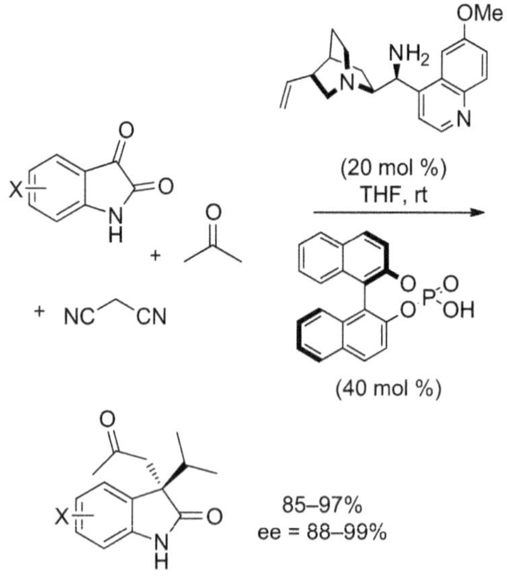

Scheme 3.199   Three-component domino Knoevenagel–Michael reaction.

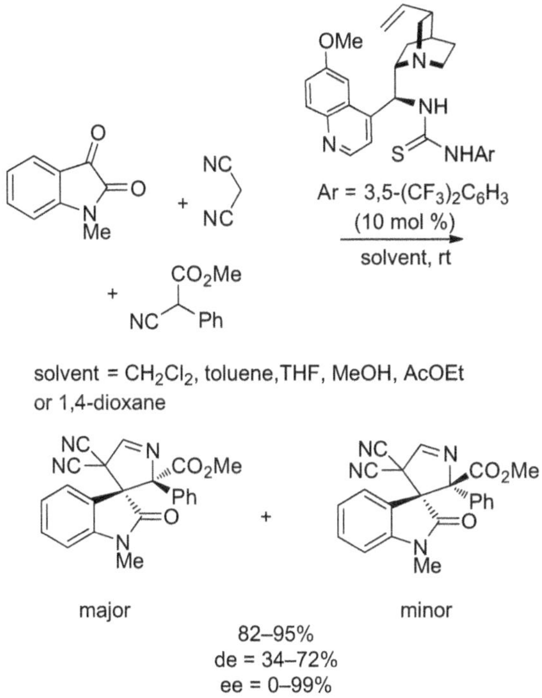

Scheme 3.200 Three-component domino Knoevenagel–Michael–cyclisation reaction.

3.3.3 Multicomponent Reactions Based on the Mannich Reaction

The Mannich reaction, discovered by Carl Mannich in 1912, is a condensation of amine derivatives, enolisable carbonyl compounds and non-enolisable aldehydes.[192] The first example of a direct organocatalytic three-component Mannich reaction was reported in 2000 by List, using L-proline as organo-catalyst,[300] and was followed by a number of excellent studies from several groups.[301] Among the work reported since 2006, Bolm and Rodriguez developed, in 2006, an L-proline-catalysed direct asymmetric Mannich reaction among cyclohexanone, formaldehyde, and various anilines.[302] Employing only 0.5 mol% of this catalyst in DMSO under microwave irradiation, the corresponding Mannich products were achieved with generally excellent enantio-selectivities, ranging from 96 to 98% ee, in combination with good to high yields (40–93%), as shown in Scheme 3.201.

In the same year, Cordova *et al.* studied the three-component Mannich reaction of dimethyl-1,3-dioxan-5-one with *p*-anisidine, and various aldehydes.[303] Among various amino acids investigated as organocatalysts, the best results were obtained by using L- or D-proline in DMSO at room

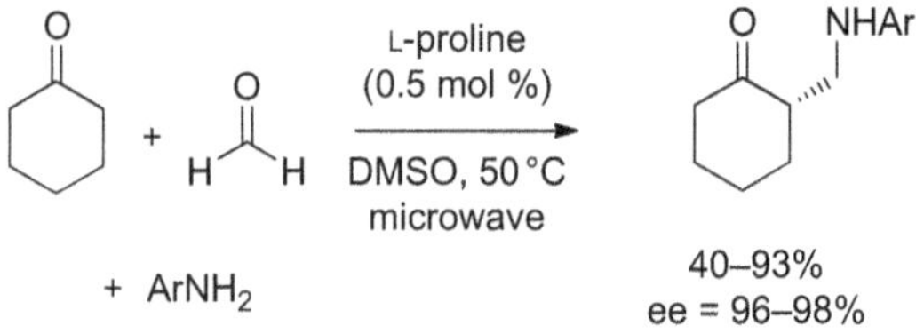

Scheme 3.201   Three-component Mannich reaction catalysed by L-proline under microwave irradiation.

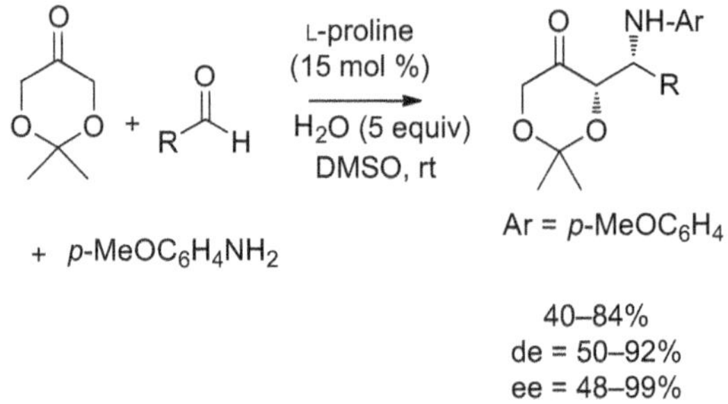

Scheme 3.202   Three-component Mannich reaction catalysed by L-proline.

temperature, which provided the protected *syn*-amino sugar products in good to high yields (40–84%), moderate to high diastereoselectivities (50–92% de), and good to excellent enantioselectivities (48–99% ee), as shown in Scheme 3.202. These authors also developed an interesting sequential Mannich–Horner–Wittig–Emmons reaction catalysed by L-proline.[304] The first step of the sequence consisted in an enantioselective three-component Mannich reaction between two different aldehydes and *p*-anisidine, which was followed, after subsequent addition of methyl diethylphosphonoacetate, by olefination to give the corresponding chiral vicinal amino alcohols in moderate to high yields (45–88%) and *syn*-diastereoselectivities (60–90% de), combined with excellent enantioselectivities (95–96% ee).

In 2006, Choudary *et al.* introduced the use of ultrasonic conditions in order to increase the yields of L-proline-catalysed Mannich reaction of hydroxyacetone, aromatic aldehydes, and *p*-anisidine.[305] Both electron-rich and electron-deficient aromatic aldehydes led, under these conditions, to the corresponding *syn*-aminoalcohols in high yields (85–98%), moderate to high *cis*-diastereoselectivities (50–92% de), and moderate to excellent enantioselectivities (66–99% ee), as shown in Scheme 3.203.

Besides proline, a number of other chiral organocatalysts have been shown to induce this type of reaction. For example, L-proline-derived tetrazole was successfully applied by Barbas *et al.* to the catalysis of the Mannich reaction of protected amino ketones with *p*-anisidine and aldehydes to give the corresponding 1,2-diamines from azido ketones, and 1,4-diamines from phthalimido

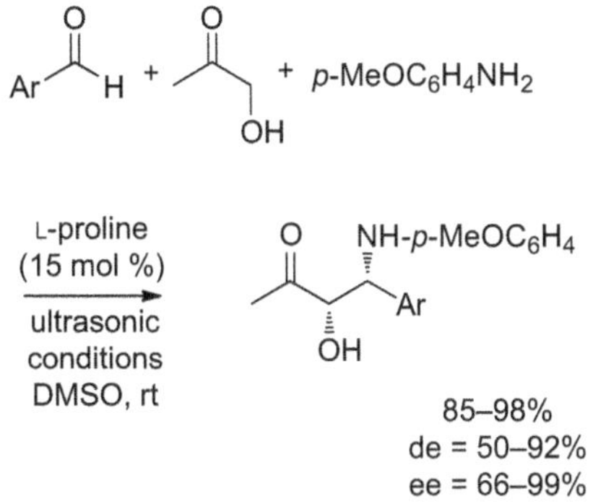

Scheme 3.203 Three-component Mannich reaction catalysed by L-proline under ultrasonic conditions.

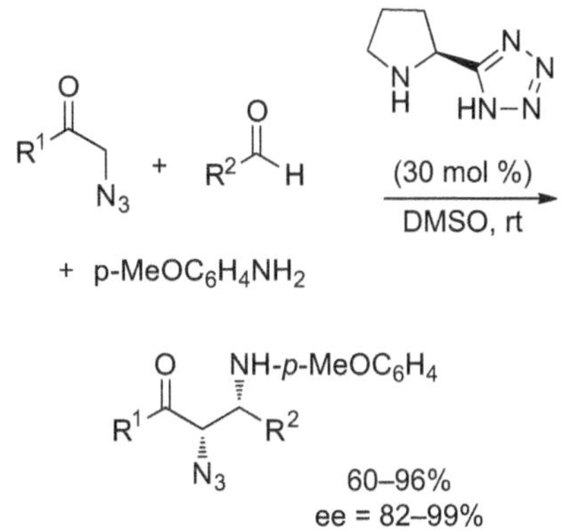

Scheme 3.204 Three-component Mannich reaction catalysed by L-proline-derived tetrazole.

ketones.[306] A range of chiral 1,2-azidoamines were achieved in good to excellent yields (60–96%), moderate to good *syn*-diastereoselectivities (40–82% de), and moderate to excellent enantioselectivities (64–99% ee), as shown in Scheme 3.204. On the other hand, a number of chiral 1,4-diamines arising from reactions of phthalimido acetone with various aldehydes and *p*-anisidine were prepared in good to high yields (53–95%), and good to excellent enantio-selectivities (57–97% ee).

Later, these authors also investigated the use of chiral linear α-amino acids, such as L-tryptophane (L-Trp) or *O-t*-Bu-L-threonine (*O-t*-Bu-L-Thr), to induce the Mannich reaction among aldehydes, α-hydroxyaldehydes, and *p*-anisidine.[307] The reaction allowed the corresponding α-hydroxy-β-aminoketones to be formed in moderate to high *anti* diastereoselectivities of up to 90% de, and enantioselectivities of up to 96% ee (Scheme 3.205). This result constituted the first direct asymmetric reaction with unmodified 1-hydroxy-2-butanone.

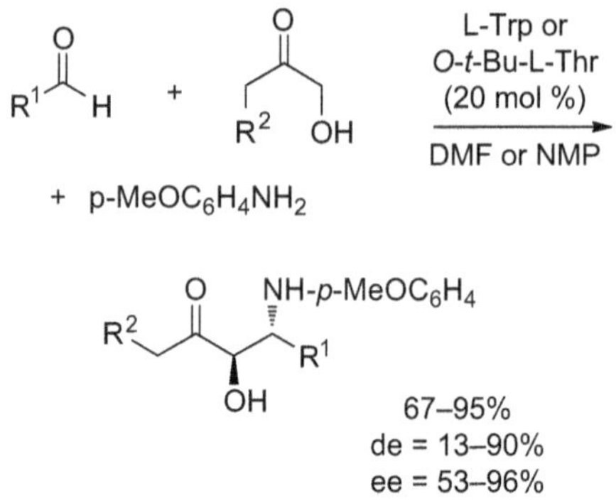

Scheme 3.205   Three-component Mannich reaction catalysed by L-tryptophane and
L-threonine derivatives.

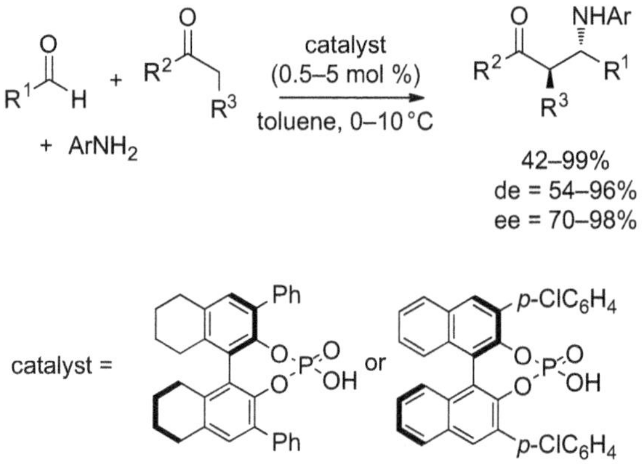

Scheme 3.206   Three-component Mannich reaction catalysed by H_8-BINOL- and
BINOL-derived phosphoric acids.

In the same year, Gong *et al.* reported the use of BINOL- and
H_8-BINOL-derived chiral phosphoric acids in the three-component Mannich
reaction among aliphatic ketones, aldehydes, and aromatic amines.[308] A
catalytic amount as low as 0.5 mol% was sufficient to promote the process,
which provided the corresponding Mannich products in moderate to excellent
yields, *anti*-diastereoselectivities, and enantioselectivities, as shown in
Scheme 3.206.

The first catalytic enantioselective Mannich reaction in purely aqueous
media was reported by Lu *et al.*, in 2007.[309] A threonine-derived organocatalyst
proved to be more effective than those derived from serine in the Mannich
reaction depicted in Scheme 3.207. The scope of the process was explored with
various aliphatic as well as aromatic aldehydes to afford the corresponding
β-amino-α-hydroxy ketones in good to high yields, and moderate to high
anti-diastereo- and enantioselectivities, as shown in Scheme 3.207. The

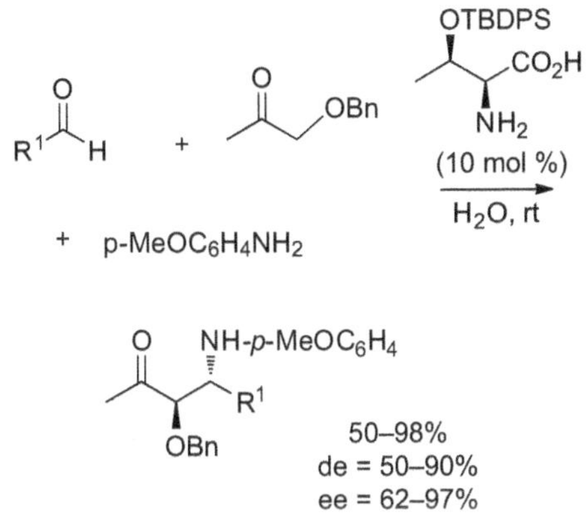

Scheme 3.207 Three-component Mannich reaction catalysed by threonine-derived catalyst.

selectivities of electron-rich aromatic aldehydes were much poorer ($\leq 62\%$ ee), and rather disappointing yields were found using aliphatic aldehydes (50–70%).

In 2008, Wu *et al.* reported a comparable Mannich reaction in aqueous media catalysed by a hydrophobic *O*-silyl serine organocatalyst.[310] The reaction among aromatic aldehydes, electron-rich aromatic amines, and a variety of carbonyl donors resulted in good enantioselectivities of up to 92% ee for the corresponding Mannich adducts. Interestingly, the application of cyclohexanone as the donor in this Mannich reaction resulted in the formation of the corresponding *syn*-product, while the use of benzyloxyacetone provided the corresponding *anti*-adduct selectively. Protected dihydroxyacetone derivatives were employed for the synthesis of amino sugars and polyhydroxylated amino acids *via* an asymmetric Mannich reaction by several groups, while the use of an unmodified dihydroxyacetone donor was not reported until 2007.[311] The main advantage of such an approach was the immensely reduced cost (>230-fold) resulting from the use of dihydroxyacetone as a direct biomimetic donor. As reported by Cordova *et al.*, Mannich reactions with dihydroxyacetone afforded the corresponding products with yields of up to 90% and excellent enantioselectivities of $>99\%$ ee. Interestingly, reactions with amino acid or peptide catalysts bearing a primary amine moiety were *anti*-selective, whereas those with proline were *syn*-selective. In order to improve the solubility of proline-derived catalysts in non-polar solvents, Carter and Yang developed, in 2009, several novel chiral proline derivatives, which were further applied to induce three-component Mannich reactions.[312] A β-amino alcohol was obtained with moderate *syn*-diastereoselectivity of 67% de when *p*-nitrobenzaldehyde was used as the aldehyde component, whereas the use of α,α-dimethoxyacetaldehyde favoured the formation of the corresponding *anti*-diastereomer in 82% de and modest enantioselectivity of 20% ee. The results are summarised in Scheme 3.208. The stereochemical outcome of the asymmetric Mannich reaction can be altered by changing the steric features at

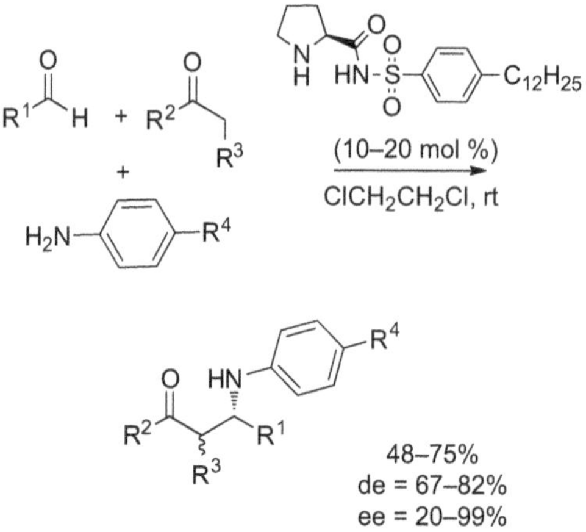

Scheme 3.208   Three-component Mannich reaction catalysed by an L-proline-derived catalyst.

the 5-position of proline. In this context, Wu *et al.* have developed proline mimetic catalysts for the asymmetric Mannich three-component reaction of hydroxyacetone.[313] For example, when the reaction of this ketone with electron-deficient aromatic aldehydes and unsubstituted anilines was induced by L-proline bearing a *trans*-carboxylic acid function at the 5-position, it provided the corresponding Mannich products in superior diastereo- and enantioselectivities of up to 88% de, and 98% ee, respectively.

In 2009, Zhu *et al.* developed an enantioselective four-component Mannich reaction catalysed by a chiral H_8-BINOL-derived phosphoric acid.[314] The process was performed among aldehydes, anilines, enecarbamates, and ethanol, which provided the corresponding aminoethers. These Mannich products were *in situ* reduced by treatment with $NaBH_3CN$ into the corresponding chiral *anti*-1,2-disubstituted 1,3-diamines, which were achieved through the two-step sequence in good to excellent yields and enantioselectivities of up to 99% ee combined with *anti*-diastereoselectivities of >90% de, as shown in Scheme 3.209. The best results were obtained with *para*-nitroaniline in combination with aliphatic as well as aromatic aldehydes.

Chiral ionic liquids have the advantages of being easily reused, non-volatile and thermally stable, and the products are easy to isolate. In this context, chiral ionic liquid catalysts have been investigated in three-component Mannich reactions.[315] As an example, a chiral ionic liquid-catalysed asymmetric Mannich reaction was reported by Wang *et al.*, in 2010.[316] This L-proline-derived ionic liquid [EMIm]$^+$[Pro]$^-$ was shown to promote the Mannich reaction of aliphatic ketones, para-substituted aromatic aldehydes, and amines to afford the corresponding (*S*)-configured β-amino carbonyls in predominantly good to excellent yields and diastereoselectivities and high enantioselectivities of up to 99% ee, as shown in Scheme 3.210. In addition, the synthesis of (*R*)-configured β-amino carbonyls from aromatic amines, aliphatic

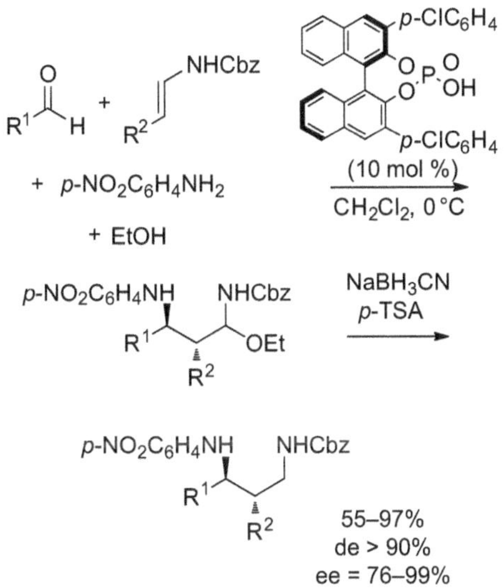

Scheme 3.209 Four-component Mannich reaction catalysed by a chiral BINOL-derived phosphoric acid.

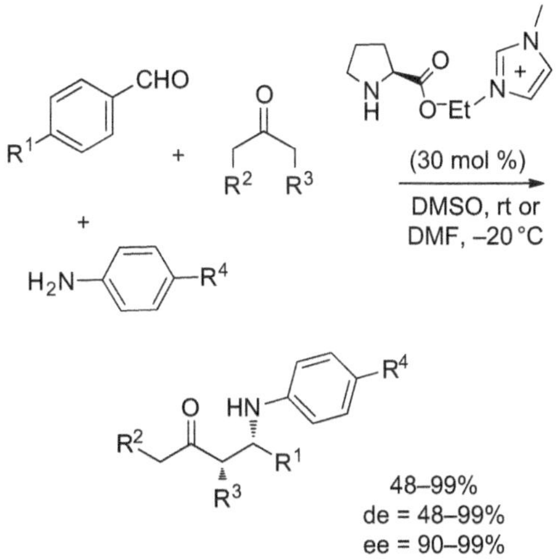

Scheme 3.210 Three-component Mannich reaction catalysed by L-proline-based chiral ionic liquid.

aldehydes and aliphatic ketones was reported by Xu *et al.*[317] By using [DEMImBF$_4$]$^+$[Pro]$^-$ chiral ionic liquid, predominantly high yields (72–96%) and moderate to excellent enantioselectivities (28–99% ee) were obtained for the corresponding (*R*)-configured β-amino carbonyls.

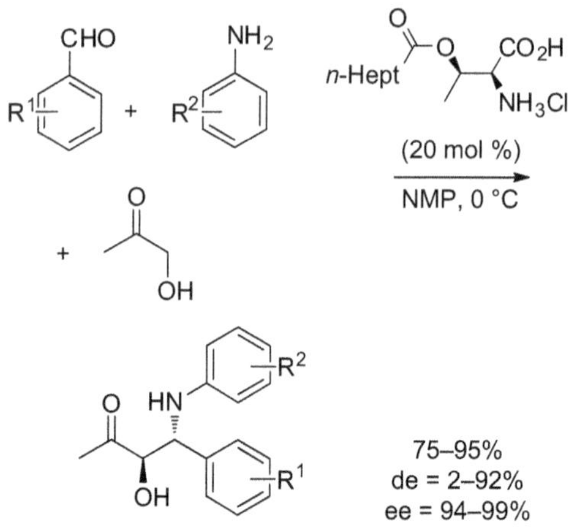

Scheme 3.211   Three-component synthesis of *anti*-1,2-amino alcohols.

In 2011, Fu *et al.* introduced other L-threonine derivatives as novel organocatalysts for enantioselective direct *anti*-Mannich reactions of hydroxyacetone with aldehydes and anilines.[318] When the reaction was promoted by an L-threonine-derived catalyst in 1-methyl-2-pyrrolidinone (NMP) as the solvent, it afforded the corresponding *anti*-1,2-amino alcohols in good yields (75–95%) and high enantioselectivities of up to 99% ee, as shown in Scheme 3.211.

In addition, Yang *et al.* have developed a highly enantioselective three-component Mannich reaction of *in situ* generated imines from 2-aminobenzoxazol and aldehydes with malonates, in the presence of a chiral cinchona alkaloid thiourea catalyst.[319] The corresponding chiral β-amino ester derivatives were produced in good to high yields and high enantioselectivities of up to >99% ee, as shown in Scheme 3.212.

A recyclable siloxy-L-serine organocatalyst was developed by Teo and Yong to induce Mannich reactions performed in ionic liquids.[320] Indeed, the reaction of cyclic as well as acyclic ketones with aldehydes and *p*-anisidine afforded, in these conditions, the corresponding β-amino ketones as mixtures of *syn*- and *anti*-diastereomers in moderate to high yields (50–91%), and enantioselectivities (55–91% ee), albeit with moderate diastereoselectivities (20–88% de). On the other hand, Tao *et al.* have developed chiral isosteviol-derived amino acids to be used in asymmetric Mannich reaction of hydroxyacetone, aromatic aldehydes, and anilines.[321] When the reaction was performed with an L-proline isosteviol-derived amino acid as catalyst, it afforded the corresponding *syn*-1,2-amino alcohols with excellent stereoselectivities of up to 98% de and >99% ee, while the use of an L-serine isosteviol-derived amino acid as catalyst allowed the corresponding *anti*-1,2-amino alcohols to be achieved in comparable excellent stereoselectivities of up to 94% de and 99% ee, as shown in Scheme 3.213. In 2012, these authors extended the scope of this methodology

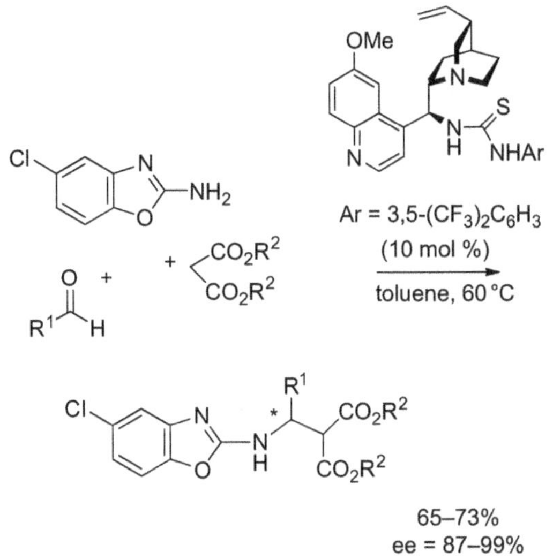

Scheme 3.212 Three-component synthesis of β-amino ester derivatives.

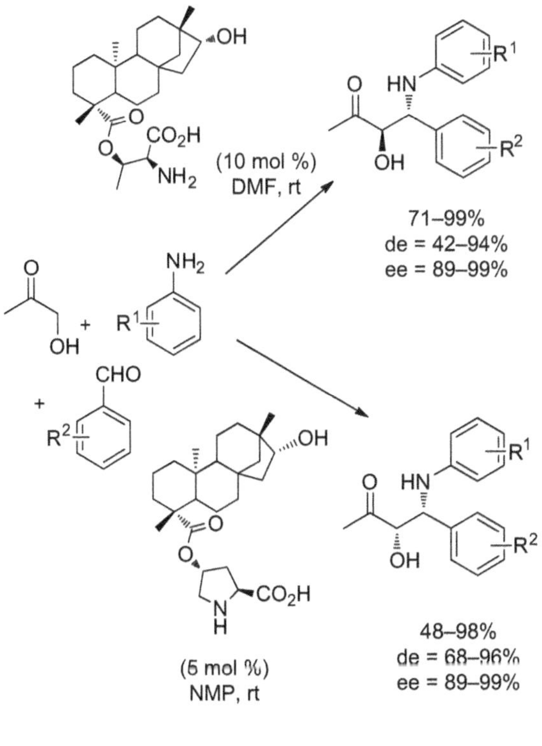

Scheme 3.213 Three-component synthesis of amino alcohols.

to cyclohexanone, which provided the corresponding chiral *syn*-Mannich products with excellent diastereo- and enantioselectivities of up to 96% de, and >99% ee, respectively, when the same L-proline isosteviol-derived amino acid was employed as catalyst.[322]

3.3.4 Multicomponent Reactions Based on the Biginelli Reaction

The Biginelli reaction is a three-component reaction allowing the synthesis of 3,4-dihydropyrimidin-2-(1*H*)-ones or -thiones by reacting urea or thiourea, a 1,3-dicarbonyl derivative, and an aldehyde. It was discovered by Pietro Biginelli in 1893.[243] The heterocyclic pyrimidinone products are known to exhibit a wide range of important pharmacological properties and make up a large family of medically relevant compounds. The asymmetric version of the Biginelli reaction is of significant contemporary interest. It must be noted that there are only few reports so far in the literature of the synthesis of chiral dihydropyrimidinones by employing organocatalysts. The first organocatalytic enantioselective Biginelli reaction was described by Gong *et al.*, in 2006.[323] Indeed, these authors reported excellent results for the Biginelli reaction of aldehydes, (thio)urea, and β-ketoesters catalysed by a chiral phosphoric acid.[324] As shown in Scheme 3.214, the corresponding Biginelli products were achieved in good to high yields and generally high enantioselectivities of up to 98% ee. In the course of developing an enantioselective synthesis of the potent melanin-concentrating hormone receptor antagonist SNAP-7941, Schauss and Goss have used the same chiral phosphoric acid to induce the key Biginelli reaction of urea, methylacetoacetate, and 3,4-difluorobenzaldehyde.[325] The required key product was produced in 96% yield and 79% ee and further converted into the expected SNAP-7941.

In 2008, Feng *et al.* investigated these reactions by employing a combination of a chiral *trans*-4-hydroxyproline-derived secondary amine and a Bronsted acid, such as 2-chloro-4-nitrobenzoic acid, as the catalyst system.[326] The dual-catalysed process was performed at 25 °C with 5 mol% of catalyst loading for each of the two catalysts in 1,4-dioxane and in the presence of an additive such as *t*-BuNH$_2$·TFA. In these conditions, the Biginelli products were obtained in

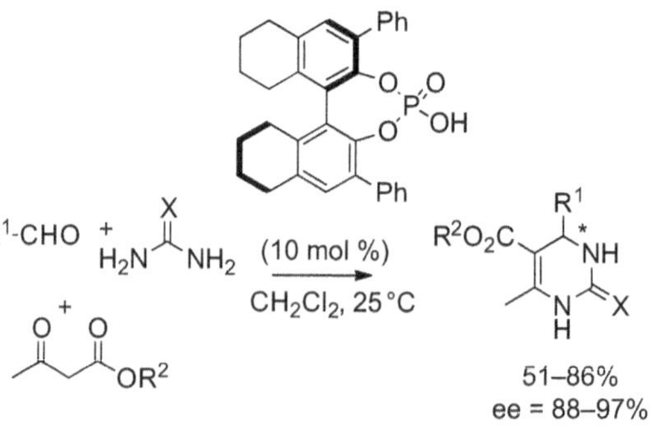

Scheme 3.214   Biginelli reaction catalysed by a chiral phosphoric acid.

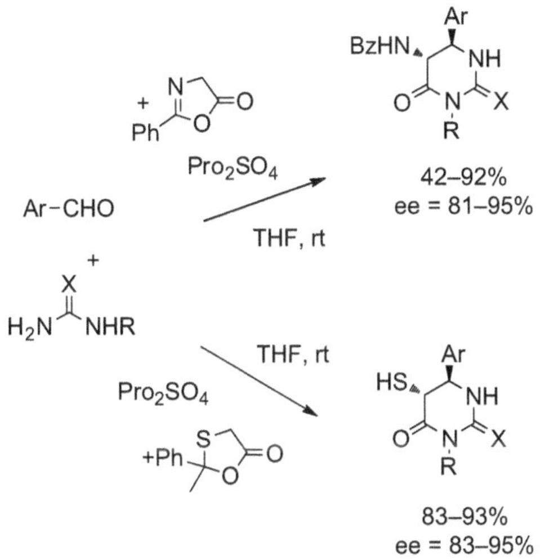

Scheme 3.215 Chiral ionic liquid-catalysed Biginelli reactions.

moderate to good yields (34–73%) and good to excellent enantioselectivities (70–98% ee). In the same year, Yadav *et al.* reported an interesting chiral ionic liquid-catalysed, efficient, and unprecedented version of the Biginelli reaction using new variants of its active methylene component, such as 2-phenyl-1,3-oxazol-5one and 2-methyl-2-phenyl-1,3-oxathiolan-5-one. The reaction of these carbonyl compounds with aromatic aldehydes and (thio)ureas afforded, when catalysed by chiral ionic liquid, such as L-prolinium sulfate (Pro$_2$SO$_4$), the corresponding 5-amino or 5-mercaptoperhydropyrimidines in good to high yields and high enantioselectivities of up to 95% ee, as shown in Scheme 3.215.[327]

Later, a diastereoselective Biginelli type reaction catalysed by L-proline was developed by Tripathi *et al.*[328] This process occurred among 3,4-dihydro-(2*H*)-pyran, aromatic aldehydes and (thio)urea, providing the corresponding hexahydropyrano pyrimidinones in almost complete relative diastereo-selectivity and in good yields, as shown in Scheme 3.216. These products constituted potent precursors for many biologically active molecules and fused oxazines, which are the lead molecules for tuberculosis chemotherapy.

In 2009, Lee *et al.* studied the efficiency of L-proline ester salts to promote the Biginelli reaction among *m*-nitrobenzaldehyde, urea, and ethyl acet-oacetate.[329] The corresponding 3,4-dihydropyrimidin-2-(1*H*)-one was obtained in 87% yield and moderate enantioselectivity of 40% ee by employing the salt of L-proline-derived *tert*-butyl ester as the catalyst. Higher enantioselectivities of up to 81% ee combined with good yields (63–88%) were reported by Wang *et al.*, using chiral substituted 5-(pyrrolidin-2-yl)tetrazoles as catalysts for the Biginelli reaction of various substituted benzaldehydes, urea (or thiourea), and a range of β-ketoesters.[330] The best results (68–81% ee) were obtained with 4-*N*-tosyl-5-(pyrrolidin-2-yl)tetrazole bearing an *n*-butyl group on the nitrogen atom. Comparable results (14–81% yields, 51–78% ee) were also reported by

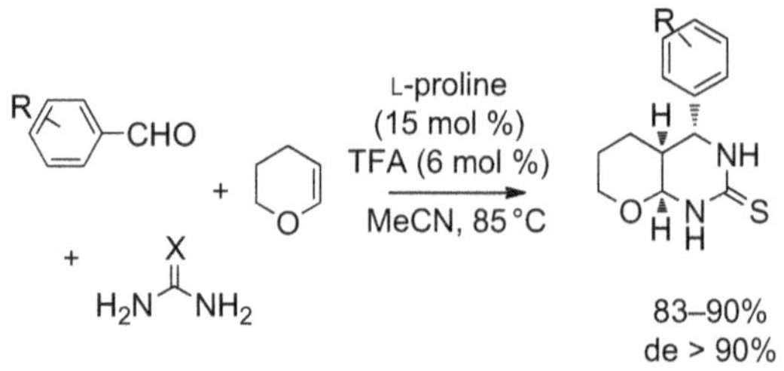

83–90%
de > 90%

Scheme 3.216   L-Proline-catalysed Biginelli type reaction.

Zhao and Ding for similar reactions, albeit catalysed by a chiral primary amine derived from quinine.[331] In addition, these authors have included aliphatic aldehydes as substrates of the Biginelli reaction, providing comparable results. For example, 43% yield combined with enantioselectivity of 72% ee was afforded for the reaction of *n*-heptanal with urea and ethyl acetoacetate. On the other hand, chiral bifunctional primary amine–thiourea catalysts derived from D-glucose have been demonstrated by Chen *et al.* to be very efficient in promoting a highly enantioselective Biginelli reaction of a range of benzaldehydes, (thio)urea, and ethyl acetoacetate.[332] The best results were obtained by using gluco-2-aminocyclohexylthiourea, depicted in Scheme 3.217, in combination with 2,4,6-trichlorobenzoic acid (TCBA) and *tert*-butylammonium trifluoroacetate as additives in dichloromethane as solvent. In these conditions, the corresponding dihydropyrimidines were achieved in moderate to high yields (51–93%) and enantioselectivities of up to 99% ee, as shown in Scheme 3.217. In order to explain the stereoselectivity of the reaction, the authors have proposed that both hydrogen-bonding interactions and enamine activation of the β-ketoester occurred in the transition states (Scheme 3.217). It must be noted that a low catalyst loading of only 5 mol% was sufficient to attain these results, however poor enantioselectivities were obtained for aliphatic aldehydes, such as *n*-butanal, which provided the corresponding product in only 15% ee. The asymmetric synthesis of dihydropyrimidines can also be catalysed by chiral bifunctional primary amine–thiourea catalysts, as reported by the Chen *et al.*[333] The activation of the intermediate imine and carbonyl component towards nucleophilic attack was effectively catalysed by the primary amine and thiourea moieties of the catalyst. The incorporation of a carbohydrate fragment in the catalyst was particularly attractive for the well-defined stereogenic centres and ubiquity. The combination of the primary amine–thiourea catalyst depicted in Scheme 3.217, 2,4,6-trichlorobenzoic acid (TCBA), and *t*-BuNH$_2$-TFA as an additive proved superior in terms of reactivity and selectivity for the catalytic enantioselective Biginelli reaction depicted below. The scope of the reaction was investigated with a variety of aromatic aldehydes to afford the corresponding dihydropyrimidines with good to high enantioselectivities, while the reaction of aliphatic aldehyde was less efficient (51% yield) and stereoselective (15% ee). A transition state was proposed by the authors in which the *Si*-face of the

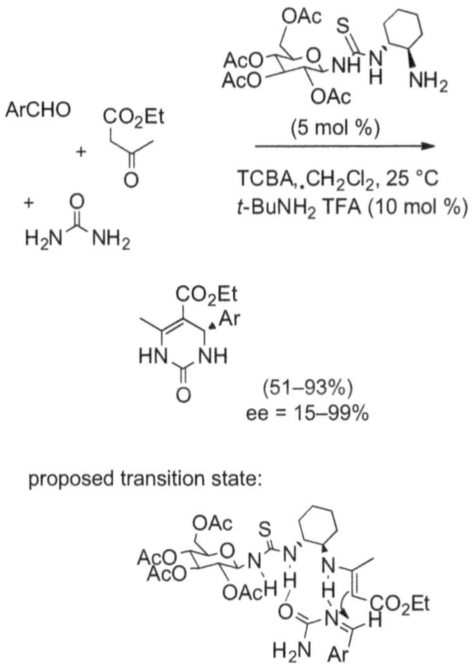

ArCHO

CO₂Et

+

O

+

$H_2N \overset{O}{\underset{}{\parallel}} NH_2$

$\xrightarrow{\text{TCBA, CH}_2\text{Cl}_2\text{, 25 °C} \atop t\text{-BuNH}_2 \text{ TFA (10 mol %)}}$

(5 mol %)

CO₂Et

Ar

HN   NH

O

(51–93%)
ee = 15–99%

proposed transition state:

Scheme 3.217 Carbohydrate-derived primary amine–thiourea-catalysed Biginelli reaction.

intermediate imine was predominately attacked by the enamine intermediate to afford the (S)-configured dihydropyrimidine, because the *Re*-face was shielded by the cyclohexane ring of the catalyst. More recently, the same authors investigated the same reaction under phase transfer conditions.[334] Indeed, by using the same catalyst at 15 mol% of catalyst loading combined with 15 mol% of TfOH and 10 mol% of *t*-BuNH₂·TFA as an additive, the reaction could be performed in aqueous media, providing the Biginelli products in good to high yields (62–93%) and high to excellent enantioselectivities (87–99% ee).

Moreover, a series of chiral phosphoric acids derived from 3,3′-disubstituted binaphthols have been investigated as catalysts for the asymmetric Biginelli reaction with thioureas by Gong *et al.*[335] The authors have shown that the size of 3,3′-substituents of the catalysts had a considerable impact on the control of the stereochemistry of the reaction. Indeed, by tuning these substituents, the stereochemistry of the reaction could be reversed, allowing the synthesis of both enantiomers of the same dihydropyrimidine. The best results were achieved by using a chiral phosphoric acid, which allowed a range of dihydropyrimidines to be produced in moderate to good yields (26–98%) and moderate to excellent enantioselectivities (26–99% ee), from the reactions of various β-ketoesters and aldehydes with thiourea. It must be noted that some aliphatic aldehydes also gave good results, such as cyclohexylcarbaldehyde, which allowed an

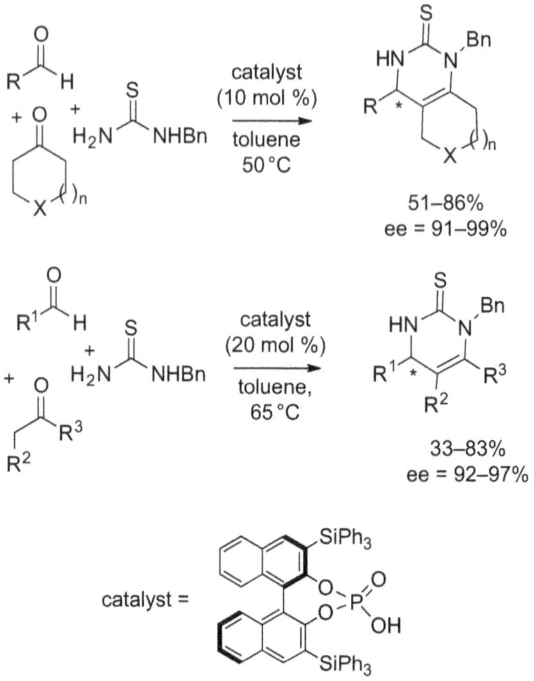

Scheme 3.218 Biginelli-like reactions catalysed by a chiral phosphoric acid.

enantioselectivity of 90% ee to be obtained in combination with 54% yield. The scope of this methodology was extended to the reaction of a range of aldehydes and thioureas with various ketones. As shown in Scheme 3.218, both cyclic and acyclic ketones provided the corresponding dihydropyrimidines in good yields and enantioselectivities of up to 99% ee, through a Biginelli-like reaction.

In 2011, Moorthy and Moorthy reported the synthesis of a series of novel sterically hindered organocatalysts derived from L-proline.[336] From a systematic investigation of these catalysts and screening studies, the chiral catalyst depicted in Scheme 3.219, containing a strong hydrogen-bonding site, was selected to be the most efficient in promoting the Biginelli condensation of aromatic as well as aliphatic aldehydes with ethyl acetoacetate and urea in a remarkably high enantioselectivity (94–99% ee), as shown in Scheme 3.219. This sterically hindered catalyst was used in combination with pentafluorobenzoic acid (PFBA) and tritylammonium trifluoroacetate as additives, allowing a range of chiral dihydropyrimidines to be produced in moderate to good yields (44–68%). The authors demonstrated that chiral prolinamides with reinforced chirality and enhanced hydrogen acidity at the C4-position and sterics built at the C2-position were excellent organocatalysts of the Biginelli reaction.

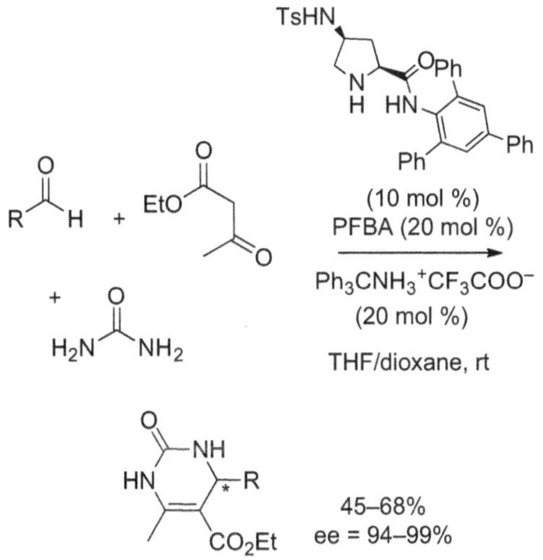

Scheme 3.219   Biginelli reaction catalysed by a sterically hindered chiral prolinamide.

3.3.5   Multicomponent Reactions Based on the Hantzsch Reaction

Another venerable and old multicomponent reaction is the so-called Hantzsch reaction, which was first reported in 1882.[242,337] It allows the synthesis of 1,4-dihydropyridines to be achieved through the reaction of enamines, aldehydes, and 1,3-dicarbonyl compounds. An asymmetric three-component Hantzsch-type reaction for the construction of dihydropyridines was developed by Jorgensen *et al.* in 2008, representing the first organocatalytic enantioselective one-pot synthesis of chiral dihydropyridines.[338] The reaction of α,β-unsaturated aldehydes, β-diketones or β-ketoesters, and primary amines afforded, by using a chiral diarylprolinol trimethylsilyl ether as catalyst, the corresponding chiral 1,4-dihydropyridines in moderate to good yields and moderate to high enantioselectivities of up to 95% ee, as shown in Scheme 3.220. The scope of the reaction was explored with α,β-unsaturated aldehydes bearing aliphatic and aromatic groups as well as heteroatoms and double bonds. Moreover, both acetoacetone and methyl acetoacetate consistently gave high enantioselectivities in combination with various substrates. The primary amine tolerated aliphatic as well as aromatic substituents, although the highest reactivity and yield were observed for aniline.

In 2008, Gong *et al.* simultaneously reported an almost identical Hantzsch-type three-component reaction for the generation of oppositely configured dihydropyridines under chiral Brønsted acid catalysis by using a chiral phosphoric acid as catalyst.[339] Advantageously, a broader scope of substrates was tolerated and the products were mostly obtained in considerably higher yields (31–93%) and enantioselectivities (66–98% ee), as shown in Scheme 3.221.

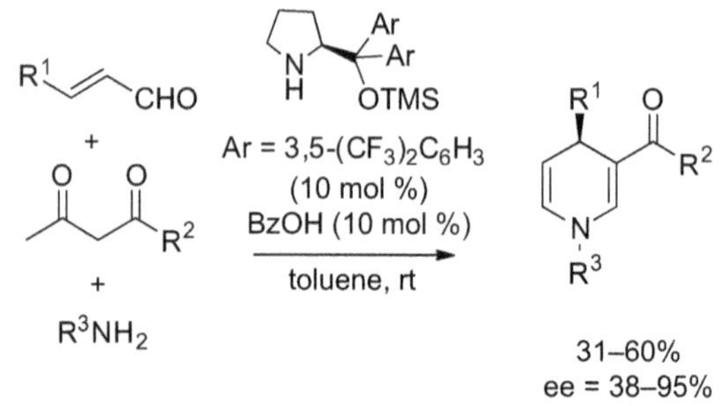

Scheme 3.220   Three-component Hantzsch-type reaction catalysed by an L-proline derivative.

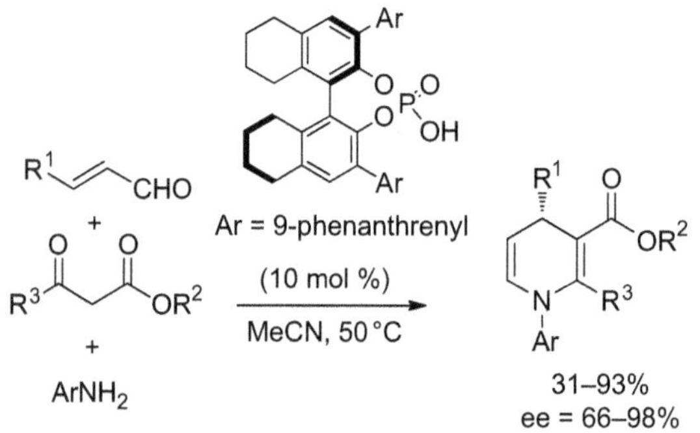

Scheme 3.221   Three-component Hantzsch-type reaction catalysed by a chiral phosphoric acid.

It was shown that both the reaction efficiency and enantioselectivity of this process were highly dependent on the substituents of the substrates. For the aniline component, electron-withdrawing substituents on the phenyl ring appeared to have a negative effect on the stereoselectivity of the reaction. Moreover, a dependence of the enantioselectivity on the ester moiety of the β-ketoester was observed.

The first highly enantioselective four-component version of the Hantzsch reaction was recently developed by Gestwicki and Evans, providing access to a range of pharmaceutically important dihydropyridines.[340] The reaction was catalysed by a BINOL-derived phosphoric acid, and occurred between ethylacetoacetate, ammonium acetate, dimedone, and an aromatic aldehyde in acetonitrile as solvent. As shown in Scheme 3.222, a series of 1,4-dihydropyridines were produced under these conditions in excellent yields and enantioselectivities in almost all cases of the substrates studied. Conversely, reactions involving aliphatic aldehydes proceeded in good yields but with no enantio-enrichment.

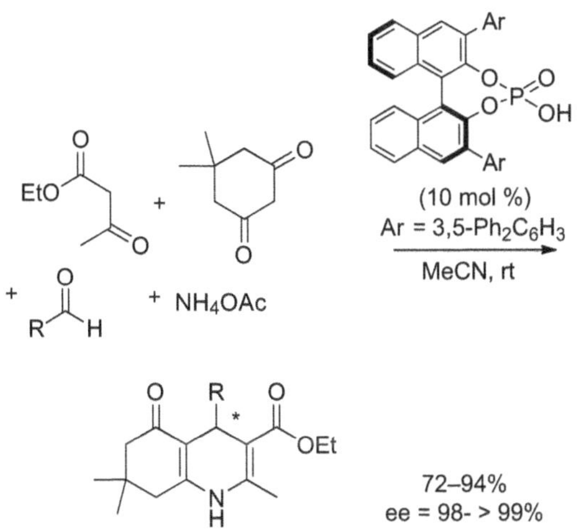

Scheme 3.222 Four-component Hantzsch reaction catalysed by a chiral phosphoric acid.

3.3.6 Multicomponent Reactions Based on the Strecker Reaction

The first multicomponecarbonyl derivatives, amines, and a cyanide source, such as hydrogen cyanident reaction, were involved in the Strecker reaction reported in 1850 by Adolf Strecker.[241a] It is a three-component coupling among carbonyl derivatives, amines, and a cyanide source, such as hydrogen cyanide, to provide α-aminonitriles which constitute potent starting materials to achieve important α-amino acids by simple hydrolysis. The mechanism of the Strecker reaction involves the initial formation of an imine from condensation of the amine component to the carbonyl component, after which addition of the cyanide component to this imine intermediate follows. Although the first enantioselective metal-catalysed Strecker reaction was reported by Kobayashi *et al.* in 2000,[341] the first organocatalytic version was not described until 2007 by Pan and List.[342] An advantage of this novel procedure was that highly toxic and volatile cyanide hydrogen was replaced by acyl cyanide. This three-component acyl-Strecker reaction among acetyl cyanide, amines, and aldehydes was catalysed by a chiral thiourea, and provided the corresponding α-amino nitriles in good to excellent yields and enantioselectivities of up to 94% ee, as shown in Scheme 3.223. Both aromatic and aliphatic aldehydes were tolerated for this process, in addition to amine inputs. Furthermore, the α-amino nitriles formed were subsequently converted into the corresponding α-amino acids and their salts through acid-mediated hydrolysis and hydrogenolysis without loss of optical purity.

In the same year, C_2-symmetrical chiral bisformamides were demonstrated by Feng *et al.* to induce the asymmetric Strecker three-component reaction.[343] The catalytic efficiency of the reaction was highly dependent on the match between the chiral centres of the diamine and proline parts of the catalyst.

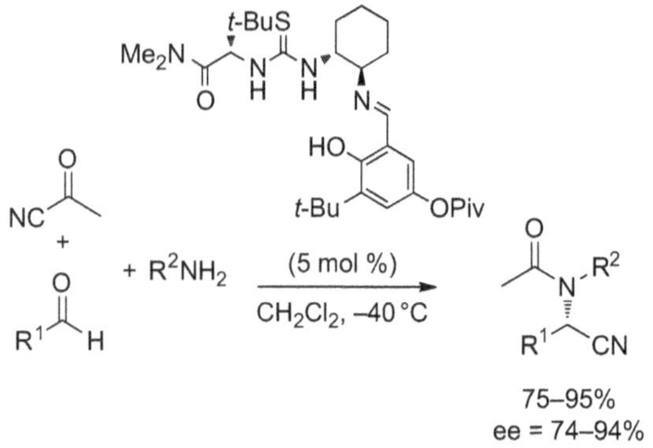

Scheme 3.223 Three-component acyl-Strecker reaction catalysed by a chiral thiourea.

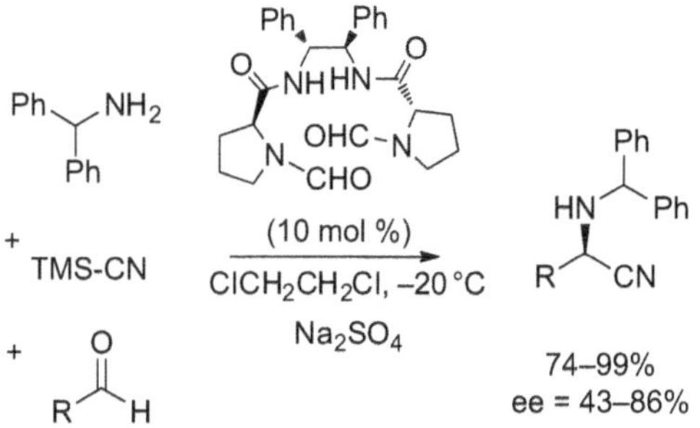

Scheme 3.224 Three-component Strecker reaction catalysed by an L-proline-derived diamine.

Disappointing enantioselectivity of 12% ee was observed when the L-proline formamide part of the catalyst was combined with (R,R)-diamine bisformamide instead of (S,S)-diamine bisformamide (61% ee). Optimisation of the reaction conditions showed that the formamide-based catalyst depicted in Scheme 3.224 was the most efficient in polar aprotic solvents, such as 1,2-dichloroethane. Indeed, the use of protic solvents, such as methanol, resulted in the formation of racemic α-aminonitriles. The scope of the process was investigated with various aromatic electron-rich as well as electron-deficient aldehydes, which allowed the corresponding chiral α-aminonitriles to be achieved in high yields of up to 99% and moderate to good enantio-selectivities (43–86% ee).

   Later, these authors developed the same asymmetric Strecker reaction induced by chiral amino acid-derived *N,N'*-dioxides as effective catalysts.[344] The combination of the Lewis basic *N*-oxide and amide moieties provided a highly efficient bifunctional organocatalyst. Once again, the catalytic efficiency was shown to be highly dependent on the match between the chiral centres of the amide and proline parts of catalyst depicted in Scheme 3.225, giving better

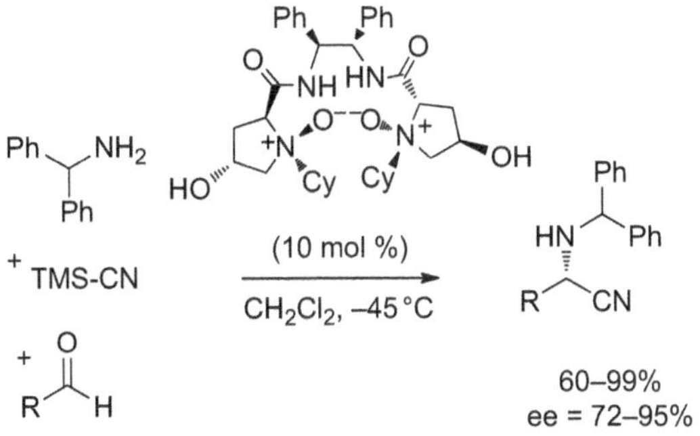

Scheme 3.225 Three-component Strecker reaction catalysed by a L-proline-derived *N,N′*-dioxide.

results for L-proline in combination with an (*S,S*)-diamide. Reduced enantio-selectivity was obtained after expanding or decreasing the size of the cyclohexyl substituent of the catalyst, while the proline-hydroxyl proved to complement the stereoselectivity. The process was compatible with a broad range of aromatic and aliphatic aldehydes, affording the corresponding α-aminonitriles in moderate to good yields and enantioselectivities of up to 95% ee, as shown in Scheme 3.225. In comparison with the C_2-symmetric chiral bisformamide of Scheme 3.224, this *N,N′*-dioxide catalyst induced the Strecker reaction with a broader scope of substrates, allowed higher enantioselectivities to be obtained, and furnished the opposite enantiomer of α-aminonitriles.

3.3.7 Multicomponent Reactions Based on the Petasis Reaction

As a multicomponent reaction, the Petasis reaction enjoys a history of nearly two decades, since it was discovered by N. A. Petasis in 1993.[345] It involves the condensation of amines, carbonyl derivatives, and aryl- or vinylboronic acids for the preparation of amine derivatives. This reaction evolved to be a powerful synthetic tool in the last decade, owing to the high accessibility of the reagents and mild reaction conditions. Its most important application is the preparation of chiral α-amino acids, using glyoxylic acid as the aldehyde component. Although remarkable success have been achieved in the asymmetric version based on the use of chiral auxiliaries, to date there are only few examples concerning the catalytic enantioselective Petasis reaction. Most of the time, these rare enantioselective versions of the Petasis reaction are limited to substrates including quinolines and ethyl glyoxylates. The first organocatalytic enantioselective Petasis reaction was reported by Schaus and Lou, in 2008.[346] The reaction among styryl boronates, secondary amines, and ethyl glyoxylate was catalysed by chiral diphenol (*S*)-VAPOL. Vaulted biaryl phenols such as (*S*)-VAPOL gave superior results for the asymmetric Petasis reaction in comparison with chiral BINOL derivatives. Advantageously, the catalyst could be recovered and recycled without loss of activity and optical purity. The scope of the process was explored with various electron-rich and electron-deficient styryl boronates, providing the corresponding α-amino acids in high yields and enantioselectivities, as shown in Scheme 3.226. Moreoever, high functional group tolerance was observed for the secondary amine.

On the other hand, Takemoto *et al.* have developed an asymmetric Petasis-type reaction of substituted quinolines, styrylboronic acids, and phenyl chloroformate catalysed by a newly designed chiral thiourea catalyst, depicted in Scheme 3.227.[347] While the addition of water resulted in a minor increase of

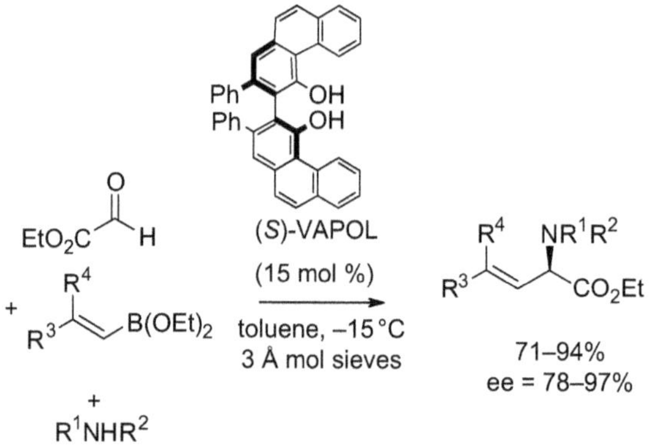

Scheme 3.226 Three-component Petasis reaction of ethyl glyoxylate, styryl boronates, and secondary amines catalysed by (S)-VAPOL.

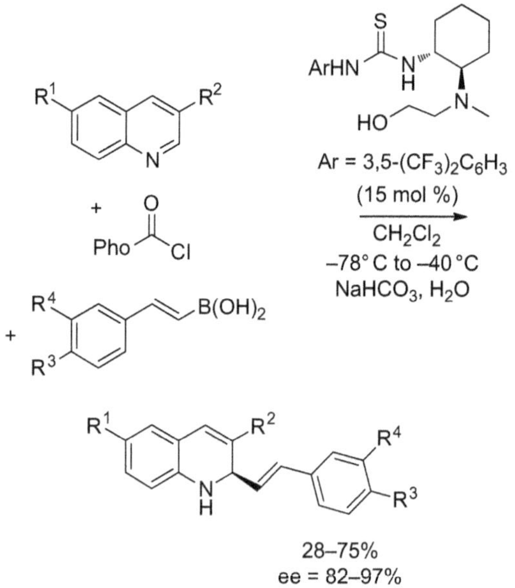

Scheme 3.227 Three-component Petasis-type reaction catalysed by a chiral thiourea catalyst.

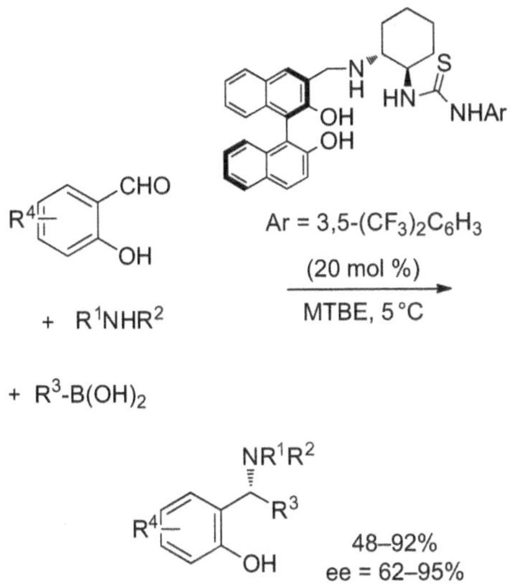

Scheme 3.228 Three-component Petasis reaction of salicylaldehydes, amines, and organoboric acids catalysed by a chiral phosphoric acid.

enantioselectivities (from 90 to 93% ees), the use of $NaHCO_3$ as an additive improved the yield of the process significantly (from 27 to 65%). Various quinolines and styrylboronic acids were tolerated to afford the corresponding dihydroquinolines in moderate yields and good enantioselectivities, as shown in Scheme 3.227. It must be noted that this Petasis-type reaction was highly selective for 1,2-addition, because the regioisomeric 1,4-adduct was not observed.

In 2012, Yuan *et al.* developed an enantioselective three-component Petasis reaction among salicylaldehydes, amines, and organoboric acids catalysed by a newly designed chiral BINOL-derived thiourea.[348] A broad range of chiral alkylaminophenols could be achieved in good yields and good to high enantioselectivities of up to 95% ee, as shown in Scheme 3.228.

3.3.8 1,3-Dipolar Cycloaddition-based Multicomponent Reactions

A 1,3-dipolar cycloaddition is a reaction between a 1,3-dipole and a dipolarophile, such as a substituted akene or alkyne, to construct five-membered heterocycles, which can bear up to four stereocentres, often with a high degree of stereocontrol.[349] The resulting heterocycles, such as isoxazolidines and pyrrolidines, constitute valuable intermediates for the synthesis of various natural products, including amino acids and alkaloids. 1,3-Dipolar cycloaddition-based multicomponent reactions usually proceed *via* an *in situ* formation of a 1,3-dipole, which then reacts with the dipolarophile component. In 2007, Cordova *et al.* described a nice highly chemo-, diastereo-, and enantioselective organocatalytic one-pot 1,3-dipolar cycloaddition-based three-component reaction among *N*-arylhydroxylamines, aldehydes, and α,β-unsaturated aldehydes.[350] This process, catalysed by chiral diphenyl-prolinol trimethylsilyl ether, represented a versatile asymmetric entry to a variety of valuable isoxazolidines in high yields and generally excellent diastereoselectivity of >92% de, associated with remarkable enantio-selectivities ranging from 94 to 99% ee, as shown in Scheme 3.229. Given that the domino aldehyde products were susceptible to instability, they were reduced *in situ* to the corresponding alcohols by treatment with NaBH₄.

As an extension of this methodology, these authors employed the same catalyst to induce an enantioselective three-component synthesis of chiral pyrrolidines.[351] The first step of the domino process consisted in imine formation between 2-aminomalonate and an aldehyde. Its corresponding azomethine ylide prepared *in situ* reacted with α,β-unsaturated aldehydes, furnishing the final products in excellent yields, diastereo- and enantio-selectivities, as shown in Scheme 3.230. Various aromatic and aliphatic enals were compatible acceptors in this process, as were a broad range of benz-aldehyde derivatives.

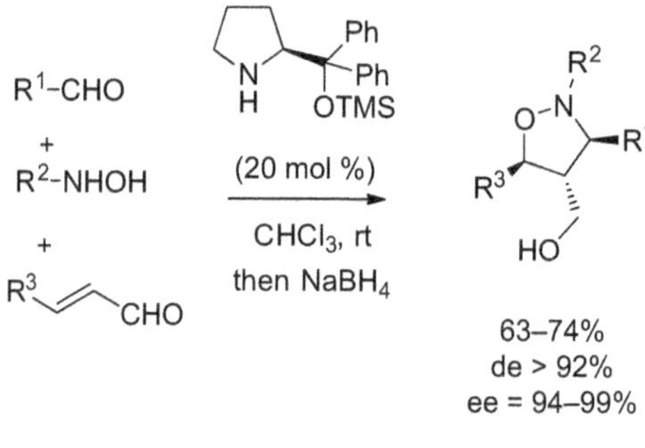

Scheme 3.229 1,3-Dipolar cycloaddition-based three-component reaction of *N*-arylhydroxylamines, aldehydes, and α,β-unsaturated aldehydes.

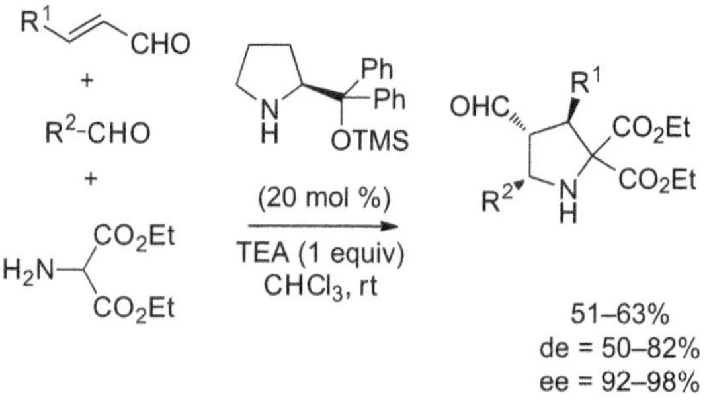

Scheme 3.230 1,3-Dipolar cycloaddition-based three-component reaction of 2-aminomalonate, aldehydes, and α,β-unsaturated aldehydes.

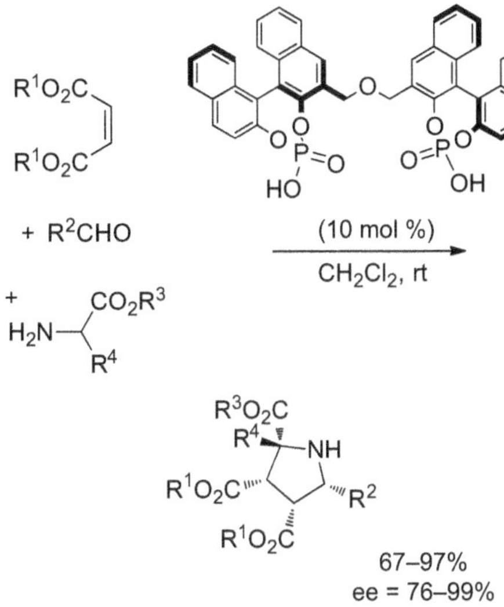

Scheme 3.231 1,3-Dipolar cycloaddition-based three-component reaction of aldehydes, dialkyl maleates, and α-amino esters.

In 2008, an enantioselective three-component reaction of aldehydes, dialkyl maleates, and α-amino esters was described by Gong *et al.*[352] As shown in Scheme 3.231, the process, catalysed by a chiral biphosphoric acid, afforded the corresponding 1,3-dipolar cycloadducts in good to excellent yields and enantioselectivities of up to 99% ee. Several aldehydes including aromatic, aliphatic, and α,β-unsaturated were compatible with this reaction. It seemed highly sensitive to the nature of the dipolarophile and suffered particularly from the presence of bulky substituents on the maleate component. By the incorporation of an unsymmetrical α-amino ester, four contiguous stereogenic centres were created with high enantioselectivities (81–97% ee).

The spiro[pyrrolidin-3,3′-oxindole] ring system constitutes the core structural element found in a large family of natural alkaloids and unnatural compounds exhibiting important biological activities. Although these are involved in elegant and creative strategies towards the construction of spirooxindole architectures, the first directly catalytic asymmetric approach to access chiral spiro[pyrrolidin-3,3′-oxindoles] was reported by Gong *et al.* only in 2009.[353] It was based on an enantioselective three-component 1,3-dipolar cycloaddition of a broad range of methyleneindolinones with aldehydes and amino diester catalysed by a chiral BINOL-derived phosphoric acid. As shown in Scheme 3.232, a series of chiral spirooxindoles were achieved in high yields with an unusual regiochemistry and excellent enantioselectivities of up to 98% ee.

Moreover, theoretical calculations disclosed that both the azomethine ylide and the methyleneindolinone are hydrogen-bonded with the catalyst, which accounted for the high enantio- and regioselectivity, and indicated that the unusual regioselectivity resulted from the stabilisation stemming from the

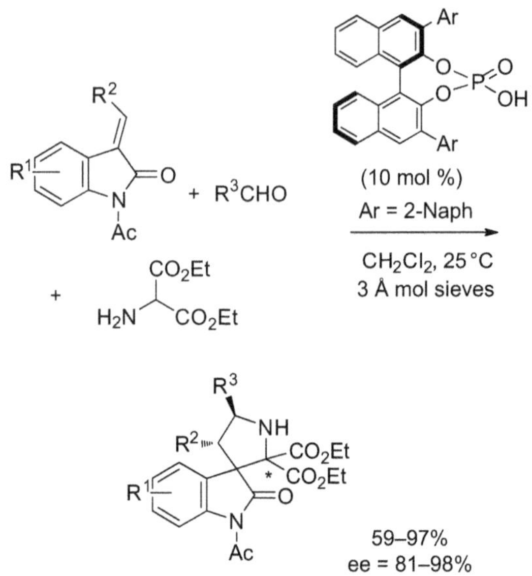

Scheme 3.232 1,3-Dipolar cycloaddition-based three-component reaction of methyleneindolinones, aldehydes, and an amino diester.

favourable π–π stacking interaction between the oxoindole ring and the conjugated esters. In this context, the authors have proposed the mechanism depicted in Scheme 3.233, in which the cycloaddition reaction started with the condensation of the aldehyde with the amino ester to give an imine intermediate. The chiral phosphoric acid and the imino ester were able to form a chiral azomethine ylide, which principally underwent a 1,2-prototropic process to form a phosphoric acid intermediate. Either the chiral azomethine ylide or this phosphoric acid intermediate reacted with the methyleneindolinones to generate iminium intermediates, wherein the hydroxyl proton and the phosphoryl oxygen of the chiral phosphoric acid would form double hydrogen-bonding interactions with the methyleneindolinone and the azomethine ylide dipole, respectively. The resulting reaction of these intermediates was a 1,3-dipolar cycloaddition reaction, regiospecifically yielding spiro[pyrrolidin-3,3′-oxindole] derivatives *via* the transition state depicted in Scheme 3.233, as suggested by DFT calculation.

In 2011, Cordova *et al.* developed the first highly diastereoselective and enantioselective organo-co-catalytic dynamic three-component reaction among aldehydes, protected α-cyanoglycine esters, and α,β-unsaturated aldehydes.[354] This powerful process afforded cyano-, formyl-, and ester-functionalised α-quaternary proline derivatives bearing four contiguous stereocentres in generally excellent enantioselectivities (93–99% ee), combined with moderate to excellent yields and diastereoselectivities (Scheme 3.234). The domino process began with the formation of an imine from the reaction between the aldehyde and the protected α-cyanoglycine ester through hydrogen-bond donating

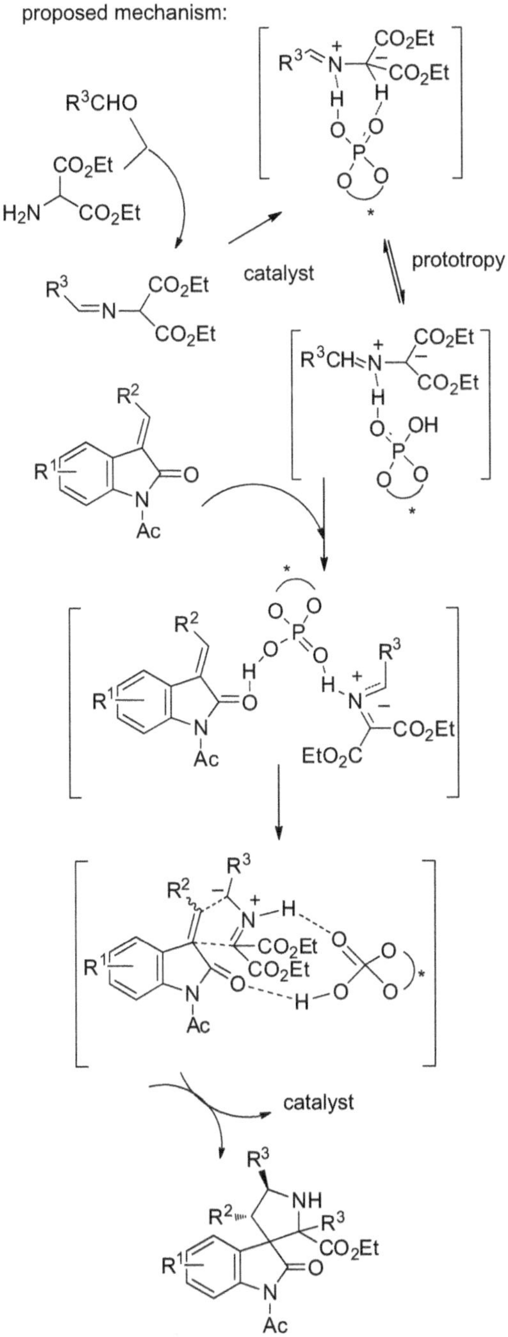

Scheme 3.233 Proposed mechanism for 1,3-dipolar cycloaddition-based three-component reaction of methyleneindolinones, aldehydes, and an amino diester.

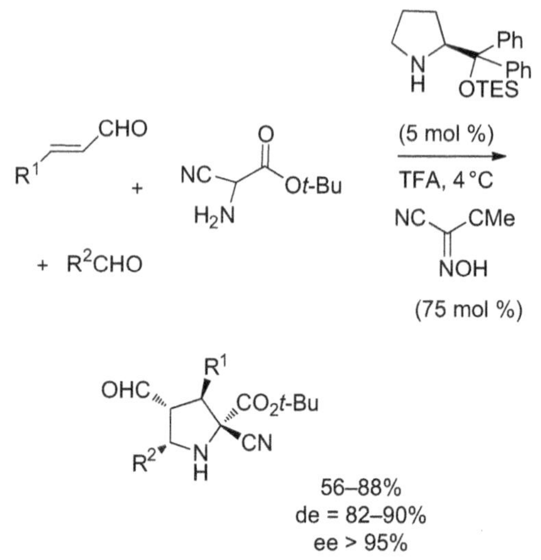

Scheme 3.234 1,3-Dipolar cycloaddition-based three-component reaction of aldehydes, protected α-cyanoglycine esters, and α,β-unsaturated aldehydes through dual organocatalysis.

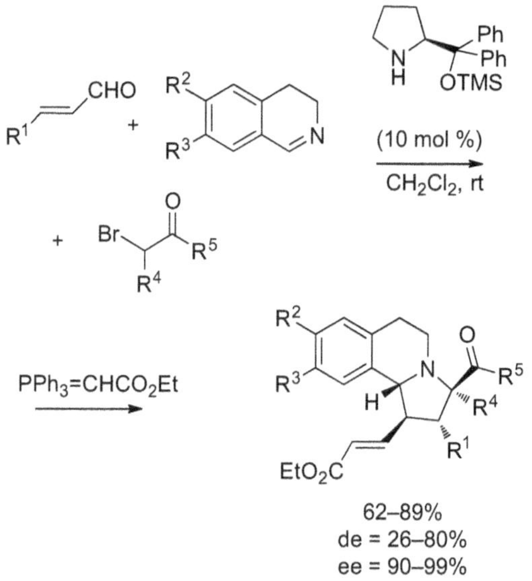

Scheme 3.235 1,3-Dipolar cycloaddition-based three-component reaction of α-bromoesters or ketones, α,β-disubstituted aldehydes, and dihydroisoquinolines.

catalysis with an achiral oxime catalyst, which then afforded the corresponding azomethine ylide. The latter subsequently underwent a 1,3-dipolar cyclo-addition with the α,β-unsaturated aldehyde activated through iminium acti-vation by chiral diphenylprolinol triethylsilyl ether catalyst to give the final product. The biomimetic cooperative combination of hydrogen-bond and iminium activation of the carbonyl components was shown to be essential to achieve chemo- and stereoselective 1,3-dipolar cycloaddition under kinetic control.

Finally, Jorgensen *et al.* described, in 2012, an asymmetric synthesis of hexahydropyrrolo-isoquinolines based on an enantioselective 1,3-dipolar cycloaddition of *in situ* generated dihydroisoquinolinium ylides from the corresponding imines and α-bromoesters or ketones to α,β-disubstituted aldehydes.[355] This domino imine-formation–1,3-dipolar cycloaddition reaction was catalysed by chiral diphenylprolinol trimethylsilyl ether in the presence of a base such as Na_2CO_3. The corresponding domino aldehyde products were further converted into the corresponding alkenes by treatment with Wittig–Horner reagent. The final pyrrolo-isoquinolines were achieved in good yields and excellent enantioselectivities along with moderate to good diastereoselectivities, as shown in Scheme 3.235.

3.3.9 Miscellaneous Multicomponent Reactions

The modern concept of a multicomponent reaction is intimately related to the reactions developed with isocyanide reagents.[356] The first isocyanide-based multicomponent reaction was discovered by Passerini, in 1921.[357] It was not until 2003 that Denmark and Fan developed an enantioselective organocatalytic version of this reaction.[358] The Ugi four-component reaction is the reaction of a carbonyl compound (usually an aldehyde), an amine, an isocyanide, and a carboxylic acid (or an alcohol) to yield α-amino acid derivatives. This reaction, first described in 1959, has been more widely studied and used than any other multicomponent reaction.[244a] The Ugi three-component is a variant of the general reaction in which either two of the usual groups are included in the same reagents or the condensation of the carbonyl compound with the amine takes place before the addition of the isocyanide and acid derivatives. Even though the imine condensation could be performed in a one-pot process and its isolation is not necessary, it must be included in this category. Although a number of diastereoselective versions of this reaction have been successfully developed, enantioselective versions remain a significant challenge. In fact, the first example of enantioselective α-addition of α-isocyanides to imines was developed by Zhu *et al.*, in 2009.[359] Indeed, these authors have demonstrated that a chiral phosphoric acid was able to catalyse the three-component reaction of aldehydes, anilines, and an α-isocyanoacetamide (Scheme 3.236), leading to the corresponding chiral 2-(1-aminoalkyl)-5-aminoxazoles in excellent yields (82–95%) and moderate to good enantioselectivities (66–90% ee), as shown in Scheme 3.236.

In 2008, Ma *et al.* reported a highly enantioselective three-component reaction for the construction of trifluoromethyl-containing compounds.[360] This domino Friedel–Crafts–aminoalkylation reaction occurred between indoles and imines generated *in situ* from trifluoroacetaldehyde methyl hemiacetal and aniline in the presence of a catalytic amount of a chiral BINOL-derived

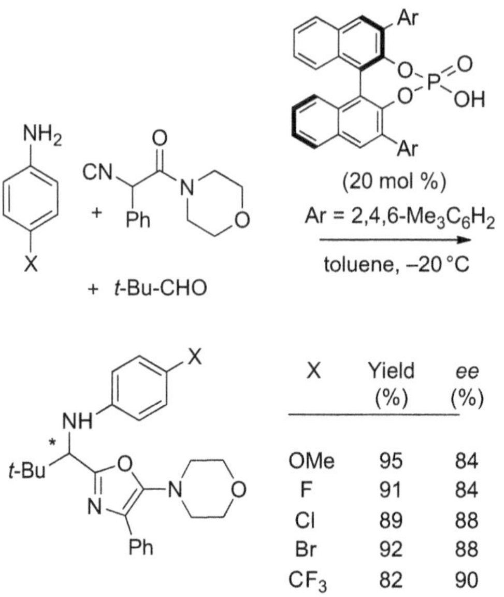

X	Yield (%)	ee (%)
OMe	95	84
F	91	84
Cl	89	88
Br	92	88
CF_3	82	90

Scheme 3.236 Three-component Ugi reaction catalysed by a chiral phosphoric acid.

phosphoric acid bearing bulky 2,4,6-triisopropylphenyl groups at the 3,3′-positions. It afforded the corresponding chiral trifluoromethyl-containing indoles in high yields and excellent enantioselectivities when using 3,4,5-trimethoxyaniline as the amine component, as shown in Scheme 3.237. Furthermore, the scope of the methodology was extended to difluoroacetaldehyde methyl hemiacetal, providing the corresponding difluoromethyl-containing indoles in quantitative yields and enantioselectivities of >93% ee.

In 2010, Enders *et al.* reported an enantioselective synthesis of polyfunctionalised 3-(cyclohexenylmethyl)-indoles bearing three stereogenic centres on the basis of a novel four-membered domino reaction, occurring among 3-(cyclohexenylmethyl)-indoles, nitroalkenes, and 2 equivalents of acrolein.[361] This novel process was based on a quadruple domino Friedel–Crafts–Michael–Michael–aldol condensation reaction catalysed by chiral diphenylprolinol trimethylsilyl ether. As shown in Scheme 3.238, the products were produced in moderate to excellent yields (23–82%) and excellent diastereo- and enantioselectivities of up to >90% de, and >99% ee, respectively. The authors have proposed the catalytic cycle depicted in Scheme 3.238. In the first step, acrolein was activated by chiral diphenylprolinol trimethylsilyl ether through the formation of the corresponding iminium ion, with which the indoles performed an intermolecular Friedel–Crafts-type reaction. The resulting enamines subsequently underwent an intermolecular Michael addition to nitroalkenes, affording novel iminium intermediates. Further hydrolysis of these intermediates led to the corresponding aldehydes, which reacted with a second equivalent of the starting iminium ion to give the

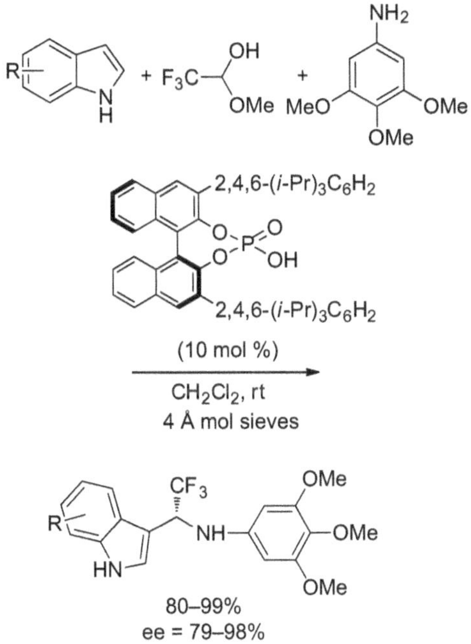

Scheme 3.237 Three-component domino Friedel–Crafts–aminoalkylation reaction.

corresponding enamine intermediates. After an intramolecular enamine-mediated aldol reaction, alcoholate intermediates were formed, which underwent dehydration and hydrolysis to afford the final indoles, while the catalyst was regenerated.

In 2012, Chen and Roy demonstrated the kinetic resolution of nitroallylic acetates *via* an enantioselective three-component coupling involving indoles, acrolein, and nitroolefin allylic acetates.[362] This domino Friedel–Crafts–Michael–elimination reaction was catalysed by the same catalyst and provided, through the iminium enamine activation mode, the corresponding densely functionalised 3-substituted indoles in moderate yields, albeit remarkable diastereo- and enantioselectivities, as shown in Scheme 3.239. It was found that the recovered nitroallylic acetates were obtained in good to excellent enantiopurity ranging from 79 to 99% ee.

A large number of organocatalysts have been shown to be highly efficient in inducing Diels–Alder reactions, providing excellent yields and stereoselectivities.[363] Indeed, prolines, imidazolidinones, and other pyrrolidine-derived reagents have been demonstrated to be capable of promoting these reactions through enamine catalysis. Hydrogen-bonding activation is also possible with a variety of reagents, such as chiral Brønsted acids and diols. The first enantioselective aza-Diels–Alder three-component reaction was reported by Cordova *et al.*, in 2005.[364] It was catalysed by L-proline, and occurred among cyclohexenones, anilines, and formaldehyde. Diversely substituted aromatic amines

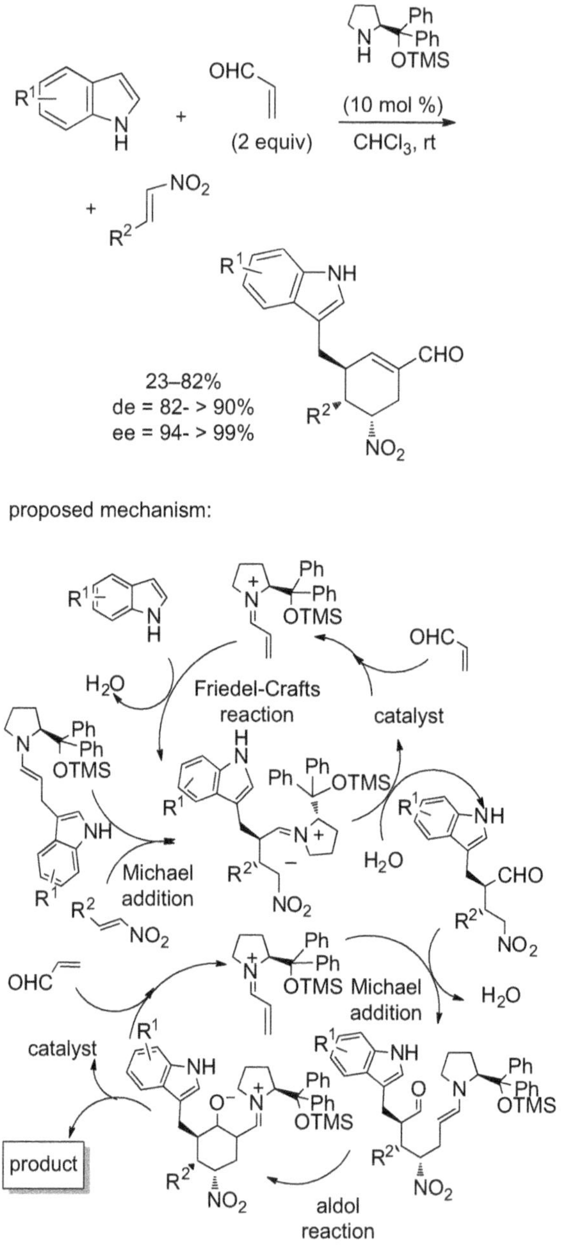

Scheme 3.238 Four-component domino Friedel–Crafts–Michael–Michael–aldol condensation reaction.

were demonstrated to be excellent substrates, providing the corresponding bicyclic amines in good yields (70–82%), and high enantioselectivities (94–99% ee), as shown in Scheme 3.240. Notably, excellent stereocontrol was achieved for the reaction of anilines bearing an electron-donating *para*-substituent.

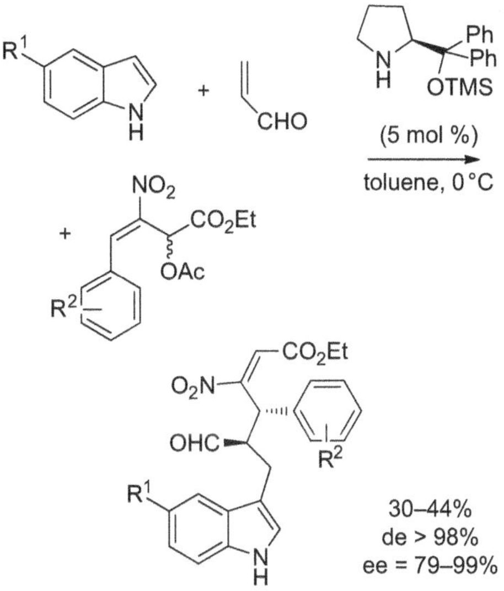

Scheme 3.239   Three-component   domino   Friedel–Crafts–Michael–elimination reaction.

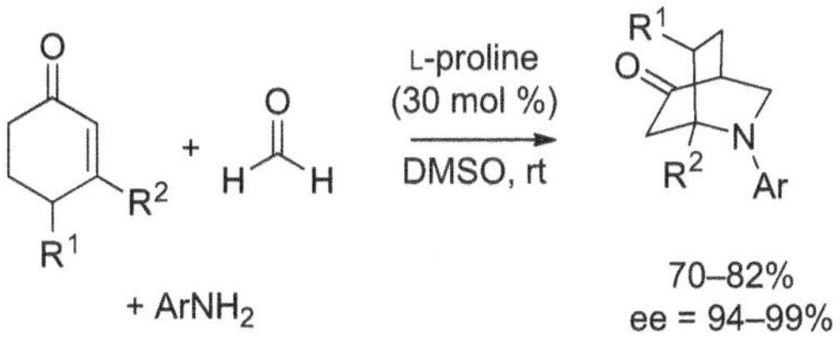

Scheme 3.240   Three-component domino imine-formation–aza-Diels–Alder reaction of formaldehyde, cyclohexenones, and aromatic amines.

To account for the stereochemical outcome of the reaction, a mechanism was proposed which involved an *in situ* enamine formation between the cyclo-hexenone derivative and L-proline. The intermediate imine formed from formaldehyde and aniline was attacked by this enamine intermediate from the less-hindered *Si*-face. Subsequently, a 6-*endo-trig*-cyclisation occurred to afford the final product. It must be noted that this approach suffered from the high catalyst loading (30 mol%), as well as the restriction to formaldehyde as the aldehyde component.

A more widely applicable procedure was reported by Gong *et al.*, who developed the first Brønsted acid-catalysed asymmetric direct aza-Diels–Alder reaction, in 2006.[365] As shown in Scheme 3.241, the process was catalysed by a chiral H$_8$-BINOL-derived phosphoric acid and was shown to be compatible with various substituted aromatic aldehydes, providing by reaction with *p*-anisidine and 2-cyclohexenone the corresponding bicyclic piperidines in moderate to good yields, moderate *endo*-selectivities of up to 74% de, and good

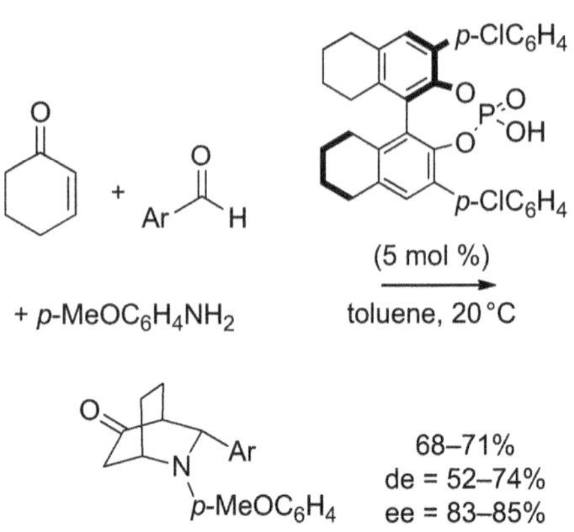

Scheme 3.241   Three-component domino imine-formation–aza-Diels–Alder reaction of aromatic aldehydes, *p*-anisidine, and cyclohexenone.

enantioselectivities of up to 85% ee. These *N*-heterocycles constituted key building blocks for a large family of biologically important compounds, such as alkaloids, amino acids, and aza-sugars.

In 2011, Jorgensen *et al.* disclosed a novel activation mode provided by organocatalysts.[366] Thus, these authors demonstrated that the merger of chiral secondary amines and polyenals generated reactive trienamine intermediates, which readily participated in stereoselective Diels–Alder reactions with different types of dienophile. The novelty of this activation strategy lay within the perfect chirality relay over a distance of up to eight bonds. An extension of this discovery was the development of the first domino multicomponent reaction, which combined trienamine catalysis with enamine activation. As shown in Scheme 3.242, chiral diphenylprolinol trimethylsilyl ether was able first to activate 2,4-hexadienal through its corresponding trienamine, which reacted with 3-olefinic oxindole employed as dienophile through a Diels–Alder cycloaddition to afford the corresponding enamine intermediate. This enamine then reacted with double-activated ethyl 2-(diethoxyphosphoryl)acrylate employed as electrophilic partner to give an 85:15 diastereomeric mixture of products in 89% yield. The non-controllable stereocentre was easily eliminated through a further acetal protection-Horner–Wadsworth–Emmons reaction sequence, forming highly functionalised cyclohexene without epimerisation of the remaining stereocentres. This chiral product was produced in 62% yield for the two steps, with an excellent enantioselectivity of 98% ee and good diastereoselectivity of 70% de through combined trienamine and enamine activation modes. It is noteworthy that a double conjugate addition sequence could also account for the products formed. However, the authors have shown by monitoring the reaction by NMR spectroscopy that no monoaddition intermediates could be observed, implying a concerted reaction mechanism.

The Povarov reaction, an inverse electron-demand aza-Diels–Alder reaction of 2-azadienes with electron-rich olefins, allows a rapid construction of poly-substituted tetrahydroquinolines.[367] Although three-component versions

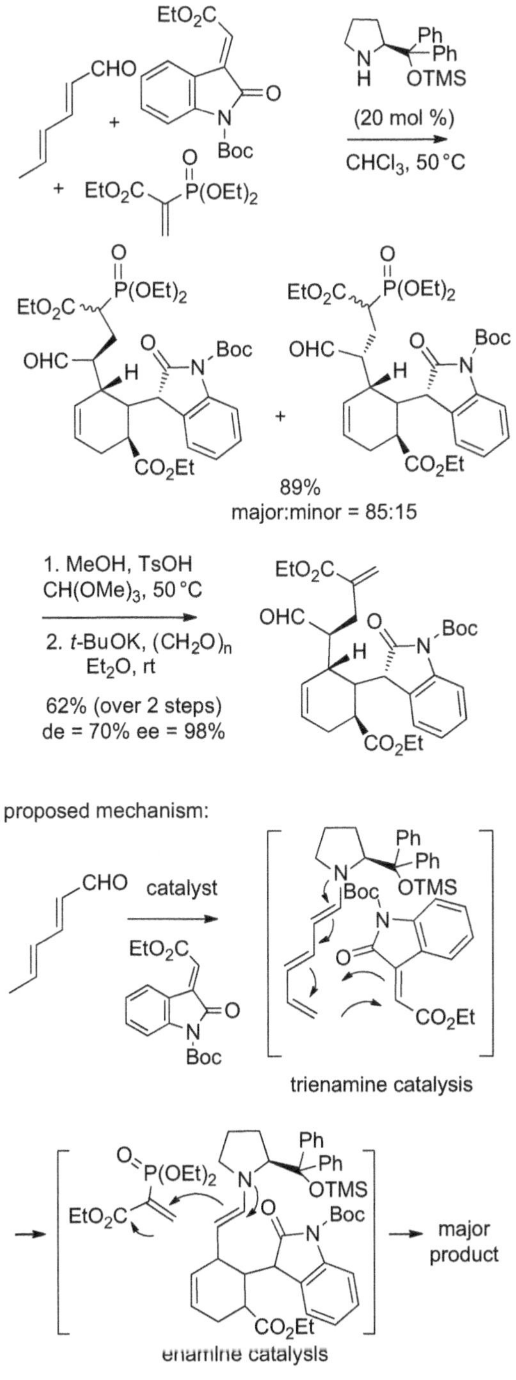

Scheme 3.242 Three-component domino Diels–Alder–Michael reaction through trienamine–enamine activation mode.

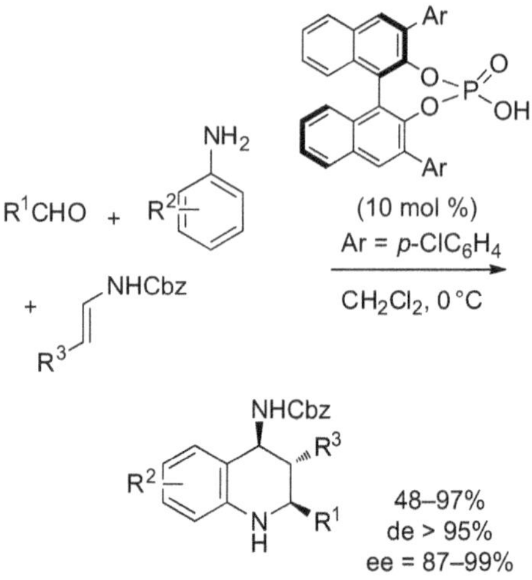

Scheme 3.243 Three-component Povarov reaction of aldehydes, anilines, and enecarbamates.

involving aniline, aldehydes, and olefins have been reported, enantioselective Povarov reactions remain rare. The first highly enantioselective example of this type of reaction was developed by Zhu *et al.*, in 2009.[368] Indeed, the reaction of a wide variety of aromatic and aliphatic aldehydes as well as anilines with benzyl-*N*-vinylcarbamates in the presence of a BINOL-derived phosphoric acid afforded the corresponding chiral (2,4-*cis*)-4-amino-2-aryl(alkyl)-tetrahydro-quinolines in good to high yields (64–90%) and remarkable enantioselectivities (92 to >99% ee), as shown in Scheme 3.243. In 2011, the same authors extended the scope of this powerful methodology to β-substituted acyclic enecarbamates which provided, by reaction with aromatic as well as aliphatic aldehydes and a wide range of anilines bearing electron-donating and electron-withdrawing groups, the corresponding 1,2,3,4-tetrahydroquinolines with three contiguous stereogenic centres in excellent diastereo- and enantioselectivities of >95% de, and up to >99% ee, respectively.[369] This remarkable three-component process afforded the corresponding *cis*-4-amino-2-aryl(alkyl)-1,2,3,4-tetrahydroquinolines in high yields and almost complete enantio-selectivities, as shown in Scheme 3.243. It must be noted that this methodology has allowed, for the first time, aliphatic aldehydes to be employed in the enantioselective Povarov reaction. In 2010, Gong *et al.* reported a three-component synthesis of the tricyclic ring system julolidine by combining the enantioselective Povarov reaction with an intramolecular hydroamination reaction.[370] In the same year, Ricci *et al.* employed another BINOL-derived phosphoric acid, (*S*)-TRIP, to induce a domino Povarov–Friedel–Crafts reaction.[371] The process began with an enantioselective two-component reaction of *N*-arylimines with 2- or 3-vinylindoles, providing the correspond-ing Povarov intermediates, which were subsequently trapped with

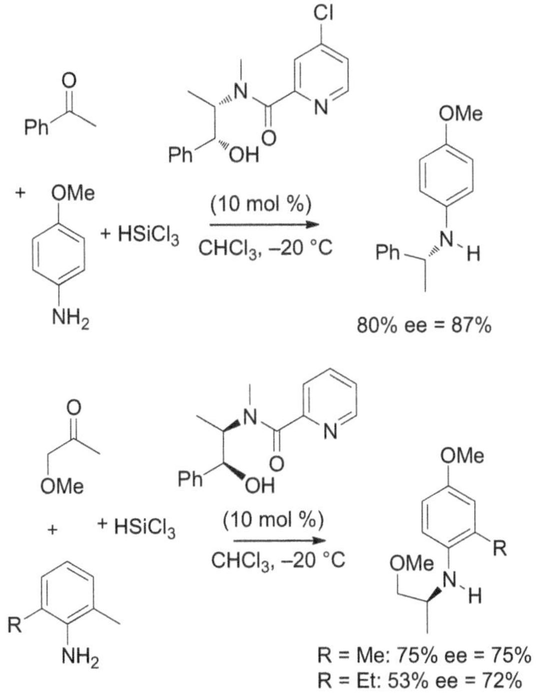

Scheme 3.244 Three-component reductive aminations of ketones.

p-methoxy-phenylamine as a nucleophile to give the corresponding indole derivatives as mixtures of diastereomers in good yields, zero to moderate diastereoselectivities, and good to excellent enantioselectivities (60–99% ee).

In another area, a convenient enantioselective three-component reductive amination of ketones was developed by Benaglia *et al.*, starting simply from a mixture of a ketone, an aryl amine, and trichlorosilane, upon catalysis with chiral picolinamides.[372] For example, the reaction of acetophenone, 4-methoxyaniline, and trichlorosilane in the presence of chiral picolinamide, depicted in Scheme 3.244, led to the corresponding chiral amine in 80% yield and a high enantioselectivity of 87% ee. The reduction of *in situ* generated *N*-aryl imines of methoxyacetone gave better results by using another catalyst depicted in Scheme 3.244 in comparable conditions, providing the corresponding chiral amines in good yields and enantioselectivities of up to 75% ee, as shown in Scheme 3.244.

An enantioselective synthesis of α-aminophosphonates by means of direct organocatalytic three-component hydrophosphonylation was successfully developed by JunAn *et al.*, in 2010.[373] Indeed, upon catalysis with a chiral phosphoric acid, the reaction among an aromatic aldehyde or a cinnamaldehyde, 4-methoxyaniline, and diisopropylphosphate afforded the corresponding α-aminophosphonates in good yields (61–91%) and moderate to good enantioselectivities (31–87% ee), as shown in Scheme 3.245.

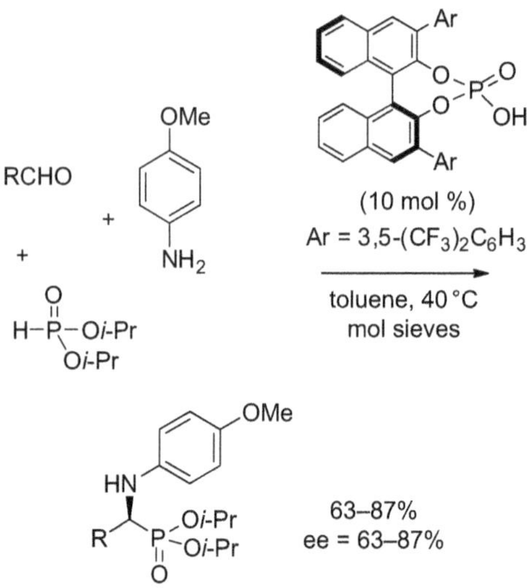

Scheme 3.245   Three-component hydrophosphonylation reaction.

In 2007, List and Zhou reported a highly enantioselective organocatalytic three-component process towards chiral *cis*-3-substituted cyclohexylamines through enamine, iminium, and Bronsted acid catalysis.[374] A domino aldol–dehydration–Michael–reductive amination reaction occurred among 2,6-diketones, *p*-alkoxy aniline, and Hantzsch ester when treated by a catalytic amount of a BINOL-derived chiral phosphoric acid. As shown in Scheme 3.246, a range of densely functionalised chiral cyclohexylamines were achieved in moderate to high yields and diastereoselectivities of up to 98% de, combined with excellent enantioselectivities of up to 96% ee.

An enantioselective pseudo-three-component domino aza-ene type reaction–cyclisation reaction was described by Terada *et al.*, in 2007.[375] It enabled the rapid construction of enantioenriched piperidine derivatives by using chiral phosphoric acids derived from BINOL as organocatalysts. The potential of this process was highlighted through its ability to achieve a rapid increase in molecular complexity from simple enecarbamates and a broad range of aldimines, while also controlling three stereogenic centres in a highly diastereo- and enantioselective manner, as shown in Scheme 3.247.

In 2010, Fréchet *et al.* reported an enantioselective three-component domino aldol–dehydration–Michael reaction performed in an aqueous buffer.[376] This novel methodology involved two different aldehydes and nitromethane as the three substrates and employed two catalysts, such as L-proline and chiral diphenylprolinol trimethylsilyl ether. In a two-phase system, aldehydes with a small size difference could be differentiated and reacted in a controlled manner to form selectively a single-cross-product. Each catalyst mediated an individual reaction step in either the aqueous or organic phase. For example, the reaction between *n*-butanal and *n*-decanal with nitromethane in aqueous buffer

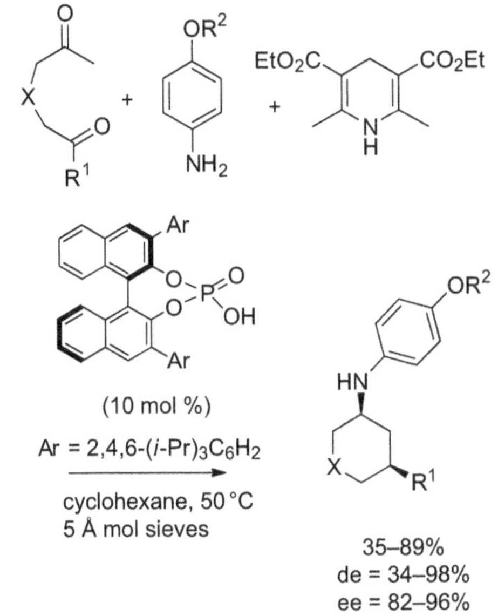

Scheme 3.246 Three-component domino aldol–dehydration–Michael–reductive amination reaction.

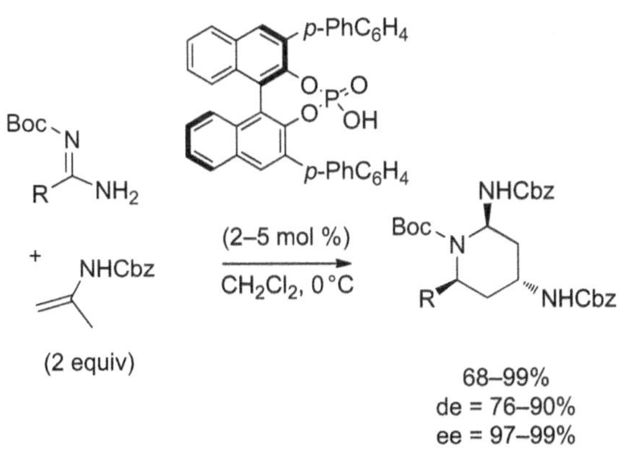

Scheme 3.247 Three-component domino aza-ene type reaction–cyclisation reaction.

catalysed by L-proline and chiral diphenylprolinol trimethylsilyl ether in the presence of oily droplets as the organic phase led selectively to a single domino product. Indeed, this major domino product arose from a two-step sequence beginning with the reaction of the relatively polar *n*-butanal with nitromethane catalysed by L-proline in the aqueous phase to give the corresponding nitroalkene intermediate through aldolisation followed by dehydration. This nitroalkene intermediate was then converted into the final product by Michael addition with *n*-decanal catalysed by chiral diphenylprolinol trimethylsilyl ether

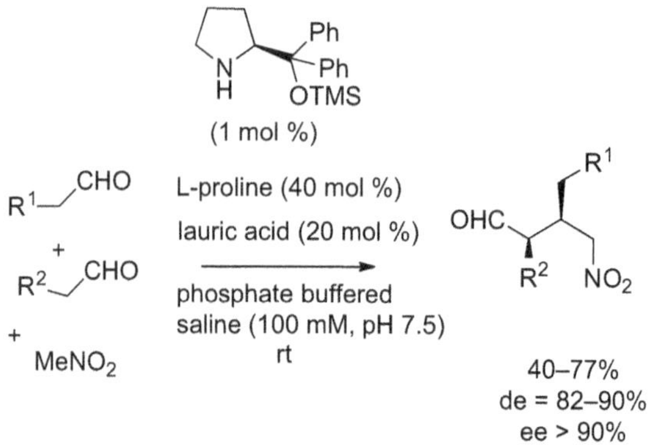

Scheme 3.248   Polarity-directed three-component reaction mediated by two catalysts in an aqueous buffer.

in the organic phase. As shown in Scheme 3.248, this methodology could be applied to various systems of aliphatic aldehydes, providing, by reaction with nitromethane, the corresponding domino products in good yields and excellent enantioselectivities of >90% ee in all cases of the substrates studied, combined with good to high diastereoselectivities of up to 90% de. This remarkable system highlighted an often-ignored approach to developing chemoselective reactions by using properties other than chemical reactivity, such as polarity, inherent to the substrates or catalysts.

3.4   Conclusions

This chapter illustrates the power of enantioselective organocatalytic domino reactions, which have quickly become a powerful, fascinating, and highly efficient tool in organic chemistry. Indeed, these beautiful one-pot reactions can be considered as one of the most influential reaction classes of the last century. The combination of asymmetric organocatalysis with the concept of domino sequences has allowed the easy attainment of high molecular complexity with very often excellent levels of stereocontrol in environmentally friendly conditions, with simple operational procedures, and advantages of savings in solvent, time, energy, and costs. With the various recently developed organo-catalytic activation modes at hand, including the concept of bifunctional organocatalysts and multicatalytic systems,[377] numerous novel cascade sequences can be envisaged, with a tendency to develop from simple domino reactions to triple and even quadruple domino reactions in cascade.[378] Major progress has been achieved in the last six years through the creation of highly enantioselective versions of existing and novel organocatalytic domino and multicomponent reactions. For example, in the area of multicomponent reactions, the first catalytic enantioselective Passerini, Biginelli, Petasis, and Hantzsch multicomponent reactions have been developed in recent years. Other improvements include significantly broadened scope, novel techniques and more environmentally benign methods, in addition to entirely novel reactions. The most progressive change is, however, the shift from transition metal catalysis towards organocatalysis. Transition metal contamination of the products is thus avoided, which is of great importance for future applications in the pharmaceutical industry. These significant improvements are expected to influence the development of novel organocatalytic enantioselective reactions to a great extent and will make them even more useful tools for the total synthesis of natural and biologically important products. Undoubtedly, the future direction in this emerging field is to continue expanding the scope of organocatalytic domino reactions through the identification of novel modes of reactivity and to apply these powerful, simple, and easy to perform strategies to the synthesis of biologically interesting molecules, including natural products.

References

1. (a) R. Noyori in *Asymmetric Catalysts in Organic Synthesis*, Wiley, New York, 1994; (b) E. N. Jacobsen, A. Pfaltz and H. Yamamoto (ed.), *Comprehensive Asymmetric Catalysis*, Springer, Berlin, 1999; (c) *Catalytic Asymmetric Synthesis*, ed. I. Ojima, Wiley-VCH, New York, 2nd edn, 2000; (d) M. Beller and C. Bolm (ed.), *Transition Metals for Organic Synthesis*, Wiley-VCH, Weinheim, 2nd edn, 2004; (e) D. J. Ramon and M. Yus, *Chem. Rev.*, 2006, **106**, 2126–2208.

2. (a) S. M. Roberts (ed.), *Biocatalysts for Fine Chemicals Synthesis*, Wiley-VCH, New York, 1999; (b) K. Drauz, H. Waldmann (ed.), *Enzyme Catalysis in Organic Synthesis*, Wiley-VCH, Weinheim, 2nd edn, 2002; (c) A. S. Bommarius and B. R. Riebel, *Biocatalysis*, Wiley-VCH, Weinheim, 2004.

3. (a) E. Knoevenagel, *Chem. Ber.*, 1896, **29**, 172–174; (b) U. Eder, G. Sauer and R. Wiechert, *Angew. Chem., Int. Ed. Engl.*, 1971, **10**, 496–497; (c) Z. G. Hajos and D. R. Parrish, *J. Org. Chem.*, 1974, **39**, 1615–1621; (d) K. A. Ahrendt, C. J. Borths and D. W. C. MacMillan, *J. Am. Chem. Soc.*, 2000, **122**, 4243–4244; (e) B. List, R. A. Lerner and C. F. Barbas, *J. Am. Chem. Soc.*, 2000, **122**, 2395–2396.

4. (a) P. I. Dalko and L. Moisan, *Angew. Chem., Int. Ed.*, 2001, **40**, 3726–3748; (b) A. Berkessel and H. Gröger in *Asymmetric Organocatalysis – From Biomimetic Concepts to Powerful Methods for Asymmetric Synthesis*, Wiley-VCH, Weinheim, 2005; (c) J. Seayad and B. List, *Org. Biomol. Chem.*, 2005, **3**, 719–724; (d) M. S. Taylor and E. N. Jacobsen, *Angew. Chem., Int. Ed.*, 2006, **45**, 1520–1543; (e) P. I. Dalko in *Enantioselective Organocatalysis*, Wiley-VCH, Weinheim, 2007; (f) P. I. Dalko, *Chimia*, 2007, **61**, 213–218; (g) H. Pellissier, *Tetrahedron*, 2007, **63**, 9267–9331; (h) A. G. Doyle and E. N. Jacobsen, *Chem. Rev.*, 2007, **107**, 5713–5743; (i) M. G. Gaunt, C. C. C. Johansson, A. McNally and N. C. Vo, *Drug Discovery Today*, 2007, **2**, 8–27; (j) *Chem. Rev.*, 2007, **107**(12), 5413–5883, Special Issue on Organocatalysis, ed. B. List; (k) D. Enders, C. Grondal and M. R. M. Hüttl, *Angew. Chem., Int. Ed.*,

2007, **46**, 1570–1581; (l) G. Guillena, D. J. Ramon and M. Yus, *Tetrahedron: Asymmetry*, 2007, **18**, 693–700; (m) D. W. C. MacMillan, *Nature*, 2008, **455**, 304–308; (n) X. Yu and W. Wang, *Chem. Asian. J.*, 2008, **3**, 516–532; (o) A. Dondoni and A. Massi, *Angew. Chem., Int. Ed.*, 2008, **47**, 4638–4660; (p) P. Melchiorre, M. Marigo, A. Carlone and G. Bartoli, *Angew. Chem., Int. Ed.*, 2008, **47**, 6138–6171; (q) F. Peng and Z. Shao, *J. Mol. Catal. A*, 2008, **285**, 1–13; (r) C. F. Barbas, *Angew. Chem., Int. Ed.*, 2008, **47**, 42–47; (s) C. Palomo, M. Oiarbide and R. Lopez, *Chem. Soc. Rev.*, 2009, **38**, 632–653; (t) S. Bertelsen and K. A. Jorgensen, *Chem. Soc. Rev.*, 2009, **38**, 2178–2189; (u) A.-N. Alba, X. Companyo, M. Viciano and R. Rios, *Curr. Org. Chem.*, 2009, **13**, 1432–1474; (v) M. Bella and T. Gasperi, *Synthesis*, 2009, 1583–1614; (w) L. Gong, Special Topic: Asymmetric Organocatalysis, *Chin. Sci. Bull.*, 2010, **55**, 17; (x) B. List, *Asymmetric Organocatalysis*, in *Top. Curr. Chem.*, 2010, **291**; (y) H. Pellissier, in *Recent Developments in Asymmetric Organocatalysis*, Royal Society of Chemistry, Cambridge, 2010; (z) *Enantioselective Organocatalysed Reactions*, ed. R. Mahrwald, Springer, Berlin, 2011, vols I and II; (aa) M. Ruiz, P. Lopez-Alvarado, G. Giorgi and J. C. Menéndez, *Chem. Soc. Rev.*, 2011, **40**, 3445–3454; (ab) F. Giacalone, M. Gruttadauria, P. Agrigento and R. Noto, *Chem. Soc. Rev.*, 2012, **41**, 2406–2447; (ac) H. Pellissier, *Adv. Synth. Catal.*, 2012, **354**, 237–294.

5. (a) R. M. Christmann de Figueiredo and M. Christmann, *Eur. J. Org. Chem.*, 2007, 2575–2600; (b) E. Marquès-Lopez and R. P. Herrera, *Nat. Prod. Rep.*, 2010, **27**, 1138–1167.

6. L. F. Tietze, G. Brasche and K. Gericke, *Domino Reactions in Organic Synthesis*, Wiley-VCH, Weinheim, 2006.

7. (a) H. Pellissier, *Tetrahedron*, 2006, **62**, 2143–2173; (b) H. Pellissier, *Tetrahedron*, 2006, **62**, 1619–1665.

8. (a) L. F. Tietze and U. Beifuss, *Angew. Chem., Int. Ed. Engl.*, 1993, **32**, 131–163; (b) L. F. Tietze, *Chem. Ind. (London)*, 1995, **12**, 453–457; (c) L. F. Tietze, *Chem. Rev.*, 1996, **96**, 115–136; (d) L. F. Tietze and M. E. Lieb, *Curr. Opin. Chem. Biol.*, 1998, **2**, 363–371.

9. (a) L. F. Tietze and A. Modi, *Med. Res. Rev.*, 2000, **20**, 304–322; (b) *Multicomponent Reactions*, ed. J. Zhu and H. Bienaymé, Wiley-VCH, Weinheim, 2005; (c) D. J. Ramon and M. Yus, *Angew. Chem., Int. Ed.*, 2005, **44**, 1602–1634; (d) R. V. A. Orru and E. Ruijter (ed.), *Synthesis of Heterocycles via Multicomponent Reactions*, in *Topics in Heterocyclic Chemistry*, Springer, Berlin, 2010, vols I and II.

10. (a) B. List, *Chem. Commun.*, 2006, 819–824; (b) A. Erkkilä, I. Majander and P. M. Pihko, *Chem. Rev.*, 2007, **107**, 5416–5470; (c) S. Mukherjee, J. W. Yang, S. Hoffmann and B. List, *Chem. Rev.*, 2007, **107**, 5471–5569; (d) X. Yu and W. Wang, *Org. Biomol. Chem.*, 2008, **6**, 2037–2046; (e) C. Palomo and A. Mielgo, *Angew. Chem., Int. Ed.*, 2006, **45**, 7876–7880.

11. T. Bui and C. F. Barbas, *Tetrahedron Lett.*, 2000, **41**, 6951–6954.

12. G. Lelais and D. W. C. MacMillan, *Aldrichimica Acta*, 2006, **39**, 79–87.
13. N. Wolgang, F. Tanaka and C. F. Barbas, *Acc. Chem. Res.*, 2004, **37**(8), 580–591.
14. T. Komnenos, *Justus Liebigs Ann. Chem.*, 1883, **218**, 145–169 .
15. P. Perlmutter, in *Conjugate Addition Reactions in Organic Synthesis*, Pergamon Press, Oxford, 1992.
16. (a) H.-C. Guo and J.-A. Ma, *Angew. Chem., Int. Ed.*, 2006, **45**, 354–366; (b) D. Almasi, D. A. Alonso and C. Najera, *Tetrahedron: Asymmetry*, 2007, **18**, 299–365; (c) J. L. Vicario, D. Badia and L. Carrillo, *Synthesis*, 2007, 2065–2092; (d) S. B. Tsogoeva, *Eur. J. Org. Chem.*, 2007, 1701–1716; (e) L. F. Tietze and A. Düfert in *Catalytic Asymmetric Conjugate Reactions* ed. A. Cordova, Wiley-VCH, Weinheim, 2010, p. 321.
17. (a) N. Halland, P. S. Aburel and K. A. Jorgensen, *Angew. Chem., Int. Ed.*, 2004, **43**, 1272–1277; (b) J. Pulkkinen, P. S. Aburel and K. A. Jorgensen, *Adv. Synth. Catal.*, 2004, **346**, 1077–1080.
18. D. Gryko, *Tetrahedron: Asymmetry*, 2005, **16**, 1377–1383.
19. J. W. Yang, M. T. Hechavarria Fonseca and B. List, *J. Am. Chem. Soc.*, 2005, **127**, 15036–15037.
20. Y. Huang, A. M. Walji, C. H. Larsen and D. W. C. MacMillan, *J. Am. Chem. Soc.*, 2005, **127**, 15051–15053.
21. L. Zu, H. Li, H. Xie, J. Wang, W. Jiang, Y. Tang and W. Wang, *Angew. Chem., Int. Ed.*, 2007, **46**, 3732–3734.
22. G.-L. Zhao, I. Ibrahem, P. Dziedzic, J. Sun, C. Bonneau and A. Cordova, *Chem. Eur. J.*, 2008, **14**, 10007–10011.
23. P. G. McGarraugh and S. E. Brenner, *Org. Lett.*, 2009, **11**, 5654–5657.
24. A. Ma and D. Ma, *Org. Lett.*, 2010, **12**, 3634–3637.
25. D. Enders, K. Lüttgen and A. A. Narine, *Synthesis*, 2007, 959–980.
26. D. Enders, A. Saint-Dizier, M. I. Lannou and A. Lenzen, *Eur. J. Org. Chem.*, 2006, 29–49.
27. (a) C. F. Nising and S. Bräse, *Chem. Soc. Rev.*, 2008, **37**, 1218–1228; (b) C. F. Nising and S. Bräse, *Chem. Soc. Rev.*, 2012, **41**, 988–999.
28. H. Li, L. Zu, H. Xie, J. Wang, W. Jiang and W. Wang, *Org. Lett.*, 2007, **9**, 1833–1835.
29. B.-C. Hong, P. Kotame, C.-W. Tsai and J.-H. Liao, *Org. Lett.*, 2010, **12**, 776–779.
30. X. Zhang, S. Zhang and W. Wang, *Angew. Chem., Int. Ed.*, 2010, **49**, 1481–1484.
31. B.-C. Hong, R. Y. Nimje and J.-H. Liao, *Org. Biomol. Chem.*, 2009, **7**, 3095–3101.
32. B.-C. Hong, C.-S. Hsu and G.-H. Lee, *Chem. Commun.*, 2012, **48**, 2385–2387.
33. B.-C. Hong, A. A. Sadani, R. Y. Nimje, N. S. Dange and G.-H. Lee, *Synthesis*, 2011, 1887–1895.
34. J.-W. Xie, W. Chen, R. Li, M. Zeng, W. Du, L. Yue, Y.-C. Chen, Y. Wu, J. Zhu and J.-G. Deng, *Angew. Chem., Int. Ed.*, 2007, **46**, 389–392.

35. (a) P. R. Schreiner, *Chem. Soc. Rev.*, 2003, **32**, 289–296; (b) P. M. Pihko, *Angew. Chem., Int. Ed.*, 2004, **43**, 2062–2064; (c) S. J. Connon, *Angew. Chem., Int. Ed.*, 2006, **45**, 3909–3912; (d) T. Akiyama, J. Itoh and K. Fuchibe, *Adv. Synth. Catal.*, 2006, **348**, 999–1010; (e) T. Akiyama, *Chem. Rev.*, 2007, **107**, 5744–5748.

36. Q. Wei and L.-Z. Gong, *Org. Lett.*, 2010, **12**, 1008–1011.

37. Q. Ren, Y. Gao and J. Wang, *Chem. Eur. J.*, 2010, **16**, 13594–13598.

38. A. Grossmann and D. Enders, *Angew. Chem., Int. Ed.*, 2011, **50**, 2–14.

39. D. Enders, O. Niemeier and A. Henseler, *Chem. Rev.*, 2007, **107**, 5606–5655.

40. X. Fang, K. Jiang, C. Xing, L. Hao and Y. R. Chi, *Angew. Chem., Int. Ed.*, 2011, **50**, 1910–1913.

41. (a) D. Roca-Lopez, D. Sadaba, I. Delso, R. P. Herrera, T. Tejero and P. Merino, *Tetrahedron: Asymmetry*, 2010, **21**, 2561–2601; (b) J. L. Vicario, D. Badia, L. Carrillo and E. Reyes (ed.), *Organocatalytic Enantioselective Conjugate Addition Reactions: A Powerful Tool for the Stereocontrolled Synthesis of Complex Molecules*, the Royal Society of Chemistry, Cambridge, 2010.

42. T. Okino, Y. Hoashi and Y. Takemoto, *J. Am. Chem. Soc.*, 2003, **125**, 12672–12673.

43. B. Tan, Z. Shi, P. Chua and G. Zhong, *Org. Lett.*, 2008, **10**, 3425–3428.

44. J. Wang, H. Xie, H. Li, L. Zu and W. Wang, *Angew. Chem., Int. Ed.*, 2008, **47**, 4177–4179.

45. (a) H. Pellissier, *Tetrahedron*, 2003, **59**, 8291–8327; (b) H. Pellissier, *Tetrahedron*, 2008, **64**, 1563–1601; (c) H. Pellissier, *Tetrahedron*, 2011, **67**, 3769–3802; (d) H. Pellissier, *Adv. Synth. Catal.*, 2011, **353**, 659–676; (e) H. Pellissier in *Chirality from Dynamic Kinetic Resolution*, Royal Society of Chemistry, Cambridge, 2011.

46. X.-F. Wang, Q.-L. Hua, Y. Cheng, X.-L. An, Q.-Q. Yang, J.-R. Chen and W.-J. Xiao, *Angew. Chem., Int. Ed.*, 2010, **49**, 8379–8381.

47. Q.-L. Hua, C. Li, X.-F. Wang, L.-Q. Lu, J.-R. Chen and W.-J. Xiao, *ACS Catalysis*, 2011, **1**, 221–226.

48. X.-F. Wang, J. An, X.-X. Zhang, F. Tan, J.-R. Chen and W.-J. Xiao, *Org. Lett.*, 2011, **13**, 808–811.

49. C. Yu, Y. Zhang, A. Song, Y. Ji and W. Wang, *Chem. Eur. J.*, 2011, **17**, 770–774.

50. S. Rajkumar, K. Shankland, G. D. Brown and A. J. A. Cobb, *Chem. Sci.*, 2012, **3**, 584–588.

51. L.-Y. Wu, G. Bencivenni, M. Mancinelli, A. Mazzanti, G. Bartoli and P. Melchiorre, *Angew. Chem., Int. Ed.*, 2009, **48**, 7196–7199.

52. H. Sunden, R. Rios, Y. Xu, L. Eriksson and A. Cordova, *Adv. Synth. Catal.*, 2007, **349**, 2549–2555.

53. L. Zu, S. Zhang, H. Xie and W. Wang, *Org. Lett.*, 2009, **11**, 1627–1630.

54. W. Wang, H. Li, J. Wang and L. Zu, *J. Am. Chem. Soc.*, 2006, **128**, 10354–10355.

55. R. Rios, H. Sunden, I. Ibrahem, G.-L. Zhao, L. Eriksson and A. Cordova, *Tetrahedron Lett.*, 2006, **47**, 8547–8551.

56. H. Sunden, I. Ibrahem, G.-L. Zhao, L. Eriksson and A. Cordova, *Chem. Eur. J.*, 2007, **13**, 574–581.

57. H. Li, J. Wang, T. E.-Nunu, L. Zu, W. Jiang, S. Wie and W. Wang, *Chem. Commun.*, 2007, 507–509.

58. S.-P. Luo, Z.-B. Li, L.-P. Wang, Y. Guo, A.-B. Xia and D.-Q. Xu, *Org. Biomol. Chem.*, 2009, **7**, 4539–4546.

59. H. Shen, K.-F. Yang, Z.-H. Shi, J.-X. Jiang, G.-Q. Lai and L.-W. Xu, *Eur. J. Org. Chem.*, 2011, 5031–5038.

60. G.-L. Zhao, J. Vesely, R. Rios, I. Ibrahem, H. Sunden and A. Cordova, *Adv. Synth. Catal.*, 2008, **350**, 237–242.

61. H. Sunden, R. Rios, I. Ibrahem, G.-L. Zhao, L. Eriksson and A. Cordova, *Adv. Synth. Catal.*, 2007, **349**, 827–832.

62. H. Li, J. Wang, H. Xie, L. Zu, W. Jiang, E. N. Duesler and W. Wang, *Org. Lett.*, 2007, **9**, 965–968.

63. Y. Yoshitomi, H. Arai, K. Makino and Y. Hamada, *Tetrahedron*, 2008, **64**, 11568–11579.

64. C. Liu, X. Zhang, R. Wang and W. Wang, *Org. Lett.*, 2010, **12**, 4948–4951.

65. S. Brandau, E. Maerten and K. A. Jorgensen, *J. Am. Chem. Soc.*, 2006, **128**, 14986–14991.

66. J. Wang, H. Li, H. Xie, L. Zu, X. Shen and W. Wang, *Angew. Chem., Int. Ed.*, 2007, **46**, 9050–9053.

67. B.-C. Hong, R. Y. Nimje, A. A. Sadani and J.-H. Liao, *Org. Lett.*, 2008, **10**, 2345–2348.

68. M. Rueping, A. Kuenkel, F. Tato and J. W. Bats, *Angew. Chem., Int. Ed.*, 2009, **48**, 3699–3702.

69. E. Reyes, G. Talavera, J. L. Vicario, D. Badia and L. Carrillo, *Angew. Chem., Int. Ed.*, 2009, **48**, 5701–5704.

70. D. Enders, C. Wang and G. Raabe, *Synthesis*, 2009, 4119–4124.

71. L. Hong, W. Sun, C. Liu, L. Wang and R. Wang, *Chem. Eur. J.*, 2010, **16**, 440–444.

72. G. Luo, S. Zhang, W. Duan and W. Wang, *Tetrahedron Lett.*, 2009, **50**, 2946–2948.

73. J. Tang, D. Q. Xu, A. B. Xia, Y. F. Wang, J. R. Jiang, S. P. Luo and Z. Y. Xu, *Adv. Synth. Catal.*, 2010, **352**, 2121–2126.

74. J.-Y. Bae, H.-J. Lee, S.-H. Youn, S.-H. Kwon and C.-W. Cho, *Org. Lett.*, 2010, **12**, 4352–4355.

75. W. Li, X. Li, W. Wu, X. Liang and J. Ye, *Chem. Commun.*, 2011, **47**, 8325–8327.

76. (a) D. Enders, C. Wang and J. W. Bats, *Synlett*, 2009, 1777–1780; (b) D. Enders, A. A. Narine, T. R. Benninghaus and G. Raabe, *Synlett*, 2007, 1667–1670.

77. (a) B.-C. Hong, N. S. Dange, C.-S. Hsu and J.-H. Miao, *Org. Lett.*, 2010, **12**, 4812–4815; (b) B.-C. Hong, N. S. Dange, C.-S. Hsu, J.-H. Liao and G.-H. Lee, *Org. Lett.*, 2011, **13**, 1338–1341.

78. (a) R. Rios, H. Sunden, I. Ibrahem, G.-L. Zhao and A. Cordova, *Tetrahedron Lett.*, 2006, **47**, 8679–8682; (b) R. Rios, H. Sunden, I. Ibrahem and A. Cordova, *Tetrahedron Lett.*, 2007, **48**, 2181–2184.

79. Y.-Q. Yuang, Z. Chai, H.-F. Wang, X.-K. Chen, H.-F. Cui, C.-W. Zheng, H. Xiao, P. Li and G. Zhao, *Chem. Eur. J.*, 2009, **15**, 13295–13298.

80. H.-F. Cui, Y.-Q. Yang, Z. Chai, P. Li, C.-W. Zheng, S.-Z. Zhu and G. Zhao, *J. Org. Chem.*, 2010, **75**, 117–122.

81. L.-L. Wang, L. Peng, J.-F. Bai, Q.-C. Huang, X.-Y. Xu and L.-X. Wang, *Chem. Commun.*, 2010, **46**, 8064–8066.

82. L. Zu, J. Wang, H. Li, H. Xie, W. Jiang and W. Wang, *J. Am. Chem. Soc.*, 2007, **129**, 1036–1037.

83. L. Zu, H. Xie, H. Li, J. Wang, W. Jiang and W. Wang, *Adv. Synth. Catal.*, 2007, **349**, 1882–1886.

84. Y. Gao, Q. Ren, H. Wu, M. Li and J. Wang, *Chem. Commun.*, 2010, **46**, 9232–9234.

85. B. Tan, N. R. Candeias and C. F. Barbas, *Nature Chem.*, 2011, **3**, 473–477.

86. X.-Q. Dong, X. Fang, H.-Y. Tao, X. Zhou and C.-J. Wang, *Chem. Commun.*, 2012, **48**, 7238–7240.

87. C.-L. Cao, X.-L. Sun, Y.-B. Kang and Y. Tang, *Org. Lett.*, 2007, **9**, 4151–4154.

88. (a) G.-L. Zhao, P. Dziedzic, F. Ullah, L. Eriksson and A. Cordova, *Tetrahedron Lett.*, 2009, **50**, 3458–3462; (b) Y. Hayashi, T. Okano, S. Aratake and D. Hazelard, *Angew. Chem., Int. Ed.*, 2007, **46**, 4922–4925.

89. M. Mariago, J. Franzen, T. B. Toulsen, W. Zhuang and K. A. Jorgensen, *J. Am. Chem. Soc.*, 2005, **127**, 6964–6965.

90. H. Sunden, I. Ibrahem and A. Cordova, *Tetrahedron Lett.*, 2006, **47**, 99–103.

91. G.-L. Zhao and A. Cordova, *Tetrahedron Lett.*, 2007, **48**, 5976–5980.

92. (a) X. Wang and B. List, *Angew. Chem., Int. Ed.*, 2008, **47**, 1119–1122; (b) X. Wang, C. M. Reisinger and B. List, *J. Am. Chem. Soc.*, 2008, **130**, 6070–6071.

93. J. Vesely, I. Ibrahem, G.-L. Zhao, R. Rios and A. Cordova, *Angew. Chem., Int. Ed.*, 2007, **46**, 778–781.

94. H. Arai, N. Sugaya, N. Sasaki, K. Makino, S. Lectard and Y. Hamada, *Tetrahedron Lett.*, 2009, **50**, 3329–3332.

95. F. Pesciaioli, F. De Vincentiis, P. Galzerano, G. Bencivenni, G. Bartoli, A. Mazzanti and P. Melchiorre, *Angew. Chem., Int. Ed.*, 2008, **47**, 8703–8706.

96. (a) I. Ibrahem, R. Rios, J. Vesely, G.-L. Zhao and A. Cordova, *Angew. Chem., Int. Ed.*, 2007, 849–851; (b) I. Ibrahem, R. Rios, J. Vesely, G.-L. Zhao and A. Cordova, *Synthesis*, 2007, 1153–1157.

97. R. Rios, I. Ibrahem, J. Vesely, H. Sunden and A. Cordova, *Tetrahedron Lett.*, 2007, **48**, 8695–8699.

98. P. T. Franke, B. Richter and K. A. Jorgensen, *Chem. Eur. J.*, 2008, **14**, 6317–6321.

99. M. Rueping, E. Sugiono and E. Merino, *Chem. Eur. J.*, 2008, **14**, 6329–6332.

100. M. Rueping, E. Sugiono and E. Merino, *Angew. Chem., Int. Ed.*, 2008, **47**, 3046–3049.

101. (a) J. Franzen and A. Fisher, *Angew. Chem., Int. Ed.*, 2009, **48**, 787–791; (b) W. Zhang and J. Franzen, *Adv. Synth. Catal.*, 2010, **352**, 499–518.

102. (a) G. Valero, J. Schimer, I. Cisarova, J. Vesely, A. Moyano and R. Rios, *Tetrahedron Lett.*, 2009, **50**, 1943–1946; (b) S. Cihalova, G. Valero, J. Schimer, M. Humpl, M. Dracinsky, A. Moyano, R. Rios and J. Vesely, *Tetrahedron*, 2011, **67**, 8942–8950.

103. Z.-Q. He, Q. Zhou, L. Wu and Y.-C. Chen, *Adv. Synth. Catal.*, 2010, **352**, 1904–1908.

104. X. Dai, X. Wu, H. Fang, L. Nie, J. Chen, H. Deng, W. Cao and G. Zhao, *Tetrahedron*, 2011, **67**, 3034–3040.

105. X. Wu, X. Dai, L. Nie, H. Fang, J. Chen, Z. Ren, W. Cao and G. Zhao, *Chem. Commun.*, 2010, **46**, 2733–2735.

106. R. R. Knowles, J. Carpenter, S. B. Blakey, A. Kayano, I. K. Mangion, C. J. Sinz and D. W. C. MacMillan, *Chem. Sci.*, 2011, **2**, 308–311.

107. A. Noole, M. Borissova, M. Lopp and T. Kanger, *J. Org. Chem.*, 2011, **6**, 1538–1545.

108. O. V. Maltsev, A. S. Kucherenko, A. L. Chimishkyan and S. G. Zlotin, *Tetrahedron: Asymmetry*, 2010, **21**, 2659–2670.

109. Y. Lee, S. W. Seo and S.-G. Kim, *Adv. Synth. Catal.*, 2011, **353**, 2671–2675.

110. M. Fernandez, E. Reyes, J. L. Vicario, D. Badia and L. Carrillo, *Adv. Synth. Catal.*, 2012, **354**, 371–376.

111. D. B. Ramachary and Y. V. Reddy, *Eur. J. Org. Chem.*, 2012, **5**, 865–887.

112. D. Enders, X. Yang, C. Wang, G. Raabe and J. Runsik, *Chem. Asian J.*, 2011, **6**, 2255–2259.

113. (a) T. Tozawa, H. Nagao, Y. Yamane and T. Mukaiyama, *Chem. Asian J.*, 2007, **2**, 123–134; (b) T. Tozawa, Y. Yamane and T. Mukaiyama, *Chem. Lett.*, 2006, **35**, 56–57.

114. J.-W. Xie, X. Huang, L.-P. Fan, D.-C. Xu, X.-S. Li, H. Su and Y.-H. Wen, *Adv. Synth. Catal.*, 2009, **351**, 3077–3082.

115. S.-L. Zhao, C.-W. Zheng, H.-F. Wang and G. Zhao, *Adv. Synth. Catal.*, 2009, **351**, 2811–2816.

116. S.-L. Zhao, C.-W. Zheng and G. Zhao, *Tetrahedron: Asymmetry*, 2009, **20**, 1046–1051.

117. M. A. Calter and J. Wang, *Org. Lett.*, 2009, **11**, 2205–2208.

118. W. Yao, L. Pan, Y. Wu and C. Ma, *Org. Lett.*, 2010, **12**, 2422–2425.

119. O. Mahè, I. Dez, V. Levacher and J.-F. Brière, *Angew. Chem., Int. Ed.*, 2010, **49**, 7072–7075.

120. K. Matoba, H. Kawai, T. Furukawa, A. Kusuda, E. Tokunaga, S. Nakamura, M. Shiro and N. Shibata, *Angew. Chem., Int. Ed.*, 2010, **49**, 5762–5766.

121. Y. Gao, Q. Ren, W.-Y. Siau and J. Wang, *Chem. Commun.*, 2011, **47**, 5819–5821.

122. Z.-P. Hu, C.-L. Lou, J.-J. Wang, C.-X. Chen and M. Yan, *J. Org. Chem.*, 2011, **76**, 3797–3804.

123. X.-M. Li, B. Wang, J.-M. Zhang and M. Yan, *Org. Lett.*, 2011, **13**, 374–377.

124. J.-j. Wang, Z.-p. Hu, C.-l. Lou, J.-l. Liu, X.-m. Li and M. Yan, *Tetrahedron*, 2011, **67**, 4578–4583.

125. G. Zhang, Y. Zhang, X. Jiang, W. Yan and R. Wang, *Org. Lett.*, 2011, **13**, 3806–3809.

126. S. Chandrasekhar, K. Mallikarjun, G. Pavankumarreddy, K. V. Rao and B. Jagadeesh, *Chem. Commun.*, 2009, 4985–4987.

127. D. Lu, Y. Li and Y. Gong, *J. Org. Chem.*, 2010, **75**, 6900–6907.

128. B.-C. Hong, P. Kotame and J.-H. Liao, *Org. Biomol. Chem.*, 2011, **9**, 382–386.

129. D. B. Ramachary, M. S. Prasad and R. Madhavachary, *Org. Biomol. Chem.*, 2011, **9**, 2715–2721.

130. D. Enders, G. Urbanietz and G. Raabe, *Synthesis*, 2011, 1905–1911.

131. D. Enders, G. Urbanietz, R. Hahn and G. Raabe, *Synthesis*, 2012, **44**, 773–782.

132. B.-C. Hong, P. Kotame and G.-H. Lee, *Org. Lett.*, 2011, **13**, 5758–5761.

133. Y. Wang, S. Zhu and D. Ma, *Org. Lett.*, 2011, **13**, 1602–1605.

134. D. Enders, C. Wang, X. Yang and G. Raabe, *Synlett*, 2011, 469–472.

135. (a) H. Xie, L. Zu, H. Li, J. Wang and W. Wang, *J. Am. Chem. Soc.*, 2007, **129**, 10886–10894; (b) R. Rios, H. Sunden, J. Vesely, G.-L. Zhao, P. Dziedzic and A. Cordova, *Adv. Synth. Catal.*, 2007, **349**, 1028–1032.

136. X. Companyo, A.-N. Alba, F. Cardenas, A. Moyano and R. Rios, *Eur. J. Org. Chem.*, 2009, 3075–3080.

137. J. Vesely, G.-L. Zhao, A. Bartoszewicz and A. Cordova, *Tetrahedron Lett.*, 2008, **49**, 4209–4212.

138. R. Rios, J. Vesely, H. Sunden, I. Ibrahem and G.-L. Cordova, *Tetrahedron Lett.*, 2007, **48**, 5835–5839.

139. V. Terrasson, A. van der Lee, R. M. de Figueiredo and J. M. Campagne, *Chem. Eur. J.*, 2010, **16**, 7875–7880.

140. Y. K. Kang, S. M. Kim and D. Y. Kim, *J. Am. Chem. Soc.*, 2010, **132**, 11847–11849.

141. H. M. Hansen, D. A. Longbottom and S. V. Ley, *Chem. Commun.*, 2006, 4838–4840.

142. V. Waschdowski, H. M. Hansen, D. A. Longbottom and S. V. Ley, *Synthesis*, 2008, 1269–1275.

143. J. Lv, J. Zhang, Z. Lin and Y. Wang, *Chem. Eur. J.*, 2009, **15**, 972–979.

144. A. Russo, S. Meninno, C. Tedesco and A. Lattanzi, *Eur. J. Org. Chem.*, 2011, 5096–5103.

145. F. Pesciaioli, P. Righi, A. Mazzanti, G. Bartoli and G. Bencivenni, *Chem. Eur. J.*, 2011, **17**, 2842–2845.

146. C. C. C. Johansson, N. Bremeyer, S. V. Ley, D. R. Owen, S. C. Smith and M. J. Gaunt, *Angew. Chem., Int. Ed.*, 2006, **45**, 6024–6028.

147. Y.-n. Xuan, S.-z. Nie, L.-t. Dong, J.-m. Zhang and M. Yan, *Org. Lett.*, 2009, **11**, 1583–1586.

148. C. Guo, M.-X. Xue, M.-K. Zhu and L.-Z. Gong, *Angew. Chem., Int. Ed.*, 2008, **47**, 3414–3417.

149. L.-P. Fan, P. Li, X.-S. Li, D.-C. Xu, M.-M. Ge, W.-D. Zhu and J.-W. Xie, *J. Org. Chem.*, 2010, **75**, 8716–8719.

150. M. Rueping, A. Parra, U. Uria, F. Besselièvre and E. Merino, *Org. Lett.*, 2010, **12**, 5680–5683.

151. X. Dou, X. Han and Y. Lu, *Chem. Eur. J.*, 2012, **18**, 85–89.

152. (a) J. Boruwa, N. Gogoi, P. P. Saikia and N. C. Barua, *Tetrahedron: Asymmetry*, 2006, **17**, 3315–3326; (b) C. Palomo, M. Oiarbide and A. Laso, *Eur. J. Org. Chem.*, 2007, 2561–2574; (c) E. Marquès-Lopez, P. Merino, T. Tejero and R. P. Herrera, *Eur. J. Org. Chem.*, 2009, 2401–2420; (d) Y. Alvarez-Casao, E. Marques-Lopez and R. P. Herrera, *Symmetry*, 2011, **3**, 220–245.

153. E. Reyes, H. Jiang, A. Milelli, P. Elsner, R. G. Hazell and K. A. Jorgensen, *Angew. Chem., Int. Ed.*, 2007, **46**, 9202–9205.

154. J. L. Garcia Ruano, V. Marcos, J. A. Suanzes, L. Marzo and J. Aleman, *Chem. Eur. J.*, 2009, **15**, 6576–6580.

155. X. Liu and Y. Lu, *Org. Biomol. Chem.*, 2010, **8**, 4063–4065.

156. Z.-X. Jia, Y.-C. Luo and P.-F. Xu, *Org. Lett.*, 2011, **13**, 832–835.

157. J. Xie, K. Yoshida, K. Takasu and Y. Takemoto, *Tetrahedron Lett.*, 2008, **49**, 6910–6913.

158. D.-Q. Xu, Y.-F. Wang, S.-P. Luo, S. Zhang, A.-G. Zhong, H. Chen and Z.-Y. Xu, *Adv. Synth. Catal.*, 2008, **350**, 2610–2616.

159. R. Dodda, J. J. Goldman, T. Mandal, C.-G. Zhao, G. A. Broker and E. R. T. Tiekink, *Adv. Synth. Catal.*, 2008, **350**, 537–541.

160. W. Zhang, S.-P. Luo, B.-L. Li, A.-B. Xia, A.-G. Zhong and D.-Q. Xu, *Chem. Asian J.*, 2009, **4**, 1834–1838.

161. B. Tan, P. J. Chua, X. Zeng, M. Lu and G. Zhong, *Org. Lett.*, 2008, **10**, 3489–3492.

162. B. Tan, P. J. Chua, Y. Li and G. Zhong, *Org. Lett.*, 2008, **10**, 2437–2440.

163. D. Ding, C.-G. Zhao, Q. Guo and H. Arman, *Tetrahedron*, 2010, **66**, 4423–4427.

164. M. Rueping, A. Kuenkel and R. Fröhlich, *Chem. Eur. J.*, 2010, **16**, 4173–4176.

165. B. Tan, Y. Lu, X. Zeng, P. J. Chua and G. Zhong, *Org. Lett.*, 2010, **12**, 2682–2685.

166. P. Chintala, S. K. Gosh, E. Long, A. D. Headley and B. Ni, *Adv. Synth. Catal.*, 2011, **353**, 2905–2909.

167. (a) A. Carlone, M. Marigo, C. North, A. Landa and K. A. Jorgensen, *Chem. Commun.*, 2006, 4928–4930; (b) S. Brandau, A. Landa, J. Franzen, M. Marigo and K. A. Jorgensen, *Angew. Chem., Int. Ed.*, 2006, **45**, 4305–4309.

168. M. Marigo, S. Bertelsen, A. Landa and K. A. Jorgensen, *J. Am. Chem. Soc.*, 2006, **128**, 5475–5479.

169. Y. Hayashi, M. Toyoshima, H. Gotoh and H. Ishikawa, *Org. Lett.*, 2009, **11**, 45–48.

170. E. Alza, S. Sayalero, X. C. Cambeiro, R. Martin-Rapun, P. O. Miranda and M. A. Pericas, *Synlett*, 2011, 464–468.

171. L. Albrecht, B. Richter, C. Vila, H. Krawczyk and K. A. Jorgensen, *Chem. Eur. J.*, 2009, **15**, 3093–3102.

172. B.-C. Hong, M.-F. Wu, H.-C. Tseng and J.-H. Liao, *Org. Lett.*, 2006, **8**, 2217–2220.

173. S. Cabrera, J. Aleman, P. Bolze, S. Bertelsen and K. A. Jorgensen, *Angew. Chem., Int. Ed.*, 2008, **47**, 121–125.

174. J. Aleman, A. Nunez, L. Marzo, V. Marcos, C. Alvarado and J. L. Garcia Ruano, *Chem. Eur. J.*, 2010, **16**, 9453–9456.

175. J. L. Vicario, S. Reboredo, D. Badia and L. Carrillo, *Angew. Chem., Int. Ed.*, 2007, **46**, 5168–5170.

176. A.-B. Xia, D.-Q. Xu, S.-P. Luo, J.-R. Jiang, J. Tang, Y.-F. Wang and Z.-Y. Xu, *Chem. Eur. J.*, 2010, **16**, 801–804.

177. H. Li, J. Zhao, L. Zeng and W. Hu, *J. Org. Chem.*, 2011, **76**, 8064–8069.

178. Q. Cai, C. Zheng, J.-W. Zhang and S.-L. You, *Angew. Chem., Int. Ed.*, 2011, **50**, 8665–8669.

179. B. Tan, G. Hernandez-Torres and C. F. Barbas, *J. Am. Chem. Soc.*, 2011, **133**, 12354–12357.

180. G.-L. Zhao, R. Rios, J. Vesely, L. Eriksson and A. Cordova, *Angew. Chem., Int. Ed.*, 2008, **47**, 8468–8472.

181. Y.-k. Liu, C. Ma, K. Jiang, T.-Y. Liu and Y.-C. Chen, *Org. Lett.*, 2009, **11**, 2848–2851.

182. S. Gogoi and C.-G. Zhao, *Tetrahedron Lett.*, 2009, **50**, 2252–2255.

183. C.-L. Cao, Y.-Y. Zhou, J. Zhou, X.-L. Sun, Y. Tang, Y.-X. Li, G.-Y. Li and J. Sun, *Chem. Eur. J.*, 2009, **15**, 11384–11389.

184. Q. Ren, Y. Gao and J. Wang, *Org. Biomol. Chem.*, 2011, **9**, 5297–5302.

185. (a) J. M. Lee, Y. Na, H. Han and S. Chang, *Chem. Soc. Rev.*, 2004, **33**, 302–312; (b) J.-C. Wasilke, S. J. Obrey, R. T. Baker and G. C. Bazan, *Chem. Rev.*, 2005, **105**, 1001–1020.

186. B. Simmons, A. M. Walji and D. W. C. MacMillan, *Angew. Chem., Int. Ed.*, 2009, **48**, 4349–4354.

187. S. P. Lathrop and T. Rovis, *J. Am. Chem. Soc.*, 2009, **131**, 13628–13630.

188. C. M. Filloux, S. P. Lathrop and T. Rovis, *Proc. Natl. Acad. Sci. USA*, 2010, **107**, 20666–20671 .

189. K. E. Ozboya and T. Rovis, *Chem. Sci.*, 2011, **2**, 1835–1838.

190. N. Shimada, B. O. Ashburn, A. K. Basak, W. F. Bow, D. A. Vicic and M. A. Tius, *Chem. Commun.*, 2010, **46**, 3774–3775.

191. D. Enders, A. Grossmann, H. Huang and G. Raabe, *Eur. J. Org. Chem.*, 2011, 4298–4301.

192. C. Mannich and W. Krosche, *Arch. Pharm.*, 1912, **250**, 647.

193. (a) M. Arend, B. Westermann and N. Risch, *Angew. Chem., Int. Ed. Engl.*, 1998, **37**, 1044–1070; (b) A. Cordova, *Acc. Chem. Res.*, 2004, **37**, 102–112; (c) J. M. M. Verkade, L. J. C. Van Hemert, P. J. L. M. Quaedflieg and F. P. J. T. Rutjes, *Chem. Soc. Rev.*, 2008, **37**, 29–41; (d) R. G. Arrayas and J. C. Carretero, *Chem. Soc. Rev.*, 2009, **38**, 1940–1948.

194. T. Itoh, M. Yokoya, K. Miyauchi, K. Nagata and A. Ohsawa, *Org. Lett.*, 2003, **5**, 4301–4304.

195. T. Itoh, M. Yokoya, K. Miyauchi, K. Nagata and A. Ohsawa, *Org. Lett.*, 2006, **8**, 1533–1535.

196. R.-G. Han, Y. Wang, Y.-Y. Li and P.-F. Xu, *Adv. Synth. Catal.*, 2008, **350**, 1474–1478.

197. H. Yang and R. G. Carter, *J. Org. Chem.*, 2009, **74**, 5151–5156.

198. D. Enders, D. P. Göddertz, C. Beceno and G. Raabe, *Adv. Synth. Catal.*, 2010, **352**, 2863–2868.

199. M. Rueping and M.-Y. Lin, *Chem. Eur. J.*, 2010, **16**, 4169–4172.

200. Q. Ren, W.-Y. Siau, Z. Du, K. Zhang and J. Wang, *Chem. Eur. J.*, 2011, **17**, 7781–7785.

201. S. Dong, X. Liu, Y. Zhang, L. Lin and X. Feng, *Org. Lett.*, 2011, **13**, 5060–5063.

202. G. Masson, C. Housseman and J. Zhu, *Angew. Chem., Int. Ed.*, 2007, **46**, 4614–4628.

203. S. Takizawa, N. Inoue, S. Hirata and H. Sasai, *Angew. Chem., Int. Ed.*, 2010, **49**, 9725–9729.

204. S. Takizawa, N. Inoue and H. Sasai, *Tetrahedron Lett.*, 2011, **52**, 377–380.

205. (a) T. Poulsen and K. A. Jorgensen, *Chem. Rev.*, 2008, **108**, 2903–2915; (b) S.-L. You, Q. Cai and M. Zeng, *Chem. Soc. Rev.*, 2009, **38**, 2190–2201.

206. L. Hong, L. Wang, W. Sun, K. Wong and R. Wang, *J. Org. Chem.*, 2009, **74**, 6881–6884.

207. Y.-J. Cao, H.-G. Cheng, L.-Q. Lu, J.-J. Zhang, Y. Cheng, J.-R. Chen and W.-J. Xiao, *Adv. Synth. Catal.*, 2011, **353**, 617–623.

208. X.-F. Wang, J.-R. Chen, Y.-J. Cao, H.-G. Cheng and W.-J. Xiao, *Org. Lett.*, 2010, **12**, 1140–1143.

209. C. C. J. Loh and D. Enders, *Angew. Chem., Int. Ed.*, 2011, **51**, 46–48.

210. O. Lozano, G. Bessley, T. Martinez Del Campo, A. L. Thompson, G. T. Giuffredi, M. Bettati, M. Walker, R. Borman and V. Gouverneur, *Angew. Chem., Int. Ed.*, 2011, **50**, 8105–8109.

211. N. Momiyama, Y. Yamamoto and H. Yamamoto, *J. Am. Chem. Soc.*, 2007, **129**, 1190–1195.

212. D. B. Ramachary and R. Sakthidevi, *Chem. Eur. J.*, 2009, **15**, 4516–4522.

213. J. Guang and C.-G. Zhao, *Tetrahedron: Asymmetry*, 2011, **22**, 1205–1211.

214. M. C. Bröhmer, E. Bourcet, M. Nieger and S. Bräse, *Chem. Eur. J.*, 2011, **17**, 13706–13711.

215. T. B. Poulsen, G. Dickmeiss, J. Overgaard and K. A. Jorgensen, *Angew. Chem., Int. Ed.*, 2008, **47**, 4687–4990.

216. (a) M. Liu, D. Zhu, Y. Lu, Y. Hou, B. Tan and G. Zhong, *Angew. Chem., Int. Ed.*, 2008, **47**, 10187–10191; (b) D. Zhu, M. Lu, P. J. Chua, B. Tan, F. Wang, X. Yang and G. Zhong, *Org. Lett.*, 2008, **10**, 4585–4588.

217. B. Han, J.-L. Li, C. Ma, S.-J. Zhang and Y.-C. Chen, *Angew. Chem., Int. Ed.*, 2008, **47**, 9971–9974.

218. Z.-Q. He, B. Han, R. Li, L. Wu and Y.-C. Chen, *Org. Biomol. Chem.*, 2010, **8**, 755–757.

219. S. Kanemasa, *Heterocycles*, 2010, **82**, 87–200.

220. M.-X. Xue, X.-M. Zhang and L.-Z. Gong, *Synlett*, 2008, 691–694.

221. (a) J. Poulin, C. M. Grisé-Bard and L. Barriault, *Chem. Soc. Rev.*, 2009, **38**, 3092–3101; (b) A. Moyano and R. Rios, *Chem. Rev.*, 2011, **111**, 4703–4832.

222. S. B. Jones, B. Simmons and D. W. C. MacMillan, *J. Am. Chem. Soc.*, 2009, **131**, 13606–13607.

223. M. E. Muratore, C. A. Holloway, A. W. Pilling, R. I. Storer, G. Trevitt and D. J. Dixon, *J. Am. Chem. Soc.*, 2009, **131**, 10796–10797.

224. J. Kaeobamrung and J. W. Bode, *Org. Lett.*, 2009, **11**, 677–680.

225. M. Rueping, A. P. Antonchick, E. Sugiono and K. Grenader, *Angew. Chem., Int. Ed.*, 2009, **48**, 908–910.

226. H. Jiang, P. Elsner, K. L. Jensen, A. Falcicchio, V. Marcos and K. A. Jorgensen, *Angew. Chem., Int. Ed.*, 2009, **48**, 6844–6848.

227. R. S. Klausen and E. N. Jacobsen, *Org. Lett.*, 2009, **11**, 887–890.

228. S. Rendler and D. W. C. MacMillan, *J. Am. Chem. Soc.*, 2010, **132**, 5027–5029.

229. S.-L. Zhang, H.-X. Xie, J. Zhu, H. Li, X.-S. Zhang, J. Li and W. Wang, *Nature Commun.*, 2011, **2**, article n°211, doi: 10:1038.

230. M. Rueping, H. Sunden, L. Hubener and E. Sugiono, *Chem. Commun.*, 2012, **48**, 2201–2203.

231. M. Rueping, J. Dufour and M. S. Maji, *Chem. Commun.*, 2012, **48**, 3406–3408.

232. M. Rueping and W. Ieawsuwan, *Chem. Commun.*, 2011, **47**, 11450–11452.

233. J. Itoh, K. Fuchibe and T. Akiyama, *Angew. Chem., Int. Ed.*, 2006, **45**, 4796–4798.

234. T. Akiyama, H. Morita and K. Fuchibe, *J. Am. Chem. Soc.*, 2006, **128**, 13070–13071.

235. M. He, J. R. Struble and J. W. Bode, *J. Am. Chem. Soc.*, 2006, **128**, 8418–8420.

236. P.-C. Chiang, J. Kaeobamrung and J. W. Bode, *J. Am. Chem. Soc.*, 2007, **129**, 3520–3521.

237. A. Chan and K. A. Scheidt, *J. Am. Chem. Soc.*, 2007, **129**, 5334–5335.

238. M. Wadamoto, E. M. Phillips, T. E. Reynolds and K. A. Scheidt, *J. Am. Chem. Soc.*, 2007, **129**, 10098–10099.

239. Z. Du, W.-Y. Siau and J. Wang, *Tetrahedron Lett.*, 2011, **52**, 6137–6141.

240. Y. Hayashi, T. Urushima, D. Sakamoto, K. Torii and H. Ishikawa, *Chem. Eur. J.*, 2011, **17**, 11715–11718.

241. (a) A. Strecker, *Ann. Chem. Pharm.*, 1850, **75**, 27–45; (b) J. Wang, X. Liu and X. Feng, *Chem. Rev.*, 2011, **111**, 6947–6983.

242. A. Hantzsch, *Justus Liebigs Ann. Chem.*, 1882, **215**, 1–82.

243. (a) P. Biginelli, *Ber. Dtsch. Chem. Ges.*, 1891, **24**, 1317–1319; (b) P. Biginelli, *Gazz. Chim. Ital.*, 1893, **23**, 360–413.

244. (a) I. Ugi, R. Meyr, U. Fetzer and C. Steinbrückner, *Angew. Chem.*, 1959, **71**, 386; (b) R. V. A. Orru and M. de Greef, *Synthesis*, 2003, 1471–1499.

245. J. M. Betancort, K. Sakthivel, R. Thayumanavan and C. F. Barbas, *Tetrahedron Lett.*, 2001, **42**, 4441–4444.

246. M. Marigo, T. Schulte, J. Franzen and K. A. Jorgensen, *J. Am. Chem. Soc.*, 2005, **127**, 15710–15711.

247. D. Enders, M. R. M. Hüttl, C. Grondal and G. Raabe, *Nature*, 2006, **441**, 861–863.

248. D. Enders, M. R. M. Hüttl, J. Runsink, G. Raabe and B. Wendt, *Angew. Chem., Int. Ed.*, 2007, **46**, 467–469.

249. A. Carlone, S. Cabrera, M. Marigo and K. A. Jorgensen, *Angew. Chem., Int. Ed.*, 2007, **46**, 1101–1104.

250. O. Penon, A. Carlone, A. Mazzanti, M. Locatelli, L. Sambri, G. Bartoli and P. Melchiorre, *Chem. Eur. J.*, 2008, **14**, 4788–4791.

251. M. C. Varela, S. M. Dixon, K. S. Lam and N. E. Schore, *Tetrahedron*, 2008, **64**, 10087–10090.

252. D. Enders, M. Jeanty and J. W. Bats, *Synlett*, 2009, 3175–3178.

253. X. Companyo, A. Zea, A.-N. R. Alba, A. Mazzanti, A. Moyano and R. Rios, *Chem. Commun.*, 2010, **46**, 6953–6955.

254. C. Cassani, X. Tian, E. C. Escudero-Adan and P. Melchiorre, *Chem. Commun.*, 2011, **47**, 233–235.

255. G. Bencivenni, L.-Y. Wu, A. Mazzanti, B. Giannichi, F. Pesciaoli, M.-P. Song, G. Bartoli and P. Melchiorre, *Angew. Chem., Int. Ed.*, 2009, **48**, 7200–7203.

256. K. Jiang, Z.-J. Jia, S. Chen, L. Wu and Y.-C. Chen, *Chem. Eur. J.*, 2010, **16**, 2852–2856.

257. P. Kotame, B.-C. Hong and J.-H. Liao, *Tetrahedron Lett.*, 2009, **50**, 704–707.

258. F.-L. Zhang, A.-W. Xu, Y.-F. Gong, M.-H. Wei and X.-L. Yang, *Chem. Eur. J.*, 2009, **15**, 6815–6818.

259. K. Jiang, Z.-J. Jia, X. Yin, L. Wu and Y.-C. Chen, *Org. Lett.*, 2010, **12**, 2766–2769.

260. S. Bertelsen, R. L. Johansen and K. A. Jorgensen, *Chem. Commun.*, 2008, 3016–3018.

261. A. Zea, A.-N. R. Alba, A. Mazzanti, A. Moyano and R. Rios, *Org. Biomol. Chem.*, 2011, **9**, 6519–6523.

262. Y.-H. Zhao, C.-W. Zheng, G. Zhao and W.-G. Cao, *Tetrahedron: Asymmetry*, 2008, **19**, 701–708.

263. J. Vesely, R. Rios, I. Ibrahem, G.-L. Zhao, L. Eriksson and A. Cordova, *Chem. Eur. J.*, 2008, **14**, 2693–2698.

264. P. Galzerano, F. Pesciaioli, A. Mazzanti, G. Bartoli and P. Melchiorre, *Angew. Chem., Int. Ed.*, 2009, **48**, 7892–7894.
265. M. Rueping and A. P. Antonchick, *Angew. Chem., Int. Ed.*, 2008, **47**, 5836–5838.
266. H. Yang and R. G. Carter, *Tetrahedron*, 2010, **66**, 4854–4859.
267. H. Yang and R. G. Carter, *Org. Lett.*, 2010, **12**, 3108–3111.
268. M. Sugiura, N. Sato, Y. Sonoda, S. Kotani and M. Nakajima, *Chem. Asian J.*, 2010, **5**, 478–481.
269. Y. Hoashi, T. Yabuta, P. Yuan, H. Miyabe and Y. Takemoto, *Tetrahedron*, 2006, **62**, 365–374.
270. Y.-K. Liu, H. Liu, W. Du, L. Yue and Y.-C. Chen, *Chem. Eur. J.*, 2008, **14**, 9873–9877.
271. L. Crovetto and R. Rios, *Synlett*, 2008, 1840–1844.
272. D. Enders, M. R. M. Hüttl, G. Raabe and J. W. Bats, *Adv. Synth. Catal.*, 2008, **350**, 267–279.
273. Y. Wang, R.-G. Han, Y.-L. Zhao, S. Yang, P.-F. Xu and D. J. Dixon, *Angew. Chem., Int. Ed.*, 2009, **48**, 9834–9838.
274. D. Enders, B. Schmid, M. Erdmann and G. Raabe, *Synthesis*, 2010, 2271–2277.
275. D. E. Bergbreiter, J. Tian and C. Hongfa, *Chem. Rev.*, 2009, **109**, 530–582.
276. T. E. Kristensen, K. Vestli, M. G. Jakobsen, F. K. Hansen and T. Hansen, *J. Org. Chem.*, 2010, **75**, 1620–1629.
277. B.-C. Hong, R.-H. Jan, C.-W. Tsai, R. Y. Nimje, J.-H. Liao and G.-H. Lee, *Org. Lett.*, 2009, **11**, 5246–5249 .
278. H. Ishikawa, T. Suzuki, H. Orita, T. Uchimaru and Y. Hayashi, *Chem. Eur. J.*, 2010, **16**, 12616–12626.
279. H. Ishikawa, M. Honma and Y. Hayashi, *Angew. Chem., Int. Ed.*, 2011, **50**, 2824–2827.
280. O. Baslé, W. Raimondi, M. del Mar Sanchez Duque, D. Bonne, T. Constantieux and J. Rodriguez, *Org. Lett.*, 2010, **12**, 5246–5249.
281. D. Enders, R. Krüll and W. Bettray, *Synthesis*, 2010, 567–572.
282. T. Urushima, D. Sakamoto, H. Ishikawa and Y. Hayashi, *Org. Lett.*, 2010, **12**, 4588–4591.
283. Y. Wang, D.-F. Yu, Y.-Z. Liu, H. Wie, Y.-C. Luo, D. J. Dixon and P.-F. Xu, *Chem. Eur. J.*, 2010, **16**, 3922–3925.
284. A. Rai, A. K. Singh, P. Singh and L. D. S. Yadav, *Tetrahedron Lett.*, 2011, **52**, 1354–1358.
285. H. Ishikawa, S. Sawano, Y. Yasui, Y. Shibata and Y. Hayashi, *Angew. Chem., Int. Ed.*, 2011, **50**, 3774–3779.
286. (a) P. Jakubec, D. M. Cockfield, M. Helliwell, J. Raftery and D. J. Dixon, *Belstein J. Org. Chem.*, 2012, **8**, 567–578; (b) P. Jakubec, M. Helliwell and D. J. Dixon, *Org. Lett.*, 2008, **10**, 4267–4270.
287. E. Knoevenagel, *Ber. Dtsch. Chem. Ges.*, 1894, **27**, 2345–2346.
288. B. List, *Angew. Chem., Int. Ed.*, 2010, **49**, 1730–1734.
289. (a) L. F. Tietze, H. Meier and H. Nutt, *Chem. Ber.*, 1989, **122**, 643–643; (b) L. F. Tietze, N. Rackelmann and I. Müller, *Chem. Eur. J.*, 2004, **10**,

2722–2731; (c) L. F. Tietze, N. Böhnke and S. Dietz, *Org. Lett.*, 2009, **11**, 2948–2950.

290. D. B. Ramachary, N. S. Chowdari and C. F. Barbas, *Angew. Chem., Int. Ed.*, 2003, **42**, 4233–4237.

291. (a) D. B. Ramachary, N. S. Chowdari and C. F. Barbas, *Synlett*, 2003, 1910–1914; (b) D. B. Ramachary, K. Anebouselvy, N. S. Chowdari and C. F. Barbas, *J. Org. Chem.*, 2004, **69**, 5838–5849.

292. J. Shi, Y. Min, M. Wang, L. Lin, X. Liu and X. Feng, *Tetrahedron*, 2011, **67**, 1781–1787.

293. D. B. Ramachary and C. F. Barbas, *Chem. Eur. J.*, 2004, **10**, 5323–5331.

294. D. B. Ramachary, Y. V. Reddy and B. V. Prakash, *Org. Biomol. Chem.*, 2008, **6**, 719–726.

295. W.-B. Chen, Z.-J. Wu, Q.-L. Pei, L.-F. Cun, X.-M. Zhang and W.-C. Yuan, *Org. Lett.*, 2010, **12**, 3132–3135.

296. F. Macaev, N. Sucman, F. Shepeli, M. Zveaghintseva and V. Pogrebnoi, *Symmetry*, 2011, **3**, 165–170.

297. D.-F. Yu, Y. Wang and P.-F. Xu, *Tetrahedron*, 2011, **67**, 3273–3277.

298. L. Liu, D. Wu, X. Li, S. Wang, H. Li, J. Li and W. Wang, *Chem. Commun.*, 2012, **48**, 1692–1694.

299. W.-T. Wei, C.-X. Chen, R.-J. Lu, J.-J. Wang, X.-J. Zhang and M. Yan, *Org. Biomol. Chem.*, 2012, **10**, 5245–5252.

300. B. List, *J. Am. Chem. Soc.*, 2000, **122**, 9336–9337.

301. (a) W. Notz, K. Sakthivel, T. Bui, G. Zhong and C. F. Barbas, *Tetrahedron Lett.*, 2001, **42**, 199–201; (b) A. Cordova, W. Notz, G. F. Zhong, J. M. Betancort and C. F. Barbas, *J. Am. Chem. Soc.*, 2002, **124**, 1842–1843; (c) Y. Hayashi, W. Tsuboi, I. Ashimine, T. Urushima, M. Shoji and K. Sakai, *Angew. Chem., Int. Ed.*, 2003, **42**, 3677–3680; (d) W. Notz, S. Watanabe, N. S. Chowdari, G. F. Zhong, J. M. Betancort, F. Tanaka and C. F. Barbas, *Adv. Synth. Catal.*, 2004, **346**, 1131–1140; (e) A. J. A. Cobb, D. M. Shaw and S. V. Ley, *Synlett*, 2004, 558–560; (f) I. Ibrahem, J. Casas and A. Cordova, *Angew. Chem., Int. Ed.*, 2004, **43**, 6528–6531; (g) A. Cordova, *Chem. Eur. J.*, 2004, **10**, 1987–1997; (h) Y. Hayashi, J. Yamaguchi, K. Hibino, T. Sumiya, T. Urushima, M. Shoji, D. Hashizume and H. Koshino, *Adv. Synth. Catal.*, 2004, **346**, 1435–1439; (i) I. Ibrahem and A. Cordova, *Tetrahedron Lett.*, 2005, **46**, 2839–2843; (j) I. Ibrahem and A. Cordova, *Tetrahedron Lett.*, 2005, **46**, 3363–3367; (k) D. Enders, C. Grondal, M. Vrettou and G. Raabe, *Angew. Chem., Int. Ed.*, 2005, **44**, 4079–4083; (l) I. Ibrahem, W. Zou, M. Engqvist, Y. M. Xu and A. Cordova, *Chem. Eur. J.*, 2005, **11**, 7024–7029.

302. B. Rodriguez and C. Bolm, *J. Org. Chem.*, 2006, **71**, 2888–2891.

303. I. Ibrahem, W. Zou, Y. Xu and A. Cordova, *Adv. Synth. Catal.*, 2006, **348**, 211–222.

304. W.-W. Liao, I. Ibrahem and A. Cordova, *Chem. Commun.*, 2006, 674–676.

305. M. L. Kantam, C. V. Rajasekhar, G. Gopikrishna, K. R. Reddy and B. M. Choudary, *Tetrahedron Lett.*, 2006, **47**, 5965 5967.

306. N. S. Chowdari, M. Ahmad, K. Albertshofer, F. Tanaka and C. F. Barbas, *Org. Lett.*, 2006, **8**, 2839–2842.
307. S. S. V. Ramasastry, H. Zhang, F. Tanake and C. F. Barbas, *J. Am. Chem. Soc.*, 2007, **129**, 288–289.
308. Q.-X. Guo, H. Liu, C. Guo, S.-W. Luo, Y. Gu and L.-Z. Gong, *J. Am. Chem. Soc.*, 2007, **129**, 3790–3791.
309. (a) L. Cheng, X. Wu and Y. Lu, *Org. Biomol. Chem.*, 2007, **5**, 1018–1020; (b) L.-W. Xu, J. Luo and Y. Lu, *Chem. Commun.*, 2009, 1807–1821.
310. Y.-C. Teo, J.-J. Lau and M.-C. Wu, *Tetrahedron: Asymmetry*, 2008, **19**, 186–190.
311. P. Dziedzic, I. Ibrahem and A. Cordova, *Tetrahedron Lett.*, 2008, **49**, 803–807.
312. H. Yang and R. G. Carter, *J. Org. Chem.*, 2009, **74**, 2246–2249.
313. (a) Q. Gu, J. J. Gong, J. Feng, X. Y. Wu and Q. L. Zhou, *Chin. J. Chem.*, 2008, **26**, 1902–1906; (b) Q. Gu, L. X. Jiang, K. Yuan, L. Zhang and X. Y. Wu, *Synth. Commun.*, 2008, **38**, 4198–4206.
314. G. Dagousset, F. Drouet, G. Masson and J. Zhu, *Org. Lett.*, 2009, **11**, 5546–5549.
315. (a) Y. L. Gu, C. Ogawa, J. Kobayashi, Y. Mori and S. Kobayashi, *Angew. Chem., Int. Ed.*, 2006, **45**, 7217–7220; (b) K. Gong, D. Fang, H. L. Wang and Z. L. Liu, *Monatsh. Chem.*, 2007, **138**, 1195–1198.
316. X. Zheng, Y.-B. Qian and Y. Wang, *Eur. J. Org. Chem.*, 2010, 515–522.
317. B. Y. Liu, D. Q. Xu, J. F. Dong and H. L. Yang, *Synth. Commun.*, 2007, **37**, 3003–3010.
318. (a) C. Wu, X. Fu, X. Ma, S. Li and C. Li, *Tetrahedron Lett.*, 2010, **51**, 5775–5777; (b) C. Wu, X. Fu and S. Li, *Tetrahedron: Asymmetry*, 2011, **22**, 1063–1073.
319. S. Bai, X. Liang, B. Song, P. S. Bhadury, D. Hu and S. Yang, *Tetrahedron: Asymmetry*, 2011, **22**, 518–523.
320. F.-F. Yong and Y.-C. Teo, *Synth. Commun.*, 2011, **41**, 1293–1300.
321. Y.-J. An, Q. Qin, C. Wang and J.-C. Tao, *Chin. J. Chem.*, 2011, **29**, 1511–1517.
322. Y.-J. An, C.-C. Wang, Z.-P. Liu and J.-C. Tao, *Helv. Chim. Acta*, 2012, **95**, 43–51.
323. X.-H. Chen, X.-Y. Yu, H. Liu, L.-F. Cun and L.-Z. Gong, *J. Am. Chem. Soc.*, 2006, **128**, 14802–14803.
324. L.-Z. Gong, X. H. Chen and X.-Y. Xu, *Chem. Eur. J.*, 2007, **13**, 8920–8926.
325. J. M. Goss and S. E. Schauss, *J. Org. Chem.*, 2008, **73**, 7651–7656.
326. J. Xin, L. Chan, Z. Hou, D. Shang, X. Liu and X. Feng, *Chem. Eur. J.*, 2008, **14**, 3177–3181.
327. L. D. S. Yadav, A. Rai, V. K. Rai and C. Awasthi, *Tetrahedron*, 2008, **64**, 1420–1429.
328. J. Pandey, N. Anand and R. P. Tripathi, *Tetrahedron*, 2009, **65**, 9350–9356.

329. J.-H. Sohn, H.-M. Choi, S. Lee, S. Joung and H.-Y. Lee, *Eur. J. Org. Chem.*, 2009, 3858–3862.

330. Y.-Y. Wu, Z. Chai, X.-Y. Liu, G. Zhao and S.-W. Wang, *Eur. J. Org. Chem.*, 2009, 904–911.

331. D. Ding and C.-G. Zhao, *Eur. J. Org. Chem.*, 2010, 3802–3805.

332. Y. Wang, H. Yang, J. Yu, Z. Miao and R. Chen, *Adv. Synth. Catal.*, 2009, **351**, 3057–3062.

333. X. Li, H. Deng, S. Z. Luo and J. P. Cheng, *Eur. J. Org. Chem.*, 2008, 4350–4356.

334. Y. Wang, J. Yu, Z. Miao and R. Chen, *Org. Biomol. Chem.*, 2011, **9**, 3050–3054.

335. N. Li, X.-H. Chen, J. Song, S.-W. Luo, W. Fan and L.-Z. Gong, *J. Am. Chem. Soc.*, 2009, **131**, 15301–15310.

336. S. Saha and J. N. Moorthy, *J. Org. Chem.*, 2011, **76**, 396–402.

337. G. Jones, in *Comprehensive Heterocyclic Chemistry II*, ed. A. R. Katritzky, C. W. Rees and E. F. V. Scriven, Pergamon, Oxford, 1996, vol. 5, pp. 167–243.

338. P. T. Franke, R. L. Johansen, S. Bertelsen and K. A. Jorgensen, *Chem. Asian J.*, 2008, **3**, 216–224.

339. J. Jiang, J. Yu, X.-X. Sun, Q.-Q. Rao and L.-Z. Gong, *Angew. Chem., Int. Ed.*, 2008, **47**, 2458–2462.

340. C. G. Evans and J. E. Gestwicki, *Org. Lett.*, 2009, **11**, 2957–2959.

341. H. Hishitani, S. Komiyama, Y. Hasegawa and S. Kobayashi, *J. Am. Chem. Soc.*, 2000, **122**, 762–766.

342. S. C. Pan and B. List, *Org. Lett.*, 2007, **6**, 1149–1151.

343. Y. Wen, Y. Xiong, L. Chang, J. Huang, X. Liu and X. Feng, *J. Org. Chem.*, 2007, **72**, 7715–7719.

344. Y. Wen, B. Gao, Y. Fu, S. Dong, X. Liu and X. Feng, *Chem. Eur. J.*, 2008, **14**, 6789–6795.

345. (a) N. A. Petasis and I. Akritopoulou, *Tetrahedron: Lett.*, 1993, **34**, 583–586; (b) R. N. Candeias, F. Montalbano, P. M. S. D. Cal and P. M. P. Gois, *Chem. Rev.*, 2010, **110**, 6169–6193.

346. S. Lou and S. E. Schaus, *J. Am. Chem. Soc.*, 2008, **130**, 6922–6923.

347. (a) Y. Yamaoka, H. Miyabe and Y. Takemoto, *J. Am. Chem. Soc.*, 2007, **129**, 6686–6687; (b) H. Miyabe and Y. Takemoto, *Bull. Chem. Soc. Jpn.*, 2008, **81**, 785–795; (c) Y. Takemoto, *Chem. Pharm. Bull.*, 2010, **58**, 593–601.

348. W.-Y. Han, Z.-J. Wu, X.-M. Zhang and W.-C. Yuan, *Org. Lett.*, 2012, **14**, 976–979.

349. H. Pellissier, *Tetrahedron*, 2007, **63**, 3235–3285.

350. R. Rios, I. Ibrahem, J. Vesely, G.-L. Zhao and A. Cordova, *Tetrahedron Lett.*, 2007, **48**, 5701–5705.

351. I. Ibrahem, R. Rios, J. Vesely and A. Cordova, *Tetrahedron Lett.*, 2007, **48**, 6252–6257.

352. X.-H. Chen, W.-Q. Zhang and L.-Z. Gong, *J. Am. Chem. Soc.*, 2008, **130**, 5652–5653.

353. X.-H. Chen, Q. Wie, S.-W. Luo, H. Xiao and L.-Z. Gong, *J. Am. Chem. Soc.*, 2009, **131**, 13819–13825.

354. S. Lin, L. Deiana, G.-L. Zhao, J. Sun and A. Cordova, *Angew. Chem., Int. Ed.*, 2011, **50**, 7624–7630.

355. A. Fraile, D. M. Scarpino Schietroma, A. Albrecht, R. L. Davis and K. A. Jorgensen, *Chem. Eur. J.*, 2012, **18**, 2773–2776.

356. (a) A. Dömling and I. Ugi, *Angew. Chem., Int. Ed.*, 2000, **39**, 3168–3210; (b) J. Zhu, *Eur. J. Org. Chem.*, 2003, 1133–1144; (c) S. S. Van Berkel, B. G. M. Bögels, M. A. Wijdeven, B. Westermann and F. P. J. T. Rutjes, *Eur. J. Org. Chem.*, 2012, 3543–3559.

357. M. Passerini and L. Simone, *Gazz. Chim. Ital.*, 1921, **51**, 126–129.

358. S. E. Denmark and Y. Fan, *J. Am. Chem. Soc.*, 2003, **125**, 7825–7827.

359. T. Yue, M.-X. Wang, D.-X. Wang, G. Masson and J. Zhu, *Angew. Chem., Int. Ed.*, 2009, **48**, 6717–6721.

360. G.-W. Zhang, L. Wang, J. Nie and J.-A. Ma, *Adv. Synth. Catal.*, 2008, **350**, 1457–1463.

361. D. Enders, C. Wang, M. Mukanova and A. Greb, *Chem. Commun.*, 2010, **46**, 2447–2449.

362. S. Roy and K. Chen, *Org. Lett.*, 2012, **14**, 2496–2499.

363. P. Merino, E. Marques-Lopez, T. Tejero and R. P. Herrera, *Synthesis*, 2010, 1–26.

364. (a) H. Sunden, I. Ibrahem, L. Eriksson and A. Cordova, *Angew. Chem., Int. Ed.*, 2005, **44**, 4877–4880; (b) I. Ibrahem, W. Zou, J. Casas, H. Sunden and A. Cordova, *Tetrahedron*, 2006, **62**, 357–364.

365. H. Liu, L.-F. Cun, A.-Q. Mi, Y.-Z. Jiang and L.-Z. Gong, *Org. Lett.*, 2006, **26**, 6023–6026.

366. Z.-J. Jia, H. Jiang, J.-L. Li, B. Gschwend, Q.-Z. Li, X. Yin, J. Grouleff, Y.-C. Chen and K. A. Jorgensen, *J. Am. Chem. Soc.*, 2011, **133**, 5053–5061.

367. L. S. Povarov, *Russ. Chem. Rev.*, 1967, **36**, 656–670.

368. H. Liu, G. Dagousset, G. Masson, P. Retailleau and J. Zhu, *J. Am. Chem. Soc.*, 2009, **131**, 4598–4599.

369. G. Dagousset, J. Zhu and G. Masson, *J. Am. Chem. Soc.*, 2011, **133**, 14804–14813.

370. C. Wang, Z.-Y. Han, H.-W. Luo and L.-Z. Gong, *Org. Lett.*, 2010, **12**, 2266–2269.

371. G. Bergonzini, L. Gramigna, A. Mazzanti, M. Fochi, L. Bernardi and A. Ricci, *Chem. Commun.*, 2010, **46**, 327–329.

372. S. Guizzetti, M. Benaglia, F. Cozzi and R. Annunziata, *Tetrahedron*, 2009, **65**, 6354–6363.

373. W. Lian, C. ShuMin, M. Wie, Z. GuangWu, N. Jing and M. JunAn, *Chin. Sc. Bull.*, 2010, **55**, 1729–1731.

374. (a) J. Zhou and B. List, *J. Am. Chem. Soc.*, 2007, **129**, 7498–7499; (b) J. Zhou and B. List, *Synlett*, 2007, 2037–2040.

375. M. Terada, K. Machioka and K. Sorimachi, *J. Am. Chem. Soc.*, 2007, **129**, 10336–10337.

376. S. T. Scroggins, Y. Chi and J. M. J. Fréchet, *Angew. Chem., Int. Ed.*, 2010, **49**, 2393–2396.

377. (a) J. Zhou, *Chem. Asian J.*, 2010, **5**, 422–434; (b) C. Zhong and X. Shi, *Eur. J. Org. Chem*, 2010, 2999–3025; (c) M. Rueping, R. M. Koenigs and I. Atodiresei, *Chem. Eur. J.*, 2010, **16**, 9350–9365; (d) Z. Shao and H. Zhang, *Chem. Soc. Rev.*, 2009, **38**, 2745–2755.

378. B. Westermann, M. Ayaz and S. S. van Berkel, *Angew. Chem., Int. Ed.*, 2010, **49**, 846–849.

General Conclusion

Major issues in modern synthetic organic chemistry, which deals with the preparation of natural products, pharmaceuticals, diagnostics, agrochemicals, and other important materials, are the improvement of efficiency, the avoidance of toxic reagents, the reduction of waste, and the responsible treatment of our resources. One of the ways to fulfil these goals is the development and use of one-pot domino processes, which consist in two or more bond-forming reactions evolving under identical conditions, in which the subsequent transformation takes place at the functionalities obtained in the former transformation and following the same principles that are found in biosynthesis in nature. These elegant reactions allow the easy attainment of high molecular complexity from simple substrates with very often high levels of stereocontrol, involve simple operational procedures, and have the advantages of savings in solvent, time, energy, and costs by avoiding the use of costly and time-consuming protection–deprotection processes, as well as purification procedures for intermediates. In particular, the use of one- and two-component, as well as multicomponent, domino reactions in asymmetric synthesis is increasing constantly, and this has rapidly become one of the most exciting fields in organic chemistry. This book updates the major progress in the general field of asymmetric one-, two-, and multicomponent domino reactions, covering the literature since the beginning of 2006. It illustrates the power and diversity of asymmetric domino reactions based on the use of chiral substrates, as well as those based on the use of chiral catalysts, which can be organocatalysts or metal catalysts.

The first chapter of the book reviews asymmetric domino reactions involving chiral substrates, including chiral auxiliaries. Of the methods available for preparing chiral compounds, asymmetric synthesis from chiral substrates still attracts a lot of attention. Indeed, it remains the method most commonly employed in the total synthesis of optically active compounds, playing an important role in medicine and materials science, as well as natural products.

RSC Catalysis Series No. 10
Asymmetric Domino Reactions
By Hélène Pellissier
© The Royal Society of Chemistry 2013
Published by the Royal Society of Chemistry, www.rsc.org

This first chapter illustrates the wide applications of domino reactions of chiral substrates in the total synthesis of biologically interesting molecules, including natural products, and that of novel chiral ligands and functional materials.

The importance of asymmetric catalysis as a tool for obtaining enantiomerically pure or enantioenriched compounds is well recognised worldwide. The second chapter of the book illustrates how much asymmetric organometallic catalysis has contributed to the development of highly efficient enantioselective domino and multicomponent reactions. The economic interest in such a combination is obvious. An explosive number of novel enantioselective metal-catalysed domino processes have been successfully developed in the last six years, allowing up to eight stereogenic centres to be generated with excellent stereoselectivities. The wide variety of these processes reflects that of the metals employed to induce them. Indeed, an increasing number of different metals, such as magnesium, scandium, titanium, ruthenium, cobalt, rhodium, iridium, nickel, palladium, platinum, copper, silver, gold, zinc, and aluminium, as well as tin, have been found to be effective catalysts.

While the end of the last century was dominated by the use of metal, and biocatalysis, a change in perception occurred during the last decade, when several reports confirmed that relatively simple organic molecules, such as proline, could be highly effective and remarkably enantioselective catalysts of a variety of fundamentally important transformations. This rediscovery has initiated an explosive growth of research activity in organocatalysis. Organocatalysts have several important advantages, because they are usually robust, inexpensive, readily available, and non-toxic. Enantioselective organocatalytic processes have grown rapidly to become one of the most fascinating and current fields in organic chemistry, with an impressive and steadily increasing number of publications regarding the applications of this type of reactions. Hence, the application of chiral organocatalysts has permitted the preparation of a number of very valuable chiral products with the exclusion of any trace of hazardous metals and with several advantages from economic and environmental points of view. The ability of organocatalysts to promote a wide range of reactions by different activation modes makes organocatalysis ideal for its application in domino reactions. The third chapter of the book deals with the combination of asymmetric organocatalysis with the concept of domino sequences. This powerful methodology allows the preparation of a number of complex molecules with very often excellent levels of stereocontrol in environmentally friendly conditions. Taking advantage of the various recently developed organocatalytic activation modes, including the concept of bifunctional organocatalysts and multicatalytic systems, numerous novel cascade sequences can be envisaged, with a trend from simple domino reactions to triple and even quadruple domino reactions in cascade.

The future direction in the field of asymmetric domino and multicomponent reactions is to continue expanding their scope through the combination of different types of reactions, the identification of novel modes of reactivity, and the employment of novel chiral catalysts, and to apply these powerful strategies to the synthesis of biologically interesting molecules, including natural products.

Subject Index

References to schemes or figures are given in *italic* type.